DATE DUE	
NOV 25 2005	
MAY 9 – 2006	

GAYLORD · PRINTED IN U.S.A.

Basic Abstract Algebra

This book is part of the Allyn and Bacon Series in Advanced Mathematics

Consulting Editor: Irving Kaplansky

Other books in the series:

    James Dugundji, *Topology*
    Horst Herrlich and George E. Strecker, *Category Theory*
    Irving Kaplansky, *Commutative Rings*
    Irving Kaplansky, *Linear Algebra and Geometry:*
        A Second Course
    Irving Kaplansky, *Set Theory and Metric Spaces*
    Ralph Kopperman, *Model Theory and Its Applications*
    Joseph J. Rotman, *The Theory of Groups:*
        An Introduction, Second Edition

Basic Abstract Algebra

Otto F. G. Schilling
Purdue University

W. Stephen Piper
*Purdue University
and the Center for Naval Analyses
an affiliate of the University of Rochester*

Allyn and Bacon, Inc.
Boston

© Copyright 1975 by Allyn and Bacon, Inc.
470 Atlantic Avenue, Boston. All rights reserved.
Printed in the United States of America.
No part of the material protected by this copyright notice may be reproduced or utilized in any form or by any means, electronic or mechanical, including photocopying, recording, or by any informational storage and retrieval system, without written permission from the copyright owner.

Library of Congress Cataloging in Publication Data

Schilling, Otto Franz Georg, 1911–1973.
    Basic abstract algebra.

    Bibliography: p.
    1. Algebra, Abstract. I. Piper, William Stephen, 1940–   joint author. II. Title.
QA162.S34   512'.02   74–4547

Contents

Preface ix

Chapter 1 • Notation and Introductory Concepts   1
    §1.1 Set Notation   1
    §1.2 Equivalence Relations and Classes   6

Chapter 2 • Arithmetic of Integers   10
    §2.1 Algebraic Properties of Integers   10
    §2.2 Analytic Properties of Integers and Induction   14
    §2.3 The Division Algorithm   21
    §2.4 Ideals in $\mathbf{Z}$ 23
    §2.5 Divisibility   25
    §2.6 Prime Numbers   31
    §2.7 Unique Factorization   33
    §2.8 Congruences Modulo m 37
    §2.9 Addition and Multiplication of Cosets Modulo m 40
    §2.10 Definitions and Examples   43
    §2.11 Simultaneous Systems of Congruences   48
    §2.12 Two Topics in Number Theory   52

Chapter 3 • Introduction to Ring Theory   55
    §3.1 Basic Elements of Ring Theory   56
    §3.2 Ring Homomorphisms   62
    §3.3 Direct Sums of Rings   71
    §3.4 Residue Class Rings of Integers   74
    §3.5 Direct Sum Decompositions of $\mathbf{Z}_m$ 77
    §3.6 Integral Domains and Fields   83

Chapter 4 • Aspects of Linear Algebra   94
    §4.1 Vector Spaces   95
    §4.2 Linear Independence and Bases   101
    §4.3 Transformations of Vector Spaces   105
    §4.4 Matrices and Linear Transformations   110
    §4.5 Determinants   121

Chapter 5 • Polynomials and Polynomial Rings   125
    §5.1 Polynomial Rings   126

Contents

- §5.2 Divisibility and Factorization of Polynomials   130
- §5.3 Residue Class Rings for Polynomials   136
- §5.4 Residue Class Fields of Irreducible Polynomials   142
- §5.5 Roots of Polynomials   146
- §5.6 The Interpolation Formula of Lagrange   150
- §5.7 Polynomial Functions   152
- §5.8 Primitive and Irreducible Polynomials   154
- §5.9 Characteristic Polynomials of Matrices   157

Chapter 6 • Group Theory   162

- §6.1 Elements of Group Theory   163
- §6.2 Subgroups and Orders of Elements   171
- §6.3 Coset Decompositions and the Theorem of Lagrange   177
- §6.4 Normal Subgroups and Factor Groups   180
- §6.5 Group Homomorphisms   184
- §6.6 Cyclic Groups   191
- §6.7 Groups of Permutations   194
- §6.8 The Isomorphism Theorems of Group Theory   201
- §6.9 Automorphisms, Center, Commutator Group   205
- §6.10 Direct Product   212
- §6.11 Homomorphisms of Abelian Groups   217

Chapter 7 • Selected Topics in Group Theory   221

- §7.1 Finitely Generated Abelian Groups   221
- §7.2 Characters of Finite Abelian Groups   230
- §7.3 Bijections of Sets   235
- §7.4 The Class Equation and Normalizers   237
- §7.5 The Elementary Theorems on Sylow Subgroups   241
- §7.6 Composition Series and the Jordan-Hölder Theorem   248
- §7.7 Groups with Operators   255
- §7.8 Modules   258

Chapter 8 • Field Theory   266

- §8.1 Algebraic Elements   266
- §8.2 Finite Fields   271
- §8.3 The Theorem of the Primitive Element   275
- §8.4 Equivalence of Fields   277
- §8.5 Counting of Isomorphisms and Separability   280
- §8.6 Prelude to Galois Theory   284
- §8.7 The Fundamental Theorem of Galois Theory   288
- §8.8 Consequences of the Fundamental Theorem   292
- §8.9 Algebraic Closure   299

Chapter 9 • Selected Topics in Field Theory   304

- §9.1 Cyclotomic Fields   305
- §9.2 Equations Solvable by Radicals   310
- §9.3 Constructions with Ruler and Compass   314
- §9.4 Trace and Norm   319
- §9.5 Theorem of the Normal Basis   325

§9.6 Hilbert's Theorem 90 and Noether's Equations   328
§9.7 Kummer or Radical Extensions   334
§9.8 Unique Factorization Domains and Elementary Symmetric Functions   341
§9.9 The Fundamental Theorem of Algebra   348
§9.10 Finite Division Rings   351
§9.11 Simple Transcendental Extensions   354
§9.12 Perfect Fields   358

Bibliography   363

Index of Mathematicians   367

Index of Notation   373

Index of Mathematical Terms   379

Preface

> *The Science of Pure Mathematics, in its modern developments, may claim to be the most original creation of the human spirit.*
>
> Alfred North Whitehead
> *Science and the Modern World*

Perspective

Why study abstract algebra? *What good is it?* The answer to such questions lies, of course, in the questing human intellect that admits no restraints upon its probes, nor limits to its adventures. Utility marches after discovery. The formal systems we are about to study have played important roles in the physical sciences and are of increasing importance in the biological and social sciences. While the algebra may be abstract, it nonetheless provides the setting or language for consideration of concrete physical and social problems.

Our approach to abstract algebra involves thinking of *abstract* as a verb form, not as an adjective. The development of algebraic structures as abstractions of the properties of integers and polynomials is shown. Thus familiar properties, observed in examples, are given a more general axiomatic setting. The intent is to involve the student in the evolution of algebraic concepts as a participant, rather than as a spectator having to assimilate lists of axiomatic descriptions. Besides the pedagogical merit of this approach, there is the mathematical merit of more closely indicating methods of mathematical research and advancement. Thus our proofs are designed to show details of arguments that later facilitate solution of exercises.

Content

This book is intended as a text for the first undergraduate and introductory graduate courses in abstract algebra. It has been so used at Purdue University. The order of topics and discussion derives from classroom experience. Students first encountering abstract algebra commonly have difficulty working with cosets of subgroups, factor groups, and residue class rings. These difficulties center on insufficient understanding of equivalence relations and classes. We address this topic in Section 1.2 and subsequently in the text refer to this initial discussion. Similarly, mathematical induction is carefully presented in Section 2.2 with subsequent references. Early chapters stress algebraic concepts and techniques so that the student may understand algebraic structures and gain facility in handling algebraic arguments.

Chapter 2 provides a careful introduction of the algebraic structure and congruence relations of integers, culminating in concrete problems in residue class rings of integers (Section 2.9) and the solution of simultaneous systems of congruences (Section 2.11). Thus, the student is provided with examples of rings and with the basic number theory necessary for questions in the theory of groups involving orders of elements, cyclic groups, and concepts like direct products.

We provide in Section 2.4 explicit proof of the equivalence of the Well-Ordering Principle and two statements of the Principle of Induction. As these equivalence proofs are often omitted from undergraduate texts, some classes may want to accept parts of this section without proof. Later we continue to state inductive arguments in detailed steps so that students may gain an appreciation of their proper use, rather than skirt the issue by blithely concluding arguments "by induction."

Rings are introduced before groups, contrary to a common sets–groups–rings–fields approach to abstract algebra. The reason is pedagogical; students are already familiar with integers and polynomials. Ready reference to integers and polynomials, where the axiomatic algebraic properties and their logical consequences are introduced, provides the student with examples that develop and test the concepts of ideals, nilpotence, idempotence, factorization, and congruence.

After discussing the algebraic structure of the integers, we abstract that structure to rings. Chapter 3 begins with the basic elements of ring theory (Sections 3.1 and 3.2): axiomatic description, subrings, ring homomorphisms, and ideals. Sections 3.4 and 3.5 consist of an optional discussion of orthogonal idempotents and ideals in residue class rings of integers and of the external and internal direct sum structure of such rings, the concept of direct sum having been introduced in Section 3.3. These sections may be omitted without loss of continuity by instructors who prefer to follow Section 3.2 immediately with the discussion of integral domains, fields, and polynomial rings (Sections 3.6, 5.1, and 5.2). Here extensive sets of exercises provide ample choice of material for further development.

The basics of linear algebra are presented in Chapter 4. This chapter is intended as a refresher for students who have had a prior course in linear algebra or as a concise source of the linear facts used in field theory (Chapters 8 and 9) and in some of the examples of groups (Chapter 6).

Chapter 5 is devoted to polynomials. The first four sections continue the ring theory of Chapter 3 with discussion of polynomial rings and residue class rings of polynomials, including Kronecker's construction of roots of irreducible polynomials. The remaining sections introduce properties of polynomials that will be useful in subsequent field theory study (especially roots of polynomials in Chapter 8 and other topics in Chapter 9).

Groups (Chapter 6) are not introduced until students have had sufficient experience in congruence theory to handle arguments involving orders of elements. The first section provides an extensive set of examples of groups from different branches of algebra. The student is urged to use these examples to verify the details of group-theoretic arguments while carrying out explicit computations with the elements of particular groups. Appropriate explicit examples of normal subgroups, quotient groups, homomorphisms, and cyclic and permutation groups appear throughout the chapter. The theorem that the image of a group homomorphism is isomorphic to its domain modulo the kernel is given special attention together with its immediate consequences (Section 6.8). Chapter 6 concludes with three optional sections designed to give the student more experience with group theory and to prepare him for subsequent topics.

For honors sections or classes seeking greater emphasis on group theory, Chapter 7 includes all the topics common to an advanced undergraduate or first graduate course: the structure of finitely generated abelian groups, group characters, the Sylow theory (Sections 7.3 through 7.5), composition series, and groups with operators and modules. These topics are presented independent of one another, so that they may be studied in any order. Most topics were chosen not only for their inherent algebraic interest but also for later reference in Chapter 9, Selected Topics in Field Theory.

Chapter 8 builds to the Galois theory. We emphasize the concept of prolongation of mappings, since it prepares the student for more advanced work in cohomological algebra and its applications. While both separable and inseparable extensions are treated, one may also conveniently limit the discussion to fields of characteristic zero.

Chapter 9 offers a broad selection of field-theoretic topics. Sections 9.2 and 9.3 on solvability by radicals and ruler and compass constructions provide answers to classical questions based on the Galois theory. Consequences of the Galois theory include the inseparable case because of its importance in modern applications of the subject, especially for abelian function fields and algebraic groups. The proof of the algebraic closure of the complex numbers follows the ideas of Lagrange and Gauss.

Introductions to individual sections serve as previews, stating briefly the principal directions of the section and often the relationship of the topics covered with those in other sections. They may involve terms about to be, but not yet, defined. These introductions should be helpful to a student re-reading the book or reviewing his course work.

As one learns best by doing, we have provided over 1400 exercises by which the student may test developing algebraic skills.

Course Selections

A variety of courses can be designed from the material offered, each tailored to the objectives of the students and instructors. We suggest the following for consideration as content of a typical

One-semester undergraduate course, with no linear algebra prerequisite	Chapters 1–2 Chapter 3, less Sections 3.4–3.5 Chapter 4, Sections 4.1–4.3 Chapter 5, Sections 5.1–5.4 Chapter 6, Sections 6.1–6.6
One-semester undergraduate course, linear algebra prerequisite	Same as above with Chapter 4 omitted, and additional topics added from Chapters 5–6
Honors undergraduate course	Chapters 1–6
First graduate course	Review Chapter 2 Chapters 3 and 5 Selection of topics from Chapter 7 Chapter 8, perhaps less Section 8.2, and only emphasizing the separable case
Second semester of above courses	Continuation into Chapters 6–7, with work from Chapter 4 as needed Topics from Chapter 9 for graduate courses.

Appreciation

No undertaking of the nature of this text can be the sole work of two individuals. Our manuscript has evolved over many semesters of trial. It owes its present existence and form to the provocative probes of colleagues and students. We wish to thank especially Michael Drazin for his helpful criticism, April Kihlstrom for her valuable comments, Bonnie Schnitta for editorial assistance, and Betty Lewis and Judy Snyder for their patience in typing and retyping the manuscript. We are also grateful to the several reviewers for their assessments, notably Thomas R. Berger of Trinity College, Connecticut, Richard L. Faber of Boston College, David Hertzig

Preface

of the University of Miami, Eugene F. Krause of Arizona State University, Joel Roberts of the University of Minnesota, and C. R. B. Wright of the University of Oregon.

Otto F. G. Schilling
W. Stephen Piper

West Lafayette, Indiana
June 1973

Otto Schilling died in June 1973 just as our manuscript was entering the production process. In reviewing the editing and composition phases of this project I have tried to be faithful to the writing done while it was a joint project, and I have been guided by the senior author's insistence on mathematical quality.

His Purdue University colleagues wrote of him: "Otto was a dedicated teacher. He tried to awaken enthusiasm for mathematics by absolute integrity and strict adherence to rigorous standards." This book, which he never saw in final form, caps the end of his long career in algebra. Otto Schilling's interests developed at Göttingen, where he was writing a thesis on algebraic number fields under Professor Emmy Noether when she was compelled to leave her native Germany. Upon her recommendation, Otto Schilling completed his thesis under Professor Helmut Hasse at Marburg. He then left Germany (in 1934) for a year at Trinity College, Cambridge and a long teaching career in the United States. His last twelve years were at Purdue University, after twenty-two years at the University of Chicago.

W. Stephen Piper

Falls Church, Virginia
June 1974

Basic Abstract Algebra

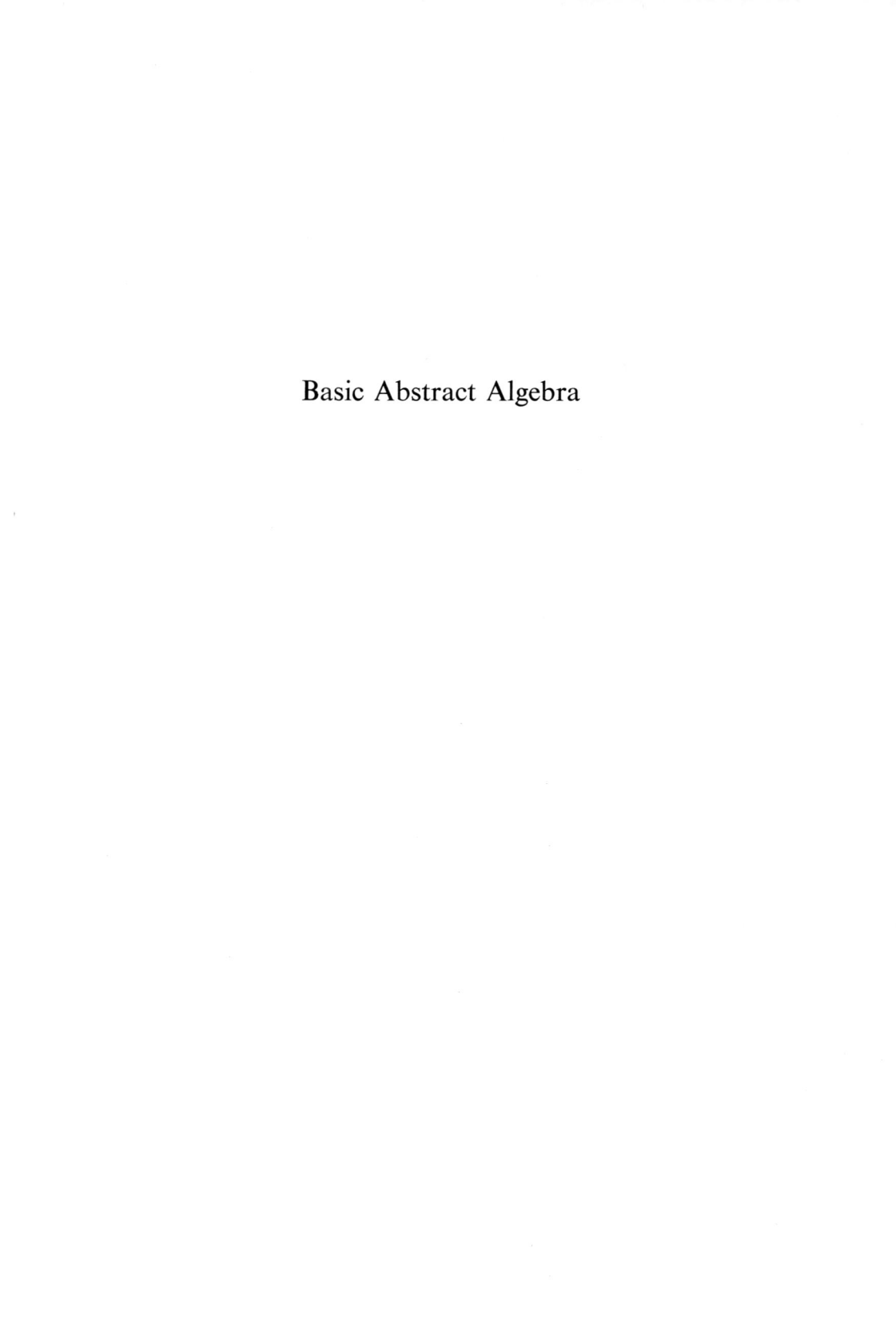

1

Notation and Introductory Concepts

A basic concept in algebra is that of equivalence relation, together with the related equivalence, congruence, or residue classes, also called cosets. After defining needed basic set theoretic notation [§1.1], we provide the definition and some examples of equivalence relations, so that the reader may focus on the concept itself before encountering residue class rings of integers modulo an integer m [§2.8], residue class rings of polynomials [§5.3], and cosets of ideals [§3.2, Exercise 3] and of groups [§6.2].

Although subsequently we reflect common usage by referring to equivalence classes by the various names of residue or congruence classes or cosets, the underlying concept is always that of the defining equivalence relation. Especially important is the fact that two equivalence classes defined by an equivalence relation are either equal or disjoint. Chapter 1 is brief, by design, so that the foundation is quickly laid for heuristic examination of the integers in Chapter 2.

§1.1 Set Notation

Throughout algebra, in fact throughout mathematics, one constantly refers to sets or collections of elements. In general we shall denote sets by capital letters $A, B, C, ..., S, ...$ and their elements by lower case letters

a, b, c, etc. If a belongs to the set or collection A, we write $a \in A$, and if a does not belong to A, we write $a \notin A$.

Membership in a set S is often denoted either by explicitly listing the elements of the set or by stating a rule for membership, as follows:

$$A = \{1, 2, 3, 4, 8, 9\} \quad \text{a set of 6 integers}$$

or

$$S = \{\text{integers } n \text{ such that } n \text{ is even}\}.$$

The latter has the following common abbreviated form:

$$S = \{\text{integers } n : n \text{ is even}\}.$$

Mathematicians have evolved certain standard symbols and terminology to aid in stating concepts. For instance, we shall make frequent use of the following symbols and set designations.

$\Rightarrow$	"implies," or "is a sufficient condition that"
$\Leftrightarrow$	"if and only if," "implies and is implied by," "is a necessary and sufficient condition," "is equivalent to," or "means the same as"
$\in$	"is an element of" or "belongs to"
N	set of positive integers (also called the set of natural numbers)
Z	set of integers
Q	set of rational numbers
R	set of real numbers
C	set of complex numbers

The standard abbreviation for "there exists" is $\exists$, and for "for all" is $\forall$, although we shall not need to use them in this text.

The set S mentioned above can also be expressed as

$$S = \{n \in \mathbf{Z} : n \text{ is even}\}.$$

The set with no elements belonging to it is called the **empty** or **null set**, designated by $\emptyset$. A set A is said to be a **subset** of a second set B if

(∗) $$a \in A \Rightarrow a \in B.$$

(Implicitly the sign $\Rightarrow$ means that this holds for *all* $a \in A$.) Translated into English words this says that if a is an element of A, then it is also an element of B. Alternatively, we can say that every element in A belongs to B. The relationship in line (∗) is denoted by

$$A \subseteq B.$$

Although many authors do not distinguish between $A \subset B$ and $A \subseteq B$, we shall reserve $A \subset B$ for the situation when A is *strictly* or *properly* contained

§1.1 Set Notation

in B, meaning that the set B contains A and also elements not belonging to A. We say then that A is a **proper subset** of B. Thus,

$$A \subseteq B \Leftrightarrow (a \in A \Rightarrow a \in B),$$
$$A \subset B \Leftrightarrow (a \in A \Rightarrow a \in B, \text{ and there is an element } b \in B \text{ such that } b \notin A).$$

For every set S,

$$\emptyset \subseteq S \quad \text{and} \quad S \subseteq S.$$

The subsets $\emptyset$ and S of a set S are commonly called **trivial subsets**; any other subset is called a **nontrivial subset**.

Many proofs involve proving that two sets consist of the same elements; that is, that they are equal. The customary technique for proving the equality of two sets A and B is to prove that each is contained in the other; that is, to show that

$$a \in A \Rightarrow a \in B \quad \text{(and hence } A \subseteq B\text{)},$$

and conversely that

$$b \in B \Rightarrow b \in A \quad \text{(and hence } B \subseteq A\text{)}.$$

For sets A and B, we make the following definitions.

Difference: $\quad B \setminus A = \{b \in B : b \notin A\}.$

Union: $\quad A \cup B = \{x : x \in A \text{ or } x \in B\}.$

(The "or" used here is inclusive and means that x belongs to A or B, or to both.)

Intersection: $\quad A \cap B = \{x : x \in A \text{ and } x \in B\}.$

Symmetric difference:
$$A \triangle B = (A \cup B) \setminus (A \cap B)$$
$$= \{a \in A : a \notin B\} \cup \{b \in B : b \notin A\}.$$

Observe that for two arbitrary sets A and B, we have

$$\emptyset \subseteq A \cap B \subseteq A \subseteq A \cup B$$

and

$$\emptyset \subseteq B \setminus A \subseteq B \subseteq A \cup B.$$

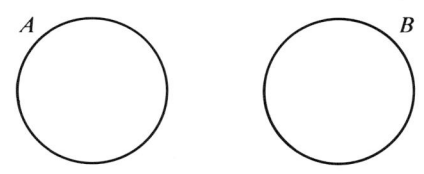

The preceding definitions are commonly illustrated by Venn diagrams, named after the English logician John Venn (1834–1923). For given sets we represent the following sets by the shaded portions of the diagrams:

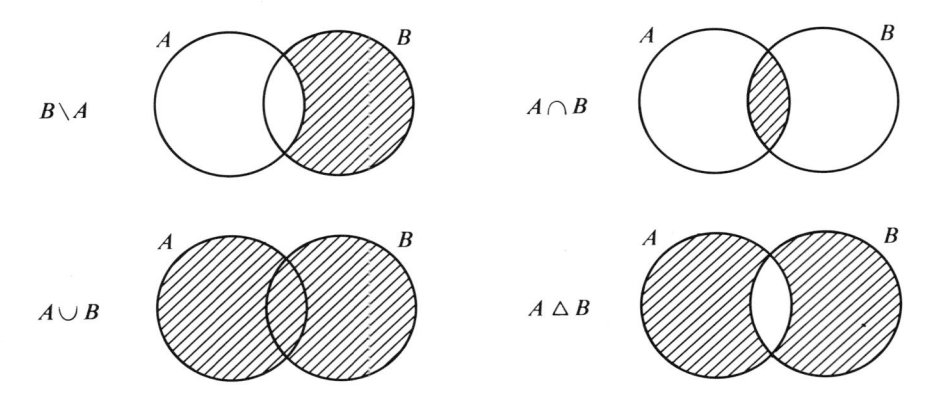

$B \setminus A$

$A \cap B$

$A \cup B$

$A \triangle B$

The definitions of union and intersection extend to a finite collection $\{A_1, ..., A_n\}$ of sets as follows.

Union: $\bigcup_{i=1}^{n} A_i = \{x : x \in A_i, \text{ for } \textit{some } i = 1, ..., n\}.$

Intersection: $\bigcap_{i=1}^{n} A_i = \{x : x \in A_i, \text{ for } \textit{all } i = 1, ..., n\}.$

With only a notational change, we can treat infinite collections of sets $\{A_i\}_{i \in I}$ indexed by an arbitrary index set I. The statement "I is an index set for a collection of sets Σ" is used in an intuitive sense, meaning simply that to each element $i \in I$ there corresponds a set A_i in the collection Σ. Also, to each set in Σ there corresponds at least one $i \in I$. The set of natural numbers is a commonly used infinite index set, but any set (finite or infinite) can serve as an index set.

REMARK. Use of an index set in referring to the union and intersection of sets in a collection Σ is convenient, but can be avoided by using

$$\bigcup_{S \in \Sigma} S \quad \text{and} \quad \bigcap_{S \in \Sigma} S$$

to denote the union and intersection, respectively, of all sets in the collection Σ.

We shall call sets A and B **disjoint** if their intersection is empty, that is, if A and B have no elements in common. We say that the sets in a collection $\{S_i\}_{i \in I}$ are **mutually disjoint** if $S_i \cap S_j = \emptyset$ for any distinct $i, j \in I$.

Given sets A and B, we often consider the set $A \times B$ of ordered pairs of elements in A and B. This set, called the **cartesian product** of A and B (after the French philosopher and mathematician René Descartes, 1596–1650), is

§1.1 Set Notation

defined to be the set

$$A \times B = \{(a, b) : a \in A, b \in B\}.$$

The pairs (a, b) are **ordered** in the sense that the first element (component) listed belongs to the first set A and the second element to the second set B. Ordered pairs (a, b), (a', b') are defined to be equal if and only if $a = a'$ and $b = b'$. The usual cartesian coordinates in the euclidean plane are ordered pairs of real numbers; the plane is simply the cartesian product $\mathbf{R}^2 = \mathbf{R} \times \mathbf{R}$ of the set of real numbers with itself (together with a distance function).

In our subsequent study of rings [§3.3] and of groups [§§6.10 and 7.1] we shall extend the concept of the cartesian product to n sets (or rings, or groups) $A_1, A_2, \ldots, A_n$:

$$A_1 \times A_2 \times \cdots \times A_n = \{(a_1, a_2, \ldots, a_n) : a_i \in A_i, i = 1, \ldots, n\},$$

the set of ordered **n-tuples** of elements $a_i \in A_i$, $i = 1, \ldots, n$.

Exercises

1. For arbitrary sets A and B, verify the following statements.
   a. $A \cup (B \setminus A) = A \cup B$
   b. $A \cap (B \setminus A) = \emptyset$
   c. $B \setminus A = B \Leftrightarrow A \cap B = \emptyset$.
2. Let $S = \{1, 2, 3, 4\}$. Write out explicitly all subsets of S. How many subsets are there? How many are proper subsets?
3. If a set has n elements, how many subsets are there? How many proper subsets?
4. For sets A, B, C verify the following statements.
   a. $A \cap (B \triangle C) = (A \cap B) \triangle (A \cap C)$
   b. $(A \triangle B) \triangle C = A \triangle (B \triangle C)$
   c. $A \triangle B = B \triangle A$
   d. $A \triangle \emptyset = A$
   e. $A \triangle A = \emptyset$
   f. $A \cap (B \cup C) = (A \cap B) \cup (A \cap C)$
   g. $A \cup (B \cap C) = (A \cup B) \cap (A \cup C)$
   h. $A \setminus (B \cap C) = (A \setminus B) \cup (A \setminus C)$
   i. $A \setminus (B \cup C) = (A \setminus B) \cap (A \setminus C)$.
5. For sets A, B, C verify that
   a. $(A \times B) \cap (A \times C) = A \times (B \cap C)$
   b. $(A \times B) \cup (A \times C) = A \times (B \cup C)$
   c. $(A \times B) \setminus (A \times C) = A \times (B \setminus C)$
   d. $(A \times B) \triangle (A \times C) = A \times (B \triangle C)$
   e. $A \times B = B \times A$ if and only if $A = B$
   f. There is a one-one correspondence between $(A \times B) \times C$ and $A \times (B \times C)$.

6. a. For subsets A, B, C of a given set S, verify that union and intersection satisfy the following statements:

   (i) $A \cup (B \cup C) = (A \cup B) \cup C$
   (ii) $A \cap (B \cap C) = (A \cap B) \cap C$.

   These are statements of the *associativity* of the union and intersection of sets. Associativity, as an algebraic concept, is discussed in §§2.1, 3.1, and 6.1, among others.

   b. State why, despite Exercise 5(f), the cartesian product of sets is not associative.

§1.2 Equivalence Relations and Classes

In this section we introduce the concept of "relatedness" of elements in a set S. For example, in the set of all living persons we are familiar with many examples of relations or relatedness. We say "a is a child of b" or "a is a sister of b" to indicate a (family) relationship. Mathematically, consider a set S, and let $\sim$ stand abstractly for a relation between two elements of S, such that for any pair a, b in S, either $a \sim b$ (read "a is in relation to b") or $a \not\sim b$ (read "a is not in relation to b").

Three common descriptive properties of relations on a set S are the following.

Reflexive property: $\qquad s \sim s \qquad$ for all $s \in S$.

Symmetric property: $s \sim s' \Rightarrow s' \sim s \qquad$ for all $s, s' \in S$.

Transitive property: $s \sim s',\ s' \sim s'' \Rightarrow s \sim s'' \qquad$ for all $s, s', s'' \in S$.

Our interest in relations is restricted to ordering relations (to be discussed in §2.2) and equivalence relations. A relation $\sim$ is called an **equivalence relation**, denoted $\approx$, if it is reflexive, symmetric, and transitive.

Several distinct equivalence relations may be defined on a particular set S. For instance, consider the fourth example below with differing values of m.

Examples of Sets with Equivalence Relations

1. In **R**, equality of real numbers is an equivalence relation.
2. For a set of triangles S, similarity of triangles is an equivalence relation on S.
3. In **R**, real numbers r, r' can be defined to be equivalent if $r - r' \in \mathbf{Z}$.
4. In **Z**, for a fixed nonzero integer m, integers a, a' can be defined to be equivalent if their difference $a - a'$ is an integral multiple of m.

Once an equivalence relation $\approx$ has been defined on a particular set S, we consider subsets of equivalent elements of S. For $s \in S$, the set

$$[s] = \{s' \in S : s' \approx s\}$$

is called the **equivalence class** or **coset** of s with respect to the relation $\approx$. (At present we shall use the term *equivalence class*; in the later study of rings [§3.2] and groups [§6.2], the term *coset* is common.) The $\approx$-equivalence class of an element $s \in S$ consists of *all* elements in S which are $\approx$-equivalent to s.

In the fourth example above,

$$[3] = \{a \in \mathbf{Z} : a - 3 \text{ is an integral multiple of } m\}.$$

In the specific case in which $m = 6$,

$$[3] = \{3 + 6k : k \in \mathbf{Z}\},$$
$$[6] = \{6k : k \in \mathbf{Z}\} = [0].$$

If $\approx$ is an equivalence relation on a set S, then every element $s \in S$ belongs to some $\approx$-equivalence class, namely $[s]$, because of the reflexive property of an equivalence relation.

We now turn to an important fact concerning equivalence classes.

Proposition. For a given equivalence relation $\approx$ on a set S, equivalence classes $[s], [a]$ are either equal or disjoint; that is,

$$[s] = [a] \quad \text{or} \quad [s] \cap [a] = \varnothing.$$

Proof. Suppose that $[s] \cap [a] \neq \varnothing$. We then show that $[s] = [a]$ by first considering $b \in [s] \cap [a]$. By definition of the equivalence classes,

$$b \in [s] \Rightarrow b \approx s, \qquad b \in [a] \Rightarrow b \approx a.$$

By symmetry and transitivity of the relation $\approx$, we have then $a \approx s$. Since $[a] = \{a' : a' \approx a\}$, again by transitivity we have $a' \approx s$ for all $a' \in [a]$. Hence $[a] \subseteq [s]$. Similarly, $[s] \subseteq [a]$, and thus $[s] = [a]$, as was to be shown.

As a corollary, whose proof is left as an exercise, we have

$$[s] = [a] \Leftrightarrow s \approx a.$$

A **partition of a set** S is a collection of mutually disjoint subsets $\{S_i\}_{i \in I}$ of S whose union is S, i.e.,

$$S_i \cap S_j = \varnothing \quad \text{unless } S_i = S_j;$$

$$S = \bigcup_{i \in I} S_i.$$

For example, the equivalence classes corresponding to an equivalence relation $\approx$ on a set S constitute a partition of S.

Conversely, given any partition $\{S_i\}_{i \in I}$ of S, indexed by some (finite or infinite) set I, we can define an equivalence relation on S as follows.

Define two elements $s, s' \in S$ to be equivalent if they belong to the same subset S_i in the partition of S. Then for each $s \in S$, $[s] = S_i$, where i is such that $s \in S_i$. Thus each equivalence relation on S defines a partition of S, and each partition gives rise to an equivalence relation.

Given an equivalence relation on S, we shall need to distinguish carefully between elements s of a set S and equivalence classes of elements of S. We have

$$s \in [s] \subseteq S.$$

Note that $[s]$ is *not an element* of S; it is a *subset* of S. However, $[s]$ is an element of the set Σ of equivalence classes in S. Thus $[s]$ will be considered both as an element and as a set. We must ascertain how it is to be considered from the context. In the same way, we should distinguish between the element $s_0 \in S$ and the subset $\{s_0\}$ of S consisting only of the single element s_0.

Exercises

1. Verify that the third and fourth examples of sets with equivalence relations satisfy the properties of an equivalence relation.
2. With reference to Example 3, show that every real number r is equivalent to some r', $0 \leq r' < 1$.
3. a. In Example 4 let m be 7 and write out all the equivalence classes.
   b. Repeat part (a) with $m = 10$.
4. Show by example that a given set might have several equivalence relations defined on it, and that in general different relations will give rise to different equivalence classes.
5. How many distinct equivalence relations can be defined on a set of three elements?
6. In a set S with equivalence relation $\approx$, prove that $[s] = [a] \Leftrightarrow s \approx a$.
7. Prove that the equivalence relations in Examples 3 and 4 satisfy the following statement:

$$x \approx y, \ x' \approx y' \Rightarrow x + x' \approx y + y'.$$

8. Prove that the equivalence relation in Example 4, but not that in Example 3, satisfies the following statement:

$$x \approx y, \ x' \approx y' \Rightarrow xx' \approx yy'.$$

9. Let $\approx$ be an equivalence relation on the finite set S. If each of the m equivalence classes with respect to $\approx$ has n distinct elements, how many elements are there in the set S? Why?
10. Consider arbitrary sets S, T and a mapping $f: S \to T$, defined for all $s \in S$. Define s' to be $\approx$-equivalent to s if $f(s) = f(s') \in T$. Prove that $\approx$ is an equivalence relation on S.

11. a. With the mapping f and relation $\approx$ defined in Exercise 10, if f satisfies the surjective property that for each $t \in T$ the set

$$\{s \in S : f(s) = t\} \neq \varnothing,$$

show that there is a one-to-one correspondence between $\approx$-equivalence classes in S and elements of T.

   b. Show that every equivalence relation $\approx$ on S determines a surjective map f to the set of $\approx$-equivalence classes of elements of S.

12. Show that the set of all lines in the cartesian plane $\mathbf{R}^2$ which are parallel to a given line constitute a partition of $\mathbf{R}^2$.

2

Arithmetic of Integers

The integers Z have properties of addition and multiplication well known to all students. We review these properties, both algebraic [§2.1] and analytic [§2.2], to include the equivalent Principles of Well-Ordering and Induction, in preparation for abstracting them to sets more general than the integers. The basic algebraic properties of interest are addition and multiplication; the analytic property of importance here is that of ordering.

Algebraic concepts of the division algorithm [§2.3], ideals [§2.4], prime element [§2.6], unique factorization [§2.7], and congruence relations [§2.8] are introduced within the context of the integers to provide rationale and experimental material for general contexts [see Chapters 3 and 5]. Unique factorization domains are studied in detail in §9.8. Ring-theoretic properties of residue classes of integers [§2.9] provide explicit examples of finite rings. We emphasize the Chinese Remainder Theorem [§2.11] because of its significance in the direct sum decomposition of Z_m presented in Chapter 3. It is also important in the theory of algebraic numbers and in algebraic geometry. The Euler φ-function and Fermat's Little Theorem are the two number theory topics presented in §2.12.

§2.1 Algebraic Properties of Integers

Familiarity with the ordinary laws for addition and multiplication of integers $a, b \in Z$, the comparison of integers ($a > b$, and equivalently $b < a$), and the properties of absolute value $|a|$ is assumed. For the sake of completeness and ease of reference, the following summaries of these axioms are given.

    I. **Equality.** The basic properties of equality (an equivalence relation) are reflexivity, symmetry, and transitivity [cf. §1.2].

    II. **Addition.** For every pair of integers a, b there exists a unique integer s called the sum of a, b and denoted by $s = a+b$. This operation of addition obeys the following axioms for all $a, b, c \in \mathbf{Z}$.

(i) Uniqueness of the sum: $\left. \begin{array}{l} a = a^* \\ b = b^* \end{array} \right\} \Rightarrow a + b = a^* + b^*.$

(We also say that the *sum is well-defined*.)
(ii) Associativity: $a + (b+c) = (a+b) + c$.
(iii) Commutativity: $a + b = b + a$.
(iv) Identity (existence of an identity): There exists an integer 0, called zero, such that $0 + a = a$ for all $a \in \mathbf{Z}$.
(v) Inverse: For each $a \in \mathbf{Z}$, there exists an integer a' such that $a + a' = 0$.

    III. **Multiplication.** For every pair of integers a, b there exists a unique integer p called the product of a, b and denoted by $p = a \cdot b$, or briefly, $p = ab$. This operation of multiplication obeys the following axioms for all $a, b, c \in \mathbf{Z}$:

(i') Uniqueness of the product: $\left. \begin{array}{l} a = a^* \\ b = b^* \end{array} \right\} \Rightarrow ab = a^*b^*.$

(We also say that the *product is well-defined*.)
(ii') Associativity: $a(bc) = (ab)c$.
(iii') Commutativity: $ab = ba$.
(iv') Identity (existence of an identity): There exists an integer 1, such that $1 \cdot a = a$, for all $a \in \mathbf{Z}$.

(v*) Cancellation: $\left. \begin{array}{l} ac = bc \\ c \neq 0 \end{array} \right\} \Rightarrow a = b.$

    The general associative laws for addition and multiplication of more than three integers are proved in the discussion on groups [§6.1]. These general laws and the proof of identities like $(a+b) + (c+d) = a + [(b+c) + d]$ are consequences of the Principle of Induction [§2.2].

    IV. **Distributive Axiom.** A connection between addition and multiplication of arbitrary integers a, b, c is provided by the distributive axiom of multiplication over addition: $a(b+c) = ab + ac$.

    Note the absence of an axiom (v'), "inverse," for multiplication analogous to the inverse axiom (v) of addition. Experience in working with integers shows that $ax = 1$ has a solution x in $\mathbf{Z}$ if and only if $a = +1$ or -1. The equation $2x = 1$ has *no integral solution*.

    We conclude this section with a series of consequences of the addition and multiplication axioms for integers; some of the proofs are left as exercises.

Property 1. The additive identity or zero element is unique. That is,
$$a + z = a \quad \text{for all } a \in \mathbf{Z} \Rightarrow z = 0.$$

Let $a = 0$. Then $0 + z = 0$, and by symmetry of the equivalence relation $=$, $0 = 0 + z$. Since $0 + z = z$, by the properties of 0, we have, by transitivity,

$$0 = 0 + z = z, \quad \text{and by symmetry } z = 0.$$

Property 2. Furthermore, $a = a + z$ for *some* $a \in \mathbf{Z}$ implies $z = 0$.

Adding to each side $a' \in \mathbf{Z}$, such that $a' + a = 0$ (i.e., an additive inverse of a), we have

$$0 = a' + a = a' + (a + z)$$
$$= (a' + a) + z = 0 + z = z.$$

Note that this result is stronger than Property 1 because of the weaker hypothesis.

Property 3. The cancellation law for addition holds; that is,

$$a + c = b + c \Rightarrow a = b.$$

Let $c' \in \mathbf{Z}$ be an additive inverse of c. Adding c' to the right of each side of the equality, we have

$$a = a + 0 = a + (c + c') = (a + c) + c'$$
$$= (b + c) + c' = b + (c + c') = b + 0 = b,$$

making use of the properties of the inverse of an element, of the 0 element, and of associativity.

Because of commutativity, we also have the left cancellation law

$$c - a = c + b \Rightarrow a = b.$$

Property 4. The additive inverse a' of an element $a \in \mathbf{Z}$ is unique.

The following proof of uniqueness of the additive inverse is typical of many uniqueness proofs in mathematics. Suppose there exist two integers a' and a'' satisfying the properties of the additive inverse; that is,

$$a + a'' = 0, \quad a + a' = 0.$$

By transitivity of the equality relation

$$a + a'' = a + a',$$

and by the law of cancellation (Property 3)

$$a'' = a'.$$

Hence the two inverses are equal.

§2.1 Algebraic Properties of Integers

COMMENT. It is common usage to let $-a$ denote the *unique* additive inverse of $a \in \mathbf{Z}$. Then $b+(-a)$ is written more simply as $b-a$. Observe that the inverse or negative of a, namely $-a$, is described as a solution of an equation involving elements of **Z**:

$$a + x = 0.$$

No appeal whatsoever is made to the ordering properties of integers.

Property 5. Every equation $a+x = b$, for $a, b \in \mathbf{Z}$, has a unique solution $x \in \mathbf{Z}$.

Adding $-a$ to each side, using associativity and the properties of 0, we have $x = -a+b$. The uniqueness of the solution follows from the cancellation law in the same way as the uniqueness of the additive inverse was proved.

Some authors cite the solvability property as an axiom in place of the existence of the inverse. The solvability and inverse axioms are equivalent, in the sense that in conjunction with associativity and identity, each implies the other.

Property 6. For any $a \in \mathbf{Z}$, $a \cdot 0 = 0$.

Since $a = a+0$, we have, using the uniqueness of multiplication,

$$aa = a(a+0) = aa + a \cdot 0, \qquad \text{by distributivity.}$$

Therefore $a \cdot 0 = 0$ by Property 2.

Property 7. The multiplicative identity is unique; that is,

$$am = a \qquad \text{for all } a \in \mathbf{Z} \Rightarrow m = 1.$$

Property 8. Furthermore, $am = a$ for *some* nonzero $a \in \mathbf{Z}$ implies $m = 1$.

Property 9. The **rule of signs** for multiplication states that for all $a, b \in \mathbf{Z}$,

$$(-a)(-b) = ab.$$

This rule is *not* dependent upon the ordering properties of the integers. It will now be proved using *only* the preceding algebraic properties.

$$
\begin{aligned}
(-a)(-b) &= (-a)(-b) + 0 && \text{property of the zero element} \\
&= (-a)(-b) + a \cdot 0 && \text{Property 6} \\
&= (-a)(-b) + a(b-b) && \text{property of inverse} \\
&= (-a)(-b) + [a(-b)+ab] && \text{distributivity, commutativity} \\
&= [(-a)+a](-b) + ab && \text{associativity, distributivity} \\
&= 0(-b) + ab && \text{property of inverse} \\
&= 0 + ab && \text{Property 6} \\
&= ab && \text{property of the zero element.}
\end{aligned}
$$

Similarly, we can prove $(-a)b = -(ab) = a(-b)$.

Property 10. The cancellation law (v*) under III (Multiplication) above is equivalent to the statement that the product of two nonzero integers is not zero.

Assuming the cancellation law we must show that
$$ab = 0, \ a \neq 0 \Rightarrow b = 0.$$
Since $ab = 0 = a \cdot 0$, by cancellation, $b = 0$. To prove the converse, note that
$$ac = bc \Rightarrow (a-b)c = 0.$$
Then $c \neq 0$ implies $a - b = 0$, or $a = b$.

The integers are an example of an algebraic structure called a ring. While rings are formally introduced in §3.1, we note now that a *ring R* is a set of elements together with two rules for composing elements (commonly called addition and multiplication) which satisfy axioms II(i) through (v); III(i'), (ii'), and (iv'); and IV.

If the multiplication in a ring R satisfies the commutative axiom III(iii') and the cancellation law III(v*), we call the ring R an *integral domain* [see §3.6] because it abstracts all of the algebraic properties of the integers.

In what follows we shall refer to **Z** as the **ring of integers**.

§2.2 Analytic Properties of Integers and Induction

The previous section emphasized addition and multiplication of integers and their governing laws. We now discuss briefly the basic rules for integers dealing with their "size" and "positiveness." These rules are generally referred to as the ordering properties of the integers. Our introduction of ordering is followed by two equivalent, fundamental principles: the Principle of Well-Ordering and the Principle of Induction (two versions of the latter are presented). The axioms of induction and well-ordering are essential in many mathematical proofs. Which of the two equivalent principles is used in a particular proof is a matter of convenience or ease of reference, so usage varies.

A word of warning is appropriate. The set of integers has rules of addition, multiplication, and ordering. These properties alone do not, however, define or categorize the integers, because other very different sets can have these same properties.

Some readers might wish to skip over the proofs of the equivalence of induction and well-ordering, as these are not commonly presented in undergraduate courses. The proofs are not so difficult as they are formal or technical.

The set of integers **Z** contains as a subset the natural numbers
$$\mathbf{N} = \{n \in \mathbf{Z} : n > 0\},$$

§2.2 Analytic Properties of Integers and Induction

where the symbol $>$ is read "greater than." The set **N** has the following properties:

Additive closure: $\qquad x, y \in \mathbf{N} \Rightarrow x + y \in \mathbf{N}.$

Multiplicative closure: $\qquad x, y \in \mathbf{N} \Rightarrow xy \in \mathbf{N}.$

The Law of Trichotomy: For $x \in \mathbf{Z}$ precisely one of the three statements
$$x \in \mathbf{N}, \qquad x = 0, \qquad -x \in \mathbf{N}$$
is true.

A simple formal consequence of these facts about **N** is that $a^2 \in \mathbf{N}$ for all nonzero integers a. To prove this assertion note that if $a \neq 0$, then either $a \in \mathbf{N}$ or $-a \in \mathbf{N}$. In the first case $a^2 \in \mathbf{N}$ in accordance with (2) above; in the second the rule of signs [§2.1, Property 9] implies that
$$a^2 = (-a)(-a) \in \mathbf{N}, \qquad \text{since } (-a) \in \mathbf{N}.$$
As a special case, $\qquad 1 = 1 \cdot 1 \in \mathbf{N}.$

Next an *ordering* of the integers is defined by writing
$$a < b \qquad (\text{equivalently, } b > a) \qquad \text{if } b - a \in \mathbf{N}$$
(i.e., if a is strictly less than b). This ordering or relation among pairs of integers satisfies:

1. $a < b$ and $b < c \Rightarrow a < c$ (transitivity).
2. $a < b \Rightarrow a + c < b + c$, for all $c \in \mathbf{Z}$.
3. $a < b \Rightarrow an < bn$, for all $n \in \mathbf{N}$.
4. $a < b \Rightarrow -b < -a$.
5. $a < b$ and $c < d \Rightarrow a + c < b + d$.

Proofs of these properties are left to the reader.

The "less than" relation or ordering of integers is transitive, but not symmetric or reflexive, and hence is *not* an equivalence relation. A similar ordering of integers is "less than or equal to," denoted by $\leq$, which is transitive and reflexive, but not symmetric.

Furthermore there is associated to each $a \in \mathbf{Z}$ its **absolute value** $|a| \in \mathbf{N} \cup \{0\}$, defined as follows:

(i) If $a \in \mathbf{N}$, then $|a| = a$.
(ii) If $a = 0$, then $|a| = |0| = 0$.
(iii) If $-a \in \mathbf{N}$, then $|a| = -a$.

Immediate consequences of the definition are that for all $a, b \in \mathbf{Z}$ the following relations hold:
$$-|a| \leq a \leq |a|, \qquad |a| = |-a|, \qquad |ab| = |a||b|.$$

The important *triangle inequality*
$$|a+b| \leq |a| + |b|$$
also holds. It is demonstrated by considering the following relations:
$$-|a| \leq a \leq |a| \quad \text{and} \quad -|b| \leq b \leq |b|.$$
Adding inequalities, we have
$$-(|a|+|b|) \leq a + b \leq |a| + |b|.$$
Hence,
$$|a+b| \leq |a| + |b|.$$

REMARK. The algebraic and analytic properties of the integers described so far are shared by other systems of numbers; for example, the rational numbers **Q**, the real numbers **R**, and real numbers of the form $a+b\sqrt{3}$, where $a, b \in \mathbf{Z}$ [see §3.6]. We now consider an ordering property that distinguishes **Z** from these other systems of numbers.

A set S is said to be **well-ordered** if every nonempty subset S' has a *least element* s_0. That is, $s_0 \leq s$ for all $s \in S'$.

Principle of Well-Ordering. The positive integers **N** form a well-ordered set.

In fact, more generally for any *fixed* $k \in \mathbf{Z}$ the subset
$$S = \{n \in \mathbf{Z} : n \geq k\}$$
of **Z** is well-ordered. Applications of the Well-Ordering Principle often involve nonnegative integers, not just the positive ones.

We accept the Principle of Well-Ordering as an axiom. After indicating some of its simple, though remarkable consequences, we show that it is equivalent to the Principle of Induction.

Proposition 1. Between 0 and 1 there are no integers.

Proof. Recall that 0 and 1 are defined to be the additive and multiplicative identities, respectively, and that we proved, using the rule of signs, that $1 > 0$. The proof that no integer m satisfies $0 < m < 1$ will be indirect. Let
$$S = \{m \in \mathbf{N} : 0 < m < 1\}$$
and suppose $S \neq \emptyset$. Since $S \subseteq \mathbf{N}$, and by assumption is nonempty, the Well-Ordering Principle ensures the existence of a least integer σ, such that $0 < \sigma < 1$. Since **N** is closed under multiplication, $\sigma^2 \in \mathbf{N}$, and further
$$0 < \sigma < 1 \Rightarrow 0 < \sigma^2 < 1 \cdot \sigma = \sigma.$$
Therefore, $\sigma^2 \in S$, but $\sigma^2 < \sigma$, contradicting the choice of σ as the least element in S. Hence S must be empty.

Thus Proposition 1 states that 1 is the *least* element in **N**. Note the following corollary.

Corollary. There are no integers between z and $z+1$, for each $z \in \mathbf{Z}$.

Proposition 2. If a set $S \subseteq \mathbf{N}$ contains 1 and if $s \in S$ implies $s+1 \in S$, then $S = \mathbf{N}$.

Proof. Again we choose an indirect argument. Let
$$S^c = \{n \in \mathbf{N} : n \notin S\}.$$
Our objective is to show that S^c is the empty set, and thus that $S = \mathbf{N}$.

Suppose that $S^c \neq \varnothing$. As a subset of **N**, S^c has a least element a by the Well-Ordering Principle. The positive integer a cannot be 1, since $1 \in S$. Therefore $a > 1$, and $a-1 \in \mathbf{N}$. Because a is the least element in S^c, $a-1 \in S$. Then by hypothesis $a = (a-1)+1$ belongs to S, and so $a \in S \cap S^c$. This intersection is empty by the definition of S^c, so that $a \notin S \cap S^c$. We have arrived at a logical contradiction and must conclude that S^c cannot be nonempty. That is, S must equal **N**.

Proposition 2 implies the following customary principle.

Principle of Induction. Assume that with each positive integer n there is associated a statement (proposition) $P(n)$ that is either true or false. Then, $P(n)$ is true for *all* positive integers n provided the following conditions hold:

(i) $P(1)$ is true.
(ii) The truth of $P(s)$, for any given $s \in \mathbf{N}$, implies the truth of $P(s+1)$.

For the proof note that the set S of positive integers s for which $P(s)$ is true satisfies the hypotheses of Proposition 2.

An alternate form of the Principle of Induction is used frequently in proofs.

Principle of Induction (*Alternate Form*). Assume, as before, that with each positive integer n there is associated a proposition $P(n)$. Then, $P(n)$ is true for *all* positive integers n provided the following conditions hold:

(i) $P(1)$ is true.
(ii) For each $m \in \mathbf{N}$, the truth of $P(s)$, for all positive $s < m$, implies the truth of $P(m)$.

Proof. The alternate form of induction can be derived directly from the Well-Ordering Principle or from the original form of the Principle of Induction. We choose the latter course, leaving the former as an exercise. If $P(n)$, for each $n \in \mathbf{N}$, is a proposition for which $P(1)$ is true and, if for each $m \in \mathbf{N}$, the truth of $P(s)$, for *all* positive $s < m$, implies that of $P(m)$, then we wish to conclude that $P(n)$ is true for all $n \in \mathbf{N}$.

Let $Q(s)$ be the proposition that $P(k)$ is true for all $k \leq s$. First note that $Q(1)$ is true, and then note that the truth of $Q(s)$ implies that of $Q(s+1)$, for all $s \in \mathbf{N}$, because the truth of $P(k)$ for all $k \leq s$ was assumed to imply the truth of $P(s+1)$, i.e., of $P(k)$ for all $k \leq s+1$. Applying the original form of induction to $Q(s)$, we conclude that $Q(n)$ is true for all $n \in \mathbf{N}$. But then $P(n)$ is true for all $n \in \mathbf{N}$, which establishes the alternate form of induction.

Conversely, if $P(n)$, for each $n \in \mathbf{N}$, is a proposition for which $P(1)$ is true and if, for any $s \in \mathbf{N}$, the truth of $P(s)$ implies that of $P(s+1)$, we can conclude, using the alternate form of induction, that $P(n)$ is true for all $n \in \mathbf{N}$. The hypotheses of the original form certainly imply those of the alternate; i.e., if the truth of $P(s)$ implies the truth of $P(s+1)$, then, *a fortiori*, the truth of $P(k)$, $k \leq s$, implies the truth of $P(s+1)$. Thus the validity of the alternate form of induction implies that of the original. Consequently the two forms are equivalent, and we shall not distinguish between them in what follows.

Proposition 3. The Principle of Induction implies the Well-Ordering Principle.

To prove this assertion we first use induction to show that 1 is the smallest positive integer or, in other words, that there are no integers between 0 and 1. The previous proof of this fact (Proposition 1) utilized the Well-Ordering Principle. Let

$$T = \{t \in \mathbf{N} : t \geq 1\}.$$

Obviously $1 \in T$, and if $n \in T$, then $n+1$, since it is greater than n, is also greater than 1 and an element of T. The Principle of Induction then implies that $T = \mathbf{N}$. Thus, $n \geq 1$ for all $n \in \mathbf{N}$, which means that 1 is the least element of $\mathbf{N}$.

We prove next that an arbitrary nonempty subset S of $\mathbf{N}$ has a least element. For each $n \in \mathbf{N}$, let $P(n)$ be the proposition that every set of positive integers containing an integer less than or equal to n has a least element. Clearly $P(1)$ is true since 1 is the least positive integer according to the preliminary observation.

Now suppose n is an integer for which $P(n)$ is true, and hence that any subset $T \subset \mathbf{N}$ containing an integer less than or equal to n possesses a least element. Consider next a set $T^* \subseteq \mathbf{N}$ which contains an element less than or equal to $n+1$. There are two possibilities to be considered. If T^* contains no integer less than $n+1$, then certainly $n+1$ is the least element of T^*. If T^* does contain an integer less than $n+1$, then it also must contain an integer less than or equal to n. Since $P(n)$ was assumed to be true, T^* has a least element. Thus, the truth of $P(n)$ implies that of $P(n+1)$. The hypotheses of the Principle of Induction are satisfied, so we conclude that $P(n)$ is true for all $n \in \mathbf{N}$.

§2.2 Analytic Properties of Integers and Induction

It is worth repeating that application of the Principle of Induction to the proof of a given proposition involves verifying that the first and second hypotheses are satisfied, and then using the conclusion of the Principle of Induction to state that the given proposition is true for all positive integers n. The first hypothesis is that $P(n)$ is true for some initial value of $n \in \mathbf{N}$. It is convenient to take this initial value to be 1, although in some particular situations it may be greater than 1. (Of course, when the induction starts at s, the result is that the statement is true for all integers $n \geq s$.) The second hypothesis is essentially that truth (assumed) for previous value(s) implies truth for the next integer value.

The exercises following this section develop several mathematical formulas with the assistance of the Principle of Induction. In proving the case for $n = s+1$, we *cannot* simply substitute into a general formula, but rather must usually express the statement for $n = s+1$ in terms of that for $n \leq s$.

As an example of the use of induction in proving formulas, consider the **Binomial Theorem** which states that for arbitrary integers a, b and for $n > 0$,

$$(a+b)^n = a^n + c_1 a^{n-1} b + \cdots + c_i a^{n-i} b^i + \cdots + b^n$$

where
$$c_i = \frac{n!}{(n-i)!(i)!} = \binom{n}{i},$$

an integer [see Exercise 10], and $n! = n(n-1)(n-2) \cdots 3 \cdot 2 \cdot 1$. Note that $c_0 = c_n = 1$. We verify that

$$\binom{s+1}{k} = \binom{s}{k-1} + \binom{s}{k},$$

and then show by induction on n that $c_i = \binom{n}{i} \in \mathbf{Z}$. Later it will be observed that the formula for $(a+b)^n$ holds for elements a and b in an arbitrary commutative ring. Then $c_i a^{n-i} b^i$ will be interpreted as the c_i-fold sum of $a^{n-i} b^i$.

For thoroughness of discussion we cite an additional fundamental property of integers. The Archimedean Principle (attributed to Archimedes of Syracuse, 287–212 B.C.), used primarily in analysis, says that for integers a and b, satisfying $0 < a < b$, there exists an integer t such that $b \leq at$.

Exercises

1. Without reference to the original form of the Principle of Induction, use the Well-Ordering Principle to prove the alternate form.
2. Prove that $n+m \neq n$ for all positive integers $n, m \in \mathbf{N}$. To start, take any $m \in \mathbf{N}$ and prove
$$\{n \in \mathbf{N} : n+m \neq n\} = \mathbf{N}.$$

3. Prove the following formulas.

   a. $\sum_{i=1}^{n} i = \dfrac{n(n+1)}{2}$

   b. $\sum_{i=1}^{n} i^2 = \dfrac{n(n+1)(2n+1)}{6}$

   c. $\sum_{i=1}^{n} i^3 = \left(\dfrac{n(n+1)}{2}\right)^2$

   d. $\sum_{i=1}^{n} i^4 = \dfrac{n(n+1)(6n^3+9n^2+n-1)}{30}$

   e. $\sum_{i=1}^{n} \dfrac{1}{i(i+1)} = \dfrac{n}{n+1}$

   f. $\sum_{i=1}^{n} (2i-1)^2 = \dfrac{n(2n-1)(2n+1)}{3}$.

4. a. Prove that $\sum_{i=1}^{n} i^{-2} < 2 - \dfrac{1}{n}$ for $n \geq 2$.

   b. Prove that $\sum_{i=0}^{n} \dfrac{1}{i!} \leq 3 - \dfrac{1}{n}$ for $n \geq 1$.

5. Prove that $x^2 > 0$ if and only if $x \neq 0$.

6. For integers $a, b_1, \ldots, b_n$, prove the following general distributive law:
$$a(b_1 + \cdots + b_n) = ab_1 + \cdots + ab_n.$$

7. Given collections of sets A_i and B_j, where $1 \leq i \leq n$ and $1 \leq j \leq m$, prove the following.

   a. $A_i \cap \left(\bigcup_{j=1}^{m} B_j\right) = \bigcup_{j=1}^{m} (A_i \cap B_j),\quad$ for fixed i.

   b. $\left(\bigcup_{i=1}^{n} A_i\right) \cap \left(\bigcup_{j=1}^{m} B_j\right) = \bigcup_{i=1}^{n} \bigcup_{j=1}^{m} (A_i \cap B_j)$.

8. Verify that
$$\binom{n+1}{i} = \binom{n}{i-1} + \binom{n}{i}$$
where
$$\binom{n}{i} = \dfrac{n!}{i!(n-i)!},$$
0! is defined to be 1, and $n! = \prod_{i=1}^{n} i$.

9. Prove the Binomial Theorem, for integers a, b:
$$(a+b)^n = \sum_{i=0}^{n} \binom{n}{i} a^{n-i} b^i.$$

The Binomial Theorem for integral powers has several independent origins. The *Precious Mirror of the Four Elements* (1303) by Chu Shih-chieh (fl. 1280–1303) begins with a diagram of the binomial coefficients through $n = 8$. In his *Algebra* the Persian poet and mathematician Omar Khayyam (c. 1050–1121/3) claimed to have found a rule for writing the sixth and higher powers of a binomial. The binomial theorem in array form for the coefficients appears in the work of al-Kashi (died c. 1436). The array of binomial coefficients for integral powers, illustrated in Figure 2.1 [cf. Exercise 8], is known in the Western world as Pascal's triangle (after Blaise Pascal, 1623–1662) although it was printed on the title page of the 1527 algebra *Rechnung* by Peter Apian (1495–1552).

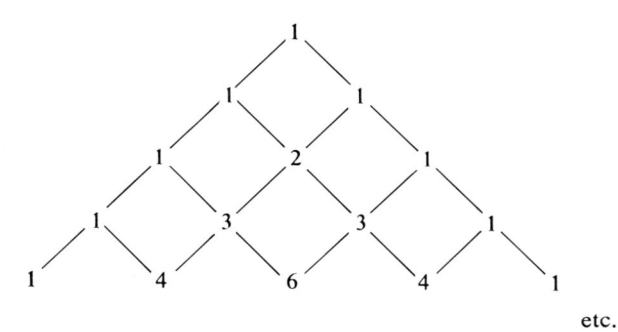

```
              1
            /   \
           1     1
          / \   / \
         1   2   1
        / \ / \ / \
       1   3   3   1
      / \ / \ / \ / \
     1   4   6   4   1
```
etc.

Figure 2.1

The corresponding theorem for nonintegral powers was discovered in 1664 or 1665 by Isaac Newton (1643–1727) and published in *Algebra* (1685) by John Wallis (1616–1703) with credit to Newton.

10. Prove that for all n, the **binomial coefficients** $\binom{n}{i}$ in Exercise 9 are integers. (The result of Exercise 8 is helpful.)
11. Prove that $\sum_{i=1}^{n} x_i^2 = 0$ for rational numbers x_i if and only if $x_i = 0$, $1 \leq i \leq n$.
12. For integers a, b prove that $a(-b) = (-a)b$. Give reasons for each step in your proof.
13. Prove that the multiplicative identity $1 \in \mathbf{Z}$ is unique.
14. Prove that $an = a$ for some nonzero $a \in \mathbf{Z}$ implies that $n = 1$.
15. Prove the Archimedean Principle for integers.
16. Let f, g be continuous real-valued functions defined on the interval $[-1, 1] \subset \mathbf{R}$, whose nth derivatives exist for all $n \in \mathbf{N}$. Find a formula for the nth derivative of the product fg and (by induction) prove its validity. (This result is due to Gottfried Wilhelm Leibniz, 1646–1716.)
17. Prove that $a^3 \leq b^3 \Leftrightarrow a \leq b$, for $a, b \in \mathbf{Z}$.
18. Prove the corollary to Proposition 1.

§2.3 The Division Algorithm

The following division algorithm for integers, often called the Euclidean Algorithm, is fundamental in algebraic number theory and group theory. Furthermore, generalizations of the division algorithm to rings other than the integers are significant in commutative algebra. (See, for example, §5.1 on polynomials.) An analogous algorithm is not valid in arbitrary rings.

Division Algorithm (for Integers). For given $a, b \in \mathbf{Z}$, $b \neq 0$, there exist *unique* integers q and r, such that

(∗) $$a = bq + r, \quad \text{where } 0 \leq r < |b|.$$

The proof will be in two parts: first the *existence* of q and r is shown, and second their *uniqueness* is demonstrated.

Case 1. Assume initially that a is nonnegative and that b is positive. Then apply the Principle of Induction to a for fixed b. Thus $P(a)$ is the statement that there exist unique $q, r \in \mathbf{Z}$ such that equation (∗) holds. We use the following steps.

    (i) For $a = 0$, take $q = r = 0$.
    (ii) For $a = 1$, if $b = 1$, take $q = 1$, $r = 0$;
           if $b > 1$, take $q = 0$, $r = 1$.
    (iii) The general inductive step consists in showing that

$$a - 1 = bq_1 + r_1, \quad \text{where } 0 \leq r_1 < b,$$

implies $\quad a = bq + r, \quad \text{where } 0 \leq r < b.$

Now, $\quad a = (a-1) + 1 = (bq_1 + r_1) + 1 = bq_1 + (r_1 + 1).$

If $r_1 + 1 = b$, let $q = q_1 + 1$, $r = 0$. Otherwise, let $q = q_1$, $r = r_1 + 1$. The second hypothesis of the Principle of Induction is satisfied [i.e., the truth of $P(a-1)$ implies the truth of $P(a)$], and we conclude that for all nonnegative a, arbitrary $b > 0$, there exist integers q, r such that $a = bq + r$.

Case 2. For the case $a \geq 0$ and $b < 0$, we have $-b > 0$ and $a = (-b)q^* + r$, $0 \leq r < -b = |b|$; therefore, let $q = -q^*$ to obtain

$$a = b(-q^*) + r = bq + r, \quad 0 \leq r < |b|.$$

Case 3. Finally, for the case $a < 0$, we have $-a > 0$, and by the preceding argument

$$-a = bq' + r', \quad \text{where } 0 \leq r' < |b|.$$

Hence $\quad a = b(-q') + (-r').$

Then if $r' = 0$, $\quad a = b(-q')$
$\qquad\qquad\qquad = bq, \quad \text{where } q = -q';$

and if $r' > 0$,

$$a = b(-q' - \varepsilon) + (|b| - r'), \quad \text{where } \varepsilon = 1 \text{ if } b > 0$$
$$\varepsilon = -1 \text{ if } b < 0$$
$$= bq + r, \quad \text{where } q = -q' - \varepsilon,\ r = |b| - r',\ \text{and } 0 \leq r < |b|.$$

To prove the *uniqueness* of q and r, assume $a = bq + r = bq^* + r^*$ where $0 \leq r, r^* < |b|$. Then $b(q - q^*) = r^* - r$, whence $|b||q - q^*| = |r^* - r|$. Now observe that the assumptions on r^* and r imply $|r^* - r| < |b|$. Also, $|q - q^*|$ is a nonnegative integer. If $|q - q^*| > 0$ (and hence at least 1),

$|b||q-q^*| \geq |b|$, but $|b||q-q^*| = |r^*-r| < |b|$. Therefore, $|q-q^*| = 0$, and hence $|r^*-r| = 0$, i.e., $q^* = q$ and $r^* = r$.

An alternate proof of the existence portion of the Division Algorithm follows from the Well-Ordering Principle. Assume $b > 0$, and let S be the set of nonnegative integers defined by

$$S = \{a-bt : t \in \mathbf{Z}, a-bt \geq 0\}.$$

If $a \geq 0$, then $a \in S$ for $t = 0$. If $a < 0$, then $ba \leq a$ for $b \geq 1$; therefore, $a - ba \in S$. Thus, in either case S is a nonempty set. Application of the Well-Ordering Principle implies that S has a least element $r \geq 0$, and $r = a - bq$, for some $q \in \mathbf{Z}$. Thus,

$$a = bq + r.$$

It remains to show that $r < b = |b|$. Suppose to the contrary that $r \geq b$; then $r = b + d$, where $0 \leq d < r$, and

$$a - bq = b + d.$$

Then
$$a - b(q+1) = d$$

is an element of S strictly less than r, contradicting the choice of r as the least element in S. Hence, r must be strictly less than b, and the existence proof is complete for positive b.

The result for $b < 0$ follows as in Case 2 of the previous proof.

§2.4 Ideals in Z

In §§2.1 and 2.3 of this chapter we considered properties of integers or pairs of integers. We now begin a more thorough examination of the algebraic structure of the integers by introducing the concept of an ideal. In §3.1 we shall consider ideals in more general rings, again as a tool for analysis of algebraic structure.

An **ideal** in the ring of integers $\mathbf{Z}$ is a nonempty subset A satisfying the following axiomatic properties:

(i) $a, b \in A \Rightarrow a + b \in A$
(ii) $a \in A$, $z \in \mathbf{Z} \Rightarrow az = za \in A$.

Example 1. The (trivial) subsets $\{0\}$ and $\mathbf{Z}$ are ideals in $\mathbf{Z}$, and are often referred to as the *trivial ideals* in $\mathbf{Z}$.

Example 2. For each $m \in \mathbf{Z}$, the set $A = \{km : k \in \mathbf{Z}\}$ is an ideal in $\mathbf{Z}$. More specifically, for $m = 6$, the set $A = \{\ldots, -12, -6, 0, 6, 12, \ldots\}$ of integral multiples of 6 is an ideal.

The reader is urged to check that A is indeed an ideal in $\mathbf{Z}$, and to write out elements of other ideals (i.e., for other values of m).

An ideal A is said to be a **principal ideal** if there exists some element $m \in A$ such that

$$A = \{km : k \in \mathbf{Z}\}.$$

In other words, there is a single element m, such that every element in A is a multiple of m. Such an ideal is commonly denoted (m) and is referred to as the **ideal generated** by m. The generator of a principal ideal is not necessarily unique. In $\mathbf{Z}$, if m generates the ideal $A = (m)$, then so does $-m$.

Theorem. Every ideal A in $\mathbf{Z}$ is principal.

Proof. For a given ideal A, we have $A = \{0\} = (0)$ or $\{0\} \subset A \subseteq \mathbf{Z}$. The last case is the one to be examined further. For a nonzero element k of A either $k > 0$ or $-k = (-1)k > 0$. In either event, the ideal A contains a positive element. Hence the set

$$P = \{a \in A : a > 0\}$$

is nonempty, and by the Well-Ordering Principle, there is a least element $m \in P$. We claim that $A = (m)$. Certainly $(m) = \{km : k \in \mathbf{Z}\} \subseteq A$, since A is an ideal. Now to prove conversely that $A \subseteq (m)$, consider an arbitrary $a \in A$. By the Division Algorithm,

$$a = mq + r, \quad \text{where } 0 \leq r < m.$$

To show that $r = 0$, observe that $mq \in A$, and hence

$$r = a - mq \in A,$$

by the properties of an ideal. We conclude that $r = 0$, since $r > 0$ leads to a contradiction, as m was chosen to be the least element in A which is positive. Thus $A \subseteq (m)$ and hence $A = (m)$, a principal ideal.

The **ideal generated by given integers** $a_1, \ldots, a_i, \ldots, a_n$, denoted $(a_1, \ldots, a_n)$, is defined to be

$$(a_1, \ldots, a_n) = \left\{ \sum_{i=1}^{n} r_i a_i : r_i \in \mathbf{Z} \right\},$$

i.e., the set of all linear combinations of the elements a_i with integral coefficients. By the preceding theorem the ideal $(a_1, \ldots, a_n)$ is principal, that is,

$$(a_1, \ldots, a_n) = (m) \quad \text{for some } m.$$

Since $m \in (a_1, \ldots, a_n)$, we can express m as an integral combination of the elements a_i:

$$m = \sum_{i=1}^{n} k_i a_i, \quad k_i \in \mathbf{Z}.$$

Of particular interest will be the case of an ideal generated by two integers a, b. Let d denote the positive generator of the ideal (a, b), assuming

that one of the integers a, b is not zero. Then
$$(d) = (a, b),$$
and there are integers s, t such that
$$d = sa + tb.$$
We leave as an exercise the proof of the following proposition.

Proposition. The set-theoretic intersection of two ideals $(a), (b)$ in **Z** is an ideal.

Exercises

1. a. Verify that the set of integral multiples of 7 is an ideal in **Z**.
    b. Verify that the set (m) of multiples of $m \in \mathbf{Z}$ is an ideal.
2. Find a (single) generator of the ideal in **Z** generated by the following integers.
    a. 2 and 3
    b. 8 and 24.
3. Find the positive generator of the smallest ideal in **Z** containing the following ideals.
    a. (4) and (18)
    b. (6) and (35).
4. If (a) and (b) are two ideals in **Z**, prove that their intersection $(a) \cap (b)$ is again an ideal in **Z**.
5. Find the positive generator of the ideal in **Z** that is the intersection of the following ideals.
    a. (2) and (3)
    b. (4) and (18)
    c. (8) and (78)
    d. (6) and (35).
6. Find q and r as in the Division Algorithm when:
    a. $a = 7, b = 12$
    b. $a = 47, b = 4$
    c. $a = 182, b = -3$
    d. $a = -189, b = -17$.
7. If a, b, c, d, x, y are integers, and $ad - bc = 1$, show that the ideals generated by x, y and by $ax + by, cx + dy$ are equal.

§2.5 Divisibility

In this section we consider the divisibility or factorization of integers. Its most significant aspect is the existence of the greatest common divisor of two integers. Proof of the existence and uniqueness follows from the Division Algorithm of §2.3 and the theorem in §2.4 that all ideals in **Z** are principal. Our discussion of factorization of integers culminates in the statement of unique factorization in §2.7.

An integer $b \neq 0$ is said to **divide** an integer a if there exists an integer q such that $a = bq$. The notational shorthand is $b | a$, read "b divides a" or

"b is a divisor (factor) of a." Similarly $b \nmid a$ means there is no $x \in \mathbf{Z}$ such that $a = bx$, and is read "b does not divide a."

Note that any integer divides zero, but zero is not a divisor of any integer. Also

$$b \mid a \Leftrightarrow a \in (b), \text{ the ideal in } \mathbf{Z} \text{ generated by } b.$$

Divisibility has the following properties.

1. For $a \neq 0$, $a \mid a$.
2. For b, c nonzero, $c \mid b$ and $b \mid a \Rightarrow c \mid a$.
3. For a, b nonzero, $a \mid b$ and $b \mid a \Rightarrow b = \pm a$.

The **greatest common divisor** (abbreviated GCD) d of integers a and b is defined as follows:

1. If $a = b = 0$, then $d = 0$.
2. If either a or $b \neq 0$, then
   (i) d is a common divisor of a and b.
   (ii) d is divisible by any other common divisor of a and b.
   (iii) $d > 0$.

Theorem. Any integers a, b have a unique GCD.

Proof. In the singular case $a = b = 0$, we define d to be 0. Thus the questions of existence and uniqueness are considered only when either a or b is different from zero.

(i) Let $A = (a, b) = \{ax + by : x, y \in \mathbf{Z}\}$ be the ideal consisting of all linear integral combinations of a and b. Then, by the theorem of §2.4, $A = (m)$, where $m = sa + tb$ with $s, t \in \mathbf{Z}$, and $m > 0$. This integer m is a common divisor of a and b. Because $a = a \cdot 1 + b \cdot 0$ and $b = a \cdot 0 + b \cdot 1$, they belong to A, whence $a = q_1 m$ and $b = q_2 m$ for some $q_1, q_2 \in \mathbf{Z}$.

(ii) If $e \mid a$, $e \mid b$, then $a = es_1$ and $b = es_2$ with $s_1, s_2 \in \mathbf{Z}$. Hence $m = es_1 s + es_2 t = e(s_1 s + s_2 t)$, which shows that $e \mid m$.

(iii) Finally $m > 0$ by the normalization of the generator of A. If δ were another GCD, then $\delta \mid m$ and $m \mid \delta$ (by definition of the GCD), and so $\delta = m$ for $\delta > 0$. Hence m is the unique GCD of a and b.

It is customary to denote the GCD of integers a and b by

$$d = (a, b).$$

Since in this expression (a, b) is interpreted as an integer, there should be no confusion with the ideal generated by a and b, also denoted (a, b), introduced in §2.4. In ideal-theoretic notation the ideals (d) and (a, b) are equal.

From part (i) of the proof, note the useful fact that the greatest common divisor $m = d = (a, b)$ of integers a, b can be expressed as a *linear*

combination of a, b. That is, given $a, b \in \mathbf{Z}$, there exist $s, t \in \mathbf{Z}$, such that
$$(a, b) = d = sa + tb.$$
The existence of such s and t will be used repeatedly in the subsequent developments of the arithmetic of integers as well as that of polynomials of one indeterminate [cf. §5.2].

If $(a, b) = 1$, then a and b are said to be **relatively prime to each other**. Thus, for relatively prime integers a, b we can find integers s, t such that
$$sa + tb = 1.$$

Proposition. For integers a, b, and m,
$$(a, m) = 1, (b, m) = 1 \Rightarrow (ab, m) = 1.$$

The proof is left as an exercise.

The GCD $d = (a_1, \ldots, a_n)$ of more than two integers, $a_1, \ldots, a_n$, not all zero, is defined as follows:

(i) $d \mid a_i$, $1 \leq i \leq n$.
(ii) If $e \mid a_i$, $1 \leq i \leq n$,   then $e \mid d$.
(iii) $d > 0$.

The GCD d is the unique positive generator of the ideal $A = (a_1, \ldots, a_n)$. Furthermore, d can be found inductively by setting
$$d_1 = (a_1, a_2), \quad d_2 = (d_1, a_3), \ldots, \quad d_i = (d_{i-1}, a_{i+1}), \ldots,$$
$$d = d_{n-1} = (d_{n-2}, a_n).$$
(Verify this statement.)

So far we have proved that two or more integers must have a unique GCD. An algorithm attributed to Euclid of Alexandria (c. 330–275 B.C.) provides for explicit computation of these GCDs. The following procedure with use of the Principle of Induction also provides an alternate proof of the existence of the GCD.

The Euclidean Algorithm. For given integers a and b, $b \neq 0$; write the following using the Division Algorithm repeatedly:

$$\begin{aligned}
a &= q_0 b + r_1, & 0 &\leq r_1 < |b|, \\
b &= q_1 r_1 + r_2, & 0 &\leq r_2 < r_1, \\
r_1 &= q_2 r_2 + r_3, & 0 &\leq r_3 < r_2, \\
&\,\,\vdots \\
r_k &= q_{k+1} r_{k+1} + r_{k+2}, & 0 &\leq r_{k+2} < r_{k+1}, \\
&\,\,\vdots \\
r_{n-1} &= q_n r_n + 0; & \text{i.e., } r_{n+1} &= 0.
\end{aligned}$$

Let n be the *smallest positive integer* such that the remainder r_{n+1} upon division of r_{n-1} by r_n is zero. Since $0 \le r_{k+2} < r_{k+1} < \cdots < r_2 < r_1 < |b|$, such an integer n must exist and can be found in a finite number (at most $|b|$) of steps. Then $r_n = (a, b)$.

Certainly, starting with the last equation of the above series of equations, we have $r_n | r_{n-1}$; hence

$$r_n | r_{n-2}, \quad \text{where } r_{n-2} = q_{n-1} r_{n-1} + r_n.$$

Ultimately, on the one hand, $r_n | b$ and $r_n | a$; hence $r_n | d$. On the other hand, $d | a$ and $d | b$ imply that $r_1 = a - q_0 b$ is divisible by d. Since $r_2 = b - q_1 r_1$, the remainder r_2 is divisible by d. Continuing in this fashion, we determine that $d | r_n$. Since $r_n > 0$, necessarily $d = r_n$.

This algorithm can be used to determine integers s and t such that $d = (a, b) = sa + tb$. From $r_{n-2} = q_{n-1} r_{n-1} + r_n$ and $r_{n-3} = q_{n-2} r_{n-2} + r_{n-1}$ we obtain

$$r_n = r_{n-2} - q_{n-1} r_{n-1},$$

$$r_n = r_{n-2} - q_{n-1}(r_{n-3} - q_{n-2} r_{n-2})$$
$$= r_{n-2}(q_{n-1} q_{n-2} + 1) + (-q_{n-1}) r_{n-3}.$$

Continuing this process, in the end we eliminate the remainders $r_{n-1}, \ldots, r_1$ in the expression for $r_n = d$. Collecting terms yields d as a linear combination of a and b.

Explicit examples will help clarify the preceding argument.

Example 1. For integers 38 and 122,

$$(38, 122) = 2 = -16 \cdot 38 + 5 \cdot 122.$$

First,
$$122 = 3 \cdot 38 + 8,$$
$$38 = 4 \cdot 8 + 6,$$
$$8 = 1 \cdot 6 + 2,$$
$$6 = 3 \cdot 2 + 0.$$

Consequently, by reversing the process, we find

$$2 = 8 - 1 \cdot 6$$
$$= 8 - 1 \cdot (38 - 4 \cdot 8) = -1 \cdot 38 + 5 \cdot 8$$
$$= -1 \cdot 38 + 5 \cdot (122 - 3 \cdot 38) = 5 \cdot 122 - 16 \cdot 38.$$

Example 2. For integers 38 and 119,

$$(38, 119) = 1 = -15 \cdot 119 + 47 \cdot 38.$$

First,
$$119 = 3 \cdot 38 + 5,$$
$$38 = 7 \cdot 5 + 3,$$
$$5 = 1 \cdot 3 + 2,$$
$$3 = 1 \cdot 2 + 1.$$

§2.5 Divisibility

Then,
$$1 = 3 - 2$$
$$= 3 - (5 - 1 \cdot 3) = 2 \cdot 3 - 5$$
$$= 2 \cdot (38 - 7 \cdot 5) - 5 = 2 \cdot 38 - 15 \cdot 5$$
$$= 2 \cdot 38 - 15 \cdot (119 - 3 \cdot 38) = 47 \cdot 38 - 15 \cdot 119.$$

We note for future use the following application of the Euclidean Algorithm:

Let n be a fixed integer greater than 1; then every positive integer a has a *unique* representation

$$a = \omega_0 + \omega_1 n + \omega_2 n^2 + \cdots + \omega_s n^s, \qquad \text{where } 0 \leq \omega_i < n,$$

called the **n-adic expansion** of a.

Many students will have encountered n-adic expansions in earlier mathematics courses concerned with bases of number systems as, for example, when considering numbers to the base 2, 7, 8, 12, 16, etc., rather than to the commonly used base 10. Most common examples are in computer applications of numbers to the base 2 (the binary number system), base 8 (octal), or base 16 (hexadecimal).

Example 3. The n-adic expansion of 38 is:

For $n = 2$: $38 = 1 \cdot 2 + 1 \cdot 2^2 + 1 \cdot 2^5$.

For $n = 8$: $38 = 6 + 4 \cdot 8$.

For $n = 16$: $38 = 6 + 2 \cdot 16$.

Example 4. The n-adic expansion of 57 is:

For $n = 3$: $57 = 1 \cdot 3 + 2 \cdot 3^3$.

For $n = 12$: $57 = 9 + 4 \cdot 12$.

Example 5. The decimal or 10-adic expansion of 412 is

$$412 = 2 + 1 \cdot 10 + 4 \cdot 10^2.$$

The proofs of existence and uniqueness of the n-adic expansion can be accomplished simultaneously by induction. If $a = 1$, let $\omega_0 = 1$ and $\omega_i = 0$ for $i > 0$. Now, to verify the second hypothesis of the Principle of Induction, assume that for all positive integers $a < k$ such a unique representation exists. To prove that k has a unique n-adic expansion, write, using the Division Algorithm,

$$k = qn + r, \qquad 0 \leq r < n, \qquad \text{where } q, r \text{ are unique.}$$

Since k, n, and r are nonnegative integers, so is q, and in fact, $q < k$. If $q = 0$, then $k \ (= r < n)$ has a unique n-adic expansion, with $\omega_0 = k$ and $\omega_i = 0$ for $i > 0$. If $q > 0$, then by the induction hypothesis,

$$q = \sigma_0 + \sigma_1 n + \sigma_2 n^2 + \cdots + \sigma_{s-1} n^{s-1}, \qquad 0 \leq \sigma_i < n.$$

Substituting this *unique* expression for q into $k = qn+r$, we have
$$\begin{aligned} k &= (\sigma_0+\sigma_1 n+\sigma_2 n^2+\cdots+\sigma_{s-1}n^{s-1})n+r \\ &= \sigma_0 n + \sigma_1 n^2 + \sigma_2 n^3 + \cdots + \sigma_{s-1}n^s + r \\ &= \omega_0 + \omega_1 n + \omega_2 n^2 + \cdots + \omega_s n^s, \end{aligned}$$
letting $\omega_0 = r$ and $\omega_i = \sigma_{i-1}$ for $i > 0$. Thus, existence of a unique n-adic expansion for all $a < k$ implies existence of a unique n-adic expansion for k. Applying induction, we conclude, for all positive integers a, the existence of a unique n-adic expansion.

Such expansions have become extremely important in the arithmetization of the theory of algebraic functions of one variable.

NOTE. The n-adic expansion of a *negative* integer need not terminate after a finite number of steps. This remark will be developed, with an example, in §2.9 following discussion of residue classes of integers.

Exercises

1. Find the greatest common divisor of each of the following pairs of integers and express it as a linear combination of the two numbers with integral coefficients.
    a. $-132, 630$ b. $44, 359$
    c. $8273, 4565$ d. $3472, 812$.

2. For arbitrary integers a, b, c prove that the following greatest common divisors are equal:
$$((a, c), b) = (a, b, c).$$

3. For integers a, b, c whose greatest common divisor is 1, prove that
$$(a+b, b, c) = 1.$$

4. Adapt the proof of the uniqueness of q and r in the Division Algorithm to prove that the coefficients ω_i, $0 \leq i \leq s$, in the n-adic expansion of a are unique.

5. Find a (single) generator of the ideal in **Z** generated by the following pairs of integers.
    a. 12 and 56   b. 8 and 78   c. 34 and 672
    d. -35 and 565   e. 48 and -1024.

6. Write out the n-adic expansion of each of the following integers for the given value of n.
    a. $7, n = 2, 3$ b. $132, n = 5$
    c. $14, n = 2$ d. $1999, n = 2, 8, 16$
    e. $1{,}000{,}000, n = 12$ f. $153, n = 7$.

7. Show that if $d = (a, b)$, the integers s, t in the expression $d = sa + tb$ are not necessarily unique. Describe the different pairs s, t.

8. a. Show that if $(a,b) = 1$, then integers s, t such that $1 = sa + tb$ also satisfy $1 = (s,b) = (s,t) = (a,t)$.
   b. Show that if $(a,b) = d$, then integers s, t such that $d = sa + tb$ also satisfy $(s,t) \mid d$. Need $d = (s,t)$?
   c. For all $s, t \in \mathbf{Z}$ show that $(a, b, as + bt) = (a,b)$.
9. Show that if $m \mid ab$ and $m \mid ac$, where $(b,c) = 1$, then $m \mid a$.
10. Prove that for integers a, b, and m,
$$(a,m) = 1 \text{ and } (b,m) = 1 \Rightarrow (ab,m) = 1.$$
11. In $\mathbf{Q}$, for a fixed nonzero integer m, we can define rational numbers $r = a/b$, $r' = a'/b'$ to be equivalent if $ab' - ba'$ is an integral multiple of m, for $a, b, a', b' \in \mathbf{Z}$; where $b, b' \neq 0$ and $(a,b) = 1$, $(a', b') = 1$. Prove that this is an equivalence relation [cf. Example 4, §1.2].
12. Prove for all $n \in \mathbf{N}$ that if the GCD of integers $a_1, \ldots, a_n$ is 1, then there exist $h_i \in \mathbf{Z}$ such that
$$h_1 a_1 + h_2 a_2 + \cdots + h_n a_n = 1.$$
13. Prove for positive integers a, b, that if $a \mid b$, then $a \leq b$.
14. a. If $(a, 4) = 2$ and $(b, 4) = 2$, prove that $(a+b, 4) = 4$.
    b. For all $n \in \mathbf{N}$, prove that $4 \nmid (n^2 + 2)$.
    c. For all $n \in \mathbf{N}$, prove that $6 \mid (n^3 - n)$.
15. In 1202 Leonardo of Pisa (c. 1180–1250), better known as Fibonacci, published a treatise, *Liber abaci*, on algebraic methods which contains the following problem:
    How many pairs of rabbits will be produced in a year, beginning with a single pair, if in every month each pair bears a new pair which becomes productive from the second month on?
    Sequences of real numbers (or integers) $\{a_n\}$ whose terms are defined recursively by $a_n = a_{n-1} + a_{n-2}$ for $n \geq 3$, where a_1, a_2 are arbitrary, are called **Fibonacci sequences**. The above problem gives rise to the particular Fibonacci sequence for which $a_1 = a_2 = 1$. Prove that the terms of this sequence satisfy
$$(a_n, a_{n+1}) = 1 \quad \text{for all } n \in \mathbf{N}.$$

§2.6 Prime Numbers

Integers a, b were defined in the preceding section to be relatively prime if their greatest common divisor $(a, b) = 1$. Now we consider properties of prime or irreducible integers, leading up to the theorem on unique factorization of integers [§2.7], which is of special significance not only for the arithmetic of integers but also ultimately in the theories of algebraic numbers and functions [see §9.8].

Any nonzero integer n has the trivial divisors $\pm 1, \pm n$. A nonzero integer p, different from ± 1, is said to be **prime** or a **prime number** if its only divisors are $\pm 1, \pm p$, that is, if it has no nontrivial divisors.

Many, but not all, authors consider only positive prime numbers. However it is more convenient for the subsequent development of irreducible elements in rings [§§5.2 and 9.8] to consider -2, -3, -5, etc., as prime numbers as well as 2, 3, 5. Obviously p is prime if and only if $-p$ is also prime.

Proposition 1. There are infinitely many positive prime numbers.

Proof. First of all, note that there exists at least one positive prime, namely $1+1 = 2$. Suppose to the contrary that the set P of *all* positive primes is finite, say $P = \{p_1, ..., p_n\}$. Let π be 1 plus the product of *all* n positive primes:

$$\pi = p_1 p_2 \cdots p_n + 1.$$

The assertion is that π is a prime not in P, which would contradict the assumption that the finite set P contains all positive primes. First, observe that $p_i \nmid \pi$, $i = 1, ..., n$. We now show that π has no proper divisors. Let D be the set of positive *proper* divisors of π. If $D \neq \emptyset$, by the Well-Ordering Principle, D has a least element s. To show that s is prime, observe that

$$s \mid \pi, \quad a \mid s, \quad a > 1 \Rightarrow a \mid \pi.$$

Therefore $a \in D$. Since s is the least positive divisor of π, a must equal s. In other words, s has no proper divisors. But $s \neq p_i$, $1 \leq i \leq n$; therefore s is a prime not in P. To avoid contradiction, we must have $D = \emptyset$, which means that π has no proper divisors, and hence is prime. Since $\pi > p_i$, $i = 1, ..., n$, π is a prime not in P. Thus P cannot then be the set of *all* positive primes.

Proposition 2. For a prime number p,

$$p \mid ab \Rightarrow p \mid a \text{ or } p \mid b.$$

This is Theorem 2 in Book VII of Euclid's *Elements*. For the proof, suppose $p \nmid a$. (If $p \mid a$, nothing has to be proved.) Then $(p, a) = 1$; thus $px + ay = 1$ with $x, y \in \mathbf{Z}$. Hence $p(bx) + (ab)y = b$, and using $p \mid ab$ or $ab = pz$ for some integer z, we find $b = p(bx + zy)$, or $p \mid b$.

Integers, other than 0 and ± 1, which are not prime are called **composite**. Note that Proposition 2 characterizes prime numbers in the sense that

$$m \mid ab \Rightarrow m \mid a \text{ or } m \mid b$$

if and only if m is prime. If m is prime, this is simply Proposition 2. A composite m can be written $m = qr$ for proper factors q, r, where $1 < |q| < |m|$, $1 < |r| < |m|$. Hence, $m \nmid q$, $m \nmid r$. In subsequent applications in ring theory it is convenient to use the above characterization of prime number, rather than our original definition that m is prime if and only if it has no proper factors.

Proposition 3. If $(c, a) = 1$ and $c \mid ab$, then $c \mid b$.

Proof. Proceed as in the proof for a prime p in Proposition 2, starting with $1 = cx + ay$. Then
$$\begin{aligned} b &= bcx + bay \\ &= bcx + czy \qquad \text{since } c \mid ab \\ &= c(bx + zy), \end{aligned}$$
which states that $c \mid b$.

Proposition 4. If $(a, c) = 1$, $a \mid m$, and $c \mid m$, then $ac \mid m$.

Proof. We have $m = ad$ and $c \mid m$; hence by Proposition 3, $c \mid d$, or $d = cc^*$. Therefore $m = acc^*$ or $ac \mid m$.

§2.7 Unique Factorization

While the following statement is sometimes referred to as the Fundamental Theorem of Arithmetic, we use the more descriptive name.

Unique Factorization Theorem. For each integer a, one of the following statements is true:

1. $a = 0$.
2. $a = \pm 1$.
3. $a = \varepsilon p_1^{\alpha_1} \cdots p_s^{\alpha_s}$, where $\varepsilon = +1$, if $a \geq 2$, and $\varepsilon = -1$, if $a \leq -2$,

and where the exponents α_i are greater than zero and the p_i are distinct positive primes. Furthermore, the integers p_i and α_i in statement 3 are uniquely determined by a.

Proof. We prove the theorem in two parts: existence and uniqueness. For the existence proof we use an indirect argument based on the Well-Ordering Principle [§2.2]. Suppose that the theorem is false, and let S be the set of positive integers $a > 1$ that cannot be expressed as the product of primes. If the theorem is false, then S must be a nonempty set, and so contain a smallest integer m. This integer m cannot be a prime, as then it would trivially be a product of primes. Therefore, m is composite with a factorization $m = bc$, where $1 < b, c < m$. Since m is the smallest integer in S, neither b nor c can belong to S; hence each is the product of primes, and therefore $m = bc$ is the product of primes and does not belong to S. This logical contradiction shows that S must be the empty set, and so the existence proof is complete.

For the uniqueness portion of the proof, suppose that an integer a can be written two ways as the product of primes:
$$a = \varepsilon_1 p_1 \cdots p_h, \qquad a = \varepsilon_2 q_1 \cdots q_k,$$

where ε_1 and ε_2 are ± 1, and $p_1, \ldots, p_h, q_1, \ldots, q_k$ are not necessarily distinct positive primes. Then $\varepsilon_1 = \varepsilon_2 = +1$ if $a > 0$, and $\varepsilon_1 = \varepsilon_2 = -1$ if $a < 0$. Next $p_1 | a$ implies $p_1 | q_1 \cdots q_k$, since $a = q_1(q_2 \cdots q_k)$. If $p_1 \neq q_1$, Proposition 2 of §2.6 implies that $p_1 | (q_2 \cdots q_k)$. Repeating this argument, if necessary, it follows that p_1 equals (at least) one of the primes $q_1, \ldots, q_k$. Relabelling these primes so that $q_1 = p_1$ we obtain, after cancellation (by axiom III(v^*) of §2.1), $p_2 \cdots p_h = q_2 \cdots q_k$. Now repeat this argument for $p_2, \ldots, p_h$. Ultimately 1 is obtained on one side after successive cancellations, hence 1 must also appear on the other side. This concludes the proof of the Unique Factorization Theorem.

REMARK. In expressing an integer a as the product of primes, we commonly collect equal prime factors and write

$$a = \pm 1 \cdot p_1^{\alpha_1} \cdots p_s^{\alpha_s},$$

where $\alpha_1, \ldots, \alpha_s$ are nonnegative integers and the p_i, $i = 1, \ldots, s$, are distinct primes.

Unique factorization provides a convenient means to determine the GCD of integers a, b. Let

(*) $\qquad a = \varepsilon p_1^{\alpha_1} \cdots p_s^{\alpha_s}, \qquad b = \varepsilon' p_1^{\beta_1} \cdots p_s^{\beta_s},$

where $\alpha_i = 0$ if $p_i \nmid a$, and similarly for β_i. Then

$$(a, b) = p_1^{\gamma_1} \cdots p_s^{\gamma_s}, \qquad \text{where } \gamma_i = \min(\alpha_i, \beta_i).$$

In addition to the GCD, we are interested in the **least common multiple** (abbreviated LCM) of two nonzero integers. For nonzero integers a, b, the LCM m is defined as follows:

(i) m is a multiple of both a and b.
(ii) m divides any integer n that is a multiple of both a and b.
(iii) $m > 0$.

It is customary to denote the LCM of a and b by $[a, b]$. For each two nonzero integers, there exists a unique positive least common multiple. The proof is left as an exercise.

If a and b have the prime factorizations in equation (*) above, then

$$[a, b] = p_1^{\delta_1} \cdots p_s^{\delta_s}, \qquad \text{where } \delta_i = \max(\alpha_i, \beta_i).$$

We have already observed that the ideal in **Z** generated by integers a and b is the same as the ideal generated by their greatest common divisor [§2.5]. Now, we conclude with two propositions concerning the LCM. The proofs are left as exercises.

Proposition 1. Consider integers a, b. Then in **Z**, the ideals $([a, b])$ and $(a) \cap (b)$ are equal.

§2.7 Unique Factorization

Proposition 2. The GCD and LCM for arbitrary nonzero integers a and b satisfy

$$|ab| = (a,b)[a,b].$$

Exercises

1. Prove that $[a, -b] = [a, b]$.
2. Repeat Exercises 1 and 5, §2.5, by comparing the unique prime factorizations of each of the integers involved.
3. Find the least common multiple of each of the following pairs of integers. [Compare Exercise 1, §2.5 and Exercise 5, §2.4.]
   a. $-132, 630$
   b. $8273, 4565$
   c. $44, 359$
   d. $3472, 812$
   e. $2, 3$
   f. $4, 18$
   g. $8, 78$
   h. $6, 35$.
4. Verify for each of the pairs of integers in Exercise 3, and prove in general, that
   $$[a,b](a,b) = |ab|$$
   for nonzero integers a, b.
5. The following statements can be proved using the prime factorization of the integers a and b.
   a. Prove that $([a,b]) = (a) \cap (b)$. Thus the set-theoretic intersection $(a) \cap (b)$ of ideals is sometimes referred to as the **least common multiple** of the ideals $(a), (b)$.
   b. Given any ideal $J \subseteq \mathbf{Z}$ for which $(a) \subseteq J$ and $(b) \subseteq J$, prove that J also contains the ideal (a, b) generated by a and b.
   c. Prove that $(a) + (b) = ((a,b))$, where the **sum of ideals** $(a), (b)$ is defined to be
   $$(a) + (b) = \{ra + sb \in \mathbf{Z}\},$$
   i.e., the set of sums of elements of (a) and of (b). The sum of ideals is also called the **greatest common divisor** of the ideals.
6. a. For integers a, b, prove that $(a) \cap (b) \supseteq (ab)$.
   b. Give examples of ideals $(a), (b)$ for which the inclusion in part (a) is an equality, and examples of ideals for which it is not.
7. Find the generator of the following ideals.
   a. $(8) \cap (5)$
   b. $(8) \cap (12)$
   c. $(8) + (5)$
   d. $(8) + (12)$
   e. $(36) \cap (27)$
   f. $(36) + (27)$.
8. For a prime number p prove that the equation $x^n = p$ has no solution of the form a/b, $a, b \in \mathbf{Z}$, for any $n > 1$.
9. For nonzero integers a, b use the Well-Ordering Principle to prove the existence of a unique positive least common multiple.

10. Prove that a positive integer of the form

$$abcabc = a\cdot 10^5 + b\cdot 10^4 + c\cdot 10^3 + a\cdot 10^2 + b\cdot 10 + c,$$

where $a \neq 0$ and $0 \leq a, b, c \leq 9$, must be divisible by 7, 11, and 13.

11. Suppose that the integers a and b are relatively prime. Let $ax_0 - by_0 = 1$ with $x_0, y_0 \in \mathbf{Z}$. Prove that $ax - by = 1$ if and only if $x = x_0 + tb$ and $y = y_0 + ta$, $t \in \mathbf{Z}$.

12. Prove that a prime integer p divides the binomial coefficients $\binom{p}{i}$, $0 < i < p$ [see Exercise 8, §2.2].

13. Prove that the integer 2 is a prime. Recall that there are no integers x or y such that $0 < x < 1 < y < 2$, by Proposition 1 and its corollary, §2.2.

14. For each prime number p, define the function $v_p(a)$ on the integers $\mathbf{Z}$ as follows:

$$v_p(0) = \infty,$$

$$v_p(a) = 0, \quad \text{if } p \nmid a,$$

$$v_p(a) = \alpha, \quad \text{if } p^\alpha \mid a, \text{ but } p^{\alpha+1} \nmid a.$$

In other words, $v_p(a)$ is the greatest integer α such that $p^\alpha \mid a$. Now given an arbitrary prime p, prove the following for $a, b \in \mathbf{Z}$.
   a. $v_p(a+b) \geq \min[v_p(a), v_p(b)]$
   b. $v_p(ab) = v_p(a) + v_p(b)$
   c. $v_p((a, b)) = \min[v_p(a), v_p(b)]$
   d. $v_p([a, b]) = \max[v_p(a), v_p(b)]$.

15. Extending the function v_p in Exercise 14 to rational numbers, define $v_p(a) = \alpha$ for $a \in \mathbf{Q}$ if $a = p^\alpha(c/d)$ where c, d are relatively prime integers, such that $p \nmid c$, $p \nmid d$. Prove the first two parts of Exercise 14 for every pair of rational numbers a and b.

16. Take $v_p(a)$, $a \in \mathbf{Q}$, as defined in Exercise 15, and define the **p-norm** of a:

$$|a|_p = (1/p)^{v_p(a)}, \qquad |0|_p = 0.$$

Now show the following.
   a. $|a+b|_p \leq |a|_p + |b|_p$
   b. $|a+b|_p \leq \max[|a|_p, |b|_p]$
   c. $|ab|_p = |a|_p |b|_p$.

17. Take $|a|_p$ as defined in Exercise 16 for $a \in \mathbf{Q}$. Prove that the classical Cauchy condition for convergence of infinite sums $\sum_{i=0}^{\infty} a_i$, $a_i \in \mathbf{Q}$, can be replaced by $\lim_{i \to \infty} |a_i|_p = 0$.

18. a. Consider arbitrary integers u, v and a prime p. Prove that if $p \mid [u, v]$, then $p \mid u$ or $p \mid v$.
    b. Extend the result of part (a) by induction to prove that if $p \mid [u_1, \ldots, u_n]$ then $p \mid u_i$ for some i, $1 \leq i \leq n$.

19. a. Prove that a necessary and sufficient condition that the equation

$$ax + by = c, \quad a, b, c \in \mathbf{Z},$$

have integral solutions x, y is that the GCD $d = (a, b)$ divide c.

b. Show further that if there exists a solution x_0, y_0, then there will be infinitely many solutions of the form

$$x = x_0 + (b/d)t, \qquad y = y_0 - (a/d)t, \qquad t \in \mathbf{Z}.$$

c. Construct explicit numerical examples of equations of the type cited in part (a). Give multiple solutions, as in part (b) for those which have solutions.

d. Suppose that $a_1, \ldots, a_n, b$ are integers. Prove that the **Diophantine equation** (after the Greek mathematician Diophantos, c. 250 A.D.)

$$a_1 x_1 + a_2 x_2 + \cdots + a_n x_n = b$$

has integral solutions if and only if the GCD d of $a_1, a_2, \ldots, a_n$ divides b.

20. Let $n = \prod_{i=1}^{s} p_i^{a_i}$ be the factorization of the positive integer n with distinct positive primes p_i. Denote by $\delta(n)$ the number of distinct positive divisors of n. Prove

$$\delta(n) = \prod_{i=1}^{s} (\alpha_i + 1).$$

First, prove $\delta(mn) = \delta(m)\delta(n)$ if $(m, n) = 1$ for positive m, n.

21. a. Use the Well-Ordering Principle to prove that every integer $n > 1$ is divisible by some positive prime p. (Note that this problem implies the existence of at least one positive prime number if we admit the existence of integers greater than 1.)

b. Using part (a), prove the existence of infinitely many positive primes. (This is the type of proof given by Euclid in his *Elements*, Book IX, Theorem 20.)

22. For integers a, b, c, d prove the following statements involving the GCD and LCM.

a. $[(a, b), c] = ([a, c], [b, c])$

b. $(ab, cd) = (a, c)(b, d)\left(\dfrac{a}{(a, c)}, \dfrac{d}{(b, d)}\right)\left(\dfrac{c}{(a, c)}, \dfrac{b}{(b, d)}\right).$

23. For a prime number p and integers a, b answer the following questions.

a. If $(a, p^2) = p$ and $(b, p^5) = p^2$, then what is $(pa + b, p^5)$?

b. If $(a, b) = p$, what is (a^2, b^5)?

§2.8 Congruences Modulo m

In this section we turn to the concept of congruence, which is all pervading in mathematics, although it is presented now only for integers. The underlying idea is to replace strict equality by a "loosened-up" concept. Rather than discussing the *equality* of integers a, b, we consider them now to be *equivalent* (or *congruent*) if they differ, not by zero, but by a multiple of a fixed integer m. Mathematicians, perhaps even Pierre de Fermat (1601–1665), made mistakes in algebraic number theory because of a lack of understanding of congruence prior to the systematic study of congruences by Carl Friedrich Gauss (1777–1855).

Application of the congruence concept permits formulation of meaningful results on the structure and explicit description of algebraic entities as, for example, in algebraic number theory [see §3.4], in the solution of polynomial equations [§5.3], and in group theory [§6.4].

Let m be a fixed integer, $m \neq 0, \pm 1$. An integer b is said to be **congruent** to an integer a **modulo** m; in symbols

$$b \equiv a \text{ modulo } m \quad \text{or} \quad b \equiv a \pmod{m},$$

if $m \mid (b-a)$, that is, $b - a = mq$ or $b = a + mq$ for some $q \in \mathbf{Z}$.

With reference to the properties of divisibility [§2.5], we can easily verify the following properties of congruence modulo m.

Property 1. Congruence modulo m is an equivalence relation [see §1.2, Example 4]. We have:

Reflexivity: $\qquad a \equiv a \pmod{m}$.

Symmetry: $\qquad a \equiv b \pmod{m} \Rightarrow b \equiv a \pmod{m}$.

Transitivity: $a \equiv b \pmod{m}$ and $b \equiv c \pmod{m} \Rightarrow a \equiv c \pmod{m}$.

Property 2. $a \equiv b \pmod{m}$ implies $a + c \equiv b + c \pmod{m}$ for all $c \in \mathbf{Z}$.

Property 3. If $a + c \equiv b + c \pmod{m}$ for some $c \in \mathbf{Z}$, then $a \equiv b \pmod{m}$.

Property 4. $a \equiv b \pmod{m}$ implies $ac \equiv bc \pmod{m}$, for all $c \in \mathbf{Z}$.

But note that $ac \equiv bc \pmod{m}$ for some $c \in \mathbf{Z}$ *does not necessarily imply* $a \equiv b \pmod{m}$. That is, a multiplicative cancellation law for integers modulo m does not hold [see §2.1, axiom III(v*)].

For example, $8 \cdot 9 \equiv 6 \cdot 9 \pmod{18}$, but $8 \not\equiv 6 \pmod{18}$ because $8 - 6$ is not a multiple of 18.

Property 5. However, $ac \equiv bc \pmod{m}$, where $(c, m) = 1$, implies $a \equiv b \pmod{m}$.

Proof. Note that $cz + my = 1$. Consequently $cz \equiv 1 \pmod{m}$, and

$$a \equiv a(cz) = (ac)z \equiv (bc)z = b(cz) \equiv b \pmod{m}.$$

An integer $a \in \mathbf{Z}$ is called a **prime residue modulo** m if $(a, m) = 1$. In particular, if p is a prime number, then $1, 2, \ldots, p-1$, and more generally $i + kp$, $1 \leq i < p$, for any $k \in \mathbf{Z}$ are prime residues modulo p.

The **coset** (also called **residue** or **congruence class**) $[a]$ modulo m of $a \in \mathbf{Z}$ is the *subset* of $\mathbf{Z}$ given by

$$[a] = \{a' \in \mathbf{Z} : a' \equiv a \pmod{m}\}.$$

We conclude this section with some comments on these cosets. In the next section we define an arithmetic for them, that is, rules for addition and multiplication of cosets.

§2.8 Congruences Modulo m

Since $a \equiv a \pmod{m}$, the coset $[a]$ is nonempty for all $a \in \mathbf{Z}$. Note that $[a]$ contains a *unique* $r \in \mathbf{Z}$ such that $0 \leq r < |m|$. Write, by the Division Algorithm [§2.3],

$$a = mq + r, \qquad 0 \leq r < |m|,$$

and observe $r \equiv a \pmod{m}$ [cf. Exercise 2, §1.2].

Thus $a' \in [a]$ means that a and a' have the same remainder r upon division by m; each $a' \in [a]$ can be written as $a' = sm + r$ for some $s \in \mathbf{Z}$.

To say that $[a] \neq [b]$ means that the residue classes $[a]$ and $[b]$, considered as subsets of $\mathbf{Z}$, have no element in common. Since congruence modulo m is an equivalence relation, the corresponding equivalence classes (i.e., the congruence classes modulo m) form a partition of $\mathbf{Z}$. By the proposition of §1.2, we then have either $[a] = [b]$ or $[a] \cap [b] = \varnothing$.

We denote the collection of distinct residue classes in $\mathbf{Z}$ modulo m by either $\mathbf{Z}/(m)$ or $\mathbf{Z}_m$, and refer to an element $a' \in [a]$ as a **representative** of the coset $[a]$. Note that there are $|m|$ "elements" in $\mathbf{Z}_m$.

Exercises

1. Write out explicitly the congruence classes in $\mathbf{Z}$ modulo m for
   a. $m = 2$
   b. $m = 3$
   c. $m = 8$
   d. $m = 10$.
2. Describe the elements of $\mathbf{Z}_m$ for
   a. $m = 4$
   b. $m = 5$.
3. Find the least positive representative of each of the following congruence classes.
   a. 5 modulo 11
   b. 5^2 modulo 11
   c. 5^8 modulo 11
   d. 628 modulo 8
   e. 17 modulo 5
   f. 641 modulo 17
   g. 42^{19} modulo 23
   h. 145 modulo 15.
4. In the text $\mathbf{Z}_m$ was defined for $m \neq 0, \pm 1$. Now extend the definition of residue classes of integers modulo m to include the cases $m = 0, \pm 1$.
   a. Describe the residue classes of $\mathbf{Z}/(1) = \mathbf{Z}/(-1)$. How many are there?
   b. What are the equivalence classes of $\mathbf{Z}$ for congruence modulo 0?
5. Prove that the congruence $ab \equiv ac \pmod{am}$ for $a \neq 0$ implies $b \equiv c$ (Compare Property 5.)
6. If $a \equiv b \pmod{m}$ and $(a, m) = 1$, prove that $(b, m) = 1$.
7. Prove for a prime p that $(a, p) > 1 \Leftrightarrow p \mid a$.
8. Let p be an odd prime. Prove that

   $$a \equiv b \pmod{p^k}, \qquad k \geq 1,$$

   implies

   $$a^p \equiv b^p \pmod{p^{k+1}}.$$

9. a. Let p be an odd prime and suppose that
$$c \equiv 1 + gp \pmod{p^2} \quad \text{with } g \in \mathbf{Z}.$$
Prove that
$$c^{p^{k-1}} \equiv 1 + gp^k \pmod{p^{k+1}} \quad \text{for all } k \geq 1.$$
  b. Prove that $c \equiv 1 + 2^2 g \pmod{2^3}$ implies
$$c^{2^{\mu-2}} \equiv 1 + g2^\mu \pmod{2^{\mu+1}} \quad \text{for } \mu \geq 2.$$
10. Prove by induction on α that for integers x and y,
$$(x+y)^{p^\alpha} \equiv x^{p^\alpha} + y^{p^\alpha} \pmod{p}$$
$$(x-y)^{p^\alpha} \equiv x^{p^\alpha} + (-y)^{p^\alpha} \pmod{p}.$$
11. Suppose that p is an odd prime. Prove that the congruences $x^2 \equiv a \pmod{p^n}$, $n > 1$, where $(a, p) = 1$, have integral solutions x if and only if the congruence $x^2 \equiv a \pmod{p}$ has an integral solution. *Hint:* The proof is by induction on n. For the inductive step try to find x_{n+1} of the form $x_{n+1} = x_n + \lambda p^n$, where $x_n^2 \equiv a \pmod{p^n}$.
12. Generalizing Exercise 11, prove that for an odd prime p,
$$x^m \equiv a \pmod{p^n}, \quad n > 1,$$
is solvable if and only if $x^m \equiv a \pmod{p}$ is solvable, where $(m, p) = (a, p) = 1$.
13. Prove that there are infinitely many positive prime numbers of the form $4n + 3$. *Hint:* Suppose there are only finitely many—say $p_1 = 3, p_2, \ldots, p_k$—and consider $m = 4p_2 \cdots p_k + 3$.

§2.9 Addition and Multiplication of Cosets Modulo m

We now define rules of addition and multiplication for cosets $[a], [b]$ modulo m, or, briefly, for elements of $\mathbf{Z}_m$ for a fixed integer m distinct from 0 and ± 1.

Define a sum $[a]+[b]$ in $\mathbf{Z}_m$ by taking $a' \in [a]$ and $b' \in [b]$ as arbitrary representatives and setting $[a]+[b] = [a'+b']$; i.e., the sum coset is the coset modulo m of the sum $a' + b'$ (in $\mathbf{Z}$) of the representatives a', b'.

Similarly, *define a product* $[a] \cdot [b]$, usually denoted simply $[a][b]$, in $\mathbf{Z}_m$ to be the coset $[a'b']$ modulo m of the product (in $\mathbf{Z}$) of the representatives a', b'.

In order that these definitions give rise to cosets $[a]+[b], [a] \cdot [b]$ *uniquely* determined by the given cosets $[a]$ and $[b]$, it is necessary to show that a different *choice of the representatives* a'' of $[a]$ and b'' of $[b]$ will lead to the *same* sum and product cosets. That is, we must prove that
$$[a'+b'] = [a''+b''] \quad \text{and} \quad [a'b'] = [a''b''].$$
For $a, a', a'' \in [a]$ and $b, b', b'' \in [b]$ we can write
$$a' \equiv a'' \equiv a \pmod{m} \quad \text{and} \quad b' \equiv b'' \equiv b \pmod{m},$$

according to the definition of cosets. Now adding congruences we have
$$a' + b' \equiv a'' + b'' \equiv a + b \pmod{m},$$
and multiplying, $\quad a'b' \equiv a''b'' \equiv ab \pmod{m}.$

Thus $[a'+b'] = [a''+b'']$ and $[a'b'] = [a''b'']$, as asserted.

We observe that the elements $[a]$ of $\mathbf{Z}_m$ satisfy the same arithmetic axioms of addition, multiplication (with the important exception of the cancellation law), and distributivity as given for elements a of $\mathbf{Z}$ in §2.1. These axioms are:

I. The sum $[a]+[b] = [a+b]$ of cosets satisfies the following properties.
   (i) Associativity: $[a]+([b]+[c]) = ([a]+[b])+[c]$.
   (ii) Commutativity: $[a]+[b] = [b]+[a]$.
   (iii) Identity: There exists a zero coset or additive identity $[0] = \{km : k \in \mathbf{Z}\}$ such that $[0]+[a] = [a]$ for all $[a] \in \mathbf{Z}_m$.
   (iv) Inverse: For each $[a] \in \mathbf{Z}_m$, there exists an element $[a'] \in \mathbf{Z}_m$ such that $[a]+[a'] = [0]$.

II. The product $[a][b] = [ab]$ of cosets satisfies the following properties.
   (i) Associativity: $[a]([b][c]) = ([a][b])[c]$.
   (ii) Commutativity: $[a][b] = [b][a]$.
   (iii) Identity: There exists an element (the class of 1 modulo m) which plays the role of multiplicative identity. That is, $[1][a] = [a][1] = [a]$.

III. The sum and product of cosets satisfy the distributivity property:
$$[a]([b]+[c]) = [a][b]+[a][c].$$

We call $\mathbf{Z}_m$ the *residue class ring of integers modulo* m. Such rings were first constructed by Adrien-Marie Legendre (1752–1833). Note carefully that in general the cancellation law *does not hold* in $\mathbf{Z}_m$. For example, let $m = 18$, then
$$[8][9] = [6][9] \quad \text{but} \quad [8] \neq [6].$$

Proofs of the preceding statements are obtained by appealing to the corresponding facts valid in $\mathbf{Z}$. For example, consider the distributive law. Let $a' \in [a]$, $b' \in [b]$ and $c' \in [c]$ be any representatives of the given cosets. Then, in $\mathbf{Z}$, $a'(b'+c') = a'b'+a'c'$. Consequently $[a'(b'+c')] = [a'b'+a'c']$ for the corresponding cosets. Hence, using the definitions of sum and product in $\mathbf{Z}_m$, we have
$$[a'(b'+c')] = [a][b'+c'] = [a]([b']+[c']) = [a]([b]+[c]),$$
and similarly,
$$[a'b'+a'c'] = [a'b']+[a'c'] = [a][b]+[a][c].$$

To illustrate the importance of the choice of coset representatives, we consider, as at the end of §2.5, the *n*-adic expansion of any $a \in \mathbf{N}$ for a fixed $n \in \mathbf{N}$. Such an expansion need not be finite if a *fixed* system of representatives for the *n* residue classes modulo n is used. For example let $n = 5$ and take 0, 1, 2, 3, and 4 to be *the* representatives of the cosets of $\mathbf{Z}_5$. Then $-1 \equiv 4 \pmod 5$, where $-1 = 4 + -1 \cdot 5$ and $-1-4 = -5 \equiv -1 \cdot 5 \pmod{5^2}$. Consequently,

$$\begin{aligned} -1 &= 4 \cdot 5^0 + -1 \cdot 5^1 = 4 \cdot 5^0 + (4 \cdot 5^0 + -1 \cdot 5)5 \\ &= 4 \cdot 5^0 + 4 \cdot 5^1 + -1 \cdot 5^2 \\ &= 4 \cdot 5^0 + 4 \cdot 5^1 + (4 \cdot 5^0 + -1 \cdot 5)5^2 \\ &= 4 \cdot 5^0 + 4 \cdot 5^1 + 4 \cdot 5^2 + -1 \cdot 5^3, \quad \text{etc.} \end{aligned}$$

Thus $-1 = \sum_{v=0}^{\infty} 4 \cdot 5^v$, an infinite sum which does not make sense for the ordinary constructions of real and complex analysis with the customary absolute value $|\ |$ as in §2.2. However, if we define $|a|_5$ to be $(1/5)^\alpha$ where $a = 5^\alpha \cdot (b/c)$, $a \in \mathbf{Q}$ and $(b,c) = 1$ for $b,c \in \mathbf{Z}$, then the infinite sum turns out to be convergent. (This discussion of $|\ |_5$ is a special case of the p-norm defined in Exercise 16, §2.7. See also Exercise 17, §2.7.)

Exercises

1. For $n = a_0 + 10a_1 + \cdots + 10^k a_k$, prove that

$$n \equiv 0 \pmod 3 \Leftrightarrow \sum_{i=0}^{k} a_i \equiv 0 \pmod 3.$$

The following problem furnishes the basis for the bookkeeper's rule of *casting out nines*, which is used for checking addition and multiplication.

2. a. Let s be the sum of the decimal digits of an integer a. Show that $a \equiv 0 \pmod 9$ if $s \equiv 0 \pmod 9$.
   b. Prove that the remainder of a sum upon division by 9 is equal to the sum (reduced modulo 9) of the remainders of the addends upon division by 9.
   c. Prove that the remainder of a product upon division by 9 is equal to the product (reduced modulo 9) of the remainders upon division by 9.
   d. Prove that if an integer m is obtained from n by permuting the digits of n, then $m \equiv n \pmod 9$.

"It is often said that ... the 'casting out of nines' is a Hindu invention, but it appears that the Greeks knew earlier of this property, without using it extensively, and that the method came into common use only with the Arabs of the eleventh century."[†]

3. Find all powers of the following.
   a. $[6]_{13}$ in $\mathbf{Z}_{13}$
   b. $[2]_{10}$ in $\mathbf{Z}_{10}$
   c. $[-9]_{11}$ in $\mathbf{Z}_{11}$
   d. $[6]_{27}$ in $\mathbf{Z}_{27}$

4. For any $a \in \mathbf{Z}$, prove that precisely one of the congruences

$$a^2 \equiv 0, \quad a^2 \equiv 1, \quad a^2 \equiv 4$$

holds modulo 8.

[†] From Carl Boyer, *A History of Mathematics*, p. 241, © 1968 by John Wiley & Sons, Inc.

5. Prove that no integer a satisfies the congruence $a^2 \equiv 17 \pmod{100}$.
6. Prove that the congruence $ax \equiv b \pmod{m}$ has a solution if $(a, m) = 1$.
7. Prove the associative and commutative laws of addition and multiplication for the arithmetic of cosets. Note that the proofs are essentially carried out in $\mathbf{Z}$ and transfered to $\mathbf{Z}_m$.
8. Match each nonzero element in $\mathbf{Z}_{11}$ with its multiplicative inverse.
9. Repeat Exercise 8 for $\mathbf{Z}_7$ and $\mathbf{Z}_{17}$.
10. What elements in $\mathbf{Z}_6$ and $\mathbf{Z}_{12}$ have multiplicative inverses?

§2.10 Definitions and Examples

In this section we consider the residue class rings $\mathbf{Z}_m$ for several values of m, explicitly giving the addition and multiplication tables for $m = 2, 3, 4,$ and 6. We also introduce ring-theoretic terminology for elements in $\mathbf{Z}_m$ with particular properties: prime residue or unit, zero divisor, nilpotent, and idempotent. These special elements and their properties are examined in specific instances in the rings $\mathbf{Z}_m$ so as to provide an elementary setting (where all computations can be executed easily) for the introduction of concepts necessary for the later study of rings, groups, and algebraic number theory, algebraic geometry, etc. These definitions, which carry over to general ring theory [§3.1] and polynomial rings [Chapter 5], provide an opportunity to gain experience in computing with elements of the residue class rings $\mathbf{Z}_m$.

The final topic of this section is the solvability theory for congruences (modulo m) of integers, which is equivalent to solving linear equations in $\mathbf{Z}_m$. Section 2.11 treats the general theory for solving simultaneous congruences.

In the following examples note that $\mathbf{Z}_m$ always satisfies the algebraic axioms II(i) through (v) and III(i') through (iv') of §2.1, but that the cancellation law (v*) is valid *only* in specific instances.

We call $[a]$ a **prime residue class modulo** m if some (and hence any) representative a' is a prime residue modulo m, that is, if $(a', m) = 1$. If m is prime, $[a]$ is a prime residue class for all a, $1 \leq a < m$. Also, for example, $[10]$ is a prime residue mod 21.

A nontrivial **divisor of zero** in $\mathbf{Z}_m$ is a residue class $[a] \neq [0]$ for which there exists a nonzero residue class $[b] \in \mathbf{Z}_m$ satisfying $[a][b] = [0]$. For example, in $\mathbf{Z}_{18}$, if $[a] = [3]$, then $[b] = [12]$ and $[b] = [6]$ satisfy $[a][b] = [0]$.

A **nilpotent element** in $\mathbf{Z}_m$ is a residue class $[a]$ satisfying

$$[a]^h = [0], \quad \text{in other words,} \quad m \mid a^h,$$

for some positive integer h. (The smallest positive h having this property is called the **index** or **exponent of nilpotency** of $[a]$.) Thus, the coset $[6] \neq [0]$ is a nilpotent element (of index 2) in $\mathbf{Z}_{18}$ because $[6]^2 = [36] = [0]$. The zero class $[0]$ is *trivially* nilpotent.

Note that divisors of zero are not *necessarily* nilpotent elements. For example, $[4] \in \mathbf{Z}_{18}$ is a divisor of zero—$[4][27] = [4][9] = [36] = [0]$—but it is not a nilpotent element because no power $[4]^h$ equals $[0]$, that is, $4^h = 2^{2h}$ is not divisible by 18 for any positive integer h. Similarly, $[3]$ is a divisor of zero, but is not nilpotent.

We now illustrate the preceding concepts in examples of addition and multiplication tables for $\mathbf{Z}_m$. For the sake of clarity in the tables all residue classes $[a]$ modulo m are denoted by overlining, viz., $\bar{a}$.

Example 1. For $m = 2$.

Addition

	$\bar{0}$	$\bar{1}$
$\bar{0}$	$\bar{0}$	$\bar{1}$
$\bar{1}$	$\bar{1}$	$\bar{0}$

Multiplication

	$\bar{0}$	$\bar{1}$
$\bar{0}$	$\bar{0}$	$\bar{0}$
$\bar{1}$	$\bar{0}$	$\bar{1}$

Note the absence of nontrivial zero divisors.

Example 2. For $m = 3$.

Addition

	$\bar{0}$	$\bar{1}$	$\bar{2}$
$\bar{0}$	$\bar{0}$	$\bar{1}$	$\bar{2}$
$\bar{1}$	$\bar{1}$	$\bar{2}$	$\bar{0}$
$\bar{2}$	$\bar{2}$	$\bar{0}$	$\bar{1}$

Multiplication

	$\bar{0}$	$\bar{1}$	$\bar{2}$
$\bar{0}$	$\bar{0}$	$\bar{0}$	$\bar{0}$
$\bar{1}$	$\bar{0}$	$\bar{1}$	$\bar{2}$
$\bar{2}$	$\bar{0}$	$\bar{2}$	$\bar{1}$

Note the absence of nontrivial zero divisors.

Example 3. For $m = 4$.

Addition

	$\bar{0}$	$\bar{1}$	$\bar{2}$	$\bar{3}$
$\bar{0}$	$\bar{0}$	$\bar{1}$	$\bar{2}$	$\bar{3}$
$\bar{1}$	$\bar{1}$	$\bar{2}$	$\bar{3}$	$\bar{0}$
$\bar{2}$	$\bar{2}$	$\bar{3}$	$\bar{0}$	$\bar{1}$
$\bar{3}$	$\bar{3}$	$\bar{0}$	$\bar{1}$	$\bar{2}$

Multiplication

	$\bar{0}$	$\bar{1}$	$\bar{2}$	$\bar{3}$
$\bar{0}$	$\bar{0}$	$\bar{0}$	$\bar{0}$	$\bar{0}$
$\bar{1}$	$\bar{0}$	$\bar{1}$	$\bar{2}$	$\bar{3}$
$\bar{2}$	$\bar{0}$	$\bar{2}$	$\bar{0}$	$\bar{2}$
$\bar{3}$	$\bar{0}$	$\bar{3}$	$\bar{2}$	$\bar{1}$

Note that the nontrivial zero divisor is nilpotent.

§2.10 Definitions and Examples

Example 4. For $m = 6$.

Addition

+	$\bar{0}$	$\bar{1}$	$\bar{2}$	$\bar{3}$	$\bar{4}$	$\bar{5}$
$\bar{0}$	$\bar{0}$	$\bar{1}$	$\bar{2}$	$\bar{3}$	$\bar{4}$	$\bar{5}$
$\bar{1}$	$\bar{1}$	$\bar{2}$	$\bar{3}$	$\bar{4}$	$\bar{5}$	$\bar{0}$
$\bar{2}$	$\bar{2}$	$\bar{3}$	$\bar{4}$	$\bar{5}$	$\bar{0}$	$\bar{1}$
$\bar{3}$	$\bar{3}$	$\bar{4}$	$\bar{5}$	$\bar{0}$	$\bar{1}$	$\bar{2}$
$\bar{4}$	$\bar{4}$	$\bar{5}$	$\bar{0}$	$\bar{1}$	$\bar{2}$	$\bar{3}$
$\bar{5}$	$\bar{5}$	$\bar{0}$	$\bar{1}$	$\bar{2}$	$\bar{3}$	$\bar{4}$

Multiplication

·	$\bar{0}$	$\bar{1}$	$\bar{2}$	$\bar{3}$	$\bar{4}$	$\bar{5}$
$\bar{0}$	$\bar{0}$	$\bar{0}$	$\bar{0}$	$\bar{0}$	$\bar{0}$	$\bar{0}$
$\bar{1}$	$\bar{0}$	$\bar{1}$	$\bar{2}$	$\bar{3}$	$\bar{4}$	$\bar{5}$
$\bar{2}$	$\bar{0}$	$\bar{2}$	$\bar{4}$	$\bar{0}$	$\bar{2}$	$\bar{4}$
$\bar{3}$	$\bar{0}$	$\bar{3}$	$\bar{0}$	$\bar{3}$	$\bar{0}$	$\bar{3}$
$\bar{4}$	$\bar{0}$	$\bar{4}$	$\bar{2}$	$\bar{0}$	$\bar{4}$	$\bar{2}$
$\bar{5}$	$\bar{0}$	$\bar{5}$	$\bar{4}$	$\bar{3}$	$\bar{2}$	$\bar{1}$

Note that no nontrivial zero divisor is nilpotent.

The following three propositions summarize and generalize the observations made in these examples.

Proposition 1. In $\mathbf{Z}_m$, a coset $[a]$ is a nontrivial zero divisor if and only if $(a, m) > 1$.

Proof. First, if $[a]$ is a zero divisor, then there exists a coset $[b] \neq [0]$, such that $m | ab$. If $(a, m) = 1$, then by Proposition 3, §2.6, $m | b$, and $[b] = [0]$, a contradiction. Hence $(a, m) \neq 1$. Now conversely, if $(a, m) = d > 1$ we may write

$$m = qd, \quad 0 < q < m, \quad \text{and} \quad a = ds, \quad 0 < s < a.$$

Then, $ms = qds = qa$, and so $[q][a] = [0]$. Since $0 < q < m$, $m \nmid q$ and so $[q] \neq 0$. Therefore $[a]$ is a zero divisor.

Proposition 2. If $m = p_1 \cdots p_h$ with distinct primes p_i, $h \geq 2$, then $\mathbf{Z}_m$ has no nonzero nilpotent elements.

Proof. Any nilpotent element is a zero divisor. For any nontrivial zero divisor $[a]$ in $\mathbf{Z}_m$, one of the prime factors $p_1, \ldots, p_h$, say p_e, must not divide a. Hence $p_e \nmid a^s$, for any $s \in \mathbf{N}$, which means that $[a]$ cannot be nilpotent, as asserted.

Proposition 3. If $m = p^\alpha$, $\alpha \geq 2$, then every divisor of zero in $\mathbf{Z}_m$ is nilpotent.

Proof. Clearly, if $[a]$ is a zero divisor, it is necessary that $p^\beta | a$, for some β, $1 \leq \beta < \alpha$, because there exists a coset $[b] \neq [0]$ such that $p^\alpha | ab$. Pick s so that $s\beta \geq \alpha$, and observe that $p^\alpha | a^s$ or equivalently, $[a]^s = [0]$.

An **idempotent** (element) of $\mathbf{Z}_m$ is a coset $[a]$ for which

$$[a]^2 = [a^2] = [a];$$

or, in other words, a solution of the equation $x^2 - x = [0]$. For example, if $m = 36$, then $[28]^2 = [28]$. In $\mathbf{Z}_{18}$, $[10]$ is an idempotent element which is also a divisor of zero: $[10][9] = [0]$. We refer to $[0], [1]$ as *trivial idempotents* in $\mathbf{Z}_m$ for any m.

A **unit** of $\mathbf{Z}_m$ is a coset $[a]$ for which there exists a (necessarily unique) coset $[b]$ such that $[a][b] = [1]$. We then speak of $[b]$ as the **multiplicative inverse** of $[a]$. While we write $[b] = [a]^{-1}$, we have *no illusions* that the coset $[a]^{-1} \in \mathbf{Z}_m$ contains proper fractions. As we show next, the units $[a]$ of $\mathbf{Z}_m$ are precisely the prime residue classes modulo m.

To wit, if $(a, m) = 1$, then $as + mt = 1$ for some $s, t \in \mathbf{Z}$ as in §2.5. Hence $as \equiv 1 \pmod{m}$, or equivalently $[a][s] = [1]$. Conversely, if $[a]$ is a unit, there exists $[b]$ such that $[a][b] = [1]$. Then $ab + mq = 1$ for some $q \in \mathbf{Z}$; hence $(a, m) = 1$.

This argument generalizes as follows to the solution of equations $[a][x] = [b]$ in $\mathbf{Z}_m$.

Proposition 4. The congruence $ax \equiv b \pmod{m}$ is solvable if and only if $d = (a, m)$ divides b. The solutions, if any, differ by multiples of $q = m/d$.

Proof. First, we prove the existence of a solution assuming that $d \mid b$. Write $d = sa + tm$ and $b = qd$. Then

$$b = qd = qsa + qtm$$

and $x = qs$ satisfies the congruence.

Second, assume that the congruence is solvable. Then $b = ax + mw$ for some $w \in \mathbf{Z}$. Hence $d \mid b$, because $d \mid a$ and $d \mid m$.

Finally, suppose that x^* is another solution of the congruence $ax \equiv b \pmod{m}$. Then $a(x - x^*) \equiv 0 \pmod{m}$. Consequently, writing $a = dv$ and $m = dq$ since $d = (a, m)$, we have $dv(x - x^*) \equiv 0 \pmod{dq}$, and $v(x - x^*) \equiv 0 \pmod{q}$. Therefore $x - x^* \equiv 0 \pmod{q}$ as asserted, because $(v, q) = 1$. Furthermore, if x^* is a solution of $ax \equiv b \pmod{m}$, then so is any $x' \equiv x^* \pmod{q}$.

Proposition 4 provides a useful tool for the study of cyclic groups in §6.6.

Exercises

1. Find all zero divisors in the residue class ring $\mathbf{Z}_m$ for
   a. $m = 20$
   b. $m = 16$
   c. $m = 7$
   d. $m = 24$.
2. Find all nilpotents in each of the rings in Exercise 1.

§2.10   Definitions and Examples   47

3. Give necessary and sufficient conditions on m for each of the following statements.
   a. No proper zero divisor of $\mathbf{Z}_m$ is nilpotent.
   b. All proper zero divisors of $\mathbf{Z}_m$ are nilpotent.
4. Find *all* x in the residue class ring $\mathbf{Z}_m$ such that $x^2 = [1]$, the multiplicative identity in $\mathbf{Z}_m$, for
   a. $m = 3$ b. $m = 8$ c. $m = 6$
   d. $m = 4$ e. $m = 10$ f. $m = 12$.
5. For each of the following congruences describe all integral solutions.
   a. $13x \equiv 9 \pmod{19}$ b. $4x \equiv -1 \pmod{21}$
   c. $2x \equiv 17 \pmod{27}$ d. $3x \equiv 36 \pmod{54}$
   e. $3x \equiv 54 \pmod{36}$ f. $12x \equiv -1 \pmod{35}$.
6. Show that any elements $[a], [b] \in \mathbf{Z}_m$ which satisfy
$$[a] + [b] = [1] \quad \text{and} \quad [a][b] = [0]$$
are idempotent. Further, show that for each nontrivial idempotent $[a] \in \mathbf{Z}_m$ there exists an idempotent $[b]$ such that the above identities are satisfied.
7. Show that the congruences
$$a + b \equiv 1 \pmod{p^2} \quad \text{and} \quad ab \equiv 0 \pmod{p^2}$$
have only the solutions $a \equiv 1$, $b \equiv 0$, or $a \equiv 0$, $b \equiv 1$. Conclude that $\mathbf{Z}_{p^\alpha}$ has no trivial idempotents for $\alpha > 1$. (The discussion at the end of §5.2 indicates that $\mathbf{Z}_p$ has no nontrivial idempotents.)
8. With reference to Exercises 6 and 7, write out explicitly all idempotents in the residue class ring $\mathbf{Z}_m$ for
   a. $m = 20$ b. $m = 70$ c. $m = 75$
   d. $m = 16$ e. $m = 24$ f. $m = 30$.
9. Prove that $[a] \in \mathbf{Z}_m$ is a nontrivial idempotent if and only if there exist m_1, m_2 such that
   (i) $m = m_1 m_2$, $(m_1, m_2) = 1$,
   (ii) $a = sm_2$,   where $sm_2 \equiv 1 \pmod{m_1}$.
10. Prove that $[a] \in \mathbf{Z}_m$ is either a unit or a zero divisor.
11. Prove that $[a] \in \mathbf{Z}_m$ is nilpotent if and only if every prime divisor of m divides a.
12. a. Prove that $\mathbf{Z}_m$ has nontrivial nilpotents if and only if $n^2 | m$ for some $n > 1$.
    b. Describe the set N of all nilpotents in the residue class ring $\mathbf{Z}_m$.
    c. Prove that N is a principal ideal and describe a generator.
13. Prove that if $[a] \in \mathbf{Z}_m$ is idempotent, then $[1-a]$ is also idempotent.
14. In the residue class ring $\mathbf{Z}_{p^\alpha}$ show that
    a. $[a]$ is a unit if $(a, p^\alpha) = 1$.
    b. $[a]$ is nilpotent if $(a, p^\alpha) \neq 1$.
15. a. If $x^2 + x + 1 \equiv 0 \pmod{m}$, prove that $[x]$ is not a zero divisor in $\mathbf{Z}_m$.
    b. If $[x]$ is nilpotent in $\mathbf{Z}_m$, prove that $[1 + x + x^2]$ is a unit in $\mathbf{Z}_m$.
    c. Prove that $x^2 + x + 1 \equiv 0 \pmod{3 \cdot 7 \cdot 8}$ has no integral solution.
    d. Find all integral solutions of $x^2 + x + 1 \equiv 0 \pmod{3 \cdot 7}$.

§2.11 Simultaneous Systems of Congruences

In §2.9 we listed the additive and multiplicative properties of the congruence classes $[a] \in \mathbf{Z}_m$. Then, Proposition 4 in the preceding §2.10 gave necessary and sufficient conditions for the solution of the equation

$$[a][x] = [b]$$

in $\mathbf{Z}_m$, or equivalently, of the congruence

$$ax \equiv b \pmod{m}.$$

In this section we take up necessary conditions and techniques for solving systems of congruence equations simultaneously and prove the so-called Chinese Remainder Theorem. This theorem on the solution of compatible systems of congruences has extremely important generalizations in the theories of algebraic numbers and of algebraic functions of one and more variables. With the Fundamental Theorem of Arithmetic (unique factorization of integers) it is one of the foundations of class field theory, a currently active area of mathematical research that involves abelian extensions (discussed further in Chapter 9). We shall make use of the theorem in §§3.4 and 5.3.

The general method for solving simultaneous congruences is best explained after examining two examples.

Example 1. Consider two congruences,

$$y \equiv 18 \pmod{7} \quad \text{and} \quad y \equiv 3 \pmod{12}.$$

Any solution y of the first congruence must be of the form

$$y = 18 + 7k.$$

Substituting this general solution of the first congruence into the second, we obtain

(*) $\qquad\qquad\qquad 18 + 7k \equiv 3 \pmod{12}$

or $\qquad\qquad\qquad 7k \equiv -15 \equiv 9 \pmod{12}.$

Since $(7, 12) = 1$, we can write $1 = 7s + 12t$. In fact,

$$1 = -5 \cdot 7 + 3 \cdot 12.$$

Now multiplying by 9, we have

$$-45 \cdot 7 = +9 - 27 \cdot 12$$

from which we see that $k = -45 \equiv 3 \pmod{12}$ satisfies congruence (*). Substituting $k = 3$ into $y = 18 + 7k$, we obtain $y = 39$ as a particular solution of the double congruence. The general solution is

$$y = 18 + 7(3 + 12q) = 39 + 7 \cdot 12q.$$

Example 2. Consider the system of congruences

$$2x \equiv -3 \pmod{7},$$
$$4x \equiv 1 \pmod{9},$$
$$x \equiv 5 \pmod{13}.$$

Following the procedure of Example 1, we solve for x satisfying the first congruence, obtaining
$$x = 2 + 7r.$$
Putting this value for x into the second congruence, we have
$$4(2+7r) \equiv 1 \pmod{9},$$
$$8 + 28r \equiv 1 \pmod{9},$$
$$r \equiv -7 \pmod{9},$$
(**) $$r \equiv 2 \pmod{9}.$$
The general solution of congruence (**) is then
$$r = 2 + 9s$$
and of the first and second congruences of the original system is
$$x = 2 + 7(2+9s) = 16 + 7\cdot 9s.$$
We repeat the process, rewriting the third congruence as
$$16 + 63s \equiv 5 \pmod{13}$$
or (***) $$11s \equiv 2 \pmod{13}.$$
Writing $1 = (11, 13)$ as $1 = 6\cdot 11 - 5\cdot 13$, and multiplying by 2, we obtain
$$12\cdot 11 \equiv 2 \pmod{13}.$$
The general solution of congruence (***) is then
$$s = 12 + 13t,$$
and of the set of three congruences is
$$x = 16 + 7\cdot 9(12 + 13t) = 772 + 7\cdot 9\cdot 13t.$$

Note that in specific cases some congruences, such as $2x \equiv -3 \pmod{7}$, can be solved at sight, $x = 2$, without resort to the formal techniques of solution, which involve expressing $1 = (2, 7)$ as a combination of 2 and 7.

With these two examples in mind we consider the Chinese Remainder Theorem proper, an existence and uniqueness statement for the solution of a general system of congruences. The name is derived from the fact that the Chinese mathematician Sun-Tsu found solutions of $x \equiv 2 \pmod 3$, $x \equiv 3 \pmod 5$, and $x \equiv 2 \pmod 7$ in the first century A.D. and subsequent Chinese mathematicians have considered similar problems.

The Chinese Remainder Theorem. Let $m_1, \ldots, m_n$ be n pairwise relatively prime integers, each greater than 1 (that is, $(m_i, m_j) = 1$, if $i \neq j$, $1 \leq i, j \leq n$) and let $a_1, \ldots, a_n$ be n arbitrary integers. Then the system of n congruences

$$x \equiv a_1 \pmod{m_1}$$
$$\cdot \quad \cdot \quad \cdot \quad \cdot \quad \cdot \quad \cdot$$
$$x \equiv a_j \pmod{m_j}$$
$$\cdot \quad \cdot \quad \cdot \quad \cdot \quad \cdot \quad \cdot$$
$$x \equiv a_n \pmod{m_n}$$

has an integral solution $x \in \mathbf{Z}$ that is *uniquely determined* modulo $m = \prod_{j=1}^{n} m_j$. In other words, if x^* is another solution, then $x \equiv x^* \pmod{m}$.

The proof is by induction on the number of congruences n in the system.

The Case $n = 2$. Consider two congruences

$$y \equiv a_1 \pmod{m_1} \quad \text{and} \quad y \equiv a_2 \pmod{m_2}$$

for arbitrary but relatively prime moduli m_1 and m_2. There exist integers s and t such that

(*) $$m_1 s + m_2 t = 1.$$

A solution y of the first congruence must have the form $y = a_1 + km_1$. Hence we must have

$$a_1 - km_1 \equiv a_2 \pmod{m_2}$$

or
$$km_1 \equiv a_2 - a_1 \pmod{m_2}.$$

Multiplying this congruence by s,

$$km_1 s \equiv s(a_2 - a_1) \pmod{m_2},$$

and making use of equation (*) to write $m_1 s = 1 - m_2 t$, we find that

$$k \equiv s(a_2 - a_1) \pmod{m_2}.$$

The general solution of this congruence is $s(a_2 - a_1) + qm_2$, and hence $y = a_1 + [s(a_2 - a_1) + qm_2]m_1$ is the general solution of the pair of congruences.

The General Case. To verify the second hypothesis of the Principle of Induction, §2.2, assume that z is a solution of the system of $n-1$ congruences

$$z \equiv a_1 \pmod{m_1}$$
$$\cdot \quad \cdot \quad \cdot \quad \cdot \quad \cdot \quad \cdot \quad \cdot$$
$$z \equiv a_{n-1} \pmod{m_{n-1}}$$

and consider the pair of congruences

$$x \equiv z \pmod{m_1 m_2 \cdots m_{n-1}} \quad \text{and} \quad x \equiv a_n \pmod{m_n}.$$

Since $(m_1 \cdots m_{n-1}, m_n) = 1$, this pair of congruences has a solution x by the proof for the case $n = 2$. This element $x \in \mathbf{Z}$ satisfies the original system of congruences, because $x \equiv z \pmod{m_1 \cdots m_{n-1}}$ implies that $x - z$ is divisible by m_j, or $x \equiv z \pmod{m_j}$, $1 \leq j < n$. Hence by the assumption on z,

$$x \equiv a_j \pmod{m_j}, \quad 1 \leq j < n.$$

In addition x satisfies $x \equiv a_n \pmod{m_n}$, so that x is a solution of the entire system. By the Principle of Induction such systems of n congruences are solvable for all n.

§2.11 Simultaneous Systems of Congruences

Finally, suppose that x^* is another solution of the given system, namely,
$$x \equiv x^* \equiv a_j \pmod{m_j}, \quad 1 \leq j \leq n.$$
Then $x - x^* \equiv 0 \pmod{m_j}$, or $m_j | (x - x^*)$. Thus
$$m | (x - x^*)$$
by Proposition 4, §2.6, so that $x \equiv x^* \pmod{m}$.

The Chinese Remainder Theorem also yields the existence of a solution of the system of congruences
$$a_i x \equiv b_i \pmod{m_i}, \quad 1 \leq i \leq n$$
where $(a_i, m_i) = 1$ and, for $i \neq j$, $(m_i, m_j) = 1$, since this system can be transformed into the system in the theorem by multiplying the ith congruence by s_i, where $a_i s_i \equiv 1 \pmod{m_i}$. The existence of such s_i is assured by the hypothesis that $(a_i, m_i) = 1$.

Exercises

1. Find all $x \in \mathbb{Z}$ which satisfy the following simultaneous congruences.
   a. $2x \equiv 8 \pmod{22}$
       $x \equiv 4 \pmod{5}$
       $x \equiv -1 \pmod{9}$
   b. $3x \equiv 9 \pmod{11}$
       $2x \equiv 8 \pmod{23}$
   c. $x \equiv 2 \pmod{5}$
       $x \equiv 3 \pmod{7}$
       $x \equiv 5 \pmod{11}$
   d. $x \equiv 1 \pmod{4}$
       $3x \equiv 1 \pmod{5}$
       $x \equiv 9 \pmod{17}$.

2. In parts (a) and (b) find an integer x that satisfies the simultaneous congruences.
   a. $x \equiv 5 \pmod{6}$
       $x \equiv 4 \pmod{5}$ Solved by Brahmegupta,
       $x \equiv 3 \pmod{4}$ seventh century A.D.
       $x \equiv 2 \pmod{3}$
   b. $x \equiv 1 \pmod{2}$
       $x \equiv 2 \pmod{3}$ Solved by Yih-Ling
       $x \equiv 5 \pmod{6}$ (died 717).
       $x \equiv 5 \pmod{12}$
   c. Find an integer x that is a multiple of 7 and has remainder 1 upon division by 2, 3, 4, 5, and 6. Solved by Ibn al-Haitam, c. 1000.
   d. Find an integer x that is a multiple of 7 and has remainders 1, 2, 3, 4, and 5 upon division by 2, 3, 4, 5, and 6, respectively. Solved by Fibonacci in 1202.

3. a. Find all $x, y \in \mathbb{Z}$ which satisfy the pair of congruences
   $$3x + 4y \equiv 2 \pmod{11}, \quad 5x + 2y \equiv 3 \pmod{11}.$$
   b. Repeat part (a) replacing 11 by 9.

4. Is the pair of congruences
$$3x + 4y \equiv 2 \pmod{7}, \qquad 5x + 2y \equiv 3 \pmod{7}$$
solvable? Justify your answer.

5. Find conditions on the integers u and v so that the congruences
$$x + 2y + 3z \equiv u \pmod{p}, \qquad 4x + 5y + 6z \equiv v \pmod{p}$$
have solutions for all primes p.

6. Prove that the system of congruences
$$x \equiv a_i \pmod{m_i}, \qquad 1 \leq i \leq n,$$
is solvable if and only if $(m_i, m_j) | (a_i - a_j)$, for all $i \neq j$, $1 \leq i, j \leq n$.

§2.12 Two Topics in Number Theory

We conclude this chapter on integers by giving Euler's formula for the number of units in $\mathbf{Z}_m$ and Fermat's Little Theorem: for a prime p any integer is congruent, modulo p, to its pth power.

For positive m the number $\varphi(m)$, called the **Euler φ-function** of m after Switzerland's great mathematician Leonhard Euler (1707–1783), is defined to be the number of integers between 1 and m which are relatively prime to m. In §3.5 we prove the following formula for $\varphi(m)$.

Euler φ-function. If m has the prime factorization
$$m = p_1^{\alpha_1} \cdots p_s^{\alpha_s},$$
where the p_i are distinct positive primes, then
$$\varphi(m) = m\left(1 - \frac{1}{p_1}\right) \cdots \left(1 - \frac{1}{p_s}\right) = \prod_{i=1}^{s} p_i^{\alpha_i - 1}(p_i - 1).$$

In particular, for a prime p,
$$\varphi(p) = p - 1 \quad \text{and} \quad \varphi(p^\alpha) = p^\alpha - p^{\alpha - 1}.$$
To verify the latter, observe that every pth integer between 1 and p is divisible by p, a total of $p^\alpha/p = p^{\alpha - 1}$ integers. Thus $p^\alpha - p^{\alpha - 1}$ are relatively prime to p^α, as asserted.

We noted in §2.10 that $[a] \in \mathbf{Z}_m$ is a unit if and only if $[a]$ is a prime residue, or equivalently, if and only if $(a, m) = 1$. Euler's φ-formula then expresses the number of units in $\mathbf{Z}_m$ as a function of m.

We make several further observations about the set $\mathbf{U}_m$ of all units of $\mathbf{Z}_m$. It is closed with respect to multiplication of cosets; that is, if $[a], [b] \in \mathbf{U}_m$, then $[a][b] \in \mathbf{U}_m$. Furthermore $[a][b] = [b][a]$, and every $[a] \in \mathbf{U}_m$ has a multiplicative inverse. In other words, $\mathbf{U}_m$ is an example of a *finite commutative group* with $\varphi(m)$ elements, a concept discussed in Chapter 6.

§2.12 Two Topics in Number Theory

The ring $\mathbf{Z}_p$ has $\varphi(p) = p-1$ units and has in common with the real numbers $\mathbf{R}$, the complex numbers $\mathbf{C}$, and the rational numbers $\mathbf{Q}$ the property that every nonzero element has a multiplicative inverse. Rings with such properties are called *fields* [see §3.6 for further details].

In addition to $([a][b])^p = [a]^p[b]^p$, we prove for $[a], [b] \in \mathbf{Z}_p$ that

$$([a]+[b])^p = [a]^p + [b]^p.$$

From the Binomial Theorem [§2.2] we have

$$([a]+[b])^p = \sum_{i=0}^{p} c_i [a]^{p-i} [b]^i,$$

where
$$c_i = \binom{p}{i} = \frac{p!}{(p-i)!(i!)} \in \mathbf{Z}.$$

Here $c_i[a]^{p-i}[b]^i$ denotes the product $[a]^{p-i}[b]^i$ added to itself c_i times in $\mathbf{Z}_p$. Observe that $p|c_i$ for $1 \leq i < p$, because $(p-i)!(i!)$ divides $p!$ and so must divide the factor $(p-1)!$ of $p!$ [see Proposition 3, §2.6, and §2.7]. Now set $[t_i] = [a]^{p-i}[b]^i$. For $1 \leq i < p$, $c_i[t_i] = [c_i t_i] = [0] \in \mathbf{Z}_p$, since $p|c_i t_i$. Consequently

$$([a]+[b])^p = [a]^p + [0] + \cdots + [0] + [b]^p.$$

Next, by a recursive argument, we can obtain the following theorem.

Fermat's Little Theorem. For all $[a] \in \mathbf{Z}_p$,

$$[a]^p = [a],$$

which is equivalent, for all $a \in \mathbf{Z}$, to

$$a^p \equiv a \pmod{p}$$

or, if $a \not\equiv 0 \pmod{p}$, to

$$a^{p-1} \equiv 1 \pmod{p}.$$

The proof of this result, stated by Fermat in 1640, was published by Euler in 1736. We leave the proof to the reader with the observation that

$$\begin{aligned}
{[a]^p = [a]} \Rightarrow [a+1]^p &= ([a]+[1])^p \\
&= [a]^p + [1]^p \\
&= [a] + [1] = [a+1].
\end{aligned}$$

Exercises

1. Prove Fermat's Little Theorem.
2. Show that the Euler function $\varphi(n)$ gives the number of distinct proper fractions between 0 and 1 whose denominators equal n.

3. a. Prove that $\varphi(2m) = \varphi(m)$ if and only if m is odd.
   b. Prove that $\varphi(n) \equiv 0 \pmod{2}$ for all $n > 2$.
4. a. Prove that $\varphi(m^2) = m\varphi(m)$.
   b. Prove that $\varphi(mn)\varphi(d) = d\varphi(m)\varphi(n)$, where $d = (m, n)$ is the GCD of m and n.
5. For which of the integers 7, 9, 13, 15, 17, 24, 33, and 36 is the Euler φ-function $\varphi(n)$ a power of 2? (We shall find in §9.3 that $\varphi(n) = 2^s$ if and only if a regular n-gon can be constructed by ruler and compass.)
6. Suppose $2^h + 1$ is a prime number p. Prove that
   a. The residue class $[2] \in \mathbf{Z}_p$ has multiplicative order $2h$; that is, $[2]^{2h} = [1]$.
   b. h is a power of 2.
   c. $p - 1 \equiv 0 \pmod{2h}$.

This problem of number theory is important in the discussion of the generalization of Gauss' construction of the regular polygon of 17 sides [§9.3].

7. For an odd integer n show that $\mathbf{Z}_n$ has an even number of prime residue classes (units).
8. Provide an alternate proof of $\varphi(p^\alpha) = p^\alpha - p^{\alpha-1}$ by using the p-adic expansion $\omega_0 + \omega_1 p + \cdots + \omega_{\alpha-1} p^{\alpha-1}$, $0 \leq \omega_i < p$, of a, $1 \leq a < p^\alpha$, as in §2.5. Note that $(a, p) = 1$ if and only if $1 \leq \omega_0 < p$.

3

Introduction to Ring Theory

Chapter 2 reviewed the associative, commutative, and distributive properties for the addition and multiplication of integers. We referred to the integers as a ring and defined residue class rings of integers. With these serving as explicit examples, we now abstract certain of the algebraic properties of the integers to more general sets and operations that will constitute rings, defined in the abstract.

Section 3.1 provides the basic definitions of ring theory. These definitions, each illustrated by one or more examples, must be mastered, almost like an alphabet. Sections 3.2 and 3.3 introduce from a ring-theoretic view the fundamental algebraic concepts of homomorphism and direct sum. Homomorphisms are discussed in general, with numerous specific examples drawn from the integers and residue class rings of integers. The discussion of direct sums in §3.3 is general, with specific direct sum considerations for the ring Z_m worked out in full in §3.5. Later the concepts of homomorphism and direct sum arise in other contexts: for vector spaces in Chapter 4 and for groups in Chapter 6.

Sections 3.4 and 3.5 have a twofold purpose. First, the complete description of the algebraic-arithmetic structure of the residue class rings Z_m, such as determination of orthogonal idempotents, subrings, and units, is given. Second, techniques for analyzing ring structures are established. They are general enough to apply to rings of polynomials [§5.2] and ultimately provide the model for generalization to advanced topics in number theory and function theory.

A natural continuation of the ring-theoretic development of §3.1 is the examination in §3.6 of rings with either of two additional properties: a multiplicative law of cancellation or existence of a multiplicative inverse for each nonzero element. Here we abstract properties of the integers and the rationals. The student is asked to examine prime, primary, and maximal ideals in the exercises for two reasons: first, they provide further experience in abstracting properties of Z and Z_m to more general rings; and second, significant theorems involving these special ideals can be proved only if restrictive assumptions, such as the chain conditions, are imposed on the rings.

The definitions, as presented and as typically applied in a first course in algebra, do not lead to important theorems, but not for any lack of significant ring and ideal theory. Such topics are customarily reserved for graduate level study. Thus we provide a basic introduction to the subject with emphasis on those aspects to be used later, first in a detailed analysis of residue class of integers [§3.4] and field theory [Chapter 8]. In Chapters 8 and 9 the proofs of the existence of algebraically closed fields and the Fundamental Theorem of Algebra utilize ideal-theoretic arguments.

§3.1 Basic Elements of Ring Theory

At this point in our discussion of the residue class ring Z_m it is helpful to have at hand some general definitions and facts from the theory of rings. More importantly, we now abstract many of the algebraic properties of the integers to define a particular algebraic structure called a ring [cf. §2.1, axioms II–IV, and §2.9].

A **ring** R is a set of elements together with two laws of composition (commonly called addition and multiplication) that associate to each *ordered* pair (a, b) of elements in R a unique third element (denoted $a+b$ for addition, ab for multiplication) in R, subject to the following axioms:

> **Addition.** Addition is associative and commutative, there exists an additive identity element (commonly denoted by 0), and each element $r \in R$ has an additive inverse r' such that $r+r' = 0$.
> **Multiplication.** Multiplication is associative, and there exists a multiplicative identity element (commonly denoted by e or 1) such that $er = re = r$ for all $r \in R$.
> **Distributivity.** Multiplication is distributive over addition:
> $$r(a+b) = ra + rb,$$
> $$(a+b)r = ar + br \qquad \text{for all } a, b, r \in R.$$

If furthermore $ab = ba$ for all elements $a, b \in R$, then the ring R is called a **commutative ring**.

In §2.1 we encountered the ring of integers **Z**, and in §2.9 the residue class rings of integers $\mathbf{Z}_m$. These constitute our basic examples of rings; other examples are given in exercises. Students familiar with matrices and linear transformations (topics treated in §4.4) will recognize that the set of $n \times n$ matrices ($n > 1$) with either integral or rational coefficients is a *noncommutative* ring with the usual rules of matrix addition and multiplication. Equivalently, the set of all endomorphisms (linear transformations) of an n-dimensional vector space over a field constitutes a *noncommutative* ring [see the end of §4.3 and Proposition 3, §4.4].

In this book all rings are assumed to be commutative unless explicitly stated to the contrary.

Because the defining algebraic axioms of a ring are abstracted from the corresponding axioms for integers, their logical consequences carry over to arbitrary rings R. We adapt from §2.1 the following properties, valid for any (commutative or noncommutative) ring R.

Property 1. The additive identity is unique.

Property 2. Furthermore, $a = a + r$, for some $a \in R$, implies $r = 0$.

Property 3. The cancellation law for addition holds.

Property 4. The additive inverse a', commonly written $-a$, of an element $a \in R$ is unique.

Property 5. Every equation $a + x = b$, for $a, b \in R$, has a unique solution $x \in R$.

Property 6. For any $a \in R$, $a \cdot 0 = 0$.

Property 7. The multiplicative identity is unique.

Property 9. The *rule of signs* for multiplication holds; that is, $(-a)(-b) = ab$ for all $a, b \in R$.

Note that Properties 8 and 10 of §2.1 involve the multiplicative law of cancellation, valid in the ring of integers, but not valid for rings in general (as was seen in §2.9 in the case of $\mathbf{Z}_{18}$). Therefore they are not included in the present list. Rings in which these two additional properties hold are called *integral domains* and will be discussed in §3.6.

Furthermore, the general associative and distributive laws for the addition and multiplication of n elements hold since the proofs require the Principle of Induction on the integral subscripts in the enumeration of the elements (and not on the ring elements themselves). Thus for example,

$$a \left(\sum_{i=1}^{n} b_i \right) = \sum_{i=1}^{n} (ab_i)$$

for $a, b_1, \ldots, b_n \in R$. We defer the proof of the general associative law to §6.1.

A nonempty subset T of a ring R is called a **subring** of R, if

- **(i)** T is closed under addition and additive inverse, as determined in R (i.e., for all $a, b \in T$, $a+b$ and $-a$ are elements in T; these imply that $0 \in T$).
- **(ii)** T is closed under the given law of multiplication of R ($a, b \in T \Rightarrow ab \in T$), and T contains a multiplicative identity e_T such that $e_T a = a e_T = a$ for all $a \in T$.

Note that it is *not* postulated that the multiplicative identity e_T of T be the same element as the multiplicative identity e in R.

Example 1. In $\mathbf{Z}_{20}$ the subset $S = \{[0], [5], [10], [15]\}$ of cosets modulo 20 satisfies the definition of a subring. Here $[5]$ is the multiplicative identity of S, while $[1]$ is the multiplicative identity of $\mathbf{Z}_{20}$. (The additive identity of a subring is always that of the ring itself.)

Example 2. The subset $\{[0], [10]\}$ of $\mathbf{Z}_{20}$ is not a subring. It has no multiplicative identity, but does satisfy the other properties of a subring.

REMARK 1. In particular note that a subring is itself a ring; that is, its elements satisfy the axiomatic properties of a ring.

REMARK 2. We use the preceding definition of subring because the ring-theoretic interpretation of systems of simultaneous congruences and its generalizations in the theory of algebraic numbers and algebraic geometry give rise to mappings of rings R into rings S for which the image of the multiplicative identity of R is *not* the multiplicative identity of S. (Some authors require that the subring have the same multiplicative identity as the ring. Others, especially those concerned with non-commutative structures, do not require a ring to have a multiplicative identity.)

It is common usage to refer to a subring $S \subseteq R$ as a **nontrivial subring** of R if S is neither R itself nor the zero subring $\{0\}$. Later, similar conventions will apply to ideals, fields, groups, and other algebraic structures.

The following paragraphs are replete with definitions, which essentially are agreements on terminology. While now they may seem a burden, their importance can be appreciated later on. The concept of an *ideal* has become extremely important in algebra since the work of Ernst Eduard Kummer (1810–1893), Leopold Kronecker (1823–1891), and J. W. R. Dedekind (1831–1916).

A nonempty subset A of a (commutative) ring R is called an **ideal** in R if

- **(i)** A is closed under addition
- **(ii)** A is closed under multiplication by elements $r \in R$; that is, if $a \in A$ and $r \in R$, then $ra \in A$.

For noncommutative rings R it is necessary to distinguish between the following types of ideals:

§3.1   Basic Elements of Ring Theory

Left ideals. $a \in A, r \in R \Rightarrow ra \in A.$
Right ideals. $a \in A, r \in R \Rightarrow ar \in A.$
Two-sided ideals. $a \in A, r \in R \Rightarrow ar \in A, ra \in A.$

For commutative rings, $ra = ar$ implies that all ideals are two-sided.

We have already encountered ideals in the ring of integers **Z** [§2.4]. In any commutative ring R, a given element a generates an ideal $(a) = \{ra : r \in R\}$, also denoted Ra. More generally, if $S = \{s_1, \ldots, s_m\}$ is any finite subset of R, the **ideal generated by** S is the subset of all elements of R of the form $r_1 s_1 + \cdots + r_m s_m$, where $r_i \in R$, $1 \leq i \leq m$. Such an ideal is commonly denoted by $(s_1, \ldots, s_m)$ or $Rs_1 + \cdots + Rs_m$.

There is a considerable arithmetic of ideals paralleling and generalizing that of the integers. For ideals A and B in a (commutative) ring R, we define terms as follows.

1. **Product:**

$$AB = \left\{ \sum_{i=1}^{n} a_i b_i : a_i \in A, b_i \in B, \text{ for arbitrary } n \in \mathbf{N} \right\};$$

that is, the set of all finite sums of products of elements in A and B.

2. **Sum:**

$$A + B = \{a+b : a \in A, b \in B\},$$

also called the **greatest common divisor** of A and B, denoted (A, B).

3. **Intersection:**

$$A \cap B = \{r \in R : r \in A \text{ and } r \in B\}$$

is simply the set-theoretic intersection. We also call $A \cap B$ the **least common multiple** of the ideals A, B, denoted $[A, B]$.

4. **Power:** The power A^k of an ideal, $k \in \mathbf{N}$, is the k-fold product of A with itself. It consists of all finite sums of k-fold products $a_1 \cdots a_k$ of elements of A.

By finite induction we can extend the concepts of product, sum, and intersection to any finite number of ideals.

Recalling the properties of GCD and LCM for integers [§§2.5 and 2.7], we note the significance of the terms greatest common divisor and least common multiple for ideals in a given ring in Proposition 1.

Proposition 1. Ideals A, B, C in a ring R satisfy the following properties.

(i) GCD properties:

$$A \subseteq (A, B).$$
$$B \subseteq (A, B).$$
$$A \subseteq C, B \subseteq C \Rightarrow (A, B) \subseteq C.$$

(ii) LCM properties:
$$[A, B] \subseteq A.$$
$$[A, B] \subseteq B.$$
$$C \subseteq A, C \subseteq B \Rightarrow C \subseteq [A, B].$$

These properties are all immediate consequences of the definitions. Less immediate is a second proposition.

Proposition 2. For ideals A, B, C in a ring R,

(i) $[A, B](A, B) \subseteq AB.$
(ii) $(A, B) = (A, C) = R \Rightarrow (A, BC) = R.$
(iii) $(A, B) = R \Rightarrow A \cap B = AB.$

Proof of (i) [cf. §2.7, Exercise 4].
$$[A, B](A, B) = (A \cap B)(A + B)$$
$$= (A \cap B)A + (A \cap B)B \subseteq BA + AB = AB.$$

Equality does not always hold in general rings.

Proof of (ii).
$$(A, B) = R \Rightarrow e = a + b, \quad \text{for some } a \in A, b \in B.$$

Similarly
$$(A, C) = R \Rightarrow e = a' + c, \quad \text{for some } a' \in A, c \in C,$$

and therefore
$$e = e^2 = \underbrace{aa' + ac + ba'}_{\in A} + \underbrace{bc}_{\in BC}.$$

Hence every element $r \in R$ is the sum of
$$r(aa' + ac + ba') \in A \quad \text{and} \quad rbc \in BC,$$

which means that $R = A + BC = (A, BC)$.

Proof of (iii). First, note that $AB \subseteq A \cap B$, since $AB \subseteq A$ and $AB \subseteq B$. To prove the reverse inclusion, consider $x \in A \cap B$. Since $A + B = R$, we can write $e = a + b$ for some $a \in A, b \in B$. Hence
$$x = xe = xa + xb = ax + xb \in AB,$$

as each summand on the right-hand side lies in AB. Therefore $A \cap B \subseteq AB$, and $A \cap B = AB$, as asserted.

As we did in the particular rings Z_m in §2.10, we consider here elements in a ring with specific properties. An element a in a ring R is:

A zero divisor if $a \neq 0$ and if there exists $b \neq 0 \in R$ such that $ab = 0$.
Nilpotent if $a^n = 0$ for some $n \in \mathbf{N}$.
Idempotent if $a^2 = a$.
A unit if there exists some $b \in R$ such that $ab = e$, the multiplicative identity of R.

Exercises

1. a. With reference to Exercise 4(a)–(e) in §1.1 verify that the set of subsets of a fixed set S constitutes a ring where intersection of sets is taken as the multiplicative law of composition and symmetric difference as the additive law.
   b. Show that, despite Exercise 4(f) in §1.1, the set of subsets of S with the laws of intersection and union does not constitute a ring.
2. Suppose that R is a ring such that $a^2 = a$ for all its elements a. Prove that $a + a = 0$ for all $a \in R$ and that $ab = ba$ for all $a, b \in R$.
3. Let a, b be elements of a commutative ring R with multiplicative identity e.
   a. Prove that $(-e)a = -a$ and $-(a-b) = b - a$.
   b. For all $m, n \in \mathbf{N}$ prove that $(ab)^n = a^n b^n$, $e^n = e$, and $a^{mn} = (a^m)^n$.
4. a. Let x be a nilpotent element in the ring R. Prove that $e + x$ and $e - x$ are units in R.
   b. If a is an idempotent element in R, prove that $e - a$ is an idempotent and a zero divisor.
5. An element $x \in R$ is called **unipotent** if $e - x$ is nilpotent. Prove that the product xy of unipotent elements is again unipotent. Prove also that every unipotent element of R is a unit of R whose inverse is unipotent.
6. Find the unipotent elements in $\mathbf{Z}_n$ for
   a. $n = 28$ b. $n = 18$ c. $n = 36$.
7. Suppose that a and b are nilpotent elements in the commutative ring R. Prove that
   a. $a + b$ is nilpotent.
   b. The nilpotent elements constitute an ideal in R.

The ideal of Exercise 7(b) is sometimes referred to as the **radical of the ring**, denoted either $\operatorname{Rad}(0)$ or $\operatorname{Rad} R$ [cf. Exercise 8, below].

8. Let A be an ideal in the commutative ring R. Define the **radical** of A to be the set

$$\operatorname{Rad} A = \{r \in R : r^n \in A, \text{ for some } n \in \mathbf{N}\}.$$

   a. Prove that $\operatorname{Rad} A$ is an ideal in R.
   b. Prove that $\operatorname{Rad}(\operatorname{Rad} A) = \operatorname{Rad} A$.

9. **a.** Describe all ideals A in $\mathbf{Z}$ for which $\text{Rad } A = A$ [cf. Proposition 2, §2.10].
   b. Describe $\text{Rad } A$ for $A = (100) \subset \mathbf{Z}$; for $A = (80) \subset \mathbf{Z}$.
10. Determine the radicals of the following rings [cf. Exercise 7(b)] where p, q are distinct primes. Distinguish the cases $n = 1$ and $n > 1$.
    a. $\mathbf{Z}_{p^n}$.   **b.** $\mathbf{Z}_{p^n q^m}$.
11. Using the definitions of product and sum of ideals, prove the following associative, commutative, and distributive laws for ideals A, B, C in a ring R.
    (i) $(AB)C = A(BC)$
    (ii) $AB = BA$
    (iii) $(A+B)C = AC + BC$.
12. Show that the *set-theoretic* union of ideals A, B in a ring R need not be an ideal.
13. **a.** Consider the ideal $A = (a_1, ..., a_n)$ generated by the elements $a_1, ..., a_n$ in a (commutative) ring R. Describe the elements of the ideal A^2 in terms of those in A.
    b. Prove that if $A = A^2$, then there exists an idempotent $b \in A$ such that $A = Rb = \{rb : r \in R\}$. In other words, prove that $A = A^2$ implies A is a principal ideal generated by an idempotent element b.
    c. If $S = \{s_1, ..., s_m\}$ is a finite subset of R, prove that the ideal generated by S is the GCD of the principal ideals
    $$S_i = s_i R = \{s_i r : r \in R\}, \quad 1 \leq i \leq m.$$
14. Define laws of composition $\#$ and $*$ on $\mathbf{N}$ by
    $$a \# b = (a, b), \quad \text{the GCD};$$
    $$a * b = [a, b], \quad \text{the LCM}.$$
    With these laws of composition is $\mathbf{N}$ a ring?

§3.2 Ring Homomorphisms

We turn now to mappings (functions) from one ring to another which are compatible with the ring structure. The root word *morphism* is commonly used in higher level algebra courses to denote a mapping from one algebraic structure (such as a ring) to another of the same kind. The prefix *homo-* means "structure-preserving" or "compatible"; other prefixes, to be introduced shortly, denote homomorphisms with special additional properties. In later chapters we shall encounter homomorphisms of other algebraic structures such as vector spaces and groups. Vector space homomorphisms are customarily called linear transformations or linear operators.

Consider rings R and S, each with its own laws of composition. A **(ring) homomorphism** φ of R into S is a single-valued mapping

$$\varphi: R \to S$$

§3.2 Ring Homomorphisms

such that for all $a, b \in R$,

(i) $\quad\quad\quad\quad\quad\quad \underset{\substack{\uparrow \\ \text{addition as} \\ \text{defined in } R}}{\varphi(a+b)} = \underset{\substack{\uparrow \\ \text{addition as} \\ \text{defined in } S}}{\varphi(a) + \varphi(b)}$

(ii) $\quad\quad\quad\quad\quad\quad \underset{\substack{\uparrow \\ \text{multiplication} \\ \text{as defined in } R}}{\varphi(ab)} = \underset{\substack{\uparrow \\ \text{multiplication} \\ \text{as defined in } S}}{\varphi(a)\varphi(b)}$

The subset of S

$$\varphi(R) = \{s \in S : s = \varphi(a), \text{ for some } a \in R\}$$
$$= \{\varphi(a) : a \in R\}$$

is a subring of S, called the **image** of R by φ. The image $\varphi(R)$ is also denoted by Im φ.

An important but simple consequence of property (i) of a homomorphism $\varphi: R \to S$ is that $\varphi(0_R) = 0_S$, where $0_R, 0_S$ are the additive identities of the rings R, S respectively. Since

$$\varphi(0_R) = \varphi(0_R + 0_R) = \varphi(0_R) + \varphi(0_R)$$

by property (i), the uniqueness of the additive identity 0_S of S implies that $\varphi(0_R) = 0_S$ [Property 2, §3.1].

The multiplicative property (ii) of such a homomorphism implies that the image $\varphi(e_R)$ of the multiplicative identity e_R of R is the multiplicative identity of the *subring* $\varphi(R) \subseteq S$. The defining equations of a ring homomorphism do *not* imply that $\varphi(e_R)$ is necessarily the multiplicative identity e_S of the ring S. (See Example 6 below.) We gather these facts in the following proposition.

Proposition 1. If $\varphi: R \to S$ is a ring homomorphism, then

(i) $\varphi(0_R) = 0_S$,
(ii) $\varphi(e_R) = e_{\varphi(R)}$.

One application of Proposition 1 is to show that some mappings from one ring to another cannot be homomorphisms. (See Example 11 below.)

Before considering further properties and special types of homomorphisms, we construct a few examples of homomorphisms, and of mappings between rings which are *not* homomorphisms. When several moduli m for the residue class rings Z_m are involved, subscripts on the cosets can clarify which ring a coset belongs to.

Example 1. The ring Z_{20} contains the subring (which happens also to be an ideal)

$$S = \{[0]_{20}, [5]_{20}, [10]_{20}, [15]_{20}\}.$$

We can define a homomorphism $\varphi: S \to \mathbf{Z}_4$ by specifying that $\varphi([5]_{20}) = [1]_4$. To complete the definition of φ, we compute:

$$\varphi([10]_{20}) = \varphi([5]_{20} + [5]_{20})$$
$$= \varphi([5]_{20}) + \varphi([5]_{20})$$
$$= [1]_4 + [1]_4 = [2]_4,$$

$$\varphi([15]_{20}) = \varphi([5]_{20}) + \varphi([10]_{20})$$
$$= \varphi([5]_{20}) + \varphi([10]_{20})$$
$$= [1]_4 + [2]_4 = [3]_4,$$

$$\varphi([0]_{20}) = \varphi([20]_{20}) = \varphi([10]_{20} + [10]_{20})$$
$$= \varphi([10]_{20}) + \varphi([10]_{20})$$
$$= [2]_4 + [2]_4 = [4]_4 = [0]_4.$$

To be assured that the mapping φ, as defined, is indeed a homomorphism, we must verify that φ respects the multiplicative structures of S and $\mathbf{Z}_4$. That is, that $\varphi([a]_{20}[b]_{20}) = \varphi([a]_{20})\varphi([b]_{20})$ for any $[a]_{20}, [b]_{20} \in S$. For example,

$$\varphi([15]_{20}[10]_{20}) = \varphi([150]_{20}) = \varphi([10]_{20})$$
$$= [2]_4 = [3]_4[2]_4$$
$$= \varphi([15]_{20})\varphi([10]_{20}).$$

The complete verification of the multiplicative property (ii) for φ follows easily from the fact that the elements of S are integral multiples of $[5]_{20}$. Since general elements $[a]_{20}, [b]_{20} \in S$ can be expressed as $[5m]_{20}, [5n]_{20}$,

$$\varphi([5m]_{20}[5n]_{20}) = \varphi([25mn]_{20} = \varphi([5mn]_{20})$$
$$= mn \cdot \varphi([5]_{20}) \quad \text{by property (i)}$$
$$= mn \cdot [1]_4 = [mn]_4$$
$$= [m]_4[n]_4$$
$$= \varphi([5m]_{20})\varphi([5n]_{20}).$$

Example 2. A similar example is the mapping ψ of the subring $S' = \{[0]_{20}, [4]_{20}, [8]_{20}, [12]_{20}, [16]_{20}\} \subset \mathbf{Z}_{20}$ to $\mathbf{Z}_5$ given by $\psi([16]_{20}) = [1]_5$ and more generally

$$\psi(n \cdot [16]_{20}) = n \cdot [1]_5 = [n]_5.$$

That ψ, so defined, is a homomorphism is left as an exercise. The proof is analogous to the discussion in Example 1.

It may seem an obvious point, but in considering whether a rule of correspondence $f: R \to S$ between two rings is a homomorphism, we must first insure that it is a *well-defined* mapping as, for instance, in the next examples.

Example 3. We *cannot* define a mapping f from $\mathbf{Z}_{37} \to \mathbf{Z}$ by setting $f([a]_{37}) = a$. This is not a well-defined function, because $[13]_{37} = [50]_{37}$ but $13 \neq 50$.

Example 4. Similarly, setting $\varphi([1]_4) = [1]_5$ cannot define a homomorphism φ from $\mathbf{Z}_4$ to $\mathbf{Z}_5$, because $[1]_4 = [5]_4$ and, if φ were a homomorphism,

$$\varphi([5]_4) = \varphi(5 \cdot [1]_4) = 5 \cdot \varphi([1]_4)$$
$$= 5 \cdot [1]_5 = [0]_5 \neq \varphi([1]_4).$$

§3.2   **Ring Homomorphisms**   65

We shall find later [§6.6] that the only homomorphism from Z_4 to Z_5 is trivial in the sense that each element of Z_4 must be mapped to $[0]_5$.

Examples 3 and 4 also illustrate the fact that for a homomorphism $\varphi: R \to S$, the original relations in R must be carried into relations valid in S. For instance, if $a+b+c=0$ in R, then $\varphi(a)+\varphi(b)+\varphi(c)$ must equal 0 in S.

Example 5.   There are homomorphisms

$$\sigma: Z_4 \to Z_{20}, \qquad \varphi': Z_{20} \to Z_4,$$

given by $\qquad \sigma([a]_4) = [5a]_{20} \quad$ and $\quad \varphi'([a]_{20}) = [a]_4.$

These mappings are independent of the choices of representatives of cosets, since

$$\sigma([a+4r]_4) = [5(a+4r)]_{20} = [5a]_{20} = \sigma([a]_4),$$
$$\varphi'([a+20s]_{20}) = [a+20s]_4 = [a]_4 = \varphi'([a]_{20}).$$

Combining Examples 1 and 5, we have the following sequence of homomorphisms:

$$S \xrightarrow{\varphi} Z_4 \xrightarrow{\sigma} Z_{20} \xrightarrow{\varphi'} Z_4,$$

where $S = \{[0]_{20}, [5]_{20}, [10]_{20}, [15]_{20}\}$.

Example 6.   For the homomorphism $\sigma: Z_4 \to Z_{20}$ in Example 5, note that $\sigma([1]_4) = [5]_{20} \neq [1]_{20}$, illustrating that a homomorphism from one ring to another need not map the multiplicative identity of the first to the multiplicative identity of the second.

Example 7.   We make two final observations about the homomorphisms $\sigma: Z_4 \to Z_{20}$ and $\varphi': Z_{20} \to Z_4$ of Example 5. First, $\operatorname{Im} \sigma = \sigma(Z_4) = \{[5]_{20}, [10]_{20}, [15]_{20}, [0]_{20}\}$ is a subring of Z_{20}, whose multiplicative identity is $[5]_{20} = \sigma([1]_4)$. Second, the subset

$$\{[a]_{20} \in Z_{20} : \varphi'([a]_{20}) = [0]_4\} = \{[a]_{20} \in Z_{20} : a \equiv 0 \pmod{4}\}$$
$$= \{[0]_{20}, [4]_{20}, [8]_{20}, [12]_{20}, [16]_{20}\}$$

is an ideal in Z_{20}. (In fact it is a subring with multiplicative identity $[16]_{20}$).

Propositions 2 and 3 below state the general case of the properties illustrated here: the image of a homomorphism $\varphi: R \to S$ is a subring of S, and the set of elements in R mapped to the additive identity of S is an ideal in R.

The following agreements on terminology pertain to special types of ring homomorphisms

$$\varphi: R \to S.$$

A homomorphism φ is said to be **onto** (alternatively, **surjective** or an **epimorphism**) if $\varphi(R) = S$; in other words, if for each $s \in S$ there exists some $r \in R$ for which $\varphi(r) = s$.

Example 8. The homomorphisms φ, ψ, and φ' of Examples 1, 2, and 5 above are onto. For instance,

$$\varphi'([0]_{20}) = [0]_4, \qquad \varphi'([2]_{20}) = [2]_4,$$
$$\varphi'([1]_{20}) = [1]_4, \qquad \varphi'([3]_{20}) = [3]_4;$$

hence there is some element in $\mathbf{Z}_{20}$ which is mapped by φ' to any given element in $\mathbf{Z}_4$.

A homomorphism φ is said to be **one-one** (alternatively, **injective** or a **monomorphism**) if $\varphi(r) = \varphi(r')$ implies $r = r'$. An equivalent description of a one-one homomorphism is that $\varphi(r) = 0_S$ implies that $r = 0_R$ since $\varphi(r')$ implies that $\varphi(r - r') = 0_S$.

Example 9. The homomorphisms φ, ψ, and σ of Examples 1, 2, and 5 are one-one. For example, consider $[x]_4 \in \mathbf{Z}_4$ such that $\sigma([x]_4) = [0]_{20}$. First, $\sigma([x]_4) = x \cdot [5]_{20} = [x \cdot 5]_{20}$; further $[x \cdot 5]_{20} = [0]_{20}$ if and only if $4 \mid x$ which means that $[x]_4 = [0]_4$. Hence σ is one-one.

A homomorphism φ is said to be **one-one, onto** (alternatively, **bijective** or an **isomorphism**) if it is both injective (one-one) and surjective (onto).

Example 10. Only the homomorphisms φ and ψ in Examples 1, 2, and 5 are isomorphisms.

Example 11. To illustrate that not all one-one, surjective mappings of rings need be isomorphisms, consider the function $T: \mathbf{Z} \to \mathbf{Z}$ defined by $T(n) = n + 1$. Since $T(0) \neq 0$, T is not a homomorphism by Proposition 1. Therefore T cannot be an isomorphism.

Similarly the one-one mapping $T': \mathbf{Z} \to \mathbf{Z}$, defined by $T'(n) = 2n$, is not a homomorphism, even though $T'(0) = 0$. The image $T'(\mathbf{Z})$ has no multiplicative identity, and so is not a subring; hence T' cannot be a ring homomorphism. More explicitly, observe, for example, that $16 = T'(8) = T'(2 \cdot 4) \neq T'(2) T'(4) = 4 \cdot 8 = 32$.

If $\varphi: R \to S$ is an isomorphism, then R and S are said to be **isomorphic rings**; this fact is denoted

$$R \cong S.$$

For given $s \in S$, the set (which may be empty)

$$\varphi^{-1}(s) = \{r \in R : \varphi(r) = s\}$$

is called the **complete inverse image** of $s \in S$ with respect to the homomorphism φ. Note that $\varphi^{-1}(s) \neq \emptyset$ for *all* $s \in S$ is equivalent to saying φ is *onto*, and that φ is *one-one* if and only if $\varphi^{-1}(s)$ has at most one element for each $s \in S$. For an isomorphism φ the inverse image $\varphi^{-1}(s)$ consists of *precisely one* element for each $s \in S$.

Proposition 2 concerns ideals and inverse images.

Proposition 2. If A^* is an ideal in the image $\varphi(R)$ of a homomorphism $\varphi: R \to S$, the complete inverse image $A = \varphi^{-1}(A^*)$ is an ideal in R.

Proof. Consider $a, b \in A$ and $r \in R$. Then
$$\varphi(a+b) = \varphi(a) + \varphi(b) \in A^*,$$
(since $\varphi(a)$, $\varphi(b)$ each belong to A^*, so does their sum, because A^* is an ideal)
$$\varphi(ra) = \varphi(r)\varphi(a) \in A^*,$$
(since $\varphi(a)$ belongs to A^*, so does the product $\varphi(r)\varphi(a)$, because A^* is an ideal) and hence $a+b$ and ra are elements in A. We conclude that A is an ideal in R.

The **kernel** of the homomorphism φ is
$$\ker \varphi = \varphi^{-1}(0_S) = \{r \in R : \varphi(r) = 0_S\}.$$
Note that φ is injective if and only if $\ker \varphi = \{0_R\}$. For an integer m, the mapping of $a \in \mathbf{Z}$ to its residue class $[a]$ in $\mathbf{Z}_m$ is a homomorphism of $\mathbf{Z}$ onto $\mathbf{Z}_m$, whose kernel is the ideal (m).

Proposition 3. If $\varphi: R \to S$ is a ring homomorphism, then $\ker \varphi$ is a (two-sided) ideal in R (even if R is noncommutative).

We present two proofs. First, since $\{0_S\}$ is an ideal in S, then the complete inverse image $\varphi^{-1}(\{0_S\}) = \ker \varphi$ is an ideal in R by Proposition 2. The second proof is by direct verification. For any $a, b \in \ker \varphi$ and any $r \in R$,
$$\varphi(a+b) = \varphi(a) + \varphi(b) = 0_S + 0_S = 0_S,$$
$$\varphi(ar) = \varphi(a)\varphi(r) = 0_S \cdot \varphi(r) = 0_S,$$
$$\varphi(ra) = \varphi(r)\varphi(a) = \varphi(r) \cdot 0_S = 0_S.$$

In the special case that $S = R$, a homomorphism $\varphi: R \to R$ is called an **endomorphism**. More commonly, an isomorphism $\varphi: R \to R$ will be called an **automorphism**.

Example 12. The mapping $\mathbf{Z}_{20} \to \mathbf{Z}_{20}$ given by $[a]_{20} \to [5a]_{20}$ is an endomorphism whose kernel $\{[4n]_{20} : n \in \mathbf{Z}\}$ is isomorphic to $\mathbf{Z}_5$.

Example 13. The mapping of the ring of complex numbers $\mathbf{C}$ given by $a + bi \to a - bi$, where $a, b \in \mathbf{R}$, is an automorphism of $\mathbf{C}$.

To complete our listing of terminology frequently used to describe homomorphisms, we introduce the term **embedding** to denote an injective homomorphism $\varphi: R \to S$ for which $\varphi(e_R) = e_S$. Not all injective homomorphisms are embeddings (consider Example 5 in which $\sigma: \mathbf{Z}_4 \to \mathbf{Z}_{20}$ is injective, and $\sigma([1]_4) = [5]_{20} \neq [1]_{20}$). Nor are all embeddings isomorphisms. For example, the identity map of the integers $\mathbf{Z}$ into the rational numbers $\mathbf{Q}$ is an embedding, but not an isomorphism, as the map is not surjective.

If $\varphi(e_R) = 0_S$, we call φ a **trivial homomorphism** because every element of R must then be mapped to the additive identity of S; $\varphi(R) = \{0_S\} \subseteq S$.

We conclude this section by outlining a fundamental property of ring homomorphisms for which we need first an equivalence concept for elements in a ring. The reader is asked to develop the details in Exercise 3.

For an ideal A in a commutative ring R, introduce a congruence (or equivalence) relation in R by defining $a \equiv b \pmod{A}$ if and only if $b - a \in A$ for a, b in R. Abstracting the construction of $\mathbf{Z}_m = \mathbf{Z}/(m)$ given in §2.8 and §2.9, we can construct a **residue class ring** R/A of R modulo A. In other words, the set of residue or equivalence classes of elements of R modulo the ideal A can be given a ring structure analogous to the ring structure of $\mathbf{Z}_m$. (The ring R/A is also called the **quotient ring** of R by A.)

By mapping each element of R to its residue (equivalence) class modulo the ideal A, we obtain a surjective homomorphism π_A from the ring R onto the residue class ring R/A. Further, the kernel of π_A is the ideal A. The homomorphism π_A commonly is called the **natural** or **canonical projection** of R onto the quotient ring R/A.

The following theorem completes this discussion. We omit the proof, preferring to provide it in a group-theoretic context in §6.8 after the reader has gained more abstract algebraic experience. The proof is not especially difficult, but since it is similar to the analogue in group theory, we present it only once.

Theorem. If $\varphi: R \to S$ is a surjective ring homomorphism with kernel A, there exists an isomorphism λ of the residue class ring R/A onto S, such that $\varphi = \lambda \circ \pi_A$.

Exercises

1. Consider subrings
$$S = \{[0], [5], [10], [15]\},$$
$$T = \{[0], [4], [8], [12], [16]\}$$
of $\mathbf{Z}_{20}$, and define the following ring homomorphisms:

$\varphi: S \to \mathbf{Z}_4$ by $\varphi([5]_{20}) = [1]_4$,
$\psi: T \to \mathbf{Z}_5$ by $\psi([16]_{20}) = [1]_5$,
$\sigma: \mathbf{Z}_4 \to \mathbf{Z}_{20}$ by $\sigma([1]_4) = [5]_{20}$,
$\tau: \mathbf{Z}_5 \to \mathbf{Z}_{20}$ by $\tau([1]_5) = [16]_{20}$,
$\varphi': \mathbf{Z}_{20} \to \mathbf{Z}_4$ by $\varphi'([a]_{20}) = [a]_4$,
$\psi': \mathbf{Z}_{20} \to \mathbf{Z}_5$ by $\psi'([a]_{20}) = [a]_5$.

a. Prove that φ, ψ, φ', and ψ' are surjective.
b. Prove that φ, ψ, σ, and τ are injective.
c. Show that only φ and ψ are isomorphisms.
d. Determine explicitly the kernel of each of these six homomorphisms.
e. Determine the complete inverse image of each element in $\mathbf{Z}_4$ and $\mathbf{Z}_{20}$ with respect to the appropriate maps φ, σ, and φ'.
f. Determine $\operatorname{Im}\sigma$ and $\operatorname{Im}\tau$.

2. a. Prove that the mapping $\varphi: \mathbf{Z} \to \mathbf{Z}_m$ defined by $\varphi(a) = [a]_m$ is a ring homomorphism.
b. What is the kernel of this mapping?
c. Is the mapping $\psi: \mathbf{Z}_{12} \to \mathbf{Z}_{36}$ defined by $\psi([a]_{12}) = [a]_{36}$ for $0 \le a < 12$ a homomorphism? Why?

3. Let A be an ideal in the commutative ring R.
a. Show that the congruence relation $a \equiv b \pmod{A}$ if and only if $b - a \in A$ for $a, b \in R$ is an equivalence relation on R.
b. Generalize the construction of $\mathbf{Z}/(m)$ in §§2.8 and 2.9 to the residue class ring R/A. That is, give the set R/A of residue (equivalence) classes of R modulo A the structure of a ring.
c. Show that the mapping $\pi_A: R \to R/A$, which maps each element of R to its residue class modulo A, is a surjective homomorphism.
d. Show that $\ker \pi_A = A$.

4. Prove that if $\varphi: R \to S$ is a surjective ring homomorphism, then $S \cong R/\ker\varphi$. (No assumption of commutativity is needed.)

5. Consider a homomorphism $\varphi: R \to S$.
a. If φ is surjective, prove that $\varphi(e_R) = e_S$.
b. Prove that $\varphi(e_R)$ is an idempotent in S.
c. If $\varphi(e_R)$ is different from 0_S and e_S, prove that S must contain nontrivial zero divisors. *Hint:* Consider $e_S - \varphi(e_R)$.

6. a. Verify that the mapping $\varphi: \mathbf{Z}_p \to \mathbf{Z}_p$ given by $\varphi([a]) = [a^p]$ is a ring homomorphism [see Fermat's Little Theorem, §2.12]. Is φ an isomorphism?
b. Is the mapping $\varphi: \mathbf{Z} \to \mathbf{Z}$ given by $\varphi(a) = a^p$ a homomorphism?

7. An ideal A in a commutative ring R is termed **dense** if $rA = \{ra : a \in A\} = \{0\}$ for any $r \in R$ implies $r = 0$.
a. Prove that if A is dense, then so is any ideal $B \supseteq A$.
b. Prove that if the ideals A, B are dense in R, then so are AB and $A \cap B$.

8. In analysis we frequently encounter the ring $\mathscr{F}$ of continuous (or differentiable) real-valued functions defined on the real line $\mathbf{R}$, where laws of composition are defined for $f, g \in \mathscr{F}$ by

$$(f+g)(x) = f(x) + g(x), \qquad (fg)(x) = f(x)g(x)$$

for all $x \in \mathbf{R}$.
a. Verify that $\mathscr{F}$ so defined is a (commutative) ring.
b. Verify for any nonempty subset S of $\mathbf{R}$ that $\mathscr{S} = \{f \in \mathscr{F} : f(x) = 0$ for all $x \in S\}$ is an ideal in $\mathscr{F}$.
c. Show that $\mathscr{F}$ has nontrivial zero divisors.

9. Consider a left ideal J in a noncommutative ring R. The set $A = A(J) = \{r \in R : rj = 0 \text{ for all } j \in J\}$ is called the **annihilator** of J.
   a. Prove that $A(J)$ is a two-sided ideal in R.
   b. With reference to Exercise 8(b), what is the annihilator of $\mathscr{S}$ when S is the set of integers $\mathbf{Z}$?

10. Use induction to prove that the sum of ideals $A_1, \ldots, A_n$ is an ideal in R for any $n \in \mathbf{N}$. The ideal
$$B = \sum_{i=1}^{n} A_i = \{a_1 + \cdots + a_n : a_i \in A_i, i = 1, \ldots, n\},$$
also denoted $B = (A_1, \ldots, A_n)$, is called the **greatest common divisor** of the ideals A_i.

11. For elements $a_i, b_j \in R$, $1 \le i \le m$, $1 \le j \le m$, prove the distributivity property that
$$\left(\sum_{i=1}^{m} a_i\right)\left(\sum_{j=1}^{n} b_j\right) = \sum_{h=1}^{m+n}\left(\sum_{i+j=h} a_i b_j\right).$$

12. Let R be a ring with laws of addition and multiplication denoted by $+$ and juxtaposition, respectively. Let e denote the multiplicative identity. Now define new laws of addition $\#$ and multiplication $*$ as follows:
$$a \# b = a + b + e, \qquad a * b = a + b + ab.$$
   a. Prove that R is a ring with respect to the composition laws $\#$ and $*$.
   b. Prove that
$$\varphi : R \to \{R, \#, *\}$$
given by $\varphi(a) = a - e$, for $a \in R$, is an isomorphism.
   c. Determine the inverse of the isomorphism φ.

13. Prove that if $\varphi : R \to S$ is a ring homomorphism and if J is an ideal in S, then $\varphi^{-1}(J)$ is an ideal in R. It is not required that $J \subseteq \text{Im}\,\varphi$. [Cf. Proposition 2.]

14. The **ring $\mathscr{Q}$ of quaternions** (due to William Rowan Hamilton, 1805–1865) can be described as the set of all formal combinations $\{a + bi + cj + dk : a, b, c, d \in \mathbf{R}\}$, i.e., a four-dimensional vector space over $\mathbf{R}$ with basis $\{1, i, j, k\}$ [cf. §§4.1 and 4.2]. Addition is defined component-wise. Multiplication of two quaternions is given by taking 1 as the multiplicative identity in $\mathscr{Q}$, requiring that all real numbers r commute with $i, j,$ and k, and extending by distributivity the following products of the basic elements:
$$i^2 = j^2 = k^2 = -1,$$
$$ij = k, \qquad jk = i, \qquad ki = j,$$
$$ji = -k, \qquad kj = -i, \qquad ik = -j.$$
Observe that $\mathscr{Q}$ is a noncommutative ring.
   a. Define the conjugate A^* of $A = a_0 + a_1 i + a_2 j + a_3 k \in \mathscr{Q}$ to be
$$A^* = a_0 - a_1 i - a_2 j - a_3 k.$$
Prove that AA^* is a positive real number for $A \ne 0$.
   b. Show that every nonzero element of $\mathscr{Q}$ is a unit.

In 1878 Georg Frobenius (1849–1917) proved that the quaternions are the only vector space (of any dimension) over **R** with a noncommutative associative product for which every nonzero element is a unit. The complex numbers are a two-dimensional vector space over **R**, but with a commutative product.

15. The ring $\mathscr{Q}$ in Exercise 14 is called the ring of (real) quaternions. Define the ring of rational quaternions by replacing real numbers in Exercise 14 with rational numbers.
    a. Prove that for each $A \neq 0$ in this new system AA^* is a positive rational number.
    b. Show that every nonzero element is a unit.
16. In Exercise 15, replace the rational numbers by the set of complex numbers $\mathbf{Q}(\sqrt{-1})$ of the form $a + b\sqrt{-1}$, where $a, b \in \mathbf{Q}$. Assume that $\sqrt{-1} \neq i, j, k$ and that $\sqrt{-1}$ commutes with i, j, k. Prove that the resulting ring contains nontrivial zero divisors.
17. An ideal J in a ring R is said to be a **prime ideal** if $ab \in J$ implies either $a \in J$ or $b \in J$. Suppose φ is a surjective homomorphism of the ring R onto the ring S, and P is a prime ideal in S. Prove that the inverse image $\varphi^{-1}(P)$ is a prime ideal in R.
18. Consider an ideal A contained in Rad R [cf. Exercise 8, §3.1]. Assume that $x \in R$ satisfies $x^2 \equiv x \pmod{A}$, i.e., x is idempotent modulo A. Prove the existence of an idempotent $y \in R$ such that $y \equiv x \pmod{A}$. *Hint:* Try $y = x + t(1 - 2x)$ with $t \in R$, requiring that $y^2 = y$, and use the *formal* expansion $t = (\frac{1}{2})[1 - (1 + 4z)^{-1/2}]$, letting $z = x^2 - x$.

§3.3 Direct Sums of Rings

We continue our general discussion of rings with consideration of two related concepts: the external direct sum of rings and the internal direct sum of ideals. Section 3.5 provides examples in $\mathbf{Z}_m$ of the direct sum discussion of the present section. Thus most of the exercises on direct sums are deferred until then.

The external direct sum is a tool to construct new rings from given ones, while internal direct sums are important for the analysis and decomposition of a given ring into algebraically "simpler" rings. The significance of these concepts will be explored within the context of residue class rings of integers in §3.5 and of polynomials in §5.3. In §6.10 we extend the concept of direct sum to groups. It should be noted too that these concepts and their applications in Chapter 7 are prototypes of important aspects of the theory of algebraic numbers and functions.

Let $R_1 = \{a_1, b_1, \ldots\}, \ldots, R_n = \{a_n, b_n, \ldots\}$ be rings with respective multiplicative identities $e_1, \ldots, e_n$, $n > 1$. Consider the set

$$R = R_1 \dotplus \cdots \dotplus R_n$$

of all n-tuples $a = (a_1, \ldots, a_n)$, $b = (b_1, \ldots, b_n), \ldots, e = (e_1, \ldots, e_n)$ of elements in $R_1, \ldots, R_n$. Define $(a_1, \ldots, a_n) = (b_1, \ldots, b_n)$ if and only if $a_i = b_i$, $1 \leq i \leq n$.

Furthermore, define rules of addition and multiplication by

$$a + b = (a_1 + b_1, \ldots, a_n + b_n),$$
$$ab = (a_1 b_1, \ldots, a_n b_n).$$

Then R is, as is easily checked, a commutative ring with e for its multiplicative identity. The ring R is called the **(external) direct sum** of the component rings $R_1, \ldots, R_n$. For each i, $1 \leq i \leq n$, the map

$$a_i \to (0, \ldots, 0, \underset{\underset{i\text{th component}}{\uparrow}}{a_i}, 0, \ldots, 0)$$

is an injective homomorphism $\varphi_i \colon R_i \to R$. Furthermore, the multiplicative identity e of R satisfies

$$e = \sum_{i=1}^{n} \varphi_i(e_i).$$

In particular, $\varphi_i(e_i) \neq e$, and $\varphi_i(R_i)$ is a subring (and also an ideal) in R.

A ring R is said to be the **(internal) direct sum** $R = A_1 \oplus \cdots \oplus A_n$ of ideals $A_1, \ldots, A_n$ if every element $r \in R$ can be expressed *uniquely* as the sum of elements $a_i \in A_i$, $1 \leq i \leq n$. (If R is a noncommutative ring we require that the ideals A_i be two-sided.)

Proposition 1. If $R = A_1 \oplus \cdots \oplus A_n$ is the (internal) direct sum of ideals, then the ideals A_i are subrings of R, $1 \leq i \leq n$.

Proof. An ideal is a subring if and only if it has a multiplicative identity. Write the multiplicative identity e of R as $e = e_1 + \cdots + e_n$, the *unique* sum of elements $e_i \in A_i$. We claim $e_i a_i = a_i$ for all $a_i \in A_i$, and hence that e_i is the multiplicative identity in A_i. The proof rests on the fact that $a_i e_j = 0$ for any $a_i \in A_i$, where $i \neq j$. Suppose to the contrary that $a_i e_j \neq 0$ for some distinct i and j. Then we would have

$$a_i = a_i e = \underset{\underset{\in A_1}{\uparrow}}{a_i e_1} + \cdots + \underset{\underset{\in A_i}{\uparrow}}{a_i e_i} + \cdots + \underset{\underset{\in A_j}{\uparrow}}{a_i e_j} + \cdots + \underset{\underset{\in A_n}{\uparrow}}{a_i e_n}$$

$$= \underset{\underset{\in A_1}{\uparrow}}{a_i e_1} + \cdots + \underset{\underset{\in A_i}{\uparrow}}{(a_i e_i + a_i e_j)} + \cdots + \underset{\underset{\in A_j}{\uparrow}}{0} + \cdots + \underset{\underset{\in A_n}{\uparrow}}{a_i e_n},$$

which are two distinct expressions for a_i as the sum of elements in the ideals A_k, contradicting the hypothesis that the sum $A_1 \oplus \cdots \oplus A_n$ is direct. Therefore we have

$$a_i = a_i e = a_i(e_1 + \cdots + e_n) = a_i e_i,$$

for all $a_i \in A_i$. Hence e_i is the multiplicative identity in A_i, which is then a subring as asserted.

§3.3 Direct Sums of Rings

We note further that $e_i e_i = e_i$ and $e_i e_j = 0$ for $i \neq j$. The elements e_i are called **orthogonal idempotents**. Such elements will be discussed in detail in §3.4 for the ring $\mathbf{Z}_m$.

We have the following consequence of this proof.

Corollary. If $a_i \in A_i$ and $a_j \in A_j$, where $i \neq j$, then $a_i a_j = 0$.

Since e_j is the multiplicative identity of A_j, we can write
$$a_i a_j = a_i(e_j a_j) = (a_i e_j) a_j = 0.$$

The terms *external* and *internal* refer to the fact that the external direct sum R of rings $R_1, \ldots, R_n$ is outside of ("external to") the rings $R_1, \ldots, R_n$, whereas expressing R as the internal direct sum of ideals $A_1, \ldots, A_n$ emphasizes the structure within R. The following proposition relates these two concepts.

Proposition 2. If $R = R_1 \dotplus \cdots \dotplus R_n$ and $\varphi_i: R_i \to R$ is the mapping of R_i to the ith component of R, then $R = \varphi_1(R_1) \oplus \cdots \oplus \varphi_n(R_n)$. Conversely, if $R = A_1 \oplus \cdots \oplus A_n$, then $R \cong A_1 \dotplus \cdots \dotplus A_n$.

The proof is not difficult, and is given in the case of $R = \mathbf{Z}_m$ in §3.5. The general proof is left as an exercise in §3.5.

Exercises

1. Show that the (external) direct sum $\mathbf{Z}_2 \dotplus \mathbf{Z}_2$ is not isomorphic to $\mathbf{Z}_4$.
2. a. Prove that $R = \mathbf{Z}_2 \dotplus \mathbf{Z}_3$ and $S = \mathbf{Z}_3 \dotplus \mathbf{Z}_2$ are isomorphic rings.
   b. Let R_1, R_2 be two distinct rings, and define $R = R_1 \dotplus R_2$ and $S = R_2 \dotplus R_1$. Prove that R and S are isomorphic rings and that $R \neq S$.
3. Prove that $\mathbf{Z}_6$ is isomorphic to the ring R of Exercise 2(a).
4. In $\mathbf{Z}_{36}$ consider the ideals A and B generated by [28] and [9], respectively.
   a. Verify that [28] and [9] are orthogonal idempotents in $\mathbf{Z}_{36}$, whose sum is [1].
   b. Prove that $\mathbf{Z}_{36} = A \oplus B$.
   c. Verify that A and B are also subrings of $\mathbf{Z}_{36}$.
5. a. Write $\mathbf{Z}_{20}$ as the (internal) direct sum of two ideals, which are also subrings of $\mathbf{Z}_{20}$.
   b. Prove that the ideals of part (a) are principal with idempotent generators.
6. Let $G = \mathbf{Z}_2 \dotplus \mathbf{Z}_2 \dotplus \mathbf{Z}_4$, and write
$$\varphi(([a]_2, [b]_2, [c]_4)) = ([c]_2, [c]_2, [a+b]_4),$$
where $a, b, c \in \mathbf{Z}$. Does φ describe a homomorphism from G to G?

7. An idempotent u in a ring R, in which $e + e \neq 0$, is termed **irreducible** if there exist no nonzero idempotents $v, w \in R$ such that $u = v + w$ and $vw = 0$. Prove that two distinct irreducible idempotents u and u_1 of R satisfy $uu_1 = 0$ (i.e., they are *orthogonal*).
8. Prove that $R = Ru \oplus Rv$ if u, v are orthogonal idempotents and $e = u + v$, where $Ru = \{ru : r \in R\}$ denotes the ideal generated by $u \in R$.

§3.4 Residue Class Rings of Integers

With the general ring-theoretic results of §§3.1 through 3.3 available we undertake a systematic examination of the residue class rings of integers $\mathbf{Z}_m$, introduced in §2.9. First we construct systems of orthogonal idempotents in $\mathbf{Z}_m$ and prove that every ideal in $\mathbf{Z}_m$ is a principal ideal. In the next section we express $\mathbf{Z}_m$ as the internal direct sum of ideals generated by these orthogonal idempotents and as the external direct sum of certain other residue class rings of integers.

In this section and the next we consider the residue class ring $\mathbf{Z}_m$ for an arbitrary, but fixed modulus m; $m \neq 0, \pm 1$. Assume throughout that m has the factorization

$$m = m_1 \cdots m_n, \quad \text{where } (m_i, m_j) = 1, \text{ for } i \neq j.$$

(It is implicitly assumed that the modulus m has at least two distinct prime divisors.)

For each i, $1 \leq i \leq n$, consider the system of congruences

$$x_i \equiv 0 \pmod{m_1}$$
$$\cdot \quad \cdot \quad \cdot \quad \cdot \quad \cdot$$
$$x_i \equiv 1 \pmod{m_i}$$
$$\cdot \quad \cdot \quad \cdot \quad \cdot \quad \cdot$$
$$x_i \equiv 0 \pmod{m_n}$$

or briefly, $\quad x_i \equiv \delta_{ij} \pmod{m_j}, \quad 1 \leq j \leq n,$

where $\delta_{ij} = 0$ for $j \neq i$, and $\delta_{ii} = 1$. (We refer to δ_{ij} as the **Kronecker delta**, after Leopold Kronecker.) Each of the n systems of congruences has a solution e_i which is unique modulo m by the Chinese Remainder Theorem of §2.11. Furthermore,

$$e_i^2 \equiv \delta_{ij}\delta_{ij} \equiv \delta_{ij} \equiv e_i \pmod{m_j}, \quad 1 \leq j \leq n.$$

Consequently $e_i^2 \equiv e_i \pmod{m}$ [cf. Proposition 4, §2.6]. In terms of elements of $\mathbf{Z}_m$, we have $[e_i]^2 = [e_i]$ and $[e_i] \neq [0]$, $[e_i] \neq [1]$; thus each coset $[e_i]$ is a nontrivial idempotent in the residue class ring $\mathbf{Z}_m$.

Moreover,

$$e_i e_h \equiv \delta_{ij}\delta_{hj} \pmod{m_j}, \quad 1 \leq i, j, h \leq n,$$

and so for $h \neq i$,

$$e_i e_h \equiv 0 \pmod{m_j}, \quad 1 \leq j \leq n.$$

§3.4   Residue Class Rings of Integers   75

Therefore, whenever $h \neq i$,
$$e_i e_h \equiv 0 \pmod{m},$$
and in $\mathbf{Z}_m$,
$$[e_i][e_h] = [0].$$
In other words, $[e_i]$ and $[e_h]$ are orthogonal idempotents in $\mathbf{Z}_m$. Finally, for $1 \leq j \leq n$,
$$e_1 + \cdots + e_i + \cdots + e_n \equiv \sum_{i=1}^{n} \delta_{ij} = \delta_{jj} = 1 \pmod{m_j}$$
and therefore
$$e_1 + \cdots + e_i + \cdots + e_n \equiv 1 \pmod{m}.$$
Equivalently in $\mathbf{Z}_m$,
$$[e_1] + \cdots + [e_i] + \cdots + [e_n] = [1].$$
Thus we have proved the following proposition.

Proposition 1. To the product decomposition $m = m_1 \cdots m_n$ of the modulus m, there corresponds an additive decomposition $[1] = \sum_{i=1}^{n} [e_i]$ of the identity $[1]$ of the residue class ring $\mathbf{Z}_m$ into mutually orthogonal idempotents $[e_i]$.

In §2.4, we demonstrated that any ideal in the ring of integers $\mathbf{Z}$ is principal. We now prove the corresponding result for the ring $\mathbf{Z}_m$.

Proposition 2. Any ideal $A^* \subseteq \mathbf{Z}_m$ is principal.

Proof. The ideals $([0])$ and $([1]) = \mathbf{Z}_m$ are the trivial cases as for $\mathbf{Z}$. Let A^* be any nontrivial ideal in $\mathbf{Z}_m$ and consider the set
$$A = \{a \in \mathbf{Z} : [a] \in A^*\}.$$
Then A, the inverse image of A^* for the coset mapping $a \to [a] \in \mathbf{Z}_m$, is an ideal of $\mathbf{Z}$ by Proposition 2, §3.2.

Since every ideal in $\mathbf{Z}$ is principal [Theorem, §2.4], we can write $A = (h)$ for some $h \in \mathbf{Z}$. Therefore $[h] \in A^*$ and $([h]) \subseteq A^*$. Conversely, for any element $[a]$ in A^*, each $b \in [a]$ belongs to A. Thus $b = hx$ for some $x \in \mathbf{Z}$, and
$$[a] = [b] = [h][x] \in ([h]).$$
In other words, $A^* \subseteq ([h])$, the ideal generated by the residue class $[h]$. Consequently $A^* = ([h])$ is a principal ideal, as asserted.

Note that for any ideal $A^* \subseteq \mathbf{Z}_m$, the corresponding ideal A in $\mathbf{Z}$ contains (m). In fact this observation is half of the proof of the following proposition.

Proposition 3. There is a one-one correspondence between ideals in $\mathbf{Z}_m$ and those in $\mathbf{Z}$ which contain (m).

To complete the proof, consider any ideal $A \subseteq \mathbf{Z}$, containing (m). The surjective homomorphism $\pi: \mathbf{Z} \to \mathbf{Z}_m$, defined by $\pi(a) = [a]$, maps A to an ideal $\pi(A) \subseteq \mathbf{Z}_m$, as is easily verified. Further, for ideals A, B, both containing (m),

$$\pi(A) = \pi(B) \Leftrightarrow A = B.$$

Example. We conclude this section with an illustration of the general discussion of the residue class rings $\mathbf{Z}_m$. Consider $\mathbf{Z}_{20}$. The modulus $m = 20$ has the factorization $20 = 2^2 \cdot 5$. Solving for orthogonal idempotents, we have

$$e_1 \equiv 1 \pmod{4} \quad \text{and} \quad e_1 \equiv 0 \pmod{5};$$

hence $\quad e_1 = 1 + 4s \equiv 0 \pmod{5}.$

A particular solution is $e_1 = 5$. Furthermore, the congruences

$$e_2 \equiv 0 \pmod{4} \quad \text{and} \quad e_2 \equiv 1 \pmod{5}$$

imply $\quad e_2 = 1 + 5u \equiv 0 \pmod{4};$

a particular solution is $e_2 = 16$. Direct computation then yields

$$e_1 e_2 = 5 \cdot 16 \equiv 0 \pmod{20},$$
$$e_1^2 = 25 \equiv 5 \equiv e_1 \pmod{20},$$
$$e_2^2 = 256 \equiv 16 \equiv e_2 \pmod{20}.$$

Furthermore, $\quad e_1 + e_2 = 21 \equiv 1 \pmod{20}.$

These congruences are reflected as equalities for the cosets in $\mathbf{Z}_{20}$ as follows:

$$[e_1][e_2] = 0, \quad [e_1] + [e_2] = [1],$$
$$[e_1]^2 = [e_1], \quad [e_2]^2 = [e_2].$$

The following observation concerning this illustration is a prelude to the next section. The multiples $[0], [5], [10], [15]$ of $[e_1] = [5]$ constitute an ideal (and subring) $S = ([5])$ in $\mathbf{Z}_{20}$, isomorphic to $\mathbf{Z}_4$. The isomorphism is given by $\varphi([5]_{20}) = [1]_4$. Similarly, $T = ([4])$ is an ideal (and subring) in $\mathbf{Z}_{20}$, isomorphic to $\mathbf{Z}_5$. The isomorphism is given by $\psi([16]_{20}) = [1]_5$ [cf. §3.2].

Exercises

1. Find all ideals in the residue class ring $\mathbf{Z}_m$ for
   a. $m = 28$ b. $m = 21$ c. $m = 15$.
2. Find orthogonal idempotents in $\mathbf{Z}_m$ for
   a. $m = 36$ b. $m = 40$ c. $m = 30$
   d. $m = 35$ e. $m = 58$ f. $m = 120$.

3. a. Find in $\mathbf{Z}_{28}$ two distinct idempotents whose sum is the multiplicative identity.
   b. Exhibit in $\mathbf{Z}_{28}$ a subring isomorphic to $\mathbf{Z}_7$.
   c. Show that every nonzero element of the subring in part (b) has a multiplicative inverse.
4. Repeat Exercise 3 with 28 replaced by 21.
5. Repeat Exercise 3(a) with 28 replaced by 15.
6. a. Let A be an ideal in a ring R and $\varphi: R \to S$ a ring homomorphism. Prove that $\varphi(A)$ is an ideal in S if φ is surjective.
   b. Show by example that $\varphi(A)$ need not be an ideal in S if φ is not surjective.
7. Verify that if A is an ideal in $\mathbf{Z}$, then $\pi(A)$ is an ideal in $\mathbf{Z}_m$, where $\pi: \mathbf{Z} \to \mathbf{Z}_m$ is the projection onto cosets modulo m, $\pi(a) = [a]_m$.

§3.5 Direct Sum Decompositions of $\mathbf{Z}_m$

This section continues the study of a general residue class ring $\mathbf{Z}_m$ as an internal and external direct sum. It entails arguments on homomorphisms which can be considered as models for similar structure problems in general ideal theory and in group theory [see §§6.10, 7.1, and 7.8].

As in §3.4 we write m as the product $m = m_1 m_2 \cdots m_n$ of $n \geq 2$ relatively prime factors. Setting

$$g_i = m/m_i = m_1 \cdots m_{i-1} m_{i+1} \cdots m_n,$$

we observe that the GCD $(g_1, \ldots, g_n) = 1$. Thus

$$1 = g_1 h_1 + \cdots + g_n h_n \quad \text{with } h_i \in \mathbf{Z},\ 1 \leq i \leq n,$$

[cf. Exercise 12, §2.5]. Since every g_j, $j \neq i$, is divisible by m_i, we have

$$1 = g_1 h_1 + \cdots + g_i h_i + \cdots + g_n h_n \equiv g_i h_i \pmod{m_i}$$

and $\qquad g_i h_i \equiv 0 \pmod{m_j} \qquad$ for every $j \neq i$.

Consequently, $g_i h_i \equiv \delta_{ij} \pmod{m_j}$ for each fixed i and all j, $1 \leq i, j \leq n$. In other words, $g_i h_i$ is a solution of the system of congruences $x_i \equiv \delta_{ij} \pmod{m_j}$, $1 \leq j \leq n$, whose solution e_i is unique modulo m according to the Chinese Remainder Theorem [§2.11]. Thus

$$g_i h_i \equiv e_i \pmod{m} \quad \text{or} \quad [g_i h_i] = [e_i] \quad \text{in } \mathbf{Z}_m.$$

Next let $B_i^* = ([e_i])$ denote the principal ideal generated in $\mathbf{Z}_m$ by the ith orthogonal idempotent $[e_i]$ of $\mathbf{Z}_m$. That is,

$$B_i^* = [g_i h_i] \mathbf{Z}_m = \{[g_i h_i][z] : [z] \in \mathbf{Z}_m\}.$$

The following proposition states properties of the ideals B_i^*, $1 \leq i \leq n$.

Proposition 1. The ideals B_i^*, $1 \leq i \leq n$, satisfy

(i) $B_i^{*2} = B_i^*$.
(ii) $B_i^* B_k^* = ([0])$, for $i \neq k$.
(iii) $B_i^* \cap B_k^* = ([0])$, for $i \neq k$.

Detailed proofs are left as exercises. Note that since $B_i^* B_k^* \subseteq B_i^* \cap B_k^*$, (ii) follows from (iii). Further B_i^* has a multiplicative identity, namely $[e_i]$. Hence $[a][e_i][e_i] = [a][e_i]$, as $[e_i]$ is idempotent, which proves (i). Since the ideal B_i^* has a multiplicative identity, Proposition 1(i) implies that it is a subring of Z_m.

Utilizing the additive decomposition of $[1] \in Z_m$ obtained in §3.4, we have for $[a] \in Z_m$,

$$[a] = [a][1] = [a]([e_1] + \cdots + [e_i] + \cdots + [e_n])$$
$$= [a][e_1] + \cdots + [a][e_i] + \cdots + [a][e_n].$$

Thus the ith "component" $[a][e_i]$ of $[a]$ lies in B_i^*; symbolically,

$$Z_m = B_1^* + \cdots + B_n^*.$$

In other words, Z_m is the sum of the ideals B_i^*, $1 \leq i \leq n$.

Proposition 2. For B_i^* defined as above, the sum $Z_m = B_1^* + \cdots + B_n^*$ is direct.

That is, for any $[a] \in Z_m$ the components $[b_i] \in B_i^*$, $1 \leq i \leq n$, in the expression

(*) $\qquad\qquad [a] = [b_1] + \cdots + [b_n]$

are uniquely determined by the coset $[a]$ [see §3.3]. To *establish the uniqueness* of representation of $[a]$ in equation (*) we must show for any other representation

$$[a] = [c_1] + \cdots + [c_n], \qquad [c_i] \in B_i^*$$

that $[b_i] = [c_i]$, $1 \leq i \leq n$. Since

$$[a] - [a] = [0] = [b_1 - c_1] + \cdots + [b_n - c_n],$$

we have $\qquad [c_2 - b_2] + \cdots + [c_n - b_n] = [b_1 - c_1] \in B_1^*,$

or equivalently $\qquad b_1 - c_1 \equiv 0 \pmod{m_2 \cdots m_n}.$

Also, $c_j - b_j \equiv 0 \pmod{m_1 \cdots m_n m_j^{-1} = g_j}$ for all j distinct from 1. This congruence implies $c_j - b_j \equiv 0 \pmod{m_1}$. Consequently

$$b_1 - c_1 = \sum_{j=2}^{n} (c_j - b_j) \in (m_1),$$

so that $(b_1 - c_1) \in (m_1) \cap (m_2 \cdots m_n)$. But

$$(m_1) \cap (m_2 \cdots m_n) = (m_1 m_2 \cdots m_n) = (m),$$

§3.5 Direct Sum Decompositions of Z_m

because of the assumption that $(m_i, m_j) = 1$ for $i \neq j$. Thus $[b_1 - c_1] = [0]$ and $[c_1] = [b_1]$. Examining in turn the differences $b_j - c_j$ on the left side, we find that $[c_j] = [b_j]$, which completes the proof.

Notational Convention. In order to simplify notation throughout the remainder of this section, we denote the residue class rings $\mathbf{Z}/(m_i) = \mathbf{Z}_{m_i}$ by Z_i, and their elements $[a]_{m_i}$ by $[a]_i$. We continue to use unmodified brackets to denote the elements $[a] = [a]_m$ in $\mathbf{Z}_m$.

Proposition 3. There exists, for each i, $1 \leq i \leq n$, an isomorphism φ_i of the residue class ring Z_i with the subring $B_i^* = ([e_i])$ of $\mathbf{Z}_m$.

Proof. To define an isomorphism

$$\varphi_i \colon Z_i \to B_1^* \subset \mathbf{Z}_m,$$

consider first a given residue class $[a]_i \in Z_i$ and any integer a in $[a]_i$. To this element $a \in \mathbf{Z}$ we associate its coset $[a] \in \mathbf{Z}_m$ and define

$$\varphi_i([a]_i) = [a][e_i] \in B_i^*.$$

(i) We first observe that the image $\varphi_i([a]_i)$ does not depend upon the choice of the representative $a \in [a]_i$. For any other representative $a' \in [a]_i$, we have $a' = a + km_i$ for some $k \in \mathbf{Z}$. Hence

$$\begin{aligned}[a'][e_i] &= [a + km_i][e_i] = [a][e_i] + [k][m_i g_i h_i]\\ &= [a][e_i] + [k][mh_i] = [a][e_i] + [k][0]\\ &= [a][e_i] = \varphi_i([a]_i),\end{aligned}$$

since $e_i \equiv g_i h_i \pmod{m}$ and $g_i m_i = m$. Thus the map φ_i is well-defined on the residue classes $[a]_i$.

(ii) Next, φ_i is a homomorphism of Z_i into $\mathbf{Z}_m$. Let a, b be representatives of $[a]_i, [b]_i$, respectively. Then

$$\underset{\text{sum in } Z_i}{[a]_i + [b]_i} = \underset{\text{sum in } \mathbf{Z}}{[a+b]_i}$$

and $\quad \varphi_i([a]_i + [b]_i) = [a+b][e_i] \quad$ the product of classes in $\mathbf{Z}_m$
$$\begin{aligned}&= ([a] + [b])[e_i]\\ &= [a][e_i] + [b][e_i]\\ &= \varphi_i([a]_i) + \varphi_i([b]_i).\end{aligned}$$

Furthermore, $\quad \underset{\text{product in } Z_i}{[a]_i [b]_i} = \underset{\text{product in } \mathbf{Z},}{[ab]_i}$

and $\quad \varphi_i([a]_i [b_i]) = [ab][e_i] \quad$ the product of classes in $\mathbf{Z}_m$
$$\begin{aligned}&= ([a][b])[e_i]\\ &= [a][e_i][b][e_i] \quad \text{since } [e_i] \text{ is an idempotent}\\ &= \varphi_i([a]_i) \varphi_i([b]_i).\end{aligned}$$

(iii) The homomorphism φ_i is surjective. Consider $[x] \in B_i^*$, pick $x \in [x]$, and take the residue class modulo m_i of x, i.e., $[x]_i$. By the definition of φ_i, $\varphi_i([x]_i) = [x]$; thus φ_i is surjective.

(iv) Finally, to show that φ_i is an isomorphism, it remains only to verify that $\ker \varphi_i$ in $\mathbf{Z}_i$ is $\{[0]_i\}$. Suppose that $\varphi_i([x]_i) = [0]$. Then any $x' \in [x]_i$ satisfies $[0] = [x'][e_i] = [x'g_i h_i]$ in $\mathbf{Z}_m$. Hence $m \mid x' g_i h_i$ and $m_i \mid x' h_i$ since $g_i = m/m_i$. From

$$1 = g_1 h_1 + \cdots + g_i h_i + \cdots + g_n h_n$$

we note that $(m_i, h_i) = 1$, and therefore $m_i \mid x'$; that is, $[x']_i = [0]_i$, as was to be shown.

NOTE. Since $[e_i]$ is an idempotent in $\mathbf{Z}_m$,

$$\varphi_i([e_i]_i) = [e_i][e_i] = [e_i].$$

Thus, the identity of $\mathbf{Z}_i$ is mapped by φ_i not to the identity of $\mathbf{Z}_m$, but to the identity of the subring $B_i^* \subset \mathbf{Z}_m$.

We now show that $\mathbf{Z}_m$ is isomorphic to an external direct sum. As a first step we define a mapping

$$\Lambda: \mathbf{Z}_m \to \mathbf{Z}_1 \dotplus \cdots \dotplus \mathbf{Z}_n$$

by selecting a representative a of $[a]$ and setting

$$\Lambda([a]) = ([a]_1, \ldots, [a]_i, \ldots, [a]_n) \in \mathbf{Z}_1 \dotplus \cdots \dotplus \mathbf{Z}_n.$$

The element $\Lambda([a])$ does not depend on the choice of the representative a. For arbitrary $a' \in [a]$, we have $a' = a + qm$ with $q \in \mathbf{Z}$ and hence

$$[a']_i = [a + qm]_i = [a]_i + [qm]_i = [a]_i,$$

since $m \equiv 0 \pmod{m_i}$, $1 \leq i \leq n$.

Proposition 4. *The map Λ defined above is an isomorphism:*

$$\Lambda: \mathbf{Z}_m \to R = \mathbf{Z}_1 \dotplus \cdots \dotplus \mathbf{Z}_n.$$

Proof. For $b \in [b]$,

$$\Lambda([a] + [b]) = \Lambda([a+b]) = ([a+b]_1, \ldots, [a+b]_n)$$
$\uparrow$ sum in $\mathbf{Z}_m$ $\uparrow$ sum in $\mathbf{Z}$

$$= ([a]_1 + [b]_1, \ldots, [a]_n + [b]_n)$$
$$= ([a]_1, \ldots, [a]_n) + ([b]_1, \ldots, [b]_n)$$
$\uparrow$ sum in R

$$= \Lambda([a]) + \Lambda([b]) \qquad \text{by the definition of } \Lambda.$$

Similarly, $\Lambda([a][b]) = \Lambda([a]) \cdot \Lambda([b])$, and so Λ is a homomorphism.

§3.5 Direct Sum Decompositions of Z_m

Furthermore, Λ is injective. Suppose there are given residue classes $[a], [b] \in Z_m$ such that $\Lambda([a]) = \Lambda([b])$. Then $[a]_i = [b]_i$, or $a - b \equiv 0 \pmod{m_i}$, for all i, $1 \leq i \leq n$. Consequently $a - b \equiv 0 \pmod{m}$ by Proposition 4, §2.6, or $[a] = [b]$.

Lastly, Λ is surjective. Given $([a_1]_1, \ldots, [a_n]_n) \in R$, let a be a solution of the n congruences $x \equiv a_i \pmod{m_i}$, $1 \leq i \leq n$ [see §2.11]. The element $a \in Z$ satisfies $[a]_i = [a_i]_i$, and so $\Lambda([a]) = ([a_1]_1, \ldots, [a_n]_n)$. Consequently Λ is an isomorphism of Z_m onto the external direct sum R as asserted.

Proposition 4 yields a proof of the formula for the Euler φ-function. The function $\varphi(m)$ is defined to be the number of positive integers less than m that are relatively prime to m, or equivalently, the number of units in the residue class ring Z_m. In particular, as stated in §2.12, for

$$m = \prod_{i=1}^{n} m_i, \quad \text{where } m_i = p_i^{\alpha_i},$$

and where the p_i are distinct positive primes, the **Euler φ-function** is given by

$$\varphi(m) = \prod_{i=1}^{n} \varphi(m_i) = \prod_{i=1}^{n} p_i^{\alpha_i}\left(1 - \frac{1}{p_i}\right)$$

$$= m \prod_{i=1}^{n} \left(1 - \frac{1}{p_i}\right).$$

For the proof we use the fact (proven below) that there is a one-one correspondence λ between the units U_m of Z_m and the cartesian product $U_1 \times \cdots \times U_n$ of units U_i in the rings Z_i, $1 \leq i \leq n$. [See §2.12.] Since U_i has $\varphi(m_i)$ elements, U_m has $\varphi(m) = \prod_{i=1}^{n} \varphi(m_i)$ elements. Writing $m_i = p_i^{\alpha_i}$, recall from §2.12 that $\varphi(m_i) = p_i^{\alpha_i}(1 - 1/p_i)$. Hence we have the formula

$$\varphi(m) = \prod_{i=1}^{n} p_i^{\alpha_i}\left(1 - \frac{1}{p_i}\right).$$

In establishing the one-one correspondence we emphasize on the one hand the *principle of localization* (passing to the examination of powers of primes), and on the other the Chinese Remainder Theorem, as the means to pass from "local" results (the structure of the units $U_i \subset Z_i$) to "global" results (the structure of the units $U_m \subset Z_m$). For $[a] \in U_m$, we know that $(a, m) = 1$, and "by localization" that $(a, m_i) = 1$ since $m_i | m$. Hence $[a]_i \in U_i$, $1 \leq i \leq n$, and so

$$\lambda([a]) = ([a]_1, \ldots, [a]_n) \in U_1 \times \cdots \times U_n.$$

Conversely, each set $([a_1]_1, \ldots, [a_n]_n)$ of n units $[a_i]_i \in Z_i$, $1 \leq i \leq n$, determines a unique unit $a \in Z_m$. The system of congruences

$$x \equiv a_i \pmod{m_i}$$

has a unique solution a modulo m by the Chinese Remainder Theorem (since the integers m_i are pairwise relatively prime). Further $(a, m) = 1$, since $(a_i, m_i) = 1$ and $a \equiv a_i \pmod{m_i}$ imply that $(a, m_i) = 1$. We conclude then that $a \in \mathbf{U}_m$, and note that

$$\lambda([a]) = ([a]_1, \ldots, [a]_n) = ([a_1]_1, \ldots, [a_n]_n).$$

Thus λ maps $\mathbf{U}_m$ onto the cartesian product of units $U_1 \times \cdots \times U_n$, which completes the proof.

Exercises

1. Express $\mathbf{Z}_m$ as the direct sum of ideals B_i^* for
   a. $m = 36$ b. $m = 35$ c. $m = 30$
   d. $m = 120$ e. $m = 28$ f. $m = 15$.
2. Show that the mapping $\varphi: \mathbf{Z}_m \to \mathbf{Z}_m$ given by $\varphi([a]) = [a][e_1] + [a][e_2]$, where $m = m_1 \cdots m_n$ ($n \geq 2$) and $[e_1], [e_2], \ldots, [e_n]$ are as defined in this section, is an endomorphism. What is the multiplicative identity of Im φ?
3. Describe explicit isomorphisms which show the following:
   a. $\mathbf{Z}_{36} \cong \mathbf{Z}_4 + \mathbf{Z}_9$ b. $\mathbf{Z}_{15} \cong \mathbf{Z}_3 + \mathbf{Z}_5$
   c. $\mathbf{Z}_{24} \cong \mathbf{Z}_3 + \mathbf{Z}_8$ d. $\mathbf{Z}_{30} \cong \mathbf{Z}_2 + \mathbf{Z}_3 + \mathbf{Z}_5$.
4. Prove directly that $\mathbf{Z}_5$ is isomorphic to a quotient ring of $\mathbf{Z}_{20}$. [See the example at the end of §3.4 and Exercise 3, §3.2.]
5. Following the notational convention of this section, let $\mathbf{Z}_m \cong Z_1 + \cdots + Z_n$. For each i, $1 \leq i \leq n$, prove that Z_i is isomorphic to a residue class ring of $\mathbf{Z}_m$.
6. Prove the properties of B_i^* cited in Proposition 1.
7. To generalize the discussion of the direct sum decomposition of $\mathbf{Z}_m$, let R be a commutative ring containing ideals $A_1, \ldots, A_n$ such that for all $i \neq j$, $1 \leq i, j \leq n$, the ideal (A_i, A_j) generated by A_i, A_j is equal to R.
   a. Describe $R/(A_1 \cdots A_n)$ as an internal direct sum.
   b. Prove that $R/(A_1 A_2 \cdots A_n)$ is isomorphic to the (external) direct sum
   $$R/A_1 + R/A_2 + \cdots + R/A_n.$$
   c. Note that $(A_1, \ldots, A_n) = R$ and prove that for any choice of $k_i \in \mathbf{N}$, $1 \leq i \leq n$,
   $$(A_1^{k_1}, \ldots, A_n^{k_n}) = R$$
   where $A_i^{k_i}$ is the k_ith power of the ideal A_i.
8. Consider a (commutative) ring R in which there are idempotents $e_1, \ldots, e_n$ (for some $n > 1$) such that
   (i) $e = e_1 + \cdots + e_n$, e the identity element in R;
   (ii) $e_i e_j = 0$ for $i \neq j$, $1 \leq i, j \leq n$.
   a. Prove that there exist ideals $A_1, \ldots, A_n$ in R such that
   $$R = A_1 \oplus \cdots \oplus A_n.$$
   b. Prove that $A_i \cap A_j = \{(0)\}$, for $i \neq j$, $1 \leq i, j \leq n$.
   c. Prove that $A_i A_i = A_i$ and that the ideal A_i is a subring, $1 \leq i \leq n$.

9. With reference to Exercise 4(b), §3.1, prove that if a ring R contains a nontrivial idempotent, then it contains two ideals A_1, A_2 such that $R = A_1 \oplus A_2$.

10. Suppose that a commutative ring R can be written as the sum of ideals $A_1, \ldots, A_n$:
$$R = A_1 + \cdots + A_n.$$
Prove that this sum is direct if and only if for all i, $2 \le i \le n$,
$$A_i \cap (A_1 + \cdots + A_{i-1}) = (0).$$
(This proof of directness is used later in a similar argument for vector spaces [§4.1] and groups [§6.10].)

11. a. Consider a divisor d of m. Prove that there is a unique ideal $A \subseteq \mathbf{Z}_m$ which has d elements.
    b. Prove that A has $\varphi(n)$ distinct elements which generate it, where $nd = m$.
    c. Conclude that the Euler φ-function satisfies
    $$\sum_{d \mid m} \varphi(d) = m,$$
    where the sum is taken over all positive divisors d of m.

The argument in Exercise 7(a) and (b) is used again in our discussion of cyclic groups [cf. Propositions 2 and 3, §6.6].

12. Prove the statement in Exercise 11(c) by induction on the number of distinct prime factors of m, i.e., by induction on r, where $m = p_1^{\alpha_1} \cdots p_r^{\alpha_r}$, $p_i \ne p_j$ for $i \ne j$.
    a. For the case $r = 1$, show that
    $$\varphi(m) = \varphi(1) + \varphi(p) + \varphi(p^2) + \cdots + \varphi(p^\alpha).$$
    b. For the inductive step, consider $m = p^\alpha n$, $(n, p) = 1$, and show that
    $$\sum_{d \mid m} \varphi(d) = \sum_{d \mid n} \varphi(d) + \sum_{d \mid n} \varphi(pd) + \sum_{d \mid n} \varphi(p^2 d) + \cdots + \sum_{d \mid n} \varphi(p^\alpha d),$$
    the sums being taken over all divisors d of m and n, as indicated.
13. Prove Proposition 2 of §3.3.

§3.6 Integral Domains and Fields

In §2.1 we listed algebraic properties of the integers. All these properties, except the cancellation law for multiplication, were included in defining the concept of a commutative ring in §3.1. A set of elements which has *all* the algebraic properties of the integers is called an integral domain.

An **integral domain** D is a commutative ring (with at least two elements) in which the cancellation law holds; that is, if $ac = bc$ for $a, b \in D$ and nonzero $c \in D$, then $a = b$.

It is customary to require that an integral domain have at least two elements, as in subsequent arguments it is awkward to consider the zero ring $R = \{0\}$ as an integral domain. This requirement is equivalent to stipulating that the multiplicative and additive identities are distinct.

In a ring the validity of the multiplicative cancellation law and the absence of zero divisors are equivalent properties, an equivalence we shall use often later on. We have already proved this equivalence for the ring of integers [Property 10, §2.1]; the proof for general rings is similar. If a is a proper (i.e., nonzero) divisor of zero, there is a nonzero element $c \in D$ such that $ac = 0 = 0 \cdot c$. Then by the cancellation law $a = 0$, a contradiction. Conversely, letting $ac = bc$ with $c \neq 0$, we obtain $(a-b)c = 0$. If there are no proper divisors of zero, then necessarily $a - b = 0$.

Examples of Integral Domains

1. The integers **Z**.
2. The rational, real, and complex numbers.
3. The residue class ring $\mathbf{Z}_p$, p a prime integer [see §2.10].
4. The so-called gaussian integers
$$\mathbf{Z}[i] = \{a + b\sqrt{-1} : a, b \in \mathbf{Z}\},$$
named for Carl Friedrich Gauss. (Test what conditions on the coefficients $a_j, b_j, j = 1, 2$, are required for the product
$$(a_1 + b_1\sqrt{-1})(a_2 + b_2\sqrt{-1})$$
of two gaussian integers to be zero.)
5. More generally, systems of real (or complex if $m < 0$) numbers of the form
$$\mathbf{Z}[\sqrt{m}] = \{a + b\sqrt{m} : a, b \in \mathbf{Z}\}.$$
6. Also, $\mathbf{Q}(\sqrt{m}) = \{a + b\sqrt{m} : a, b \in \mathbf{Q}\}$.
7. The set of all polynomials $p(x)$ with integral (also rational, real, or complex) coefficients. (Polynomials will be formally introduced in §5.1.)

Since the axiomatic properties of an integral domain are an abstraction of the algebraic properties of integers, Properties 1 through 10 in §2.1 generalize to integral domains, as do their proofs [cf. §3.1].

Note that nonzero elements a in an integral domain do not necessarily have multiplicative inverses in the domain. For example, in **Z** no integer times 2 is equal to 1.

Also, in $\mathbf{Z}[\sqrt{2}]$ there is no element $u + v\sqrt{2}$ with $u, v \in \mathbf{Z}$ such that $(3 + \sqrt{2})(u + v\sqrt{2}) = 1$. We have, for the real number $(3 + \sqrt{2})^{-1}$,

$$\frac{1}{3 + \sqrt{2}} = \frac{1}{3 + \sqrt{2}} \frac{3 - \sqrt{2}}{3 - \sqrt{2}} = \frac{3 - \sqrt{2}}{7} = 3/7 - (1/7)\sqrt{2}.$$

That is, the real number $(3+\sqrt{2})^{-1}$ belongs to $\mathbf{Q}(\sqrt{2})$, but not to the given integral domain with coefficients a, b in $\mathbf{Z}$.

An alternate proof follows from the fact that $(3+\sqrt{2})(u+v\sqrt{2}) = 1$ implies

$$3u + 2v = 1 \quad \text{and} \quad 3v + u = 0.$$

This pair of linear equations has a unique rational, but no integral, solution.

An integral domain F, containing at least two elements, in which every nonzero element (every element different from the additive identity) has a multiplicative inverse, is called a **field**.

The integral domains in Examples 2, 3, and 6 above are fields. The trivial ring $\{0\}$ is not considered a field.

A field can also be described as a commutative ring F (with at least two elements) in which each nonzero element has a multiplicative inverse. The equivalence of the two definitions is immediate since an element $r \in F$ having a multiplicative inverse r^{-1} cannot be a zero divisor. If $r \in F$ were to be a divisor of zero, there would exist a nonzero element $r' \in F$ such that

$$rr' = 0.$$

Multiplying by r^{-1}, we reach the contradiction $r' = 0$. Therefore, as asserted, r is not a zero divisor.

Proposition 1. *A finite integral domain D is necessarily a field.*

Proof. Let $a_1, \ldots, a_n$ be the distinct elements of D. For any $a \neq 0$ the n products $aa_1, \ldots, aa_n$ are distinct, because if $aa_i = aa_j$ for some pair $i \neq j$, then $a_i = a_j$ by the cancellation law. Since $1 \in D$ is one of the products aa_k, the element a_k is the multiplicative inverse of a. Thus, each nonzero element has a multiplicative inverse.

It is known—from experience—that every nonzero integer has an inverse in the ring (field) of rational numbers $\mathbf{Q}$. We now show that for every integral domain D we can construct a field $Q(D)$, called the **field of quotients** of D, that contains an isomorphic image of D. [See §3.2 for the definition of ring isomorphism.] The proof of the existence of the quotient field $Q(D)$ involves ordered pairs (a, b) of elements of D, where $b = 0$. Such pairs are simply an abstraction of the rationals, written a/b, for integers a and $b \neq 0$.

Theorem. *For a given integral domain D there exists a field $Q(D)$, and an embedding $\lambda: D \to Q(D)$.*

Proof. (i) Consider the totality of pairs of elements in D, such as (a, b), (a^*, b^*), (c, d), (c^*, d^*), in which the second components $b, b^*, d, d^*, \ldots$, are *different from zero*. In this set of pairs introduce the relation $\approx$:

$$(a, b) \approx (a^*, b^*) \quad \text{if } ab^* = ba^* \text{ in } D,$$

an equivalence relation in the sense of §1.2. Specifically,

$$(a, b) \approx (a, b) \qquad \text{because } ab = ba,$$

$$(a, b) \approx (c, d) \Rightarrow (c, d) \approx (a, b)$$
$$\text{because } ad = bc \Leftrightarrow cb = da,$$

$$(a, b) \approx (c, d) \text{ and } (c, d) \approx (e, f) \Rightarrow (a, b) \approx (e, f)$$
$$\text{because } ad = bc \text{ and } cf = de \text{ imply}$$
$$adf = bcf = bde \text{ and } af = be.$$

Since $\approx$ is an equivalence relation, each pair (a, b) belongs to one and only one equivalence class $[a, b]$. Let $Q(D)$ denote this set of equivalence classes of pairs of elements in D.

(ii) In $Q(D)$ we introduce addition and multiplication as follows:

$$[a, b] + [c, d] = [a^*d^* + c^*b^*, b^*d^*]$$
$$[a, b][c, d] = [a^*c^*, b^*d^*]$$

where $(a^*, b^*) \in [a, b]$ and $(c^*, d^*) \in [c, d]$. These definitions are a direct carry-over of the rules for sum and product of fractions $a/b \in \mathbf{Q}$.

It appears at first glance that these sums and products depend on the choice of the representatives (a^*, b^*) and (c^*, d^*) of the classes $[a, b]$ and $[c, d]$, i.e., that they may not be single-valued functions of the classes. Thus it *must be proved* that these sums and products are independent of the choice of representatives.

To show that the sum is well-defined we verify for $(a, b) \approx (a^*, b^*)$ and $(c, d) \approx (c^*, d^*)$ that

$$(ad + bc, bd) \approx (a^*d^* + b^*c^*, b^*d^*),$$

and for the product that

$$(ac, bd) \approx (a^*c^*, b^*d^*).$$

In other words we must verify that

$$(ad + bc)b^*d^* = (a^*d^* + b^*c^*)bd,$$
$$(ac)(b^*d^*) = (bd)(a^*c^*),$$

using $ab^* = a^*b$ and $cd^* = c^*d$.

These and subsequent verifications are *carried out in the integral domain D*. We find

$$(ad + bc)b^*d^* = adb^*d^* + bcb^*d^* = (a^*d^* + b^*c^*)bd,$$

and
$$(ac)(b^*d^*) = a^*bcd^* = (bd)(a^*c^*)$$

by the distributive, associative, and commutative laws in D. Thus the definitions of addition and multiplication of equivalence classes in terms of class representatives are indeed independent of the choice of representatives.

§3.6 Integral Domains and Fields

(iii) The associative, commutative, and distributive laws are easily verified for the sum and product of the elements of $Q(D)$, i.e., the equivalence classes of pairs of elements in D. The element $z = [0, 1]$ satisfies $z + [a, b] = [a, b]$ (z is then the additive identity of $Q(D)$), and $\varepsilon = [1, 1]$ satisfies $\varepsilon[a, b] = [a, b]$ for all $[a, b] \in Q(D)$ (ε is then the multiplicative identity of $Q(D)$). Each element $[a, b] \in Q(D)$ has an additive inverse $[-a, b]$. This completes the verification of the additive properties of a field for $Q(D)$.

For $[a, b] \neq z$—that is, $a \neq 0$—we have $[b, a] \in Q(D)$, and therefore $[a, b][b, a] = [ab, ab] = [1, 1] = \varepsilon$. Thus $Q(D)$ is a field. In particular the elements $[a, 1]$, $a \neq 0$, in D have the multiplicative inverses $[1, a]$ in $Q(D)$.

(iv) Next we define a mapping $\lambda: D \to Q(D)$ by $\lambda(a) = [a, 1]$, which we show to be an injective homomorphism. First, for all $a, b \in D$,

$$\lambda(a+b) = [a+b, 1] = [a, 1] + [b, 1] = \lambda(a) + \lambda(b)$$

and
$$\lambda(ab) = [ab, 1] = [a, 1][b, 1] = \lambda(a)\lambda(b).$$

Second, $\lambda(a) = [a, 1] = [0, 1] = z$ implies that a is zero. Hence λ is injective.

Finally, as an immediate consequence of its definition, λ is an embedding since $\lambda(1) = [1, 1] = \varepsilon$, the multiplicative identity in $Q(D)$.

We should observe that constant use was made of the assumption that D is an integral domain, specifically that the "denominators" bd and b^*d^* of the sums and products are different from 0 in D. It is customary to identify (i.e., equate) the images $\lambda(a) = [a, 1]$ of the elements a in D with the elements in D. With this understanding we say that D is embedded in its field of quotients $Q(D)$. In particular $Q(\mathbf{Z})$ is identified with the field $\mathbf{Q}$ of all "fractions" of elementary (down-to-earth) algebra.

As with subrings [§3.1], particular subsets of a field are called subfields. A nonempty subset K of a field F is called a **subfield** of F, if

(i) K is closed under addition and additive inverse, i.e., for all $a, b \in K$, we have $a + b$ and $-a \in K$.

(ii) K is closed under multiplication and multiplicative inverse (for all nonzero elements), i.e., for all $a, b \in K$, the product $ab \in K$, and for $a \neq 0$, the inverse a^{-1} of a in F belongs to K.

Alternatively, a subfield of a field F is a subring that contains the multiplicative inverse of each of its elements different from the additive identity. Again, $\{0\}$ is not considered a subfield. Of course, a subfield F' of a field F is itself a field.

The rational numbers $\mathbf{Q}$ are a subfield of the real numbers $\mathbf{R}$. Both $\mathbf{Q}$ and $\mathbf{R}$ are subfields of the complex numbers $\mathbf{C}$.

We now turn to what amounts to a proof of the uniqueness of the field of quotients $Q(D)$ of a given integral domain D.

Proposition 2. If D is an integral domain contained in a field F, then F contains an isomorphic image of the field of quotients $Q(D)$.

Proof. First observe that the multiplicative identities of F and D are equal, as F has no zero divisors. Then let a/b denote the quotient of elements $a, b \in D$, $b \neq 0$ ($a/b = ab^{-1}$ in F), and define a mapping

$$\varphi([a, b]) = \frac{a}{b}$$

from $Q(D)$ into F. Using the definitions of sum and product in $Q(D)$ and the rules for sums and products of elements a/b and c/d in F, we note immediately that φ is a single-valued homomorphism on $[a, b]$. Observe here that if $(a^*, b^*) \in [a, b]$, then $a^*/b^* = a/b$ in F. Furthermore,

$$\varphi([a, b] + [c, d]) = \frac{a}{b} + \frac{c}{d},$$

since $[a, b] + [c, d] = [ad + bc, bd]$ has the image $(ad + bc)/(bd)$ equal to $a/b + c/d$ in F. Also

$$\varphi([a, b][c, d]) = \left(\frac{a}{b}\right)\left(\frac{c}{d}\right)$$

and

$$\varphi([a, b]^{-1}) = \frac{b}{a} \quad \text{if } [a, b] \neq z.$$

Finally, φ is a one-one mapping because $\varphi([a, b]) = \varphi([c, d])$ means $a/b = c/d$, and thus $ad = bc$ or $[a, b] = [c, d]$. Therefore the set of images $\varphi(Q(D))$ is a subfield K of F. This subfield K is contained in every subfield L of F containing D; *it is uniquely determined by D as the intersection of all subfields L of F containing D.*

To conclude this section we consider the prime field of a field F. The properties of this smallest subfield contained in a field F reflect significant algebraic and arithmetic distinctions between fields, as we shall note in later sections. The **prime field** P of a field F is defined to be the intersection of all subfields L of F. Thus P is a subfield of F which contains no subfields.

Proposition 3. The prime field P of F is isomorphic either to the field of rational numbers $\mathbf{Q}$ or to the residue class field $\mathbf{Z}_p$, where the modulus p is uniquely determined by F.

Said another way, any field F has a subfield isomorphic to $\mathbf{Q}$ or to $\mathbf{Z}_p$. In the first case the field F is said to have **characteristic 0** (or ∞) denoted char $F = 0$; in the second case F is said to have **characteristic p**, denoted char $F = p$.

Proof. Let e denote the multiplicative identity of F and A the additive subset of F generated by e; that is,

$$A = \{0\} \cup \{n \cdot e : n \in \mathbf{N}\} \cup \{(-n) \cdot e : n \in \mathbf{N}\},$$

where $n \cdot e = e + \cdots + e$, n times, and $(-n) \cdot e = n \cdot (-e)$. The associative law of addition in F implies that $(r \cdot e) + (s \cdot e) = (r+s) \cdot e$; and the distributive law for multiplication in F implies that $(r \cdot e)(s \cdot e) = (rs) \cdot e$, for any $r, s \in \mathbf{Z}$. Consequently A is an integral domain, since it is a subset of the field F and hence has no zero divisors. We show next that A is isomorphic either to $\mathbf{Z}$ or to $\mathbf{Z}_p$, for some prime p.

Define a mapping $\pi: \mathbf{Z} \to F$ by setting $\pi(1) = e$, and more generally $\pi(z) = z \cdot e$ for all $z \in \mathbf{Z}$. That π is a homomorphism from $\mathbf{Z}$ onto the subring A of F follows immediately from its definition and the description of the elements in A. The ideal

$$\ker \pi = \{z \in \mathbf{Z} : \pi(z) = z \cdot e = 0 \in A\}$$
$$= (m) \quad \text{for some } m \in \mathbf{Z},$$

as every ideal in $\mathbf{Z}$ is principal. We now show that the integer m can only be either 0 or a prime number p. (The ideal $(m) = (1) = \mathbf{Z}$ is excluded because $\pi(1) = e$, and hence $1 \notin \ker \pi$.)

Case 1. If $(m) = 0$, then $\mathbf{Z}$ is isomorphic to A. Hence $Q(A) \cong \mathbf{Q}$, and the prime field P of F is isomorphic to $\mathbf{Q}$. In this case the characteristic of F (and of the subfield $Q(A)$) is zero.

Case 2. If $(m) \neq 0$, then m must be a prime p, for otherwise $m = m_1 m_2$ where $1 < m_1, m_2 < m$ implies that $\pi(m) = \pi(m_1)\pi(m_2) = 0$. But $\pi(m_1)$ and $\pi(m_2)$ are distinct from 0 since m_1, m_2 do not belong to $(m) = \ker \pi$. Thus $\pi(m_1), \pi(m_2)$ would be zero divisors, but a field has no nontrivial zero divisors.

Now define a mapping $\rho: \mathbf{Z}_p \to A$ by $\rho([a]) = a \cdot e$, where a represents the equivalence class $[a]$. Since $p \cdot e = 0$, we have $(a + rp) \cdot e = a \cdot e$ for all $r \in \mathbf{Z}$; hence the definition of ρ is independent of the choice of $a \in [a]$. That ρ is indeed an isomorphism follows immediately from its definition. In this case, since $\mathbf{Z}_p$ is a field, A is itself a field, the prime field of F. Hence char $F = p$.

NOTE. In a field of characteristic p the mapping $a \to \varphi_p(a) = a^p$ is a monomorphism of the field into (not necessarily onto) itself. We have

$$\varphi_p(a+b) = (a+b)^p = a^p + \binom{p}{1} \cdot (a^{p-1}b) + \cdots$$
$$+ \binom{p}{i} \cdot (a^{p-i}b^i) + \cdots + \binom{p}{p-1} \cdot (ab^{p-1}) + b^p$$
$$= a^p + b^p = \varphi_p(a) + \varphi_p(b)$$

since p divides the binomial coefficients $\binom{p}{i}$, $1 \leq i < p$. For the product, the equation $\varphi_p(ab) = \varphi_p(a)\varphi_p(b)$ holds trivially. The kernel of φ_p, being an ideal in the field F, is zero, for if $\varphi_p(a) = 0$ for $a \neq 0$, then $\varphi_p(e) = \varphi_p(aa^{-1}) = \varphi_p(a)\varphi_p(a^{-1}) = 0$, a contradiction.

If, in particular, F is a finite field, then $\{a^p : a \in F\} = F$, and φ_p is called the **Frobenius automorphism** of F. Furthermore, if $F = \mathbf{Z}_p$, then $[a]^p = [a]$ for all cosets $[a]$ of $\mathbf{Z}_p$ [cf. §2.12]. Finite fields will be considered in more detail in §8.2.

Exercises

1. Let a, b, c, d be elements of an integral domain D. Giving reasons for each step of your argument, verify the equalities
$$(a-b) - (c-d) = (a+d) - (b+c),$$
$$(a-b)(c-d) = (ac+bd) - (ad+bc),$$
where $a-b$ is defined to be the sum $a+(-b)$ and $-b$ is the unique solution of the equation $x+b = 0$.

2. a. In an integral domain D, prove that if $a^u = b^u$ and $a^v = b^v$, where $(u, v) = 1$, then $a = b$.
   b. Construct an explicit example for some elements a, b in some integral domain D to show that part (a) is false when $(u, v) \neq 1$.

3. Prove that $R = \{a+b\sqrt{7} : a, b \in \mathbf{Z}\}$ is an integral domain.

4. Find the field of quotients of the integral domain R of Exercise 3. Show that it is isomorphic to the field $\mathbf{Q}[\sqrt{7}] = \{x+y\sqrt{7} : x, y \in \mathbf{Q}\}$.

5. Find all units in the rings:
   a. $\mathbf{Q}[\sqrt{5}]$ b. $\mathbf{Q}[\sqrt{-1}]$
   c. $\mathbf{Z}[\sqrt{-1}]$ d. $\mathbf{Z}[\sqrt{-3}]$
   e. $\mathbf{Q}[\sqrt{-m}]$, $m \in \mathbf{N}$, $m \neq 1, 3$ and not a square
   f. $\mathbf{Z} + 3i \cdot \mathbf{Z} = \{a + (3i)b : a, b \in \mathbf{Z}\}$.

6. Prove that the ring $\mathbf{Z}[\sqrt{2}]$ has infinitely many units. (In fact, the group of units of this ring is isomorphic to the (external) direct sum of the additive groups $\mathbf{Z}_2$ and $\mathbf{Z}$. See §§6.1 and 6.10.)

7. Let F be a finite field of characteristic p. Prove that the mapping given by $\varphi_p(a) = a^p$ is an isomorphism of F onto itself. For which $x \in F$ does $\varphi_p(x) = x$? Do these elements form a subfield of F? Why?

8. In §3.1 we did not require that the multiplicative identity e' of a subring R' in a ring R be the same as the multiplicative identity e of R. Prove that if R is an integral domain then $e = e'$.

9. Let $\varphi : R \to S$ be a ring homomorphism such that $\operatorname{Im} \varphi \neq \{0_S\}$. If S is an integral domain, prove that $\varphi(e_R) = e_S$, where e_R, e_S are the multiplicative identities of R and S, respectively.

10. Let R be a commutative ring in which the only ideals are (0) and R. Prove that R must be a field.

In §3.2, we defined for an ideal A in a commutative ring R the residue class ring (also called quotient or difference ring) R/A of equivalence classes of elements of R modulo A. Exercises 11–17 develop related properties of the ideal A and of the quotient ring R/A. Note carefully the distinction between a quotient ring and the ring (field) of quotients.

Integral Domains and Fields

11. Prove that R/A is an integral domain if and only if for any $a, b \in R$, for which $ab \in A$, either $a \in A$ or $b \in A$. (Such an ideal A is called a **prime ideal**.)
12. If R is the ring of integers and $A = (m)$, $m \in \mathbf{Z}$, for what m is $\mathbf{Z}/(m) = \mathbf{Z}_m$ an integral domain?
13. Prove that R/A has only nilpotent zero divisors if and only if for any $a, b \in R$, for which $ab \in A$, either $a \in A$ or $b^n \in A$, for some $n \in \mathbf{N}$. (Such an ideal A is called a **primary ideal**.)
14. As in Exercise 12, for what m does $\mathbf{Z}_m$ have only nonzero nilpotent elements for zero divisors? In other words, describe all primary ideals in $\mathbf{Z}_m$ [cf. Proposition 3, §2.10].
15. Prove that R/A is a field if and only if no proper ideal of R properly contains A; that is, any ideal A' for which $A \subset A' \subseteq R$ must be R itself. (Such an ideal A is called a **maximal ideal**.)
16. Find all maximal ideals in $\mathbf{Z}$.
17. Conclude from Exercises 11 and 15 that in any ring R every maximal ideal is prime, and from Exercise 16 that in $\mathbf{Z}$ every prime ideal is maximal.
18. Let φ be a ring homomorphism from a ring R into a field F. Prove that $\ker \varphi$ is a prime ideal in R.
19. For a given maximal ideal M in a ring R, prove that the residue class rings R/M^n, $n \in \mathbf{N}$, each contain a unique maximal ideal.
20. Let $\mathscr{F}$ be the ring of real-valued continuous functions defined on the interval $[-1, 1] \subset \mathbf{R}$ [cf. Exercise 8, §3.2]. Prove that for a fixed x_0 in this interval the set $\mathscr{S} = \{f \in \mathscr{F} : f(x_0) = 0\}$ is a maximal ideal.
21. a. In §3.3 we defined the concept of the (external) direct sum R of rings $R_1, \ldots, R_n$. Observe that the cartesian product
$$\mathbf{R}^2 = \mathbf{R} + \mathbf{R} = \{(x, y) : x, y \in \mathbf{R}\},$$
is a ring, but not a field.
   b. Prove by examples that the (external) direct sum of integral domains is not an integral domain.
22. a. Prove that R/A has no nonzero nilpotent elements if and only if $A = \text{Rad } A$. (Such an ideal A is called a **semiprime ideal**.)
    b. Describe all semiprime ideals in $\mathbf{Z}$.
    c. Prove that a prime ideal in a ring is also a semiprime ideal.
23. a. Using the notation in the construction of the field of rational numbers $\mathbf{Q}$ as the quotient field of the integral domain $\mathbf{Z}$, define $[a, b] > 0$ if $a'b' > 0$, where (a', b') represents the class $[a, b]$. Prove that an ordering [see §2.2] of $\mathbf{Q}$ is defined and that the customary rules for the absolute value hold.
    b. Prove that this ordering of $\mathbf{Q}$ extends the customary ordering of $\mathbf{Z}$ ($b > a \Rightarrow b - a \in \mathbf{N}$) and that the Archimedian Principle [§2.2] holds in $\mathbf{Q}$.
24. Prove that any ring isomorphism $\varphi : \mathbf{Z} \to \mathbf{Z}$ must be the identity map.
25. Repeat Exercise 24 for $\varphi : \mathbf{Q} \to \mathbf{Q}$. However, exhibit an isomorphism of the complex numbers $\mathbf{C}$, which is not the identity map.

26. Determine the smallest integral domain containing $1/3$ and $-1/21$ in the field $\mathbf{Q}$.

27. For a prime number p, define
$$R_p = \{a/b \in \mathbf{Q} : a, b \in \mathbf{Z}, (a, b) = 1, (b, p) = 1\}.$$
Prove the following statements.
  a. R_p is a ring.
  b. Every ideal $A \subseteq R_p$ is principal and is equal to
$$p^\alpha R_p = \{p^\alpha c : c \in R_p\},$$
where $\alpha = \min_{x \in A} v_p(x)$, as defined in Exercise 14, §2.7.
  c. The residue class ring R_p/pR_p is isomorphic to $\mathbf{Z}_p$.
  d. $\bigcap_p R_p = \mathbf{Z}$, where the intersection is taken over all prime integers p.

28. With reference to Exercise 27, prove that $R_p/p^\alpha R_p \cong \mathbf{Z}_{p^\alpha}$.

29. In a commutative ring R let N denote the ideal of nilpotent elements [cf. Exercise 7, §3.1]. Prove that the residue class ring R/N has no nonzero nilpotent elements.

30. Find a field F and two nonzero elements $x, y \in F$ such that $x^2 + y^2 = 0$.

31. Let φ be a (ring) homomorphism of a field F to a ring R such that $\varphi(e) \neq 0$. Prove that $\ker \varphi = 0$ and hence that φ is a monomorphism of F into (not necessarily onto) R.

32. Consider $\mathbf{N}$ with the customary properties of addition and ordering.
  a. Construct the integral domain of integers analogous to the construction of the field of rational numbers $\mathbf{Q}$ from $\mathbf{Z}$. Thus let $[a, b]$ be the set of all pairs of numbers u, v in $\mathbf{N}$ such that $a + v = u + b$ for a, b in $\mathbf{N}$.
  b. Show that an equivalence relation is defined. Then define sum and product for these equivalence classes. What role does the class $[1, 1]$ play?
  c. What is the multiplicative unit in $\mathbf{Z}$ thus constructed?
  d. Is the map $a \to [a+1, 1]$ an embedding of $\mathbf{N}$ into $\mathbf{Z}$?

33. Consider $\mathbf{N}$ with the customary properties of addition and ordering. Prove that the ordering for $\mathbf{N}$ can be extended to an ordering of $\mathbf{Z}$, constructed as in Exercise 32, by defining $[a, b] > [1, 1]$ to mean $a > b$.

34. Associate to each element $a \in \mathbf{N}$ a symbol $-a$; also introduce a symbol 0, defining $0 + x = x + 0 = x$ for all $x \in S$.
  a. For $a, b \in \mathbf{N}$, define
    (i) $a + (-b) = (-b) + a = c$, if $a = b + c$ with $c \in \mathbf{N}$.
    (ii) $(-a) + b = b + (-a) = -c \in S$, if $c \neq 0$;
    and $(-a) + b = b + (-a) = 0$, if $a = b$.
    (iii) $(-a) + (-b) = -(a+b)$.
  Verify that the set S has all the customary properties of addition; in particular, that there exists in S an element t such that $s + t = t + s = 0$ for every element $s \in S$.
  b. Next define a product in S as follows:
    (iv) $0s = s0 = 0$ for all $s \in S$.
    (v) $(-a)b = b(-a) = -(ab)$ and $(-a)(-b) = ab$ for all $a, b \in \mathbf{N}$.
  Now prove that S is an integral domain and that S is isomorphic as a ring to the ring $\mathbf{Z}$, constructed in Exercise 32.

35. An integral domain D is said to be **well-ordered** if there is an ordering relation $<$ on D such that every nonempty set of positive (>0) elements in D contains a least element. Prove that any well-ordered integral domain D is isomorphic to the ring of integers [see §2.2].
36. If F is a field, prove that the field of quotients $Q(F)$ is isomorphic to F.
37. In an integral domain D, consider a subset S such that $0 \notin S$ and $s, s' \in S$ imply that $ss' \in S$. Let D_S be the subset of the field of quotients $F = Q(S)$ that consists of the elements a/s, $a \in D$ and $s \in S$. Prove the following statements.
    a. If A is an ideal in D, then the set
    $$A \cdot D_S = \{a/s : a \in A, s \in S\}$$
    is an ideal in D_S.
    b. For ideals A, B in D,
    $$(A+B) \cdot D_S = A \cdot D_S + B \cdot D_S$$
    $$(A \cap B) \cdot D_S \subseteq (A \cdot D_S) \cap (B \cdot D_S).$$
    c. If J is any ideal in D_S, then $(J \cap D) \cdot D_S = J$.
    d. For any ideal $A \subseteq D$, we have $A \subseteq (A \cdot D_S) \cap D$.
    e. For any ideal $A \subseteq D \subseteq F$, the ideal $(A \cdot D_S) \cap D$ in F consists of all elements $a \in A$ for which there exists an $s \in S$ such that $sa \in A$.
38. With reference to Exercise 37, let A' and B' be ideals in D_S. Prove the following statements.
    a. $(A' \cap D) D_S \subseteq A'$
    b. $[(A' \cap D) D_S] \cap D = A' \cap D$
    c. $(A' + B') \cap D \supseteq (A' \cap D) + (B' \cap D)$
    d. $(A'B') \cap D \supseteq (A' \cap D)(B' \cap D)$
    e. $\mathrm{Rad}(A' \cap D) = (\mathrm{Rad}\, A') \cap D$.
39. Let P be a prime ideal in a ring R and let $R_P = \{a/b : a \in R, b \notin P\}$. Prove that the *extended ideal* PR_P is a maximal ideal in R_P and that all nonunits of R_P form an ideal. Furthermore, prove that PR_P contains all proper ideals of R_P.
40. With the notation of Exercise 39, prove that the extended ideal $AR_P \neq R_P$ if and only if $A \subseteq P$.
41. Let D be the set of all functions from $\mathbf{N}$ to $\mathbf{Z}$, and for $f, g \in D$ define
    $$(f+g)(n) = f(n) + g(n)$$
    $$(f * g)(n) = \sum_{d \mid n} f(d) g\left(\frac{n}{d}\right),$$
    the sum being taken over all divisors d of n. Prove that D with these laws of composition is an integral domain.

4

Aspects of Linear Algebra

In this chapter we pause in our development of abstract algebra to consider the linear algebra topics of vector space, linear independence, basis, matrix, and determinant. The material here exemplifies some of the algebraic concepts already encountered and is necessary for subsequent considerations of polynomials, groups, and field extensions. Linear algebra is often presented as a subject separate from abstract algebra, but our intent is to indicate their interrelations. This chapter is by no means a complete presentation of linear algebra.

Building on our study of the arithmetic properties of integers [§2.1] and then of fields [§3.6], we define an algebraic structure called a vector space. This concept generalizes to that of a module in §7.7, and some of the aspects of vector spaces will carry over directly to modules. Vector spaces are used in §8.1 in the treatment of field extensions leading to the Galois theory. Then in §9.4, in preparation for the results of Hilbert and Noether on certain field extensions, we need to consider linear transformations, their representation by matrices, and determinants.

Some students will already have studied vector spaces and matrices and can proceed directly to Chapter 5. Others may wish to peruse this chapter, reviewing the principal statements, and then subsequently use these sections for reference. However this chapter is intended primarily for those students with no previous course in linear algebra. While not a substitute for a full course in linear algebra, it offers a comprehensive introduction.

§4.1 Vector Spaces

From analytic geometry and calculus we recall the concept of "vector," usually considered then as a pair of real numbers (a_1, a_2) or as a triple (a_1, a_2, a_3). In either case a vector was geometrically described as a line segment with both direction and magnitude. By geometric argument, usually by means of a parallelogram of forces, the following algebraic properties of vectors $A = (a_1, a_2, a_3)$ and $B = (b_1, b_2, b_3)$ were demonstrated:

$$A \pm B = (a_1 \pm b_1, a_2 \pm b_2, a_3 \pm b_3),$$

and for $r \in \mathbf{R}$,

$$r \cdot A = rA = (ra_1, ra_2, ra_3).$$

Hence $0 \cdot A = (0, 0, 0)$, the zero vector, often denoted simply by 0, and $1 \cdot A = A$.

But the importance of vectors is not limited to two- or three-dimensional euclidean space. In fact their value is that they provide a convenient means of expressing many interacting aspects of problems in fields such as physics, linear programming, statistics, and other areas of mathematical applications, as well as in advanced algebra. We begin with an axiomatic description of a vector space together with its associated field of scalars. The elements of a vector space are called **vectors**.

A **vector space** (or **linear space**) V **over a field** F is a set of elements together with two laws of composition (commonly called *vector addition* and *multiplication by scalars*). The law of vector addition associates to each ordered pair (A, B) in the cartesian product $V \times V$ a unique element $A + B \in V$, and the law of multiplication by scalars associates to each ordered pair (r, A) in $F \times V$ a unique element $rA \in V$, subject to the following axiomatic properties:

Addition. Vector addition is associative and commutative; there exists an additive identity element (denoted by 0); and each element $A \in V$ has an additive inverse B, such that $A + B = 0$. (This inverse B is designated $-A$.)

Multiplication. Multiplication by scalars (that is, multiplication of elements in V by those in F) is associative,

$$r(r'A) = (rr')A \quad \text{for } A \in V \text{ and } r, r' \in F;$$

and if 1 is the multiplicative identity in F, then $1 \cdot A = A$ for all $A \in V$.

Distributivity. Multiplication by scalars and vector addition satisfy the two distributive laws for all $A, B \in V$ and $r, r' \in F$:

$$r(A + B) = rA + rB,$$
$$(r + r')A = rA + r'A.$$

Because the defining algebraic axioms of a vector space over a field resemble those of the integers [§2.1] and of a ring [§3.1], we can easily derive the following logical consequences.

Property 1. The additive identity (called the zero vector 0) is unique.

Property 2. Furthermore, $A = A + B$ for some $B \in V$ implies $B = 0$.

Property 3. The cancellation law for addition holds.

Property 4. The additive inverse $-A$ of an element $A \in V$ is unique.

Property 5. Every equation $A + X = B$, for $A, B \in V$, has a unique solution $X \in V$.

Property 6. For any $r \in F$, $r \cdot 0 = 0$.

Property 7. The identity for multiplication by scalars is unique. That is,
$$rA = A, \quad \text{for all } A \in V \Rightarrow r = 1 \in F.$$

In fact, the next property provides a stronger statement.

Property 8. If $rA = A$ for some $A \neq 0$ in V, then $r = 1 \in F$.

Property 9. For any $r \in F$, and any $A \in V$, the additive inverse $-(rA)$ of rA is $(-r)A$.

Most commonly in geometry and analysis we consider vector spaces V over the field of real numbers **R**, and also over **C** and **Q**. The examples below are of vector spaces over these fields.

Example 1. Let V be the set of ordered n-tuples $(a_1, ..., a_n)$ of real numbers $a_i \in \mathbf{R}$, $1 \leq i \leq n$, where addition of vectors and multiplication by scalars are defined component-wise:
$$(a_1, ..., a_n) + (b_1, ..., b_n) = (a_1 + b_1, ..., a_n + b_n),$$
$$r(a_1, ..., a_n) = (ra_1, ..., ra_n).$$
We should check that these laws of composition satisfy the axiomatic properties in the definition of a vector space. (The proof is the same as for pairs of real numbers, i.e., vectors in the euclidean or cartesian plane.) Thus, V is a vector space over **R**. It is commonly denoted by $\mathbf{R}^n$.

Example 2. To obtain a vector space over **Q** or **C**, Example 1 is modified to require that the components a_i, $1 \leq i \leq n$, of the n-tuple $(a_1, ..., a_n)$ belong to **Q** or **C**, respectively.

Example 3. Let $\mathscr{F}$ denote the set of all real-valued functions defined on the real number line **R**. That is, $\mathscr{F}$ is the set of all functions $f: \mathbf{R} \to \mathbf{R}$. Define laws of composition as follows for $f, g \in \mathscr{F}$, and $r \in \mathbf{R}$, for all $x \in R$:

$f + g$ is the function which maps x to $f(x) + g(x)$;

rf is the function which maps x to $rf(x)$.

Again, after verifying that these laws of composition satisfy the axiomatic properties in the definition of a vector space, we conclude that $\mathscr{F}$ is a vector space over **R**.

We shall encounter further examples of vector spaces later: vector spaces of matrices in §4.4, vector spaces of polynomials in §5.1, and algebraic field extensions in §8.1.

In speaking of a vector space V we must also have in mind the associated field of scalars F. Thus a vector space involves two sets: the set of vectors V and the set of scalars F. To make this association explicit we denote a vector space V over a field F by V/F. We use $\mathbf{R}^n$ to denote the vector space over $\mathbf{R}$ of n-tuples of real numbers, and more generally F^n, the vector space over a field F of n-tuples of elements of F.

While we shall usually refer to the general case of an arbitrary field of scalars F, it suffices for our present purposes to think of F as the field $\mathbf{R}$.

A subset W of a vector space V/F is called a **subspace** of V if

  (i) W is closed under vector addition and additive inverse, as determined in V; that is,

$$A, B \in W \Rightarrow \begin{cases} A + B \in W \\ -A \in W. \end{cases}$$

(Hence $0 \in W$.)

  (ii) W is closed under multiplication by scalars; i.e.,

$$A \in W, \, r \in F \Rightarrow rA \in W.$$

Note that a subspace W of a vector space V/F is itself a vector space over F.

For example, to verify that

$$W = \{(a,b,c) \in \mathbf{R}^3 : a = 3c\} = \{(3c,b,c) : b, c \in \mathbf{R}\}$$

is a subspace of $\mathbf{R}^3$, we note that

$$(3c,b,c) + (3c',b',c') = (3c+3c', b+b', c+c') = (3(c+c'), b+b', c+c') \in W,$$

and $\quad -(3c,b,c) = (-3c, -b, -c) = (3(-c), -b, -c) \in W,$

and finally for any $r \in \mathbf{R}$

$$r(3c,b,c) = (r(3c), rb, rc) = (3(rc), rb, rc) \in W.$$

As a second example consider the set W of 3-tuples $(x,y,z) \in \mathbf{R}^3$ whose components satisfy the system of linear homogeneous equations

$$3x - y + z = 0,$$
$$x + 2y - 5z = 0.$$

The set W is a vector space over $\mathbf{R}$, since if x, y, z and x', y', z' are solutions, then so are $x+x'$, $y+y'$, $z+z'$, and rx, ry, rz, for all $r \in \mathbf{R}$.

By a **linear combination** of the elements of a finite subset $S = \{A_1, \ldots, A_n\}$ of a vector space V/F, we mean an element in V of the form

$$r_1 A_1 + \cdots + r_n A_n,$$

where $r_1, \ldots, r_n \in F$. Such a combination is called **nontrivial** if at least one of the coefficients r_i, $1 \leq i \leq n$, is distinct from zero. The set W of all linear combinations of elements in S with coefficients in F is a subspace of V, called the subspace **generated** or **spanned by** S. We sometimes refer to W as the **span** of S, denoted $W = \text{Span}(S)$.

Proposition 1. For a finite set $S = \{A_1, \ldots, A_n\} \subseteq V/F$, let

$$W = \left\{ \sum_{i=1}^{n} r_i A_i : r_i \in F, \ 1 \leq i \leq n \right\}.$$

Then

(i) W is a subspace of V.

(ii) W is the smallest subspace of V that contains S.

By "smallest" we mean here that no proper subset of W both is a subspace of V and contains S. The proof is left as an exercise.

As an example, consider the vector space $\mathbf{R}^4$ of 4-tuples of real numbers and the subset

$$S = \{(0,0,0,1), (4,0,1,1), (1,0,1,1)\}.$$

Then
$$\begin{aligned} W &= \{a(0,0,0,1) + b(4,0,1,1) + c(1,0,1,1) : a, b, c \in \mathbf{R}\} \\ &= \{(4b+c, 0, b+c, a+b+c) : a, b, c \in \mathbf{R}\} \\ &= \{(u, 0, v, w) : u, v, w \in \mathbf{R}\}. \quad \text{(Why?)} \end{aligned}$$

Similarly in $\mathbf{R}^3$, the span of $S = \{(1,0,1), (0,0,3), (2,0,5)\}$ is

$$\begin{aligned} W &= \{a(1,0,1) + b(0,0,3) + c(2,0,5) : a, b, c \in \mathbf{R}\} \\ &= \{(a+2c, 0, a+3b+5c) : a, b, c \in \mathbf{R}\} \\ &= \{(u, 0, v) : u, v \in \mathbf{R}\}. \quad \text{(Why?)} \end{aligned}$$

Now given subspaces W_1 and W_2 of a vector space V/F, we can define two additional subspaces:

the set-theoretic **intersection** $W_1 \cap W_2$;
the **sum** $W_1 + W_2$, where

$$W_1 + W_2 = \{A \in V : A = A_1 + A_2, \ A_i \in W_i, \ i = 1, 2\}.$$

The definitions of intersection and sum of two subspaces easily extend to any finite number of subspaces $W_1, \ldots, W_m$ in V. A sum $U = W_1 + \cdots + W_m$ of $m \geq 2$ subspaces $W_i \subseteq V$, $1 \leq i \leq m$, is called an **(internal) direct sum**, denoted

$$U = W_1 \oplus \cdots \oplus W_m,$$

if each vector $u \in U$ can be expressed *uniquely* as a sum $u = w_1 + \cdots + w_m$ of elements $w_i \in W_i$.

NOTE. It is not required that the subspaces W_i be distinct.

§4.1 Vector Spaces

Proposition 2. Consider subspaces $W_1, \ldots, W_m$ of V. Then the sum $U = W_1 + \cdots + W_m$ is direct if and only if $W_i \cap (W_1 + \cdots + W_{i-1}) = \{0\}$ for all i, $2 \leq i \leq m$.

Proof. If for some i the intersection $W_i \cap (W_1 + \cdots + W_{i-1})$ contains a nonzero vector u, then this vector $u \in W_1 + \cdots + W_m = U$ has two distinct expressions,

$$u = \underset{\in W_1}{w_1} + \underset{\in W_2}{w_2} + \cdots + \underset{\in W_{i-1}}{w_{i-1}} + 0 + 0 + \cdots + 0$$

$$u = 0 + 0 + \cdots + 0 + \underset{\in W_i}{v} + 0 + \cdots + 0$$

as the sum of elements of W_j, $1 \leq j \leq m$, so that U cannot be a direct sum.

Conversely, suppose that for all i, $2 \leq i \leq m$,

$$W_i \cap (W_1 + \cdots + W_{i-1}) = \{0\}.$$

To prove that the sum U is direct consider a vector $u \in U$ which has two distinct expressions as the sum of elements in W_j, $1 \leq j \leq m$:

(∗)
$$u = w_1 + \cdots + w_m,$$
$$u = w'_1 + \cdots + w'_m.$$

Then
$$(w_1 - w'_1) + \cdots + (w_m - w'_m) = 0.$$

Let i be the greatest subscript for which $w_i - w'_i \neq 0$. Since we assume the expressions (∗) are distinct, this subscript i is at least 2. Now,

$$w'_i - w_i = (w_1 - w'_1) + \cdots + (w_{i-1} - w'_{i-1})$$

and $w'_i - w_i$ is a nonzero vector in $W_1 + \cdots + W_{i-1}$. Hence

$$w'_i - w_i \in W_i \cap (W_1 + \cdots + W_{i-1}).$$

But as this intersection is the 0 subspace, $w'_i - w_i = 0$, a contradiction. Thus since u cannot have two distinct expressions (∗), the sum U is direct, as asserted.

A similar argument proves the analogous statement for the direct sum of rings [§3.3] and the direct product of groups [§6.10].

We say that a vector space V/F is **finitely generated** if there is some finite set of vectors $S = \{A_1, \ldots, A_m\}$ such that every vector $A \in V$ can be written in the form

$$A = r_1 A_1 + \cdots + r_m A_m;$$

that is, $V = \text{Span}(S)$. The coefficients $r_1, \ldots, r_m \in F$ are *not* required to be unique.

For example, $\mathbf{R}^2$ is a finitely generated vector space since every vector $(a, b) \in \mathbf{R}^2$ can be written as a linear combination of the vectors $\{(0, 1), (1, 1), (1, 2)\}$:

$$(a, b) = (b-a) \cdot (0, 1) + a \cdot (1, 1) + 0 \cdot (1, 2).$$

But also,
$$(a, b) = (b-2a) \cdot (0, 1) + 0 \cdot (1, 1) + a \cdot (1, 2).$$

Alternatively we can conclude that $\mathbf{R}^2$ is finitely generated by observing that every $(a, b) \in \mathbf{R}^2$ can be written as a linear combination of the vectors $\{(0, 1), (1, 1)\}$.

Exercises

1. Verify that the axiomatic properties are satisfied for the sets in Examples 1 and 3.
2. Prove Proposition 1.
3. In euclidean three-space $\mathbf{R}^3$, describe the subspaces containing the following sets of vectors.
   a. $\{(0, 0, 1), (0, 1, 1), (0, 4, 2)\}$
   b. $\{(1, 1, 1), (2, 1, 1)\}$
   c. $\{(4, 2, 1)\}$
   d. $\{(1, 2, 1), (1, 3, 1), (1, 0, 0)\}$
   e. $\{(1, 2, 1), (1, 3, 1), (1, 0, 0), (0, 5, 0)\}$.
4. Verify, for subspaces W_1, W_2 of a vector space V, that $W_1 \cap W_2$ and $W_1 + W_2$ are again subspaces of V.
5. In $\mathbf{R}^3$ demonstrate that the set-theoretic union of subspaces need not be a subspace.
6. Prove Property 7.
7. Prove that the set of triples $(x, y, z) \in \mathbf{R}^3$, satisfying the homogeneous system of linear equations

$$3x - y + z = 0$$
$$x + 2y - 5z = 0$$

   is a vector space over $\mathbf{R}$.
8. Determine whether the following subsets of $\mathbf{R}^3$ are subspaces.
   a. $W_1 = \{(a, b, 0) : a, b \in \mathbf{R}\}$
   b. $W_2 = \{(x, y, z) : 3x - 4y - 5z = 0\}$
   c. $W_3 = \{(6a - 7b, c - 4b, a + b - c) : a, b, c \in \mathbf{R}\}$
   d. $W_4 = \{(x, y, z) : x, y \in \mathbf{R}\}$
   e. $W_5 = \{(x, y, z) : x \geq y\}$.
9. Determine the intersection of the subspaces W_1 and W_2 in Exercise 8.
10. Prove that the span of a finite subset $S = \{A_1, \ldots, A_m\}$ contained in a vector space V/F is the intersection of all subspaces $U \subseteq V$ which contain S.
11. Consider subspaces $W_1, W_2 \subseteq V$ and vectors $a, b \in V$. Define $a + W_1 = \{a + w : w \in W_1\}$. Prove that $a + W_1 \subseteq b + W_2$ if and only if $W_1 \subseteq W_2$ and $a - b \in W_2$.

§4.2 Linear Independence and Bases

In defining a finitely generated vector space [at the end of §4.1] we made no statement concerning the number of elements in a generating set. In fact we showed by example that a given vector space could have several finite generating sets with differing numbers of elements. By introducing the fundamental concept of linear independence we are able to specify minimal generating sets (called bases) for a given finitely generated vector space and thus to define the dimension of such a vector space. Subsequently in our studies of polynomials and field extensions we shall encounter the similar concepts of algebraic independence and integral dependence.

A finite subset $S = \{A_1, \ldots, A_m\} \subseteq V/F$ is said to be **linearly dependent** if there exist m scalars $r_1, \ldots, r_m$, not all of which are zero, such that

$$r_1 A_1 + \cdots + r_m A_m = 0.$$

If no such scalars exist we call the subset S **linearly independent**.

The terms "linear dependence" and "linear independence" may be better understood when restated as in the following proposition.

Proposition 1. A finite set of vectors $S = \{A_1, \ldots, A_m\}$, $m \geq 2$, is linearly dependent if and only if at least one of them can be expressed as a linear combination of the remaining ones.

In this phraseology a finite set of at least two vectors is linearly independent if and only if none of the vectors can be written as a linear combination of the others. Also note that $\{0\}$ is a linearly dependent set, while $\{A\}$ is linearly independent if $A \neq 0$.

Proof. Suppose S is a linearly dependent set. By definition there exist scalars $r_1, \ldots, r_m \in F$, not all zero, such that

$$r_1 A_1 + \cdots + r_m A_m = 0.$$

For some subscript i, $r_i \neq 0$, and

$$-r_i A_i = r_1 A_1 + \cdots + r_{i-1} A_{i-1} + r_{i+1} A_{i+1} + \cdots + r_m A_m.$$

Since $-r_i$ is a nonzero element of F, it has a multiplicative inverse, and

$$A_i = s_1 A_1 + \cdots + s_{i-1} A_{i-1} + s_{i+1} A_{i+1} + \cdots + s_m A_m,$$

where $s_j = -r_j r_i^{-1} \in F$, $1 \leq j \leq m$, $j \neq i$.

Conversely, if one of S, say A_i, is a linear combination of the others, i.e., if

$$A_i = r_1 A_1 + \cdots + r_{i-1} A_{i-1} + r_{i+1} A_{i+1} + \cdots + r_m A_m,$$

then

$$r_1 A_1 + \cdots + r_{i-1} A_{i-1} + (-1) A_i + r_{i+1} A_{i+1} + \cdots + r_m A_m = 0.$$

We thus have a nontrivial linear combination of the elements of S equal to zero. Hence S is linearly dependent, which completes the proof.

Three further remarks remain. First, any finite set of vectors containing the zero vector is linearly dependent. Second, any finite set of vectors is either linearly independent or linearly dependent. Third, the most common means of proving the linear independence of a set of vectors $\{A_1, \ldots, A_m\} \subset V$ is to consider a general linear combination

$$r_1 A_1 + \cdots + r_m A_m = 0$$

and then to prove that *each* of the coefficients $r_1, \ldots, r_m$ *must* be zero.

For example, to show that $\{(0,0,1), (0,2,1), (1,-3,1)\}$ is linearly independent in $\mathbf{R}^3$, consider real numbers a, b, c, for which

$$a(0,0,1) + b(0,2,1) + c(1,-3,1) = (0,0,0).$$

This one vector equation translates into three scalar equations:

$$a + b + c = 0, \quad 2b - 3c = 0, \quad c = 0.$$

Since $c = 0$, so must b, and hence a. Therefore the vectors are linearly independent as asserted—no nontrivial linear combination of them can equal the zero vector.

To show that the set of vectors $\{(0,1), (1,1), (1,2)\}$ in $\mathbf{R}^2$ is linearly dependent, we can determine real numbers a, b, c, not all zero, such that

$$a(0,1) + b(1,1) + c(1,2) = (0,0).$$

That is, we find a, b, c, not all zero, satisfying

$$b + c = 0, \quad a + b + 2c = 0.$$

One solution is $a = 3$, $b = 3$, $c = -3$. An alternative proof of dependence would be to write one of the vectors as a linear combination of the two remaining ones. For instance,

$$(1,2) = (0,1) + (1,1).$$

The concepts of linear independence and span of a set are combined in defining a basis of a *finitely generated* vector space V/F. A finite subset S of a nonzero space V/F is called a **basis** of V if

(i) S is a linearly independent set;
(ii) the span of S equals V; i.e., each $v \in V$ can be expressed as some linear combination of elements in S.

The fundamental properties of a basis are stated in the following two theorems.

Theorem 1. Every finitely generated vector space V/F has a basis.

Proof. By hypothesis, V/F has a finite generating set $S^* = \{A_1, \ldots, A_m\}$. If S^* is a linearly independent set, then it is a basis, and nothing is to be proved. If S^* is a linearly dependent set, we can write by Proposition 1

$$A_i = s_1 A_1 + \cdots + s_{i-1} A_{i-1} + s_{i+1} A_{i+1} + \cdots + s_m A_m$$

for some i, $1 \leq i \leq m$, and some $s_1, ..., s_{i-1}, s_{i+1}, ..., s_m \in F$. The span of
$$S' = \{A_1, ..., A_{i-1}, A_{i+1}, ..., A_m\}$$
equals $\text{Span}(S^*) = V$. If S' is a linearly independent set, it is a basis for V/F. Otherwise repeat the above process, deleting a vector A_j dependent upon those remaining, and in at most $m-1$ steps obtain a linearly independent set S whose span is V. Hence S is a basis, as desired.

For the second theorem we need the following lemma.

Steinitz Exchange Lemma. If vectors $B_1, ..., B_n$ are linearly independent in a vector space V/F generated by vectors $A_1, ..., A_m$, then $n \leq m$.

Proof. We have already noted that no linearly independent set can contain the zero vector. Hence $B_i \neq 0$, $1 \leq i \leq n$. Since the vectors $A_1, ..., A_m$ generate V, the vector B_1 can be written as a linear combination
$$B_1 = r_1 A_1 + \cdots + r_m A_m.$$
Since $B_1 \neq 0$, at least one of the coefficients $r_1, ..., r_m$ is nonzero, say r_{i_1}. Then
$$A_{i_1} = \frac{1}{r_{i_1}}(B_1 - r_1 A_1 - \cdots - r_{i_1-1} A_{i_1-1} - r_{i_1+1} A_{i_1+1} - \cdots - r_m A_m),$$
from which we conclude that the set
$$\{A_1, ..., A_{i_1-1}, B_1, A_{i_1+1}, ..., A_m\}$$
generates V. Now write B_2 as a linear combination of these vectors:
$$B_2 = r'_1 A_1 + \cdots + r'_{i_1-1} A_{i_1-1} + r'_{i_1+1} A_{i_1+1} + \cdots + r'_m A_m + s_1 B_1.$$
At least one of the coefficients r'_i is nonzero, as B_1, B_2 are linearly independent. Continuing in this fashion we ultimately replace n vectors $A_{i_1}, ..., A_{i_n}$ by the linearly independent vectors $B_1, ..., B_n$ to obtain a new set
$$\{B_1, ..., B_n\} \cup [\{A_1, ..., A_m\} \setminus \{A_{i_1}, ..., A_{i_n}\}]$$
which generates V. Because the B_i are linearly independent, at the jth step the linear combination
$$B_j = r''_1 A_1 + \cdots + r''_m A_m + s'_1 B_1 + \cdots + s'_{j-1} B_{j-1},$$
where $r''_{i_1} = \cdots = r''_{i_{j-1}} = 0$, has at least one nonzero coefficient r''_i of A_i. We conclude then that $n \leq m$, as we cannot exhaust $\{A_1, ..., A_m\}$ before completing the n exchanges.

The preceding result is due to Ernst Steinitz (1871–1928), who was one of the originators of what today is called "abstract algebra." His paper, *Algebraische Theorie der Körper* (*Jour.f.d. r.u.a. Math.*, **137** (1910), 167–309), presents the foundations of this branch of mathematics. Its introduction clearly demonstrates that

Steinitz was motivated by critical analysis and insight, and not by abstraction *per se*.

Theorem 2. Any two bases of a finitely generated vector space V/F have the same number of elements.

Proof. If $\{A_1, \ldots, A_m\}$ and $\{B_1, \ldots, B_n\}$ are two bases of V/F, then as both generate V and are linearly independent, we apply the Steinitz Exchange Lemma twice to obtain $n \leq m$ and $m \leq n$.

Thus any finitely generated vector space V/F has a basis, and any two bases have the same number of elements. In other words, the number of elements in a basis for a given finitely generated vector space V/F is unique, even if V has many bases. This unique number is called the **dimension** of V/F, denoted $\dim_F V$, or simply $\dim V$, if the scalar field F is understood.

We conclude this section with a characteristic property of bases.

Proposition 2. If $S = \{A_1, \ldots, A_n\}$ is a basis for V/F, then each vector in V can be expressed *uniquely* as a linear combination of elements in S.

Proof. Consider an element $v \in V$ having two expressions as linear combinations of elements of S:

(∗)
$$v = r_1 A_1 + \cdots + r_n A_n,$$
$$v = s_1 A_1 + \cdots + s_n A_n.$$

Subtracting, we obtain

$$(r_1 - s_1) A_1 + \cdots + (r_n - s_n) A_n = 0.$$

Since the A_i are linearly independent (by hypothesis), the coefficients $r_i - s_i = 0$, $1 \leq i \leq n$, and thus the two expressions (∗) are not distinct.

As a corollary, we obtain the converse to Proposition 2. If S is a spanning, linearly dependent set in V/F, then some $v \in V$ has at least two distinct expressions as linear combinations of elements of S. [See the example at the end of §4.1.]

Exercises

1. If W is a subspace of a finitely generated vector space, prove that W is finitely generated (by elements in W).
2. If V is an n-dimensional vector space, prove that V can be written as the direct sum of n one-dimensional subspaces.
3. If W is a subspace of a finite dimensional vector space V, prove that $\dim W \leq \dim V$, where equality holds only if $W = V$.

4. Consider a nontrivial subspace W of a finitely generated vector space V.
   a. Prove there exists a subspace W' such that V equals the direct sum $W \oplus W'$.
   b. Is the subspace W' unique? Why?
5. In $\mathbf{R}^2$, let W be the subspace generated by the vector $(1, 5)$. Find W' as in Exercise 4.
6. Consider a finitely generated vector space V and a finite subset S.
   a. If S is a linearly dependent subset of V and generates V, prove that S contains a basis of V.
   b. If S is a linearly independent subset of V, prove that there exists a basis S' of V such that $S \subseteq S'$.
   c. If S is a basis for V, prove that no proper subset of S generates V, and that no subset S' of V which properly contains S is linearly independent.
7. Determine all of the linearly independent subsets of the following sets of vectors.
   a. $\{(1, 1), (2, 1), (5, 1), (4, 3)\} \subset \mathbf{R}^2$
   b. $\{(1, 1, 1), (5, 0, 1), (4, -1, -1), (0, 2, 1)\} \subset \mathbf{R}^3$.
8. Show that any vector space V/F has at least two distinct bases if $\dim V \geq 2$, or if $\dim V = 1$ and F has more than two elements.
9. State whether the following sets are linearly dependent or independent.
   a. $\{(1, 2), (1, 3), (1, 4)\} \subset \mathbf{R}^2$
   b. $\{(1, 2, 1), (1, 3, 2), (1, 4, 3)\} \subset \mathbf{R}^3$
   c. $\{(1, 2, 1), (1, -, 1), (3, -5, 3)\} \subset \mathbf{R}^3$
   d. $\{(5, -3, 1), (-2, 4, 0), (12, -10, 2)\} \subset \mathbf{R}^3$.
10. The set of vectors $\{(1, -1, 0), (2, 1, 1), (0, 3, 1), (0, -1, 1)\}$ spans $\mathbf{R}^3$. Find a subset which is a basis.
11. Prove the following standard result concerning vector space dimensions. If S and T are finite dimensional subspaces of a vector space V, prove that
$$\dim(S+T) + \dim(S \cap T) = \dim S + \dim T.$$

§4.3 Transformations of Vector Spaces

Section 4.1 offered several examples of vector spaces, including $\mathbf{R}^n$. We now consider isomorphisms of vector spaces and prove that any n-dimensional vector space over $\mathbf{R}$ is isomorphic to $\mathbf{R}^n$. In fact, for any field F, every n-dimensional vector space V/F is isomorphic to F^n.

More generally we consider mappings from one vector space U to another V, over the same field F, which respect the linear space structure. Such mappings are called linear transformations, although we might also think of them as homomorphisms of vector spaces. Specifically, a **linear transformation** (or operator) is a mapping $\varphi: U \to V$, such that, for all $A, B \in U$ and all $r \in F$,

(i) $\varphi(A+B) = \varphi(A) + \varphi(B)$,
(ii) $\varphi(rA) = r\varphi(A)$.

Vector spaces U/F and V/F are said to be **isomorphic** if there exists a linear transformation $\varphi: U \to V$, such that

(iii) $\varphi(A) = 0 \Leftrightarrow A = 0$,
(iv) $V = \{\varphi(A) : A \in U\}$.

In other words, a vector space isomorphism is injective, surjective, and compatible with the operations of vector addition and multiplication by scalars [properties (i) and (ii) above].

The properties of linearity (i) and (ii) in the definition are most significant in that, for a linear transformation $\varphi: U \to V$, once the images $\varphi(A_1), \ldots, \varphi(A_m)$ for a basis $A_1, \ldots, A_m$ of U have been specified, the image $\varphi(A)$ is determined for each $A \in U$:

$$A \in U \Rightarrow A = r_1 A_1 + \cdots + r_m A_m$$

with unique scalars $r_1, \ldots, r_m \in F$, and so

$$\varphi(A) = r_1 \varphi(A_1) + \cdots + r_m \varphi(A_m).$$

This concept is illustrated by the mapping $\varphi: \mathbf{R}^2 \to \mathbf{R}^2$, given by $\varphi((1,0)) = (3,4)$, $\varphi((0,1)) = (-3,5)$. Hence

$$\begin{aligned}\varphi((a,b)) &= a\varphi((1,0)) + b\varphi((0,1)) \\ &= a(3,4) + b(-3,5) = (3(a-b), 4a+5b).\end{aligned}$$

Theorem. If V/F is an n-dimensional vector space, then there exists at least one isomorphism $\varphi: V \to F^n$.

Proof. By Theorem 1, §4.2, V/F has a basis $\{A_1, \ldots, A_n\}$. Now for $A \in V$ there exist *unique* scalars $r_1, \ldots, r_n \in F$ such that $A = r_1 A_1 + \cdots + r_n A_n$. Define a mapping $\varphi: V \to F^n$ by setting

$$\varphi(A) = \varphi(r_1 A_1 + \cdots + r_n A_n) = (r_1, \ldots, r_n).$$

Certainly the mapping φ depends upon the choice of basis for V/F. But once a basis has been chosen, the mapping φ is well-defined, because the expression of vectors in V in terms of a given basis is unique. It remains to verify that φ is an isomorphism.

Consider also $B = s_1 A_1 + \cdots + s_n A_n$, where $s_1, \ldots, s_n \in F$, and $c \in F$. Then

$$\begin{aligned}\varphi(A+B) &= \varphi(r_1 A_1 + \cdots + r_n A_n + s_1 A_1 + \cdots + s_n A_n) \\ &= \varphi((r_1 + s_1) A_1 + \cdots + (r_n + s_n) A_n) \\ &= (r_1 + s_1, \ldots, r_n + s_n) \\ &= (r_1, \ldots, r_n) + (s_1, \ldots, s_n) = \varphi(A) + \varphi(B),\end{aligned}$$

$$\begin{aligned}\varphi(cA) &= \varphi(c(r_1 A_1 + \cdots + r_n A_n)) \\ &= \varphi(cr_1 A_1 + \cdots + cr_n A_n) \\ &= (cr_1, \ldots, cr_n) \\ &= c(r_1, \ldots, r_n) = c\varphi(A),\end{aligned}$$

$$\begin{aligned}\varphi(A) &= \varphi(r_1 A_1 + \cdots + r_n A_n) = (r_1, \ldots, r_n) \\ &= 0 \Leftrightarrow r_1 = \cdots = r_n = 0 \Leftrightarrow A = 0.\end{aligned}$$

§4.3 Transformations of Vector Spaces

For any $(r_1, \ldots, r_n) \in F^n$, consider $A = r_1 A_1 + \cdots + r_n A_n$. Then $\varphi(A) = (r_1, \ldots, r_n)$, and hence φ is surjective. This completes the proof that the mapping φ, defined above, is an isomorphism.

Replacing F by $\mathbf{R}$ in the theorem yields the following corollary.

Corollary 1. Any n-dimensional vector space over $\mathbf{R}$ is isomorphic to $\mathbf{R}^n$.

Corollary 2. Any two finite dimensional vector spaces over F are isomorphic if and only if they have equal dimensions.

We conclude this section by considering the **set of all linear transformations** from one vector space U/F to another V/F. This set, denoted $\mathrm{Hom}_F(U, V)$, is itself a vector space over F with the following laws of composition. [As usual, two linear transformations φ, ψ are said to be equal if $\varphi(A) = \psi(A)$ for all $A \in U$.] For $\varphi, \psi \in \mathrm{Hom}_F(U, V)$ and $r \in F$ we define the *sum* $\varphi + \psi$ and the *product by a scalar* $r\varphi$ to be the mappings in $\mathrm{Hom}_F(U, V)$ given for *all* $A \in U$ by

$$(\varphi + \psi)(A) = \varphi(A) + \psi(A),$$
$$(r\varphi)(A) = r\varphi(A).$$

We must verify first that $\varphi + \psi$ and $r\varphi$ are indeed linear transformations from U to V, and then that, with these laws of composition, the set $\mathrm{Hom}_F(U, V)$ is a vector space over F. We verify that $\varphi + \psi \in \mathrm{Hom}_F(U, V)$ and leave the remaining proofs as exercises [cf. the axioms for a vector space, §4.1]. Consider for $A, B \in U$, $r \in F$,

$$\begin{aligned}(\varphi + \psi)(A + B) &= \varphi(A + B) + \psi(A + B) \\ &= \varphi(A) + \varphi(B) + \psi(A) + \psi(B) \\ &= (\varphi + \psi)(A) + (\varphi + \psi)(B)\end{aligned}$$

and

$$\begin{aligned}(\varphi + \psi)(rA) &= \varphi(rA) + \psi(rA) \\ &= r\varphi(A) + r\psi(A) \\ &= r[(\varphi + \psi)(A)];\end{aligned}$$

hence $\varphi + \psi \in \mathrm{Hom}_F(U, V)$. (Justify each step of the preceding computations.)

Proposition. For vector spaces U/F and V/F of respective dimensions n and m, $\mathrm{Hom}_F(U, V)$ has dimension mn.

Proof. Let $\{A_1, \ldots, A_n\}$ and $\{B_1, \ldots, B_m\}$ be bases for U and V, respectively. We prove the proposition by displaying a basis of mn elements for $\mathrm{Hom}_F(U, V)$. For each i, $1 \le j \le n$, and j, $1 \le j \le m$, define

$$\varphi_{ij} \colon U \to V$$

by

$$\varphi_{ij}(A_i) = B_j,$$
$$\varphi_{ij}(A_h) = 0, \quad \text{for } h \ne i.$$

First we verify that the mappings φ_{ij} generate $\operatorname{Hom}_F(U,V)$. To this end consider any linear transformation $\varphi: U \to V$. Since φ is determined by its action on the chosen basis elements $\{A_1, \ldots, A_n\}$, we can write for each i, $1 \le i \le n$,

$$\varphi(A_i) = \sum_j r_{ij} B_j, \qquad 1 \le j \le m.$$

But then
$$\varphi = \sum_{i,j} r_{ij} \varphi_{ij}, \qquad 1 \le i \le n, \ 1 \le j \le m,$$

as claimed, since for each basis vector A_i, $1 \le i \le n$,

$$\sum_{i,j} r_{ij} \varphi_{ij}(A_i) = \sum_j r_{ij} B_j = \varphi(A_i), \qquad 1 \le i \le n, \ 1 \le j \le m.$$

To ascertain that the transformations φ_{ij} are linearly independent, consider a linear combination

$$\psi = \sum_{i,j} r_{ij} \varphi_{ij} = 0, \qquad 1 \le i \le n, \ 1 \le j \le m.$$

Since ψ is the zero linear transformation, we have $\psi(A_k) = 0$, for each basis vector A_k. That is, for each k, $1 \le k \le n$,

$$0 = \sum_{i,j} r_{ij} \varphi_{ij}(A_k), \qquad 1 \le i \le n, \ 1 \le j \le m,$$
$$= \sum_j r_{kj} B_j, \qquad 1 \le j \le m.$$

But the vectors B_j, forming a basis of V, are linearly independent. Therefore any linear combination of them that equals zero must have zero coefficients; that is, $r_{kj} = 0$, $1 \le j \le m$, for each k. Thus the mappings φ_{ij}, $1 \le i \le n$, $1 \le j \le m$, are linearly independent and so constitute a basis of mn elements of $\operatorname{Hom}_F(U,V)$.

When $U = V$ we refer to $\operatorname{Hom}_F(V,V)$, also denoted $\operatorname{End}_F(V)$, as the **ring of endomorphisms** of the vector space V/F. Here we can introduce a ring structure because we can compose two mappings $\varphi, \psi \in \operatorname{End}_F(V)$, as well as add them. The composition operation $\varphi \circ \psi$ is defined by

$$(\varphi \circ \psi)(A) = \varphi(\psi(A)), \qquad \text{for all } A \in V.$$

It is left to the reader to verify that $\operatorname{End}_F(V)$ with the operations $+$ and $\circ$ is a ring (noncommutative, if $\dim V \ge 2$).

Exercises

1. If the vector space V/F has dimension greater than 1, prove there exist at least two distinct isomorphisms $\varphi: V \to F^n$.
2. Prove that for each nonzero $c \in F$, the mapping given by $\varphi_c(A) = cA$ is an isomorphism of V/F with itself.

§4.3 Transformations of Vector Spaces

3. For $n > 1$, prove that each of the following mappings of F^n to itself is an isomorphism.
   a. $\tau(a_1, \ldots, a_i, \ldots, a_j, \ldots, a_n) = (a_1, \ldots, a_j, \ldots, a_i, \ldots, a_n)$
      (i.e., the ith and jth components are interchanged) for given i, j, $i \neq j$.
   b. $\alpha_c(a_1, \ldots, a_i, \ldots, a_j, \ldots, a_n) = (a_1, \ldots, a_i, \ldots, a_j + ca_i, \ldots, a_n)$
      (i.e., c times the ith component is added to the jth) for $c \in F$ and given i, j, $i \neq j$.
   c. $\mu_c(a_1, \ldots, a_i, \ldots, a_n) = (a_1, \ldots, ca_i, a_n)$
      (i.e., the ith component is multiplied by $c \in F$) for $c \neq 0$ and given i.

4. Given vector spaces U/F, V/F, and W/F and linear transformations $\varphi: U \to V$ and $\psi: V \to W$, prove that the composite $\psi \circ \varphi$ defined by
$$(\psi \circ \varphi)(A) = \psi(\varphi(A)), \quad \text{for all } A \in U,$$
is a linear transformation from U to W.

5. Given $\varphi \in \operatorname{Hom}_F(U, V)$ and $r \in F$, prove that $r\varphi$, defined by $(r\varphi)(A) = r\varphi(A)$ for all $A \in U$, belongs to $\operatorname{Hom}_F(U, V)$.

6. Verify that $\operatorname{Hom}_F(U, V)$ with the composition laws of vector addition and multiplication by scalars, as given in this section, is a vector space over F. That is, verify the following.
   a. $\varphi + \psi = \psi + \varphi$
   b. $r(\varphi + \psi) = r\varphi + r\psi$
   c. $(\varphi + \psi) + \lambda = \varphi + (\psi + \lambda)$
   d. $(r + r')\varphi = r\varphi + r'\varphi$
   e. $r(r'\varphi) = (rr')\varphi$
   f. $1 \cdot \varphi = \varphi$.

7. Give an explicit basis for
   a. $\operatorname{Hom}_{\mathbf{R}}(\mathbf{R}^2, \mathbf{R}^3)$
   b. $\operatorname{Hom}_{\mathbf{Q}}(\mathbf{Q}^3, \mathbf{Q}^3)$.

8. Show by an example that the product law of composition of mappings for $\operatorname{End}_F(V)$ is noncommutative when $V = \mathbf{R}^2$.

9. a. Prove that $\operatorname{End}_F(V)$, with sum and product as defined in the text, is a (noncommutative) ring if $\dim V \geq 2$.
   b. Describe the units in $\operatorname{End}_F(V)$.

10. If a linear transformation $\varphi: \mathbf{R}^2 \to \mathbf{R}^3$ satisfies $\varphi((1, 3)) = (4, 1, 2)$ and $\varphi((1, 1)) = (4, -1, 0)$, find the image of the following.
    a. $(1, 0)$
    b. $(2, 1)$
    c. $(4, 4)$
    d. $(3, 4)$
    e. $(-1, -1)$
    f. $(2, 3)$.

11. If a mapping $\varphi: \mathbf{R}^2 \to \mathbf{R}^2$ maps $(1, 0)$ to $(4, 8)$, $(1, 1)$ to $(-5, 3)$, and $(0, 1)$ to $(0, 1)$, can it be a linear transformation?

12. As for rings, define the **kernel** $\ker \varphi$ of a linear transformation $\varphi: U \to V$ to be the subset of vectors $A \in U$ for which $\varphi(A) = 0$. Prove that $\ker \varphi$ is a subspace of U.

13. a. If $\varphi: U \to V$ is a linear transformation defined on an n-dimensional vector space U with a k-dimensional kernel ($k \leq n$), define the **quotient space** $U/\ker \varphi$ and find its dimension [cf. the discussion of $\mathbf{Z}_m$ in §2.9].
    b. Prove that $\operatorname{Im} \varphi = \{B \in V : B = \varphi(A) \text{ for some } A \in U\}$ is a subspace of V. Find its dimension.
    c. Prove that $\operatorname{Im} \varphi$ and $U/\ker \varphi$ are isomorphic vector spaces.

14. Considering F/F as a (one-dimensional) vector space over itself, prove that $\text{Hom}_F(U, F)$ is isomorphic to U. Choose bases for U and F, and describe one for $\text{Hom}_F(U, F)$. (The vector space $\text{Hom}_F(U, F)$ is called the **dual space** of U, commonly denoted U^*; its elements are called **linear functionals**.)

15. Let $\varphi: U \to V$ be a linear transformation which is an isomorphism.
    a. Define the inverse mapping from V to U.
    b. Show that the inverse is a linear transformation.

16. Consider an idempotent transformation $\varphi \in \text{End}_F(V)$, that is, $\varphi^2 = \varphi$. Such a transformation is commonly called a **projection**. Prove that $V = \ker \varphi \oplus \text{Im}\, \varphi$.

17. Projections (idempotents) $\varphi_1, \ldots, \varphi_h \in \text{End}_F(V)$ are said to be **orthogonal** if $\varphi_i \circ \varphi_j = 0$ for all $i \neq j$, $1 \leq i, j \leq h$.
    a. Let $\varphi_1, \ldots, \varphi_h$ be orthogonal idempotents such that $\varphi_1 + \cdots + \varphi_h = 1_V$, the identity map of V onto itself. Prove that there exist subspaces U_i of V such that $V = U_1 \oplus \cdots \oplus U_h$. *Hint:* Consider $U_i = \text{Im}\, \varphi_i$.
    b. Prove conversely that if V is the (internal) direct sum of subspaces $U_1, \ldots, U_h$, there exists a system of orthogonal idempotents $\varphi_1, \ldots, \varphi_h \in \text{End}_F(V)$ such that $\varphi_1 + \cdots + \varphi_h = 1_V$.

§4.4 Matrices and Linear Transformations

Matrices provide a convenient shorthand notation for describing linear transformations of finite dimensional vector spaces. Indeed this is how they arose (in 1858) in the work of Arthur Cayley (1821–1895) at Cambridge, culminating in his development of matrix algebras. While Cayley had considered matrices in the abstract, in 1925 Werner Heisenberg recognized the noncommutative product of matrices as necessary for his significant developments in quantum mechanics. Matrices are used to deal computationally with problems involving linear transformations. Besides applications in linear algebra, they arise in the theories of differential and integral equations, in group theory, and in linear programming as means of representing systems of linear equations.

Here our concern is with the simple properties of matrices and the representation by matrices of the linear transformations introduced in §4.3, which will be used in §9.4 in the study of algebraic field extensions. In §4.5 we shall discuss briefly the related subject of determinants.

A **matrix** with coefficients (also entries or components) in a field F is an ordered rectangular array

$$A = \begin{bmatrix} a_{11} & \cdots & a_{1n} \\ \vdots & & \vdots \\ a_{m1} & \cdots & a_{mn} \end{bmatrix}$$

of elements $a_{ij} \in F$, $1 \leq i \leq m$, $1 \leq j \leq n$. We call the array A an $m \times n$ matrix

§4.4   Matrices and Linear Transformations

because it has m rows and n columns. In abbreviated form it is denoted

$$A = [a_{ij}], \quad 1 \le i \le m,\ 1 \le j \le n,$$

where a_{ij} is the entry in the ith row and jth column of the array.

The n-tuple $(a_{i1}, a_{i2}, \ldots, a_{in})$ of elements in A is commonly referred to as the ith **row** of the matrix A. Analogously, we shall refer to the m-tuple $(a_{1j}, a_{2j}, \ldots, a_{mj})$ as the jth **column**.

Two $m \times n$ matrices are said to be *equal* if their ijth coefficients are equal for all i, j.

The set $F_{m,n}$ of $m \times n$ matrices with entries in F is a vector space over F with laws of composition given as follows.

(i) *Vector addition.* The sum of matrices $A = [a_{ij}]$ and $B = [b_{ij}]$ is defined to be the matrix $A + B$ having $a_{ij} + b_{ij}$ for its ijth coefficient.

(ii) *Multiplication by scalars.* The product of $A = [a_{ij}]$ by $r \in F$ is defined to be the matrix $r \cdot A = rA$ whose ijth coefficient is ra_{ij}.

Direct computations verify that the axiomatic properties of a vector space [§4.1] are satisfied. The zero matrix (additive identity) is simply the $m \times n$ array of zeros.

For example, for real matrices (matrices with components in **R**) we have

$$\begin{bmatrix} 1 & 3 \\ -4 & 8 \end{bmatrix} + \begin{bmatrix} 1 & -2 \\ 5 & -1 \end{bmatrix} = \begin{bmatrix} 2 & 1 \\ 1 & 7 \end{bmatrix}$$

and

$$5 \cdot \begin{bmatrix} 1 & 3 \\ -4 & 8 \end{bmatrix} = \begin{bmatrix} 5 & 15 \\ -20 & 40 \end{bmatrix}.$$

Special matrices of significance are those which for fixed i, j have all their entries 0, *except* for a 1 in the ith row, jth column (the ijth position), where 1 denotes the multiplicative identity of F. These matrices will be denoted E_{ij}, $1 \le i \le m$, $1 \le j \le n$. For any matrix $A = [a_{ij}] \in F_{m,n}$, we can write

$$A = \sum_{i,j} a_{ij} E_{ij}, \quad 1 \le i \le m,\ 1 \le j \le n.$$

Thus the mn matrices E_{ij} generate the vector space $F_{m,n}$. But further they are linearly independent, since any linear combination

$$\sum_{i,j} r_{ij} E_{ij} = 0 \quad \text{(the zero matrix)}$$

implies $r_{ij} = 0$ for all i, j. Hence the mn matrices E_{ij} constitute a basis for $F_{m,n}$. Collecting these results, we have proved the following proposition.

Proposition 1. *The set $F_{m,n}$ of $m \times n$ matrices with entries in F is an mn-dimensional vector space over F.*

Example 1. The space of matrices $\mathbf{R}_{2,3}$ has a basis

$$E_{11} = \begin{bmatrix} 1 & 0 & 0 \\ 0 & 0 & 0 \end{bmatrix} \quad E_{12} = \begin{bmatrix} 0 & 1 & 0 \\ 0 & 0 & 0 \end{bmatrix} \quad E_{13} = \begin{bmatrix} 0 & 0 & 1 \\ 0 & 0 & 0 \end{bmatrix}$$

$$E_{21} = \begin{bmatrix} 0 & 0 & 0 \\ 1 & 0 & 0 \end{bmatrix} \quad E_{22} = \begin{bmatrix} 0 & 0 & 0 \\ 0 & 1 & 0 \end{bmatrix} \quad E_{23} = \begin{bmatrix} 0 & 0 & 0 \\ 0 & 0 & 1 \end{bmatrix}$$

and

$$\begin{bmatrix} a_{11} & a_{12} & a_{13} \\ a_{21} & a_{22} & a_{23} \end{bmatrix} = \sum_{i,j} a_{ij} E_{ij}, \quad i = 1, 2, \ j = 1, 2, 3.$$

Given an $m \times n$ matrix A, we define its **transpose** ^tA to be the $n \times m$ matrix obtained from A by making the ith row of A the ith column of ^tA. If $A = [a_{ij}]$, then the ijth entry b_{ij} in $^tA = [b_{ij}]$ is a_{ji}. As an example, for

$$A = \begin{bmatrix} a_{11} & a_{12} & a_{13} \\ a_{21} & a_{22} & a_{23} \end{bmatrix}, \quad {}^tA = \begin{bmatrix} a_{11} & a_{21} \\ a_{12} & a_{22} \\ a_{13} & a_{23} \end{bmatrix}.$$

Furthermore, the mapping $\varphi \colon F_{m,n} \to F_{n,m}$ given by $\varphi(A) = {}^tA$, for $A \in F_{m,n}$, is a (vector space) isomorphism. Note that $^t({}^tA) = A$.

A second matrix operation useful in the representation theory of linear transformations by matrices is that of matrix multiplication. To define the **product of matrices** A, B we require that the number of columns of A be equal to the number of rows of B. Thus we shall multiply $m \times n$ by $n \times p$ matrices.

The product of matrices

$$A = [a_{ij}], \quad 1 \le i \le m, \ 1 \le j \le n$$
$$B = [b_{jk}], \quad 1 \le j \le n, \ 1 \le k \le p$$

is defined to be the $m \times p$ matrix

$$AB = C = [c_{ik}], \quad 1 \le i \le m, \ 1 \le k \le p$$

where $c_{ik} = \sum_{j=1}^{n} a_{ij} b_{jk}$. That is, the ikth coefficient of the product is the sum of ordered products of elements in the ith row of A with those in the kth column of B.

This notationally somewhat complex rule may be seen more clearly in an example:

$$\begin{bmatrix} 1 & 0 & 5 & 4 \\ 3 & 8 & -1 & 5 \end{bmatrix} \cdot \begin{bmatrix} 1 & 2 & 3 \\ -4 & 0 & 1 \\ 2 & -1 & -1 \\ 0 & 1 & 8 \end{bmatrix} = \begin{bmatrix} c_{11} & c_{12} & c_{13} \\ c_{21} & c_{22} & c_{23} \end{bmatrix}$$

$$= \begin{bmatrix} 11 & 1 & 30 \\ -31 & 12 & 58 \end{bmatrix},$$

where

$$c_{11} = 1\cdot 1 + 0\cdot(-4) + 5\cdot 2 + 4\cdot 0 = 11,$$
$$c_{12} = 1\cdot 2 + 0\cdot 0 + 5\cdot(-1) + 4\cdot 1 = 1,$$
$$c_{13} = 1\cdot 3 + 0\cdot 1 + 5\cdot(-1) + 4\cdot 8 = 30,$$
$$c_{21} = 3\cdot 1 + 8\cdot(-4) + (-1)\cdot 2 + 5\cdot 0 = -31, \text{ etc.}$$

The product of matrices provides a single-valued mapping from the cartesian product $F_{m,n} \times F_{n,p}$ to $F_{m,p}$, satisfying the following properties for $A, B \in F_{m,n}$, $C, D \in F_{n,p}$, $E \in F_{p,q}$, and $r \in F$.

(i) $$(A+B)C = AC + BC$$
$\quad\quad\quad\quad\quad\quad\quad\;\;\uparrow\quad\quad\quad\quad\;\;\;\uparrow$
$\quad\quad\quad\quad\quad\quad\text{sum in } F_{m,n}\quad\text{sum in } F_{m,p}$

(ii) $$A(C+D) = AC + AD$$
$\quad\quad\quad\quad\quad\quad\quad\;\;\uparrow\quad\quad\quad\quad\;\;\;\uparrow$
$\quad\quad\quad\quad\quad\quad\text{sum in } F_{n,p}\quad\text{sum in } F_{m,p}$

(iii) $$(rA)C = A(rC)$$

(iv) $$(AC)E = A(CE)$$

We leave aside verifications of these properties, which are straightforward, although at times computationally tedious, consequences of the definition of product.

Note that this product is noncommutative. Indeed if A and B are $m \times n$ and $n \times p$ matrices, respectively, where $m \neq p$, then BA is not even defined. If both A and B are $n \times n$ matrices, then AB need not equal BA, as is seen in the simple example

$$\begin{bmatrix} 0 & 0 \\ 1 & 2 \end{bmatrix} \begin{bmatrix} 1 & 0 \\ 1 & 1 \end{bmatrix} = \begin{bmatrix} 0 & 0 \\ 3 & 2 \end{bmatrix},$$

but

$$\begin{bmatrix} 1 & 0 \\ 1 & 1 \end{bmatrix} \begin{bmatrix} 0 & 0 \\ 1 & 2 \end{bmatrix} = \begin{bmatrix} 0 & 0 \\ 1 & 2 \end{bmatrix}.$$

The set $F_{n,n}$ of $n \times n$ matrices with entries in F has the matrix $I_n = [\delta_{ij}]$—with the Kronecker delta—for its multiplicative identity element.

We turn now to the interrelation between matrices and linear transformations. Consider vector spaces U/F and V/F with respective dimensions n and m. In the proposition of §4.3 we proved that $\dim \text{Hom}_F(U,V) = mn$ and in Proposition 1 above that $\dim F_{m,n} = mn$. Then by Corollary 2, §4.3, these two vector spaces over F are isomorphic. Thus we have proved the following proposition.

Proposition 2. With vector spaces U, V as described above,
$$\text{Hom}_F(U,V) \cong F_{m,n}.$$

To define such an isomorphism explicitly, first choose bases $\mathscr{A} = \{A_1, \ldots, A_n\}$ and $\mathscr{B} = \{B_1, \ldots, B_m\}$ of U and V respectively. In terms of these bases a linear transformation $\varphi: U \to V$ is described by

$$\varphi(A_1) = a_{11} B_1 + \cdots + a_{m1} B_m = \sum_{i=1}^{m} a_{i1} B_i,$$

$$\cdots\cdots\cdots\cdots\cdots\cdots\cdots\cdots\cdots\cdots\cdots\cdots$$

(∗) $$\varphi(A_j) = a_{1j} B_1 + \cdots + a_{mj} B_m = \sum_{i=1}^{m} a_{ij} B_i,$$

$$\cdots\cdots\cdots\cdots\cdots\cdots\cdots\cdots\cdots\cdots\cdots\cdots$$

$$\varphi(A_n) = a_{1n} B_1 + \cdots + a_{mn} B_m = \sum_{i=1}^{m} a_{in} B_i.$$

Thus, when $A = r_1 A_1 + \cdots + r_n A_n$,

$$\varphi(A) = \sum_j r_j \varphi(A_j) = \sum_j r_j \left(\sum_i a_{ij} B_i \right)$$

$$= \sum_i \left(\sum_j r_j a_{ij} \right) B_i = \sum_i s_i B_i, \quad 1 \le i \le m, \; 1 \le j \le n,$$

where the coefficients s_i are given by the matrix product

$$\begin{bmatrix} s_1 \\ \vdots \\ s_m \end{bmatrix} = \begin{bmatrix} a_{11} & \cdots & a_{1n} \\ \vdots & & \vdots \\ a_{m1} & \cdots & a_{mn} \end{bmatrix} \begin{bmatrix} r_1 \\ \vdots \\ r_n \end{bmatrix} = M_\varphi \cdot \begin{bmatrix} r_1 \\ \vdots \\ r_n \end{bmatrix}.$$

Alternatively, using the transpose of one-column matrices to save space, we have

(∗∗) $$^t[s_1, \ldots, s_m] = M_\varphi \cdot {}^t[r_1, \ldots, r_n].$$

Now to the linear transformation φ associate the $m \times n$ matrix

$$M_\varphi = \begin{bmatrix} a_{11} & \cdots & a_{1n} \\ \vdots & & \vdots \\ a_{m1} & \cdots & a_{mn} \end{bmatrix},$$

the *transpose* of the array of coefficients in (∗) above. We speak of $M_\varphi = [a_{ij}]$ as the **matrix representation of the linear transformation** φ *with respect to the given bases* of U and V. With a different choice of bases we get a different matrix representation of φ, and a different isomorphism $\operatorname{Hom}_F(U, V) \to F_{m,n}$.

Example 2. To illustrate this representation, consider vector spaces $U = V = \mathbf{R}^2$. Choose $\{(1,0), (0,1)\}$ as a basis for U and $\{(1,2), (3,4)\}$ as a basis for V. Define a linear transformation $\varphi: U \to V$ by

$$\varphi((1,0)) = 4(1,2) - 3(3,4),$$

$$\varphi((0,1)) = 2(1,2) + 5(3,4)$$

§4.4 Matrices and Linear Transformations

and to φ associate the matrix

$$M_\varphi = \begin{bmatrix} 4 & 2 \\ -3 & 5 \end{bmatrix}.$$

In representing a linear transformation $\varphi: V \to V$, we would use the same choice of basis for V when considered as the domain as for V considered as the range.

To emphasize the importance of the basis in the representation M_φ, replace the previous basis of U by $\{(1,1), (2,-3)\}$ and compute the representing matrix N_φ. First,

$$\varphi((1,1)) = \varphi((1,0)) + \varphi((0,1))$$
$$= [4(1,2) - 3(3,4)] + [2(1,2) + 5(3,4)]$$
$$= 6(1,2) + 2(3,4),$$

$$\varphi((2,-3)) = 2[4(1,2) - 3(3,4)] - 3[2(1,2) + 5(3,4)]$$
$$= 2(1,2) - 21(3,4).$$

Thus,
$$N_\varphi = \begin{bmatrix} 6 & 2 \\ 2 & -21 \end{bmatrix} \neq M_\varphi.$$

If further, $\psi \in \text{Hom}_F(V, W)$ is represented by the $p \times m$ matrix $M_\psi = [b_{ki}]$ with respect to bases $\mathscr{B}$ of V and $\mathscr{C} = \{C_1, \ldots, C_p\}$ of W, then we show that $\psi \circ \varphi \in \text{Hom}_F(U, W)$ is represented with respect to the bases $\mathscr{A}, \mathscr{C}$ of U, W, respectively, by the $p \times n$ matrix

$$M_{\psi \circ \varphi} = [b_{ki}][a_{ij}] = M_\psi \cdot M_\varphi.$$

For the proof we examine the following diagram:

$$U \longrightarrow \varphi(U) \subseteq V$$
$$V \longrightarrow \psi(V) \subseteq W$$
$$U \longrightarrow \psi \circ \varphi(U) \subseteq W$$

and relate the effect of the maps $\varphi, \psi \circ \varphi$ on a vector $A = \sum_{j=1}^n r_j A_j$ to the bases $\mathscr{A}, \mathscr{B}, \mathscr{C}$ of U, V, W, respectively. Therefore, expressing the linear transformation $\psi: V \to W$ by

$$\psi\left(\sum_{\gamma=1}^m s_\gamma B_\gamma\right) = \sum_{\mu=1}^p t_\mu C_\mu$$

and writing the analogue of (**) for ψ, we have for the components of $\psi(\varphi(A))$

$${}^t[s_1, \ldots, s_m] = M_\varphi \cdot {}^t[r_1, \ldots, r_n],$$
$${}^t[t_1, \ldots, t_p] = M_\psi \cdot {}^t[s_1, \ldots, s_m].$$

Hence by the associativity of the matrix product,

$${}^t[t_1, \ldots, t_p] = M_\psi(M_\varphi \cdot {}^t[r_1, \ldots, r_n])$$
$$= (M_\psi \cdot M_\varphi) \cdot {}^t[r_1, \ldots, r_n].$$

Since, for each vector $A \in U$, the image $(\psi \circ \varphi)(A) = \psi(\varphi(A))$ has coordinates $t_1, \ldots, t_p$ relative to the basis $\mathscr{C}$ of W, we can conclude that $M_{\psi \circ \varphi} = M_\psi \cdot M_\varphi$. (Be certain to note however that M_φ is the matrix representation of φ with respect to the bases $\mathscr{A}$ and $\mathscr{B}$, that M_ψ is that of ψ

with respect to the bases $\mathscr{B}$ and $\mathscr{C}$, and that $M_{\psi \circ \varphi}$ is that of $\psi \circ \varphi$ with respect to the bases $\mathscr{A}$ and $\mathscr{C}$.)

Now let $U = V = W$ and $\mathscr{A} = \mathscr{B} = \mathscr{C}$. Then the equation $M_{\psi \circ \varphi} = M_\psi \cdot M_\varphi$ implies, in addition to the (vector space) isomorphism of Proposition 2, another isomorphism.

Proposition 3. For an n-dimensional vector space V/F the rings $\text{End}_F(V)$ and $F_{n,n}$ are isomorphic.

As in Example 2, let U equal $\mathbf{R}^2$ with basis $\{A_1 = (1, 0), A_2 = (0, 1)\}$, and V equal $\mathbf{R}^2$ with basis $\{B_1 = (1, 2), B_2 = (3, 4)\}$. Considering as before the mapping $\varphi: U \to V$ represented by

$$M_\varphi = \begin{bmatrix} 4 & 2 \\ -3 & 5 \end{bmatrix},$$

we compute for $A = r_1 A_1 + r_2 A_2 \in U$ the image $\varphi(A) = s_1 B_1 + s_2 B_2$:

$$\begin{bmatrix} s_1 \\ s_2 \end{bmatrix} = \begin{bmatrix} 4 & 2 \\ -3 & 5 \end{bmatrix} \begin{bmatrix} r_1 \\ r_2 \end{bmatrix} = \begin{bmatrix} 4r_1 + 2r_2 \\ -3r_1 + 5r_2 \end{bmatrix};$$

or $\qquad \varphi(A) = (4r_1 + 2r_2) B_1 + (-3r_1 + 5r_2) B_2.$

And if the transformation $\psi: V \to V$ is represented by

$$M_\psi = \begin{bmatrix} 1 & -1 \\ -1 & 1 \end{bmatrix},$$

then $\psi(\varphi(A)) = t_1 C_1 + t_2 C_2$ has coefficients given by

$$\begin{bmatrix} t_1 \\ t_2 \end{bmatrix} = \begin{bmatrix} 1 & -1 \\ -1 & 1 \end{bmatrix} \begin{bmatrix} 4 & 2 \\ -3 & 5 \end{bmatrix} \begin{bmatrix} r_1 \\ r_2 \end{bmatrix}$$

$$= \begin{bmatrix} 7 & -3 \\ -7 & 3 \end{bmatrix} \begin{bmatrix} r_1 \\ r_2 \end{bmatrix} = \begin{bmatrix} 7r_1 - 3r_2 \\ -7r_1 + 3r_2 \end{bmatrix},$$

or $\qquad \psi(\varphi(A)) = (7r_1 - 3r_2) B_1 + (-7r_1 + 3r_2) B_2.$

The relationship, given below in Proposition 4, between matrices M_φ and N_φ representing a given linear transformation $\varphi: V \to V$ with respect to different bases $\mathscr{A} = \{A_1, ..., A_n\}$ and $\mathscr{B} = \{B_1, ..., B_n\}$ of V/F will be needed in §5.9. To this end we now examine the relationship between the components of a vector $A \in V$ relative to two bases $\mathscr{A} = \{A_1, ..., A_n\}$ and $\mathscr{B} = \{B_1, ..., B_n\}$ of V.

The basis vectors B_i can be expressed in terms of $A_1, ..., A_n$ as

$$B_i = \sum_{j=1}^{n} p_{ji} A_j$$

and the basis vectors A_j in terms of $B_1, ..., B_n$ as

$$A_j = \sum_{k=1}^{n} q_{kj} B_k.$$

§4.4 Matrices and Linear Transformations

Set $P = [p_{ji}]$ and $Q = [q_{kj}]$. Then, since

$$B_i = \sum_j p_{ji}\left(\sum_k q_{kj} B_k\right) = \sum_k \left(\sum_j q_{kj} p_{ji}\right) B_k = \delta_{ik} B_k, \quad 1 \leq j, k \leq n,$$

we have
$$\sum_{j=1}^n q_{kj} p_{ji} = \delta_{ik},$$

which is to say that
$$QP = [q_{kj}][p_{ji}] = I_n.$$

Similarly, $PQ = I_n$.

If the products PQ, QP of $n \times n$ matrices P, Q are the $n \times n$ identity matrix I_n, we call P the **inverse** of Q (denoted p^{-1}), and conversely. Matrices in $F_{n,n}$ which have multiplicative inverses (i.e., the units of the ring $F_{n,n}$) are called **nonsingular**.

Now let
$$A = r_1 A_1 + \cdots + r_n A_n$$
$$= u_1 B_1 + \cdots + u_n B_n$$
$$= u_1 \left(\sum_{j=1}^n p_{j1} A_j\right) + \cdots + u_n \left(\sum_{j=1}^n p_{jn} A_j\right)$$
$$= \sum_{j=1}^n \left(\sum_{i=1}^n p_{ji} u_i\right) A_j = \sum_{j=1}^n r_j A_j.$$

Thus $r_j = \sum_{i=1}^n p_{ji} u_i$, and setting $P = [p_{ji}]$ we have
$$^t[r_1, \ldots, r_n] = P \cdot {}^t[u_1, \ldots, u_n].$$

With these preparations we are ready to establish the following relationship between two matrices that represent a linear transformation φ with respect to different bases.

Proposition 4. Let M_φ and N_φ be $n \times n$ matrices representing a linear transformation $\varphi: V \to V$ with respect to different bases $\mathscr{A}, \mathscr{B}$ of V/F. Then there exists a nonsingular $n \times n$ matrix P such that $N_\varphi = P^{-1} M_\varphi P$.

Proof. Expressing $A \in V$ as $A = r_1 A_1 + \cdots + r_n A_n = u_1 B_1 + \cdots + u_n B_n$ relative to the bases $\mathscr{A}$ and $\mathscr{B}$, we obtain, as in the discussion after Proposition 2, the following equations for the components of the image $\varphi(A)$:

$$^t[s_1, \ldots, s_n] = M_\varphi \cdot {}^t[r_1, \ldots, r_n],$$
$$^t[v_1, \ldots, v_n] = N_\varphi \cdot {}^t[u_1, \ldots, u_n].$$

Since
$$^t[r_1, \ldots, r_n] = P \cdot {}^t[u_1, \ldots, u_n],$$

and similarly

$$^t[v_1,\ldots,v_n] = Q \cdot {}^t[s_1,\ldots,s_n] = P^{-1} \cdot {}^t[s_1,\ldots,s_n],$$

we have

$$^t[s_1,\ldots,s_n] = M_\varphi \cdot {}^t[r_1,\ldots,r_n] = M_\varphi P \cdot {}^t[u_1,\ldots,u_n]$$

and

$$^t[v_1,\ldots,v_n] = P^{-1} \cdot {}^t[s_1,\ldots,s_n] = P^{-1} M_\varphi P \cdot {}^t[u_1,\ldots,u_n]$$
$$= N_\varphi \cdot {}^t[u_1,\ldots,u_n].$$

This last equality is valid for all choices of $u_1,\ldots,u_n$, i.e., for all $A \in V$, and therefore $N_\varphi = P^{-1} M_\varphi P$, as was to be shown.

Example 3. Again we return to our earlier example to explicate the computations:

$$B_1 = (1,2) = A_1 + 2A_2,$$
$$B_2 = (3,4) = 3A_1 + 4A_2,$$
$$A_1 = (1,0) = -2B_1 + B_2,$$
$$A_2 = (0,1) = \tfrac{3}{2}B_1 - \tfrac{1}{2}B_2,$$

and

$$P = \begin{bmatrix} p_{11} & p_{12} \\ p_{21} & p_{22} \end{bmatrix} = \begin{bmatrix} 1 & 3 \\ 2 & 4 \end{bmatrix},$$

$$P^{-1} = \begin{bmatrix} q_{11} & q_{12} \\ q_{21} & q_{22} \end{bmatrix} = \begin{bmatrix} -2 & \tfrac{3}{2} \\ 1 & -\tfrac{1}{2} \end{bmatrix}.$$

Further, $\varphi: \mathbf{R}^2 \to \mathbf{R}^2$ defined by $\varphi((a,b)) = (2a+b, -a+4b)$ is represented with respect to the basis $\mathscr{A} = \{A_1, A_2\}$ by

$$M_\varphi = \begin{bmatrix} 2 & 1 \\ -1 & 4 \end{bmatrix}$$

and with respect to the basis $\mathscr{B} = \{B_1, B_2\}$ by

$$N_\varphi = P^{-1} M_\varphi P = \begin{bmatrix} -2 & \tfrac{3}{2} \\ 1 & -\tfrac{1}{2} \end{bmatrix} \begin{bmatrix} 2 & 1 \\ -1 & 4 \end{bmatrix} \begin{bmatrix} 1 & 3 \\ 2 & 4 \end{bmatrix}$$
$$= \begin{bmatrix} \tfrac{5}{2} & -\tfrac{1}{2} \\ \tfrac{1}{2} & \tfrac{7}{2} \end{bmatrix}.$$

Exercises

1. a. Prove that the map $\varphi: F_{m,n} \to F_{n,m}$ given by $\varphi(A) = {}^t A$ is a (*vector space*) isomorphism.
   b. Prove that the map $\varphi: F_{n,n} \to F_{n,n}$ given by $\varphi(A) = {}^t A$ is not a *ring* isomorphism for $n \geq 2$.

2. Evaluate the following matrix products.

   a. $\begin{bmatrix} 1 & 2 & 0 & 1 \\ 1 & 8 & -1 & 0 \end{bmatrix} \begin{bmatrix} 1 & -1 & 0 \\ 0 & 1 & 4 \\ 5 & -3 & 7 \\ 4 & 6 & 0 \end{bmatrix}$

   b. $\begin{bmatrix} 4 & 8 & 6 \\ 5 & 3 & 1 \\ 0 & 4 & 5 \end{bmatrix} \begin{bmatrix} 1 & 0 & 3 \\ -1 & 8 & 0 \\ 2 & 9 & 0 \end{bmatrix}$

   c. $\begin{bmatrix} 1 & 2 \\ 5 & 8 \end{bmatrix} \begin{bmatrix} -1 & 3 \\ 2 & 0 \end{bmatrix}$

   d. $\begin{bmatrix} 0 & 0 & 1 \\ 0 & 1 & 0 \\ 0 & 0 & 1 \end{bmatrix} \begin{bmatrix} 4 & 5 & 8 \\ 9 & 6 & 1 \\ 3 & 2 & 5 \end{bmatrix}$.

3. Prove associativity of the matrix product. That is, for matrices A, B, C of respective sizes $m \times n$, $n \times p$, $p \times q$, prove that $(AB)C = A(BC)$.

4. Prove distributivity of matrix multiplication over matrix addition. That is, for matrices A, B, C of respective sizes $m \times n$, $n \times p$, $n \times p$, prove that $A(B+C) = AB+AC$.

5. Evaluate the following matrix products

   a. $\begin{bmatrix} 2 & 4 \\ -1 & 5 \end{bmatrix} \begin{bmatrix} 4 & 8 \\ 1 & 6 \end{bmatrix} \begin{bmatrix} -1 & 3 \\ 2 & -4 \end{bmatrix}$

   b. $\begin{bmatrix} 2 & 4 \\ -1 & 5 \end{bmatrix} \left(\begin{bmatrix} 4 & 8 \\ 1 & 6 \end{bmatrix} + \begin{bmatrix} -1 & 3 \\ 2 & -4 \end{bmatrix} \right)$

   c. $\left(\begin{bmatrix} 2 & 4 \\ -1 & 5 \end{bmatrix} + \begin{bmatrix} 4 & -8 \\ 1 & -6 \end{bmatrix} \right) \begin{bmatrix} -1 & 3 \\ 2 & -4 \end{bmatrix}$.

6. Determine the representing matrix M_φ with respect to the usual (canonical) basis $\{(1,0,0),(0,1,0),(0,0,1)\}$ of $\mathbf{R}^3$ for each of the linear transformations $\varphi \colon \mathbf{R}^3 \to \mathbf{R}^3$ described as follows.
   a. $\varphi((x,y,z)) = (x,y,0)$
   b. $\varphi((x,y,z)) = (x,2y,-z)$
   c. $\varphi((x,y,z)) = (x+y+z, y+z, z)$
   d. $\varphi((x,y,z)) = (5x,5y,5z)$
   e. $\varphi((x,y,z)) = (0,y,0)$
   f. $\varphi((x,y,z)) = (0,x+z,0)$
   g. $\varphi((x,y,z)) = (z,y,x)$.

7. Determine the matrix representing the composition of the following mappings in Exercise 6.
   a. The mapping in (a) followed by that in (f).
   b. The mapping in (f) followed by that in (a).
   c. The mapping in (f) followed by that in (f).
   d. The mapping in (c) followed by that in (a).
   e. The mapping in (g) followed by that in (c).

8. Determine the representing matrix for each of the transformations of $\mathbf{R}^3$ given in Exercise 6 with respect to the basis $\{(1, 1, 2), (2, 1, 0), (2, 0, 0)\}$ of $\mathbf{R}^3$.

9. Determine the representing matrix M_φ with respect to the basis $\{(1, 0), (0, 1)\}$ of $\mathbf{R}^2$ for the linear transformation $\varphi: \mathbf{R}^2 \to \mathbf{R}^2$, where φ is rotation through the following angles in radians.
   a. $\pi/6$
   b. $\pi/4$
   c. $3\pi/2$
   d. $2\pi/3$.

10. Repeat Exercise 9, expressing the matrices with respect to the basis $\{(1, 0), (1, 1)\}$ of $\mathbf{R}^2$.

11. In Exercise 9 use matrices to verify that four successive rotations through $\pi/6$ radians equal one through $2\pi/3$ radians.

12. The canonical (customary) basis of $\mathbf{R}^n$ is $\{E_i = (\delta_{i1}, \ldots, \delta_{in}) : 1 \le i \le n\}$. In terms of this basis, give the matrix representation of each of the linear transformations $\varphi: \mathbf{R}^n \to \mathbf{R}^n$ described as follows.
   a. For given i, j, $1 \le i < j \le n$,
   $$\varphi(E_i) = E_j, \qquad \varphi(E_j) = E_i$$
   $$\varphi(E_k) = E_k, \qquad 1 \le k \le n, \ k \ne i, j.$$
   b. For given i, $1 \le i \le n$, and nonzero $c \in F$,
   $$\varphi(E_i) = cE_i, \qquad \varphi(E_k) = E_k, \qquad 1 \le k \le n, \ k \ne i.$$
   c. For given i, j, $1 \le i, j \le n$, and $c \in F$,
   $$\varphi(E_j) = E_j + cE_i,$$
   $$\varphi(E_k) = E_k, \qquad 1 \le k \le n, \ k \ne j.$$
   d. Prove that the linear transformations in parts (a), (b), and (c) are isomorphisms.

13. Prove that an $n \times n$ matrix A is nonsingular if and only if its rows (or columns) are linearly independent. *Hint:* Consider the dimension of the image of the linear transformation determined by A.

14. In the ring of 2×2 real matrices prove that
   a. $\begin{bmatrix} 0 & 3 \\ 0 & 1 \end{bmatrix}$ is an idempotent zero divisor.
   b. $\begin{bmatrix} 2 & 1 \\ 4 & 2 \end{bmatrix}$ and $\begin{bmatrix} 1 & -1 \\ 2 & -2 \end{bmatrix}$ are zero divisors.
   c. $\begin{bmatrix} 2 & 2 \\ 4 & 2 \end{bmatrix}$ is a unit.
   d. $\begin{bmatrix} 0 & 0 \\ 1 & 0 \end{bmatrix}$ and $\begin{bmatrix} 8 & -8 \\ 8 & -8 \end{bmatrix}$ are nilpotent elements.

15. Let R be the set of 2×2 matrices with integral components.
   a. Prove that R is a noncommutative ring.
   b. Describe all units in R.

c. Give examples of matrices which are not units and are not zero divisors in R.
d. Describe the nontrivial two-sided ideals in R.

16. Let R be the noncommutative ring of 2×2 matrices with rational components.
    a. Describe all units in R.
    b. Prove that if a matrix $\alpha \in R$ is not a unit, then it is a zero divisor.
    c. Show that R has nontrivial one-sided ideals, but has no nontrivial two-sided ideals.

17. a. Describe all units in the noncommutative ring R of 3×3 matrices with rational components.
    b. Give examples of idempotent and nilpotent elements in R.

18. a. Give an example of a noncommutative ring in which the sum of two nonzero nilpotent elements is a unit [cf. Exercise 7(a), §3.1].
    b. In the ring of 2×2 matrices with coefficients in Z_2, prove that the identity matrix cannot be written as the sum of two nilpotent matrices. Verify, however, that the following matrices are nilpotent and sum to the identity:

$$\begin{bmatrix} 1 & 1 \\ 1 & 1 \end{bmatrix}, \quad \begin{bmatrix} 0 & 1 \\ 0 & 0 \end{bmatrix}, \quad \begin{bmatrix} 0 & 0 \\ 1 & 0 \end{bmatrix}.$$

§4.5 Determinants

We conclude our present discussion of linear algebra with a summary of the properties of the determinant function, which maps the ring of $n \times n$ matrices $F_{n,n}$ into its field of coefficients F. In §§5.9 and 9.4 we utilize determinants to define the characteristic polynomial of a linear transformation.

The earliest use in the Western world of what we call determinants apparently was by Leibniz in a 1693 letter to G. F. A. de L'Hospital (1661–1704). Later Gabriel Cramer (1704–1752) and Colin Maclaurin (1698–1746) independently developed the theory of determinants related to the solution of systems of two, three, and four linear equations. Maclaurin's *Treatise of Algebra* appeared posthumously in 1748, and Cramer published his widely known rule for such solutions in 1750. Analogous results were known to the Japanese mathematician Seki Kowa (or Seki Takakusa, 1642–1708). One of the modern axiomatic treatments of determinants is due to Karl Theodor Wilhelm Weierstrass (1815–1897), with earlier discussions by Augustin-Louis Cauchy (1789–1857), William Rowan Hamilton, Pierre-Simon de Laplace (1749–1827), and Alexandre Theophile Vandermonde (1735–1796).

Consider an $n \times n$ matrix $A = [a_{ij}] = [A^1, ..., A^n] \in F_{n,n}$, where A^j denotes the jth column of A. The **determinant** of A, denoted

$$\det(A) = |A| = \det([A^1, ..., A^n]),$$

is the element of F, determined from A in accordance with the following axioms.

I. Linearity with respect to the columns A^j
   (i) If
$$A = [A^1, ..., A^j, ..., A^n],$$
$$B = [A^1, ..., A^{j-1}, B^j, A^{j+1}, ..., A^n],$$
$$C = [A^1, ..., A^{j-1}, A^j + B^j, A^{j+1}, ..., A^n],$$
   then $\det(A) + \det(B) = \det(C)$.

   (ii) For all $c \in F$,
$$\det([A^1, ..., cA^j, ..., A^n]) = c \det([A^1, ..., A^j, ..., A^n]).$$

II. If two adjacent columns A^j, A^{j+1} are equal, for some j, $1 \leq j < n$, then $\det(A) = 0$.

III. For the $n \times n$ identity matrix I_n, $\det(I_n) = 1$.

The determinant function is uniquely determined by the preceding axiomatic properties from which the following consequences can be derived. Proofs of these properties are omitted; some require consideration of permutations of n symbols, a concept to be introduced in §6.7.

Property 1. If the matrix B is obtained from the matrix
$$A = [A^1, ..., A^i, ..., A^j, ..., A^n]$$
by interchanging the ith and jth columns, then $\det(B) = -\det(A)$.

Property 2. If any two columns of the matrix A are proportional (i.e., if one is a scalar multiple of the other), then $\det(A) = 0$.

Property 3. If the matrix B is obtained from the matrix
$$A = [A^1, ..., A^i, ..., A^j, ..., A^n]$$
by adding c times the ith column to the jth column, $i \neq j$, then $\det(A) = \det(B)$.

Using properties of permutations [§6.7], we can show another property.

Property 4. The $\det([A^1, ..., A^n])$ is given explicitly in terms of its n^2 entries a_{ij} by
$$\det(A) = \sum_\pi \operatorname{sgn}(\pi) a_{\pi(1), 1} a_{\pi(2), 2} \cdots a_{\pi(n), n}.$$

The sum is taken over all of the $n!$ permutations π of $\{1, ..., n\}$, and $\operatorname{sgn}(\pi) = \pm 1$ denotes the sign (parity) of the permutation [cf. Exercise 6, §6.7].

The formulas for determinants of 2×2 and 3×3 matrices are easily stated:

$$\begin{vmatrix} a_{11} & a_{12} \\ a_{21} & a_{22} \end{vmatrix} = a_{11}a_{22} - a_{12}a_{21},$$

$$\begin{vmatrix} a_{11} & a_{12} & a_{13} \\ a_{21} & a_{22} & a_{23} \\ a_{31} & a_{32} & a_{33} \end{vmatrix} = a_{11}a_{22}a_{33} + a_{12}a_{23}a_{31} + a_{13}a_{32}a_{21} - a_{13}a_{22}a_{31} \\ - a_{23}a_{32}a_{11} - a_{33}a_{21}a_{12}.$$

For larger values of n, the following formula, called the **Laplace expansion by minors**, reduces the determinant of an $n \times n$ matrix A to a linear combination of the determinants of n $(n-1) \times (n-1)$ matrices.

Property 5. For any fixed i, $1 \leq i \leq n$,

$$\det(A) = (-1)^{i+1} a_{i1} \det(A_{i1}) + \cdots + (-1)^{i+n} a_{in} \det(A_{in})$$

where the $(n-1) \times (n-1)$ matrix A_{ij} is obtained from the $n \times n$ matrix $A = [a_{ij}]$ by deleting its ith row and jth column. The matrix A_{ij} is called the ijth **minor** of A.

We complete our list of common properties of determinants with five additional facts.

Property 6. $\det(A) = \det({}^t A)$.

Property 7. $\det(AB) = \det(A)\det(B) = \det(BA)$.

Property 8. If A is a nonsingular matrix (i.e., if it has a multiplicative inverse A^{-1}), then $\det(A^{-1}) = [\det(A)]^{-1}$.

In Exercise 13, §4.4, we noted that a matrix is nonsingular if and only if its columns are linearly independent. From this consideration we can derive Property 9.

Property 9. A matrix is nonsingular if and only if its determinant is nonzero.

The following statement on homogeneous systems of linear equations is a consequence of Property 9. The system of n equations in n unknowns $x_1, \ldots, x_n$

$$\sum_{1 \leq j \leq n} a_{ij} x_j = 0, \quad 1 \leq i \leq n$$

has a nontrivial solution (i.e., not all $x_i = 0$) if and only if $\det([a_{ij}]) = 0$.

Property 10. $\det(cI_n) = c^n$.

The Laplace expansion of a determinant in Property 5 involved the $(n-1)\times(n-1)$ minors A_{ij} of a matrix A. The proof of the Cayley-Hamilton Theorem in §5.9 utilizes the (classical) **adjoint matrix** $\text{Adj}(A)$ of A, also defined in terms of minors:

$$\text{Adj}(A) = [\alpha_{ij}], \quad \text{where } \alpha_{ij} = (-1)^{i+j}\det(A_{ji}).$$

The component α_{ij} is commonly called the ijth **cofactor** of A.

Finally, A and its adjoint satisfy

(*) $$A \cdot \text{Adj}(A) = \text{Adj}(A)\cdot A = (\det(A))I_n.$$

Adjoints provide a convenient means of computing the inverse of a given (nonsingular) $n \times n$ matrix A, especially those for small n.

5

Polynomials and Polynomial Rings

While the reader no doubt has encountered polynomials with integer, rational, real, or complex coefficients earlier in his mathematical studies, he now will see them treated constructively where the coefficients lie in an arbitrary field. Emphasis is placed upon arithmetic properties analogous to those in the domain of integers, again an example of the extension of familiar concepts. In §§5.1 and 5.2, we parallel the previous discussion of the ring of integers and the residue class rings thereof. Special attention is given to the congruence relations used to develop the theory of formal derivatives and the study of multiple roots [§5.3]. Thus in part we prepare for the theory of inseparable field extensions [§5.5 and Chapter 8].

Kronecker's construction of the roots of polynomial equations in §5.4 will be applied to the proof of the existence of splitting fields. This is necessary preparation for the study of field extensions and the Galois theory in Chapter 8. Gauss' Lemma on Primitive Polynomials and Eisenstein's Irreducibility Criterion are discussed in §5.8. Aspects of the latter will be used to prove that the field of complex numbers admits only linear irreducible polynomials—the Fundamental Theorem of Algebra [§9.9].

This chapter ends with an examination of very special polynomials: the characteristic polynomials of a matrix or, equivalently, of a linear transformation. These facts are needed for the study of field extensions in Chapter 9.

§5.1 Polynomial Rings

Polynomials are introduced in the calculus as polynomial functions

$$t \to f(t) = a_0 + a_1 t + \cdots + a_n t^n \in \mathbf{R}$$

for all $t \in \mathbf{R}$ and given $a_i \in \mathbf{R}$, $0 \leq i \leq n$, for some $n \geq 0$. For reasons to become clearer later, we present a different approach, equivalent for the field $\mathbf{R}$ but applicable to fields F different from $\mathbf{R}$, e.g., $\mathbf{Z}_p$. Polynomial functions will be treated in §5.7.

We consider infinite vectors or tuples $\alpha = (a_0, a_1, \ldots, a_i, \ldots)$, $\beta = (b_0, b_1, \ldots, b_i, \ldots)$, and $\gamma = (c_0, c_1, \ldots, c_i, \ldots)$ with components a_i, b_i, c_i in a field F, such that a_i, b_i, and c_i are zero for all i larger than some positive integer N, depending on α, β, and γ. (That is, all but a finite number of the components a_i, b_i, and c_i are zero.) As for finite n-tuples, it is agreed that $\alpha = \beta$ if and only if $a_i = b_i$ for all subscripts i. We define

$$\alpha + \beta = (a_0 + b_0, a_1 + b_1, \ldots) \quad \text{and} \quad r\alpha = (ra_0, ra_1, \ldots)$$

for $r \in F$. A quick check shows that a vector space of infinite dimension over F is obtained (since no finite set generates the entire space).

Next, recall the distributive law of multiplication for polynomial functions

$$\left(\sum_{i=0}^{n} a_i t^i\right)\left(\sum_{j=0}^{m} b_j t^j\right) = \sum_{h=0}^{n+m} \left(\sum_{i+j=h} a_i b_j\right) t^h, \quad \text{for every } t \in \mathbf{R},$$

where $0 \leq i \leq n$ and $0 \leq j \leq m$. As an explicit example,

$$(6 + 3t + t^2)(4 + t^2 - t^3) = 24 + 12t + 10t^2 - 3t^3 - 2t^4 - t^5.$$

Using this distributive law as a model, we define the product $\alpha\beta$ to be the infinite vector $(d_0, d_1, \ldots, d_h, \ldots)$ whose components d_h are given by the formulas

$$d_h = \sum_{i+j=h} a_i b_j, \quad 0 \leq i, j \leq h,$$

for $h = 0, 1, 2, \ldots$. Note that the summation involved in defining each of the sums d_h is finite, and that only a finite number of the d_h are distinct from zero.

Simple juggling of indices and summations verifies the following algebraic properties for infinite vectors α, β, γ:

(i) $\quad\quad\quad\quad\quad\quad\quad\quad \alpha\beta = \beta\alpha$
(ii) $\quad\quad\quad\quad\quad\quad\quad\quad \alpha(\beta\gamma) = (\alpha\beta)\gamma$
(iii) $\quad\quad\quad\quad\quad\quad\quad\quad \alpha(\beta + \gamma) = \alpha\beta + \alpha\gamma$

(iv) $\quad(1, 0, \ldots)\alpha = \alpha$
(v) $\quad(0, \ldots, 0, 1, 0, \ldots)(0, \ldots, 0, 1, 0, \ldots) = (0, \ldots, 0, 1, 0, \ldots),$
$\qquad\quad\uparrow \qquad\qquad\quad\uparrow \qquad\qquad\quad\uparrow$
$\quad\;$ ith component $\quad\;\;$ jth component $\quad\;\;$ $(i+j)$th component

where the enumeration of components begins with zero.

Property (i) follows from the commutativity of the elements $a_i, b_j \in F$, because $\sum_{i+j=h} a_i b_j = \sum_{i+j=h} b_j a_i$. Similarly, property (iii) follows from the distributivity of multiplication over addition in F, $a_i(b_j+c_j) = a_i b_j + a_i c_j$, and the commutativity of addition in F.

Property (ii) requires more careful attention to the summation notation. The ηth component of $\alpha(\beta\gamma)$ is

$$\sum_{i+h=\eta} a_i \left(\sum_{j+k=h} b_j c_k \right) = \sum_{i+h=\eta} \sum_{j+k=h} a_i b_j c_k$$
$$= \sum_{k+h=\eta} \sum_{i+j=h} a_i b_j c_k = \sum_{k+h=\eta} \left(\sum_{i+j=h} a_i b_j \right) c_k,$$

which is also the ηth component of $(\alpha\beta)\gamma$. Note the use of the associative and distributive properties of elements in F. Property (iv) is a simple consequence of the definition of the sums d_h in the product of $(1, 0, \ldots, 0, \ldots)$ and α.

In other words, the infinite vectors $\alpha, \beta, \ldots$ form a set obeying the same rules of algebra as the usual polynomial functions (the case for $F = \mathbf{R}$). For this reason and for the sake of notational simplicity we identify (i.e., equate) the infinite tuple $(a_0, 0, \ldots)$ with the element $a_0 \in F$; the tuple $(0, 0, \ldots, 1, 0, \ldots)$, where 1 is in the $(i+1)$st position, with x^i (or t^i, etc.); and $(a_0, a_1, \ldots, a_n, 0, \ldots)$ with $a_0 + a_1 x + \cdots + a_n x^n = \sum_{i=0}^{n} a_i x^i$. The resulting set $F[x]$ (or $F[t]$, etc.) is called **the ring of polynomials in the indeterminate** x (or t, etc.) **with coefficients in the field** F. Coefficient fields common in our subsequent discussion are $\mathbf{Q}, \mathbf{R}, \mathbf{C}$, or $\mathbf{Z}_p$, p a prime number.

The principal properties of polynomial rings are summarized below. Consider two polynomials $f(x) = \sum_{i=0}^{k} a_i x^i$ and $g(x) = \sum_{i=0}^{k} b_i x^i$ in the indeterminate x with coefficients a_i, b_i in F. Here some or all of the coefficients a_i, b_i may be 0. Note that terms like $0x^m = 0$, m a positive integer, can be added to any polynomial without changing it. The elements $a \in F$ are identified with the polynomials $a + 0x + \cdots + 0x^k$. They are called **constant polynomials**, or more briefly, "constants." The polynomial $(0, 0, \ldots)$ is called the **zero polynomial**.

Property 1 (*Equality of Polynomials*). Polynomials $f(x)$ and $g(x)$ are equal if and only if $a_i = b_i$, for all i, $0 \leq i \leq k$.

Property 2 (*Addition of Polynomials*)

$$f(x) + g(x) = (a_0 + b_0) + (a_1 + b_1)x + \cdots + (a_k + b_k)x^k.$$

Property 3 (*Product of Polynomials*)

$$f(x)g(x) = d_0 + d_1 x + \cdots + d_h x^h + \cdots + d_{2k} x^{2k},$$

where $d_h = a_0 b_h + a_1 b_{h-1} + \cdots + a_h b_0 = \sum_{i=0}^{h} a_i b_{h-i}, \quad 0 \leq h \leq 2k.$

As indicated in its construction, $F[x]$ is a *commutative* ring.

The **degree**, denoted $\deg f(x)$, of a nonzero polynomial $f(x)$ is the greatest integer n for which $a_n \neq 0$ in the expression of $f(x)$ as $f(x) = a_0 + a_1 x + \cdots + a_k x^k$. For convenience, in stating the properties of the degree of a polynomial we assign to the element $0 \in F \subset F[x]$ the symbol $-\infty$ as degree. It is agreed that

$$-\infty + -\infty = -\infty$$

$$-\infty + k = -\infty \quad \text{for every integer } k$$

$$-\infty < k.$$

If $\deg f(x) = n$, then a_n is called the **leading coefficient**. A **monic polynomial** is one whose leading coefficient is 1.

Property 4. The degree of polynomials satisfies the following conditions.

(i) $\deg[f(x)g(x)] = \deg f(x) + \deg g(x).$

(ii) $\deg a = 0$ for $a \neq 0$ in F.

(iii) $\deg[f(x) + g(x)] \leq \max(\deg f(x), \deg g(x));$ i.e., the larger of $\deg f(x)$ and $\deg g(x)$.

Proposition. The polynomial ring $F[x]$ is an integral domain.

Proof. To verify that $F[x]$ has no proper zero divisors (or equivalently, that the multiplicative cancellation law holds, cf. §3.6), suppose that for $f(x) = \sum_{i=0}^{n} a_i x^i$ and $g(x) = \sum_{j=0}^{m} b_j x^j$, where $a_n \neq 0$, $b_m \neq 0$, we had $f(x)g(x) = 0$. Then each of the coefficients of powers of x in the product would be zero. In particular, $a_n b_m = 0$, and since the coefficients lie in the field F, then either $a_n = 0$, or $b_m = 0$, a contradiction. Thus, $F[x]$ has no proper zero divisors.

More generally, we can consider rings of polynomials with coefficients in an integral domain D. Nowhere in the preceding discussion was the existence of a multiplicative inverse used. In particular, $\mathbf{Z}[x]$ is a ring, and by the proof of the above proposition, an integral domain.

The construction of the quotient field of an integral domain [Theorem, §3.6] can be applied to $F[x]$. The quotient field $Q(F[x]) = F(x)$ is called the **field of rational functions** of the indeterminate x with coefficients in F.

Its elements are usually identified with the "quotients" $a(x)/b(x)$ of polynomials $a(x), b(x)$ in $F[x]$, with $b(x) \neq 0$.

Having defined in detail polynomials in one indeterminate x, we conclude this section with the definition of the **ring of polynomials** $F[x_1, \ldots, x_n]$ **in several indeterminates** $x_1, \ldots, x_n$. There are two common (and equivalent) definitions. First, for all $i \in \mathbf{N}$, inductively define $F[x_1, \ldots, x_{i+1}]$ to be the ring of polynomials $F[x_1, \ldots, x_i][x_{i+1}]$ in the indeterminate x_{i+1} with coefficients in the ring $F[x_1, \ldots, x_i]$. A second definition views $F[x_1, \ldots, x_n]$ as the set of finite sums of monomials of the form $ax_1^{i_1} \cdots x_n^{i_n}$, where $a \in F$ and $i_1, \ldots, i_n$ are nonnegative integers, together with the customary algebraic rules of commutativity, associativity, etc.

Polynomials in two indeterminates occur in some exercises, in the discussion of formal derivatives in §5.5, and in §9.8.

Exercises

1. Prove that the distributive law holds in $F[x]$.
2. Prove that the ideal generated by x in $\mathbf{Z}[x]$ is prime, but not maximal.
3. a. Prove that the ideal generated by x in $\mathbf{Q}[x]$ is maximal (and hence prime).
   b. Prove that the ideal generated by $x^2 + 2$ in $\mathbf{Q}[x]$ is a prime and maximal ideal.
4. Prove that not all ideals in $\mathbf{Z}[x]$ are principal.
5. Prove that $F[x, y]$ contains ideals which are not principal.
6. Let F be a field. Consider the set $F[[x]]$ consisting of all formal power series $a_0 + a_1 x + \cdots + a_n x^n + \cdots$, $a_i \in F$. Prove that $F[[x]]$ is an integral domain if sum and product of power series are defined in the same manner as in the polynomial ring $F[x]$. Find all units of $F[[x]]$, and prove that the elements of the quotient field of $F[[x]]$ have the form $x^{-m}(a_0 + a_1 x + \cdots)$ with $a_0 \neq 0$.
7. In the polynomial ring $F[x, y]$, suppose that $g(x, y) = g_0(x) + g_1(x) y + \cdots + g(x) y^n$ such that $g_0(0) = g(0, 0) = 0$ and $g_1(0) = \partial g(x, y)/\partial y \neq 0$ for $(x, y) = (0, 0)$. Prove that there exists a unique power series $y = f(x) = a_1 x + \cdots$ for which $g(x, y) = 0$ in $F[[x]]$.
8. Let $K \supseteq F$ be fields and let $a_1, \ldots, a_n \in K$. Consider the set A of polynomials $f(x_1, \ldots, x)$ in the polynomial ring $F[x_1, \ldots, x_n]$ for which $f(a_1, \ldots, a_n) = 0$.
   a. Prove that A is an ideal in $F[x_1, \ldots, x_n]$.
   b. Prove that A is a prime ideal in $F[x_1, \ldots, x_n]$.
9. For an arbitrary ring R, prove that $R[x]$ is an integral domain if and only if R is an integral domain.
10. Prove that for each $a \in F$ the mapping $\psi_a: F[x] \to F$ given by $\psi_a(f(x)) = f(a)$ is a ring homomorphism.
11. Prove Properties 4(i) and 4(iii) for the degree of polynomials.

§5.2 Divisibility and Factorization of Polynomials

Most of the proofs of the following statements concerning the divisibility and factorization of polynomials are modeled after the proofs for the corresponding statements for integers. As a general rule, we replace the absolute value $|a|$ for $a \in \mathbf{Z}$ by the degree $\deg f(x)$ for $f(x) \in F[x]$. This holds especially for applications of the Principle of Induction. Note that the proof of the division algorithm for polynomials requires that the coefficients be in a field, *not* just an integral domain.

Proposition 1. Given a polynomial $a(x)$ and a nonzero polynomial $b(x)$ in $F[x]$, there exist unique polynomials $q(x)$ and $r(x)$ in $F[x]$ such that

$$a(x) = q(x)b(x) + r(x),$$

where either $r(x) = 0$ or $0 \leq \deg r(x) < \deg b(x)$.

Proof. [See the Division Algorithm, §2.3.] If $\deg a(x) < \deg b(x)$, then set $q(x) = 0$ and $r(x) = a(x)$. Otherwise, write

$$a(x) = a_0 + a_1 x + \cdots + a_n x^n,$$
$$b(x) = b_0 + b_1 x + \cdots + b_m x^m, \qquad b_m \neq 0.$$

If $n = m$ and $a_n \neq 0$, then the degree of $a(x) - (a_n b_m^{-1}) b(x)$ is *less* than m. Set $q(x) = a_n b_m^{-1}$, and $r(x)$ equal to the preceding difference.

More generally, if $n > m$, induction on the degree n of $a(x)$ is made. Assume the existence statement for polynomials of degree less than n. The degree of $c(x) = a(x) - (a_n b_m^{-1}) x^{n-m} b(x)$ is less than n; thus by the induction hypothesis there exist polynomials $q_1(x)$ and $r(x)$ for which $c(x) = q_1(x) b(x) + r(x)$, where $\deg r(x) < \deg b(x)$ or $r(x) = 0$. Consequently,

$$a(x) = q_1(x) b(x) + r(x) + (a_n b_m^{-1}) x^{n-m} b(x)$$
$$= [q_1(x) + (a_n b_m^{-1}) x^{n-m}] b(x) + r(x)$$
$$= q(x) b(x) + r(x),$$

where $q(x) = q_1(x) + (a_n b_m^{-1}) x^{n-m}$ and $r(x) = 0$ or $\deg r(x) < \deg b(x)$. The Principle of Induction implies now that the *existence* statement holds for all $a(x)$ and any given polynomial nonzero $b(x)$.

The *uniqueness* of $q(x)$ and $r(x)$ is proved as follows. Suppose

$$a(x) = q(x)b(x) + r(x)$$
$$= q^*(x)b(x) + r^*(x).$$

Then, $$[q(x) - q^*(x)] b(x) = r^*(x) - r(x).$$

Since $\deg[r^*(x) - r(x)] < \deg b(x)$, properties of the degree imply that $q(x) - q^*(x)$ must be 0. Hence also $r^*(x) - r(x) = 0$.

The striking analogy between the division algorithm for polynomials and that for integers is the basis for further parallel definitions, statements, and theorems. The subsequent results are obtained by replacing arguments involving the absolute value used for integers by the degree of polynomials. (For example, compare the statement of the Division Algorithm [§2.3] with Proposition 1 above, or the theorem of §2.4 with Proposition 2 below.)

Proposition 2. All ideals A in $F[x]$ are principal; if $A \neq \{0\}$, then $A = (a(x))$, where $a(x)$ is a monic polynomial.

Proof. The zero ideal (0) is principal, and its generator is the zero polynomial. Now for $A \neq (0)$, the set

$$S = \{\deg f(x) : f(x) \neq 0, \ f(x) \in A\}$$

is a nonempty set of nonnegative integers. By the Well-Ordering Principle S has a least element k, and so A contains a polynomial $c(x)$ of degree k. Multiply $c(x)$ by the multiplicative inverse of its leading coefficient, thereby obtaining a monic polynomial $a(x)$ of degree k, also belonging to A.

To prove that $a(x)$ is the desired generator of A, write

$$f(x) = q(x)a(x) + r(x)$$

for arbitrary $f(x) \in A$. Hence $r(x) = f(x) - q(x)a(x) \in A$, and $\deg r(x) < \deg a(x)$. Since no nonzero polynomial in A has degree less than k, we conclude that $r(x) = 0$. Therefore $a(x) | f(x)$, as claimed.

If $\deg a(x) = 0$, then $a(x) = 1$ (the monic constant polynomial), and so $(a(x)) = (1) = F[x]$. Similarly, for all nonzero $c \in F$, $(c) = F[x]$.

As in an arbitrary ring [§3.1], elements $u \in F[x]$ for which there exist polynomials v such that $uv = 1$ are called **units** of the ring $F[x]$. Using $0 = \deg 1 = \deg(uv) = \deg u + \deg v$, we note that *the units of $F[x]$ are the nonzero constants* of F.

A polynomial $g(x)$ is called a **divisor** of $f(x)$, if $f(x) = g(x)h(x)$ with $h(x) \in F[x]$; for this we use the notation $g(x) | f(x)$ [cf. §2.5].

The **greatest common divisor** $d(x)$ of two polynomials $a(x)$ and $b(x)$, denoted $d(x) = (a(x), b(x))$, is defined as follows:

1. If $a(x) = b(x) = 0$, then $d(x) = 0$.
2. If either $a(x)$ or $b(x) \neq 0$, then

   (i) $d(x)$ is to be a common divisor of $a(x)$ and $b(x)$,
   (ii) any other common divisor $e(x)$ of $a(x)$ and $b(x)$ is to divide $d(x)$,
   (iii) $d(x)$ is to be monic.

The existence and uniqueness proofs of the greatest common divisor are modeled after the corresponding proofs for integers. The ideal A, consisting of all linear combinations $a(x)f(x)+b(x)g(x)$ in $F[x]$, has a unique monic generator $d(x)$, by Proposition 2 above, which can be shown to be the desired greatest common divisor [cf. §2.5].

Proposition 3. If $d(x) = (a(x), b(x))$, there exist polynomials $s(x), t(x) \in F[x]$ such that $d(x) = s(x)a(x)+t(x)b(x)$.

Polynomials $a(x)$ and $b(x)$ are said to be **relatively prime** if $A = (a(x), b(x)) = 1$. In other words, polynomials $a(x), b(x)$ are relatively prime if and only if there exist polynomials $s(x), t(x) \in F[x]$ such that $a(x)s(x)+b(x)t(x) = 1$.

A polynomial $p(x)$ of $F(x)$ is called an **irreducible** or **prime** polynomial, if it has only trivial divisors; i.e., if $f(x)|p(x)$, then either $f(x) = c$, a nonzero constant, or $f(x) = cp(x)$. [Compare the definition of prime number in §2.6.]

Theorem 1 (*Unique Factorization of Polynomials*). Each nonzero polynomial $f(x)$ can be written uniquely (apart from arrangement of the factors) as a product

$$f(x) = \delta p_1(x)^{\alpha_1} \cdots p_s(x)^{\alpha_s},$$

where δ is the coefficient of the highest nonzero power of $f(x)$ and the polynomials $p_i(x)$ are monic irreducible polynomials whose multiplicities α_i are positive integers. ($p(x)^0 = 1$ as for integers.)

Because the proof is formally the same as that for the Fundamental Theorem of Arithmetic in §2.7, we do not repeat it here. Note that the properties of prime integers carry over to irreducible polynomials. In particular, compare Proposition 2, §2.6, with the following.

Proposition 4. A polynomial $p(x) \in F[x]$ is irreducible if and only if

$$p(x)|a(x)b(x) \Rightarrow p(x)|a(x) \text{ or } p(x)|b(x).$$

We illustrate now the determination of the greatest common divisors of two polynomials. Note the similarity of this procedure to that in §2.5 for finding the GCD (a, b) of integers a, b.

Example. Consider the polynomials

$$a(x) = x^4 - x^2 + x - 3, \quad b(x) = x^2 - x + 1.$$

To show that $(a(x), b(x)) = 1$, and that

$$1 = \left(\frac{x-3}{7}\right)a(x) + \left(\frac{-x^3+2x^2+4x-2}{7}\right)b(x),$$

§5.2 Divisibility and Factorization of Polynomials

we carry out the long division of polynomials.

$$
\begin{array}{r}
x^2 + x - 1 \\
x^2 - x + 1 \overline{\smash{)}\, x^4 - x^2 + x - 3} \\
\underline{x^4 - x^3 + x^2 } \\
x^3 - 2x^2 + x \\
\underline{x^3 - x^2 + x } \\
-x^2 - 3 \\
\underline{-x^2 + x - 1} \\
-x - 2
\end{array}
$$

Thus $a(x) = x^4 - x^2 + x - 3 = (x^2 + x - 1)b(x) - (x + 2)$. Dividing $b(x) = x^2 - x + 1$ by $-x - 2$, we find

$$x^2 - x + 1 = (-x + 3)(-x - 2) + 7.$$

The remainder 7 is not the greatest common divisor, since it is not a monic polynomial. However, since it is constant, it indicates that $(a(x), b(x)) = 1$.

Reversing the process, we obtain

$$
\begin{aligned}
7 &= b(x) - (-x + 3)(-x - 2) \\
&= b(x) - (-x + 3)[a(x) - (x^2 + x - 1)b(x)] \\
&= (x - 3)a(x) + [1 + (x - 3)(-x^2 - x + 1)]b(x) \\
&= (x - 3)a(x) + (-x^3 + 2x^2 + 4x - 2)b(x).
\end{aligned}
$$

Dividing this last equation by 7 yields the desired expression for the GCD 1.

In §2.6, we proved that there exist infinitely many prime integers. An analogous proof yields the corresponding result for irreducible (monic) polynomials. More simply, for infinite fields, the polynomials $x - a$, for $a \in F$, are irreducible, monic, and infinite in number.

We conclude this section with a discussion of the number of roots of a polynomial equation. Consider the ring $D[x]$ of polynomials in one indeterminate with coefficients in an integral domain D. (In subsequent applications of this theorem we shall use a field F in place of the domain D.)

An element $a \in D$ is called a **zero** or **root** of the polynomial

$$p(x) = \sum_{i=0}^{n} c_i x^i \in D[x] \quad \text{if } p(a) = \sum_{i=0}^{n} c_i a^i = 0.$$

Theorem 2. If $p(x) \in D[x]$ has degree n, then $p(x)$ has at most n roots in D.

The proof is by induction on n, using the following technical lemma.

Lemma. For a polynomial $p(x) \in D[x]$, a is a root of $p(x)$ (i.e., $p(a) = 0$) if and only if $(x - a) \mid p(x)$ in $D[x]$.

Proof of Lemma. By the Division Algorithm for Polynomials (Proposition 1 above), there exist unique polynomials $q(x)$ and $r(x)$ such that
$$p(x) = q(x)(x-a) + r(x)$$
where $r(x)$ is a constant polynomial. Since the map $D[x] \to D$ given by $f(x) \to f(a)$, $a \in D$, is a ring homomorphism [cf. Exercise 10, §5.1],
$$p(a) = q(a)(a-a) + r(a) = r(a).$$
Thus $p(a) = 0$ if and only if the constant polynomial $r(x) = r(a)$ is the zero polynomial; that is, if and only if $(x-a) | p(x)$.

Proof of Theorem. Observe that a linear polynomial equation
$$c_1 x + c_0 = 0$$
has at most one root in D. (The existence of a root depends upon whether $c_1 | c_0$ in D.) Suppose, as the induction hypothesis, that every polynomial of degree $n-1$ has at most $n-1$ roots in D. Consider an arbitrary polynomial $p(x)$ of degree n. If $p(x)$ has no roots in D, the statement is satisfied, as $n > 0$. Therefore consider a polynomial $p(x)$ having at least one root in D, say a. Then by the lemma, $(x-a) | p(x)$. In particular,
$$p(x) = (x-a)q(x),$$
where the degree of $q(x)$ is $n-1$. For any root c of $p(x)$,
$$0 = p(c) = (c-a)q(c).$$
Since $c-a$ and $q(c)$ are elements of the integral domain D, either $c-a = 0$ or $q(c) = 0$. By the induction hypothesis, $q(x)$ has at most $n-1$ roots. Hence, $p(x)$ has at most $1 + (n-1) = n$ roots.

The necessity of the hypothesis that the coefficients lie in an integral domain is evidenced by the following example:
$$x^2 - [4] = [0] \quad \text{with coefficients in } \mathbf{Z}_{15}$$
has roots [2], [7], [8], [13]. Examples for other polynomials over rings $\mathbf{Z}_m$, m not a prime, are easy to construct. The idempotents in $\mathbf{Z}_m$ are the roots of the polynomial equation
$$x^2 - x = [0] \quad \text{with coefficients in } \mathbf{Z}_m.$$

Exercises

1. Prove that the ideal in $\mathbf{Q}[x]$ generated by $x^2 + 2$ is maximal.
2. Find the greatest common divisor of polynomials $p(x), q(x)$ in $\mathbf{Q}[x]$, and express it as a linear combination of $p(x)$ and $q(x)$ with coefficients in $\mathbf{Q}[x]$, where:
   a. $p(x) = x^3 + 7x - 3$
      $q(x) = x^2 + 5$
   b. $p(x) = x^2 + x + 1$
      $q(x) = x^4 + x^3 + x^2 + x + 1$
   c. $p(x) = x^4 - 3x^3 + 3x - 1$
      $q(x) = x^3 - 5x + 7$.

§5.2 Divisibility and Factorization of Polynomials

3. Find the greatest common divisor of $x^6 - 1$ and $x^3 - 1$ in $F[x]$, F an arbitrary field. Generalize this special result for arbitrary exponents m, n of x.

4. Find a polynomial $q(x) \in \mathbf{Q}[x]$ for which
$$(x^2 + 1)q(x) \equiv 1 \pmod{x^3 + 1}.$$

5. Prove that the polynomial $x^4 + 2x^2 + 2 \in \mathbf{Q}[x]$ is irreducible.

6. Let a and b be nonzero relatively prime polynomials in $\mathbf{Q}[x]$. Prove that the rational function $1/ab$ can be written as $u/a + v/b$ for some polynomials $u, v \in \mathbf{Q}[x]$.

7. Suppose that $f(x) \in F[x]$, F a field. Prove that
$$f(x) = (x - a)q_a(x) + f(a), \quad \text{for every } a \in F$$
where $q_a(x) \in F[x]$. Thus, give an alternate proof that $f(a) = 0$ if and only if $(x - a) | f(x)$.

8. Prove the *Rational Root Theorem*: If $f(x) = \sum_{i=0}^{n} a_i x^i$ is a polynomial with integral coefficients, prove that any rational root r/s, $(r, s) = 1$, of $f(x)$ must be such that $r | a_0$ and $s | a_n$.

9. Prove the *Integral Root Theorem*: Any rational root a of a monic polynomial $f(x) \in \mathbf{Z}[x]$ must be an integral divisor of the constant term of $f(x)$.

10. Consider polynomials $f(x), g(x) \in F[x]$ whose degrees are at most n. Prove that $f(x) = g(x)$ if $f(a) = g(a)$ for $n + 1$ *distinct* elements $a \in F$.

11. Let $a_0, a_1, \ldots, a_n$ be $n + 1$ distinct elements of the field. Show that the polynomial
$$f(x) = \sum_{i=0}^{n} b_i \left(\prod_{j=0,\ j \ne i}^{n} (x - a_j)(a_i - a_j)^{-1} \right)$$
satisfies $f(a_i) = b_i$, $0 \le i \le n$, where the elements $b_0, b_1, \ldots, b_n$ are in F [cf. the Lagrange Interpolation Formula, §5.6].

12. In $\mathbf{Z}_p[x]$, p a prime number, consider the polynomials
$$a(x) = x^3 + [7]_p x - [3]_p,$$
$$b(x) = x^2 + [5]_p.$$
   a. For which p are $a(x), b(x)$ relatively prime?
   b. For which p is the greatest common divisor $d(x) = (a(x), b(x))$ the residue class of
$$q(x) = (x^3 + 7x - 3, x^2 + 5) \in \mathbf{Q}[x]$$
   taken modulo p?

13. Let $a(x) = x^3 + [1]_5 x + [3]_5$ and $b(x) = x^2 + [4]_5$ in $\mathbf{Z}_5[x]$. Find their greatest common divisor and express it as a linear combination of $a(x)$ and $b(x)$ with coefficients in $\mathbf{Z}_5[x]$.

14. Prove that $x^3 + [3]_5 x + [2]_5$ is irreducible in $\mathbf{Z}_5[x]$.

15. Prove that $x^3 + [3]_4 x + [2]_4$ is reducible in $\mathbf{Z}_4[x]$. Find all of its zeros in $\mathbf{Z}_4[x]$.

16. Determine whether the polynomial $x^3 - [9]_{11} \in \mathbb{Z}_{11}[x]$ is irreducible.
17. Determine all roots of the following:
    a. $x^2 - [1]$ in $\mathbb{Z}_{12}$
    b. $x^2 + [1]$ in $\mathbb{Z}_{10}$.
18. Prove that the polynomial $x^3 - x + [2]_3$ is irreducible in $\mathbb{Z}_3[x]$.
19. Find all irreducible polynomials of degree less than 3 in $\mathbb{Z}_5[x]$.
20. Find all $x \in \mathbb{Z}_{12}$ satisfying the equation
$$x^3 - [2]_{12} x^2 - [3]_{12} x = [0]_{12}.$$
    Express the problem in terms of congruences of integers.
21. Consider $x^2 + 7x + 8 \in \mathbb{Z}[x]$. Find all primes p such that:
    a. $x^2 + [7]_p x + [8]_p$ is reducible in $\mathbb{Z}_p[x]$.
    b. $x^2 + [7]_p x + [8]_p$ is irreducible in $\mathbb{Z}_p[x]$.
22. Prove that the congruence $x^3 - 9 \equiv 0 \pmod{31}$ has no integral solution.
23. Prove that the equation $x^2 = [1]_{15}$ has precisely 4 zeros in $\mathbb{Z}_{15}$.
24. a. Find all polynomials $z \in \mathbb{Q}[x]$ such that
$$(x^2 + 1) z \equiv x \pmod{x^4 - x + 1}.$$
    b. Prove that this congruence is not solvable if the coefficients are taken in $F = \mathbb{Z}_5$.
    c. Prove that this congruence is solvable if the coefficients are taken in $F = \mathbb{Z}_p$, p a prime, $p \neq 5$.
25. Write out the proof of Theorem 1.
26. Following the proof of Proposition 1, §2.6, prove that there exist many irreducible polynomials.

§5.3 Residue Class Rings for Polynomials

We continue the discussion of analogues to the arithmetic of integers.

In what follows let $m = m(x)$ be a fixed element (of positive degree) in the polynomial ring $R = F[x]$. It will play the same role in the following discussion that the integer m did in §2.8 and subsequent sections. For notational simplicity we frequently write f for $f(x) \in F[x]$, etc.

If a, b are polynomials in R, then

$$b \equiv a \pmod{m}, \quad m \neq 0,$$

is to mean $m \mid (b-a)$, or $b - a \in (m)$, where $(m) = mR$ is the ideal in R generated by m. Also, $(m) \neq R$ because $\deg m \geq 1$. Thus congruence here has the same meaning as in §2.8.

If m were to equal 0, then congruence modulo m would be equality in R. If $m = 1$ or another nonzero constant, then all polynomials are congruent to each other. Thus, to avoid such singular cases, we assume once and for all that the modulus m is a polynomial of positive degree.

§5.3 Residue Class Rings for Polynomials

All general statements of §2.8 carry over to the polynomial ring R. We observe especially the following analogue of Property 5, §2.8.

Proposition 1. For polynomials $a, b, c, m \in R$,

$$ac \equiv bc \pmod{m}, \quad (c, m) = 1 \Rightarrow a \equiv b \pmod{m}.$$

Furthermore, if a is a prime residue modulo m, then so are the polynomials $a' = a + qm$, $q \in R$. That is, $(a', m) = 1$ for all $a' \equiv a \pmod{m}$. If p is an irreducible polynomial, then all nonzero polynomials (including the nonzero constants) of degree less than $\deg p$ are prime residues modulo p. Thus, if $\deg f < \deg p$, there exists a polynomial g such that $fg \equiv 1 \pmod{p}$.

For a fixed modulus m the **residue class modulo** m (or coset modulo m), denoted $[a]$, of a polynomial $a \in R$ consists of all polynomials a' satisfying $a' \equiv a \pmod{m}$. That is, $[a] = \{a + rm : r \in R\}$. As a consequence of the Division Algorithm for Polynomials, each residue class $[a]$ contains a *unique* polynomial $r \in R$ which either is equal to 0, in which case $[a] = [0] = mR = (m)$, or has degree less than $\deg m$. Note especially that the residue classes, determined by the elements c, d of F, are distinct if and only if c and d are.

The results for integers in §2.8 through §2.10 hold *mutatis mutandis* for polynomials. In particular, Proposition 2 is a special case of the result in Exercise 3, §3.2. Its proof, which we omit, is an exact analogue of the argument in §2.9 that $\mathbf{Z}_m$ satisfies the ring axioms.

Proposition 2. The residue classes of elements of $R = F[x]$ modulo m constitute a ring $R_m = R/(m)$.

Note that if m is a reducible polynomial then R_m has zero divisors, and hence is not an integral domain. We shall prove in §5.4 that R_m is a field when m is an irreducible polynomial. By Exercise 15, §3.6, this is equivalent to saying that (m) is a maximal ideal in R for an irreducible polynomial m.

Let us consider the following examples of arithmetic operations in R_m, or equivalently of the solution of congruences modulo m.

Example 1. Determination of the polynomial $f(x)$ of smallest degree in x satisfying the congruence

$$(x^2 - x + 1) \cdot f(x) \equiv x^2 - 3 \pmod{x^4 - x^2 + x - 3}.$$

First, $(x^2 - x + 1, x^4 - x^3 + x - 3) = 1$ and

$$1 = \tfrac{1}{7}(x - 3)(x^4 - x^3 + x - 3) + \tfrac{1}{7}(-x^3 + 2x^2 + 4x - 2)(x^2 - x + 1)$$

from the example of §5.2. Then setting $y = \tfrac{1}{7}(-x^3 + 2x^2 + 4x - 2)$ and taking all congruences modulo $x^4 - x^2 + x - 3$, we have

$$y \cdot (x^2 - x + 1) \equiv 1$$

and
$$f(x) \cdot y \cdot (x^2 - x + 1) \equiv y \cdot (x^2 - 3).$$

Thus
$$f(x) \equiv [(-x^3+2x^2+4x-2)/7](x^2-3)$$
$$\equiv [(-x^5+2x^4+4x^3-2x^2+3x^3-6x^2-12x+6)/7]$$
$$\equiv [(-x^5+2x^4+7x^3-8x^2-12x+6)/7].$$

Next, to reduce $f(x)$ modulo x^4-x^2+x-3 we carry out the division

$$
\begin{array}{r}
-x+2 \\
x^4-x^2+x-3 \overline{\smash{\big)}\,-x^5+2x^4+7x^3-8x^2-12x+6}\\
\underline{-x^5 +x^3-x^2+3x}\\
2x^4+6x^3-7x^2-15x+6\\
\underline{2x^4-2x^2+2x-6}\\
6x^3-5x^2-17x+12.
\end{array}
$$

So,
$$-x^5+2x^4+7x^3-8x^2-12x+6$$
$$=(-x+2)(x^4-x^2+x-3)+6x^3-5x^2-17x+12.$$

Therefore, $\quad f(x) \equiv [(6x^3-5x^2-17x+12)/7] \pmod{x^4-x^2+x-3}$,

where the cubic polynomial on the right side of this congruence is the polynomial of minimal degree in x which solves the given congruence.

Example 2. Following the discussion in §3.4, we now determine orthogonal idempotents in the residue class ring R_m, where $R = \mathbf{Q}[x]$ and $m = x^2(x+1)(x-1)$.

The first idempotent is a solution e_1 of the simultaneous system of congruences

(*)
$$y \equiv 1 \pmod{x^2},$$
$$y \equiv 0 \pmod{x+1},$$
$$y \equiv 0 \pmod{x-1}.$$

To determine e_1, note first that the general solution of the last two congruences is

$$y = \alpha(x+1)(x-1) = \alpha(x^2-1), \quad \alpha \in \mathbf{Q}[x].$$

(Recall that $(x+1)(x-1)$ divides y since $(x+1)|y$, $(x-1)|y$, and $(x+1,x-1) = 1$.) Now it remains only to determine $\alpha \in \mathbf{Q}[x]$ for which

$$\alpha(x^2-1) \equiv 1 \pmod{x^2} \quad \text{or} \quad -\alpha \equiv 1 \pmod{x^2}.$$

A particular solution is $\alpha = -1$. Therefore $e_1 = -1(x^2-1) = 1-x^2$ solves the system (*) and is the first of our desired set of orthogonal idempotents.

Next, we determine e_2 from the system of congruences

$$w \equiv 0 \pmod{x^2},$$
$$w \equiv 1 \pmod{x+1},$$
$$w \equiv 0 \pmod{x-1}.$$

Again note that the general solution of the first and third congruences is $w = \beta x^2(x-1)$ for $\beta \in \mathbf{Q}[x]$. Thus we need only find a polynomial β satisfying

$$\beta x^2(x-1) \equiv 1 \pmod{x+1}.$$

§5.3 Residue Class Rings for Polynomials

Noting that $x \equiv -1 \pmod{x+1}$, we rewrite this congruence as

$$\beta(-2) \equiv 1 \pmod{x+1},$$

which has the solution $\beta = -\frac{1}{2} \in \mathbf{Q}[x]$. Hence $e_2 = -\frac{1}{2}x^2(x-1)$.

The third idempotent e_3 is a solution of the system of congruences

$$z \equiv 0 \pmod{x^2},$$
$$z \equiv 0 \pmod{x+1},$$
$$z \equiv 1 \pmod{x-1}.$$

As before, we need only solve for γ such that

$$\gamma x^2(x+1) \equiv 1 \pmod{x-1}.$$

Since $x \equiv 1 \pmod{x-1}$, this congruence becomes

$$2\gamma \equiv 1 \pmod{x-1},$$

which has the solution $\gamma = \frac{1}{2}$. Thus $e_3 = \frac{1}{2}x^2(x+1)$.

As in §3.4 we easily verify that these polynomials e_1, e_2, e_3 are orthogonal idempotents in R_m because for $i \neq j$

$$[e_i][e_j] = [0] \quad \text{and} \quad [e_i][e_i] = [e_i]$$

and that
$$[1] = [e_1] + [e_2] + [e_3]$$
$$= [1-x^2] - \tfrac{1}{2}[x^2(x-1)] + \tfrac{1}{2}[x^2(x+1)].$$

As for the ring $\mathbf{Z}_m$, the residue class ring R_m, $m = \prod_{i=1}^{n} p_i^{\alpha_i}$, $n > 1$, can be written as a direct sum of subrings (ideals) B_i^*, isomorphic to $R_{p_i^{\alpha_i}}$ [see §3.5].

We have seen earlier [Exercise 3(c), §3.2] that the canonical or natural projection mapping

$$\lambda: R \to R_m, \qquad \lambda(f) = [f],$$

of each polynomial $f \in F[x] = R$ to its equivalence class $[f]$ modulo the given polynomial m is a surjective ring homomorphism. We consider now the *restriction* of the mapping λ defined on $F[x]$ to the subset $F \subset F[x]$. That is, we limit our attention to the effect of λ on F, rather than on all of R.

Proposition 3. The restriction of the projection $\lambda: R \to R_m$ to F is a monomorphism of F into R_m.

Proof. For $c, d \in F$, we have

$$\lambda(c+d) = \lambda(c) + \lambda(d)$$
$$\underset{\text{sum in } F}{\uparrow} \quad \underset{\text{sum in } R_m}{\uparrow}$$

and

$$\lambda(c \cdot d) = \lambda(c) \cdot \lambda(d).$$
$$\underset{\text{product in } F}{\uparrow} \quad \underset{\text{product in } R_m}{\uparrow}$$

Further, λ is a monomorphism of F since $\lambda(c) = [0]$ in R_m means that $c \in (m)$. Because c is a constant, and is divisible by the polynomial m (of positive degree), it must necessarily equal 0. Therefore, $\ker \lambda$ equals the zero element 0 of F. (An alternate argument for the injectivity is that $\ker \lambda$ is an ideal in F. But since a field has only trivial ideals, $\ker \lambda \neq F$ implies $\ker \lambda = (0)$.]

Thus identifying F with its image $\lambda(F)$ under λ, we speak of F as a subfield of R_m. In particular, if $\deg m = 1$, then

$$\lambda(F) = \{\lambda(c) : c \in F\} = R_m.$$

An observation important in the general theory of fields [see Chapter 8] is that R_m can be considered as a *vector space* over F. To this end we define the sum of the "vectors" $[a]$ and $[b]$ to be the sum $[a+b]$ of the cosets in R_m. Multiplication by scalars $c \in F$ is defined by

$$c[a] = [c][a] = [ca].$$

The customary axioms for a vector space, §4.1, are easily verified.

Since each coset $[a] \in R_m$ can be represented *uniquely* by a polynomial which either is 0 or has degree less than $n = \deg m$, it follows that the cosets $[1], [x], [x^2], \ldots, [x^{n-1}]$ constitute a *basis* of R_m over its field of scalars F [see §4.2]. For, if there were a proper dependence relation

$$a_0[1] + a_1[x] + \cdots + a_{n-1}[x]^{n-1} = [0]$$

(that is, if not all coefficients were 0), then the residue class

$$[a_0 + a_1 x + \cdots + a_{n-1} x^{n-1}]$$

would equal $[0]$. This would imply that

$$a_0 + a_1 x + \cdots + a_{n-1} x^{n-1} \in (m),$$

contradicting the fact that m is the monic polynomial of least degree n in (m). Hence $\dim_F R_m = \deg m$.

Exercises

1. Describe the cosets of $F[x]$ modulo $(x^3 + 1)$ when
   a. $F = \mathbf{Q}$ b. $F = \mathbf{Z}_2$.
2. Prove that $x^2 - 2 \in \mathbf{Z}[x]$ is an irreducible polynomial, but that $x^2 - [2] \in \mathbf{Z}_2[x]$ is reducible. Hence conclude that if a polynomial is irreducible in $\mathbf{Z}[x]$ it need not be irreducible in $\mathbf{Z}_p[x]$, p a prime.
3. Prove that the congruence $fg \equiv 1 \pmod{p}$ always has a solution g for $f, p \in F[x]$, when p is an irreducible polynomial not dividing f.
4. a. Find nontrivial idempotent elements in the residue class ring $\mathbf{Q}[x]/(x^2(x^2+1))$.
   b. Show that $\mathbf{Q}[x]/(x(x+1)(x^2+3))$ has no proper nilpotent elements.

§5.3   Residue Class Rings for Polynomials   141

5. a. Write the multiplicative unit of the ring $\mathbf{Q}[x]/(x(x+1)(x-1)^2)$ as a sum of three distinct idempotent elements.
    b. Write $\mathbf{Q}[x]/(m)$ as the internal direct sum of three ideals and the external direct sum of three rings when $m = x^2(x+1)(x-1) \in \mathbf{Q}[x]$ [cf. §3.5].
6. Find a polynomial $y \in \mathbf{Q}[x]$ of degree less than 4 for which $(x+1)y \equiv x^3+1 \pmod{x^3+3x-1}$.
7. Find a polynomial $y \in \mathbf{Z}_3[x]$ for which $(x^2+[1]_3)y \equiv [1]_3 \pmod{x^3+x+[1]_3}$.
8. Find inverses of the residue classes of the polynomials $x+1$ and x^2+3 in the residue class ring $\mathbf{Q}[x]/(x^5-1)$.
9. Let $F = \mathbf{Q}(\sqrt[3]{7})$ where $\sqrt[3]{7}$ is a solution of the equation $x^3 = 7$ in $\mathbf{C}$. Write $(1+\sqrt[3]{7})(1-4(\sqrt[3]{7})^2)$ and $(1+\sqrt[3]{7})^{-1}$ as polynomials in 1, $\sqrt[3]{7}$, $(\sqrt[3]{7})^2$ with coefficients in $\mathbf{Q}$. [See §3.6, Example 6 for the definition of $\mathbf{Q}(\sqrt[3]{7})$.]
10. Prove that the polynomial x^2-2 is irreducible in $\mathbf{Q}(\sqrt{3})[x]$ and $\mathbf{Q}(\sqrt[3]{3})[x]$.
11. Let F be a subfield of a field K, and φ the ring homomorphism of $F[x] \to K$ given by $\varphi(f(x)) = f(a)$ for some fixed $a \in K$. Prove that $\ker \varphi$ either is 0 or is generated by a monic irreducible polynomial $m_a(x) \in F[x]$.
12. Prove Proposition 2.
13. Consider polynomials $f(x), m(x) \in F[x]$, where F is an arbitrary field and $\deg m \geq 1$. Prove that there exist unique polynomials $h_0, h_1, \ldots, h_r \in F[x]$ for which
    a. $\deg h_i \leq \deg m$, $\quad 0 \leq i \leq r$,
    b. $f = \sum_{i=0}^r h_i m^i$.
    (Cf. the n-adic expansion for integers at the end of §2.5.) The expression in (b) is called the **m-adic expansion** of f. Especially important is the case where $m(x)$ is an irreducible polynomial $p(x)$.
14. Determine the p-adic expansion of a polynomial $f(x) \in \mathbf{Q}[x]$ for the following polynomials $p(x)$ and $f(x)$.
    a. $p(x) = x+2$ $\qquad$ b. $p(x) = x^2+2$
        $f(x) = x^3+1$ $\qquad\qquad\;\; f(x) = x^4+x$.
15. Consider the polynomial $p = p(x) = x-a$ in $F[x]$. Prove that the cosets $[f(x)]$ of R_{p^α} are of the form
    $$[\omega_0 + \omega_1(x-a) + \cdots + \omega_{\alpha-1}(x-a)^{\alpha-1}] = \sum_{i=0}^{\alpha-1} \omega_i([x-a])^i,$$
    where the coefficients $\omega_i \in F$ are unique.

These expressions are the polynomial analogue of the n-adic expansion introduced in §2.5. This type of analogy with truncated power series expansions of holomorphic functions of a complex variable led Kurt Hensel (1861–1941) to his theory of p-adic numbers, a theory which provides a powerful tool for recent developments in the theory of algebraic numbers and algebraic functions. Isaac Newton used such expansions in his studies of algebraic plane curves.

16. Consider an irreducible polynomial p in $F[x]$. Suppose that a/p^n is a proper fraction in $F(x)$, where $a \in F[x]$. Prove that there exist unique polynomials $a_0, \ldots, a_n$ such that

$$a/p^n = a_n + a_{n-1}/p + \cdots + a_0/p^n$$

where $\deg a_i < \deg p$, $0 \leq i \leq n$.

17. Consider $m = x^2 - [5]_{11} \in \mathbf{Z}_{11}[x]$.
   a. Prove that m is reducible.
   b. Show that the residue class ring $\mathbf{Z}_{11}[x]/(m)$ is the (internal) direct sum of two fields isomorphic to $\mathbf{Z}_{11}$.

§5.4 Residue Class Fields of Irreducible Polynomials

We turn now to a topic of great significance in the theory of equations and fields [see Chapter 8]. Recall that in §2.12 we found that the residue class ring $\mathbf{Z}_p$, p a prime integer, is a field; that is, every nonzero residue class has a multiplicative inverse. The initial discussion in this section is directed toward a similar question. Specifically, if R is the ring of polynomials in x over a field F and p is an irreducible polynomial, then the residue class ring R_p is a field. Moreover, Proposition 2 shows that, roughly speaking, for an irreducible polynomial $p \in R$, the residue class field R_p contains a zero of p.

Proposition 1. If p is an irreducible polynomial in $R = F[x]$, then R_p is a field.

Proof. To prove that each $[a] \neq [0] \in R_p$ has a multiplicative inverse, pick in $[a]$ the representative a' having degree less than $\deg p$, as in §5.3. Then $(a', p) = 1$, and $a'b + pq = 1$ for some polynomials b and q. Consequently $a'b \equiv 1 \pmod{p}$, or

$$[a'][b] = [a][b] = [1],$$

as asserted. Thus R_p is a field. It contains the field $\lambda(F)$, which is isomorphic to F. (Here λ is the natural projection of R onto R_p given by $\lambda(a) = [a]$ for all $a \in R$, as in §5.3.)

Proposition 2 (*Kronecker's Construction*). If $p \in R = F[x]$ is an irreducible polynomial, then the field R_p contains a root of p.

The construction of roots of irreducible polynomials is due to Leopold Kronecker. As in Proposition 3, §5.3, where we identified the field F with $\lambda(F) \subseteq R_p$, we now associate to the polynomial $p(x) \in R = F[x]$ a unique polynomial $p(t) \in R_p[t]$, where t is another indeterminate over R_p. We shall prove that the residue class (or coset) $[x] \in R_p$ is a zero of the polynomial $p(t) \in R_p[t]$. First, if

$$p(x) = a_n x^n + \cdots + a_1 x + a_0,$$

§5.4 Residue Class Fields of Irreducible Polynomials

define $p(t)$ to be the polynomial

$$p(t) = a_n t^n + \cdots + a_1 t + a_0.$$

Considering a_i as an element of R_p—that is, identifying $\lambda(a_i) = [a_i] \in R_p$ with $a_i \in R$— we may view $p(t)$ as an element of $R_p[t]$. Second,

$$\lambda(p(x)) = [p(x)] = a_n[x]^n + \cdots + a_1[x] + a_0 = [0] \in R_p.$$

(Again we write a_i for $\lambda(a_i) = [a_i]$.) Hence $p(t) \in R_p[t]$ has a root in R_p, namely $[x]$. Thus given an irreducible polynomial $p \in R = F[x]$, we have constructed a field $R_p \supseteq R$ in which p has a root.

Kronecker's construction is admittedly rather subtle. The following examples should serve to illustrate the argument.

Example 1. Let $F = \mathbf{Q}$ and $p = x^2 - a$ where a is not the square of a rational number. Then R_p is a field and $\dim_\mathbf{Q} R_p = 2$. Also note that $[x]^2 = [a]$ for the residue class $[x] \in R_p$. In other words, $[x]$ satisfies the quadratic equation $t^2 = [a]$;

$$t^2 - [a] = (t - [x])(t + [x])$$

in the polynomial ring $R_p[t]$.

Specifically, let $a = 2$. Then $p = x^2 - 2$ is an irreducible polynomial in $R = \mathbf{Q}[x]$. The map

$$\lambda: \mathbf{Q}[x] \to \mathbf{Q}[x]/(x^2 - 2) = R_p$$

takes

$$f = a_{2n} x^{2n} + a_{2n-1} x^{2n-1} + \cdots + a_1 x + a_0 \in \mathbf{Q}[x]$$

to its equivalence class

$$\lambda(f) = [f] = a_{2n}[x^2]^n + a_{2n-1}[x^2]^{n-1}[x] + \cdots + a_1[x] + a_0,$$

where $\lambda(a_i) = [a_i]$ has been identified with $a_i \in \mathbf{Q}$. In R_p we have $[x]^2 = [2] = 2$ since $x^2 \equiv 2 \pmod{x^2 - 2}$. Hence

$$[f] = a_{2n} 2^n + a_{2n-1} 2^{n-1}[x] + \cdots + a_1[x] + a_0,$$

and so can be expressed as $b[x] + c$ with coefficients $b, c \in \mathbf{Q}$. Furthermore, extending λ to a mapping from $R[t] = \mathbf{Q}[x, t]$ to $R_p[t]$ by defining $\lambda(t) = t$, we obtain (again using the fact that $[x]^2 = [2]$ in R_p)

$$\lambda(t^2 - 2) = t^2 - [x]^2 = (t - [x])(t + [x])$$
$$= (t - \sqrt{2})(t + \sqrt{2}),$$

where we conveniently write $[x]$ as $\sqrt{2}$.

In §3.6 we noted that the set

$$\{a + b\sqrt{2} : a, b \in \mathbf{Q}\}$$

is a field; it can be described as the field obtained by adjoining the element $\sqrt{2}$ to **Q**. By this, we mean that $\mathbf{Q}(\sqrt{2})$ is the smallest field containing both **Q** and $\sqrt{2}$; smallest is used in the sense that $\mathbf{Q}(\sqrt{2})$ is a subfield of F for *any* field F such that $\mathbf{Q} \subset F$, $\sqrt{2} \in F$.

The significant point to be made here is that the field $\mathbf{Q}(\sqrt{2})$ is isomorphic to the residue class field of $\mathbf{Q}[x]$ modulo a particular irreducible (over **Q**) polynomial, namely $x^2 - 2$. The isomorphism

$$\mu: \mathbf{Q}(\sqrt{2}) \to R_p$$

is given by

$$\mu(a + b\sqrt{2}) = a + b[x] \in R_p.$$

This same polynomial, $x^2 - 2$, while irreducible over **Q**, factors over $\mathbf{Q}(\sqrt{2})$ as follows:

$$x^2 - 2 = (x + \sqrt{2})(x - \sqrt{2}).$$

Example 2. As a second, but very similar example, let $F = \mathbf{R}$ and p be the irreducible polynomial $p = x^2 + 1$. Then the residue class field R_p of $F[x]$ modulo p is isomorphic to **C**, which is often defined as

$$\mathbf{C} = \{a + bi : a, b \in \mathbf{R}\},$$

where $i^2 = -1$, or $i = \sqrt{-1}$. The isomorphism

$$\mu: \mathbf{C} \to R_p$$

is given by

$$\mu(a + bi) = [a] + [b][x] \in R_p.$$

As an historical note, it is interesting that Augustin-Louis Cauchy often viewed the complex numbers algebraically as equivalence classes of real polynomials modulo $x^2 + 1$.

Example 3. Consider $F = \mathbf{Z}_3$ and $m = x^2 + [2]_3 x + [2]_3$. Then m is an irreducible polynomial in F, because no residue class $[a]$ of $\mathbf{Z}_3$, $a \in \mathbf{Z}$, satisfies $[a]^2 + [2][a] + [2] = [0]$. Consequently the residue class ring R_m is a field. Since $\dim_F R_m = 2$, we see that R_m is a field of $3^2 = 9$ elements.

We are now ready to prove the theorem that any polynomial in $F[x]$ splits into linear factors in some sufficiently large field. This theorem lays the foundations for the general theory of equations [see the Galois theory in §§8.6 and 8.7]. The theorem of §5.2 states that a polynomial $f(x)$ of degree n in $F[x]$ has at most n roots in F. The following significant theorem yields the existence of a field Ω in which $f(x)$ has precisely n (not necessarily distinct) roots.

Theorem. Consider $f(x) = \sum_{i=0}^{n} a_i x^i \in F[x]$. There exists a field Ω and an injective homomorphism $\mu: F \to \Omega$ such that $\mu f(x) = \sum_{i=0}^{n} (\mu a_i) \bar{x}^i \in \Omega[\bar{x}]$

factors completely. That is,

$$\mu f(x) = \prod_{i=1}^{n}(\bar{x}-\alpha_i^*), \qquad \alpha_i^* \in \Omega,$$

where $\mu x = \bar{x}$, an indeterminate over Ω.

Proof. We proceed by induction on n, noting that there is nothing to prove when $n = 1$. As the induction hypothesis, suppose that for any polynomial $g(x)$ of degree $m-1$ with coefficients in any field $K \supseteq F$ there exists a field Ω and an injective homomorphism $\mu': K \to \Omega$ with the properties in the statement of the theorem.

Now consider a polynomial $f(x) \in F[x]$ of degree m. If $f(x)$ factors completely in $F[x]$, then nothing has to be proved, and we can take $\Omega = F$. Therefore assume that $f(x)$ has at least one nonlinear irreducible (over F) factor $f_1(x)$. Applying the Kronecker construction of Proposition 2, we obtain a root α of $f(x)$ in the field $K_1 = F[x]/(f_1(x))$, where $\alpha = [x]$, the coset of x with respect to the ideal $(f_1(x))$.

As in §5.3 let λ denote the projection mapping to the congruence classes; for notational simplicity we write λF for $\lambda(F)$ and λa for $\lambda(a)$, etc. Since $f_1(\alpha) = 0$ in K and $K \supset \lambda F \cong F$, the image $\lambda f(x) = \sum_{i=0}^{m}(\lambda a_i) y^i \in K[y]$, where $\lambda x = y$, an indeterminate over K, has the factor $(y-\alpha)$:

$$\lambda f(x) = (y-\alpha)g(y) \in K[y].$$

The polynomial $g(y)$ has degree $m-1$. Therefore by the induction hypothesis there exists a field Ω and an injective homomorphism $\mu': K \to \Omega$ such that $\mu'g(y)$ factors completely in $\Omega[\bar{x}]$.

To complete the proof define $\mu: F \to \Omega$ by $\mu = \mu' \circ \lambda$. Then, with $\bar{x} = \mu'y$,

$$\mu f(x) = \mu' \circ \lambda(f(x))$$
$$= \mu'\left[\sum_{i=0}^{m}(\lambda a_i)y^i\right] = \sum_{i=0}^{m}(\mu a_i)\bar{x}^i \in \Omega[\bar{x}]$$
$$= (\bar{x}-\alpha^*)\prod_{i=1}^{m-1}(\bar{x}-\alpha_i^*),$$

where $\alpha^* = \mu'\alpha$ and the α_i^*, $1 \leq i < m$, are the roots in Ω of $g(y) \in K[y]$.

It is customary to identify the isomorphic image μF of F with the given field F, and to identify the indeterminate $\bar{x}$ with x. Then the theorem states that there exists a field $\Omega \supseteq F$ such that $f(x)$ factors completely in $\Omega[x]$.

The Fundamental Theorem of Algebra, to be discussed in §9.9, says that any polynomial with complex (and hence real, rational, or integral) coefficients splits in the field of complex numbers **C**. But **C** will not be a useful splitting field for our subsequent considerations as it has infinite dimension (as a vector space) over **Q**, whereas the splitting field Ω just constructed for $f(x) \in \mathbf{Q}[x]$ is a finite dimensional vector space over **Q**. Furthermore the theorem is applicable for fields of nonzero characteristic, while the Fundamental Theorem of Algebra certainly is not.

Exercises

1. Consider $F = \mathbf{Z}_{13}$.
   a. Prove that the polynomial $x^2 - [5]_{13} \in F[x]$ is irreducible.
   b. Furthermore, show that the residue class ring $F[x]/(x^2 - [5]_{13})$ is a field K having $13^2 = 169$ elements.
2. Prove that the polynomial $x^2 + [1]_7 \in \mathbf{Z}_7[x]$ is irreducible. Let t denote the residue class of x in $\mathbf{Z}_7[x]/(x^2 + [1]_7)$.
   a. Express the elements $[1]_7/(t + [1]_7)$ and $([2]_7 + [3]_7 t)([4]_7 + [6]_7 t)$ as polynomials of degree less than 2 in t with coefficients in $\mathbf{Z}_7$.
   b. Show that $\mathbf{Z}_7[x]/(x^2 + [1]_7)$ is a field of 49 elements.
3. Let $R = \mathbf{Q}[x, y]$. Prove that the residue class ring R/A, where $A = (x^2 - 2, y^2 + 3)$, is a field.
4. Construct a splitting field (i.e., a field in which each polynomial splits into a product of linear factors) of each of the polynomials $f(x) \in \mathbf{Q}[x]$.
   a. $f(x) = x^2 - 3$ b. $f(x) = x^3 - 2$
   c. $f(x) = (x^2 - 3)(x^3 - 2)$ d. $f(x) = x^4 + 1$
   e. $f(x) = x(x^5 + 7)$ f. $f(x) = (x^5 + 7)^3(x^4 + 1)$.
5. Determine the dimension of each of the splitting fields in Exercise 4 considered as a vector space over $\mathbf{Q}$.
6. Prove the theorem of this section by successively constructing fields in which irreducible factors of $f(x)$ split off at least one root. *Hint:* Apply Kronecker's construction to an irreducible factor $f_1(x)$ of $f(x)$ to obtain a field K_1 in which $f(x)$ has a root α_1. Then consider an irreducible factor, if any, of $f(x_1) \in K_1[x_1]$, where x_1 is an indeterminate over K_1.
7. Verify that the mapping μ in Example 1 is an isomorphism.
8. a. Show that the ideal $A = (y^2 - x, x^2 - y) \subset \mathbf{R}[x]$ is not a prime ideal. Find prime and maximal ideals in $\mathbf{R}[x]$ that contain A.
   b. Repeat part (a) with $A = (y^2 - x - 1, y - 1 - 3x)$.
9. Prove that $(y^2 - x, y - 4x + 3)$ is a maximal ideal in $\mathbf{R}[x]$.

§5.5 Roots of Polynomials

This section discusses the existence and properties of roots of polynomials over a field in preparation for the field theory in Chapter 8. The formal derivative of such polynomials, which involves congruences of polynomials, is needed for this discussion.

Let $F[x]$ be the polynomial ring in the indeterminate x with coefficients in the field F. Further let τ be an indeterminate over the ring $R = F[x]$. The typical element of $R[\tau]$ is

$$\sum_{i=0}^{n} h_i(x) \tau^i, \text{ where } h_i(x) \in R, \ 0 \leq i \leq n.$$

Such an element can be rewritten as

$$\sum_{i,j} a_{ij} x^j \tau^i, \quad 0 \leq i \leq n, \ 0 \leq j \leq m$$

§5.5 Roots of Polynomials

with coefficients $a_{ij} \in F$, where

$$\sum_{j=0}^{m_i} a_{ij} x^j = h_i(x), \qquad 0 \le i \le n,$$

and $m = \max m_i$. Conversely, a polynomial in two indeterminates

$$\sum_{i,j} b_{ij} x^j \tau^i = \sum_i \left(\sum_j b_{ij} x^j \right) \tau^i = \sum_i k_i(x) \tau^i$$

with $k_i(x) = \sum_j b_{ij} x^j$ can be considered as an element of $R[\tau]$.

Now for $f(x) = \sum_{i=0}^{n} a_i x^i \in R$ consider $f(x+\tau) = \sum_{i=0}^{n} a_i (x+\tau)^i$. This element of $R[\tau]$ can be expanded, using the Binomial Theorem, in the form

$$f(x+\tau) = f(x) + \tau f_1(x) + \tau^2 f_2(x) + \cdots + \tau^n f_n(x)$$

where the polynomials $f_i(x)$ are uniquely determined by $f(x)$. Next consider in $R[\tau]$ the principal ideal

$$(\tau^2) = \{\tau^2 h(x,\tau) : h(x,\tau) \in R[\tau]\}.$$

Then $$f(x+\tau) \equiv f(x) + \tau f_1(x) \pmod{\tau^2}.$$

The uniquely determined polynomial $f_1(x)$ will be seen to have the formal properties of a derivative and henceforth will be denoted by $f'(x)$. We call $f'(x)$ the **formal derivative** of $f(x)$.

To verify the (formal) properties of the derivative consider the following congruence for another polynomial $g(x)$:

$$g(x+\tau) \equiv g(x) + g'(x)\tau \pmod{\tau^2}.$$

Then, modulo τ^2,

$$(f+g)(x+\tau) \equiv (f+g)(x) + [(f+g)(x)]'\tau$$
$$\equiv f(x) + g(x) + [f(x)+g(x)]'\tau$$

and
$$(f+g)(x+\tau) = f(x+\tau) + g(x+\tau)$$
$$\equiv [f(x)+f'(x)\tau] + [g(x)+g'(x)\tau]$$
$$\equiv [f(x)+g(x)] + [f'(x)+g'(x)]\tau.$$

Consequently $[f(x)+g(x)]' = f'(x)+g'(x)$ since the coefficient of τ in the congruence is uniquely determined. Next, modulo τ^2,

$$(fg)(x+\tau) \equiv (fg)(x) + [(fg)(x)]'\tau$$
$$\equiv f(x)g(x) + [f(x)g(x)]'\tau$$

and
$$(fg)(x+\tau) = f(x+\tau)g(x+\tau)$$
$$\equiv [f(x)+f'(x)\tau][g(x)+g'(x)\tau]$$
$$\equiv f(x)g(x) + [f'(x)g(x)+f(x)g'(x)]\tau.$$

Consequently $[f(x)g(x)]' = f'(x)g(x) + f(x)g'(x)$.

Finally, $c' = 0$ for $c \in F$, since $c \equiv c + 0 \cdot \tau \pmod{\tau^2}$. Also $(x^n)' = n \cdot x^{n-1}$, since $(x+\tau)^n \equiv x^n + n \cdot x^{n-1} \tau \pmod{\tau^2}$, according to the Binomial Theorem. Here, as at the end of §2.2, $n \cdot x^{n-1}$ denotes the n-fold sum of x^{n-1}.

NOTE. If F has prime characteristic p and $m \equiv 0 \pmod{p}$, then the derivative of the nonconstant polynomial x^m is 0.

Next we use the (formal) derivative as a tool to examine polynomials for multiple roots. (Recall from Theorem 2 of §5.2 that a polynomial of degree n has at most n distinct zeros in F.)

A zero c in a field $K \supseteq F$ of a polynomial $f(x) \in F[x]$ is said to be **α-fold**, or to have **multiplicity** α, if

$$f(x) \equiv 0 \pmod{(x-c)^\alpha},$$

but $\qquad f(x) \not\equiv 0 \pmod{(x-c)^{\alpha+1}} \qquad$ in $K[x]$.

This is equivalent to saying that $f(x) = (x-c)^\alpha g(x)$ for some $g(x) \in K[x]$ relatively prime to $x-c$ in $K[x]$.

Proposition 1. If c is an α-fold zero of $f(x)$, then c is a zero of multiplicity at least $\alpha - 1$ of the derivative $f'(x)$.

Proof. Let $f(x) = (x-c)^\alpha g(x)$. Then

$$\begin{aligned} f'(x) &= \alpha(x-c)^{\alpha-1} g(x) + (x-c)^\alpha g'(x) \\ &= (x-c)^{\alpha-1} [\alpha g(x) + (x-c) g'(x)]; \end{aligned}$$

i.e., $(x-c)^{\alpha-1} \mid f'(x)$.

Note that $(x-c)^\alpha \mid f(x)$ can happen if F has prime characteristic p. Consider $F = \mathbf{Z}_p$ and $f(x) = x^p - a$, $a \in F$. Then $f'(x) = px^{p-1} = 0$, thus $(x-c)^p \mid f'(x)$, since $a = c^p$ with $c \in \mathbf{Z}_p$ according to §2.12; i.e., $(x^p - a) = (x-c)^p$.

Proposition 2. An *irreducible* polynomial $f(x)$ in $F[x]$ can have multiple roots in a field K containing F only if $f'(x) = 0$.

Proof. From the previous discussion if $c \in K$ is a multiple zero of $f(x) \in F[x]$, then $f'(c) = 0$. Therefore $x - c$ divides the GCD $(f(x), f'(x))$ computed in $F[x]$. But since $f(x)$ is assumed to be irreducible and since $\deg f' < \deg f$, we have $(f(x), f'(x)) = 1$ *unless* $f'(x)$ is the zero polynomial. Consequently the assumption that $c \in K$ is a multiple zero of $f(x)$ implies necessarily that $f'(x)$ is the zero polynomial.

Now suppose an irreducible polynomial $f(x) = \sum_{i=0}^{n} a_i x^i$ has multiple zeros. Then its derivative

$$(*) \qquad f'(x) = \sum_{i=1}^{n} i \cdot a_i x^{i-1}$$

is the zero polynomial; hence each coefficient $i \cdot a_i$ must be 0. Thus, *if* char $F = 0$, we must have $a_i = 0$ for $i > 0$. This fact implies that there are no nonconstant irreducible polynomials $f(x)$ in $F[x]$ with multiple roots in any field $K \supseteq F$.

However, *if char $F = p > 0$* [see §3.6], then the derivative of any polynomial $f(x) \in F[x]$, whose only nonzero coefficients are a_i, $i \equiv 0 \pmod{p}$, is zero. If an irreducible polynomial $f(x) \in F[x]$ has multiple roots, its derivative, given in equation (∗) above, must be identically zero; that is $f(x)$ must be of the form

$$f(x) = \sum_{i=0}^{n} a_i x^i, \quad \text{where } a_i \neq 0 \text{ only if } p \mid i.$$

Such a polynomial can be expressed as a polynomial $g(x^p) \in F[x^p]$, or $g(y) \in F[y]$, where $y = x^p$. Specifically,

$$g(x^p) = \sum_{j=0}^{m} b_j x^{pj} = \sum_{j=0}^{m} b_j y^j.$$

Furthermore, $g(y) \in F[y] \subset F[x]$ is an irreducible polynomial in y if $f(x)$ is irreducible. It may happen that $g(y)$ also has multiple zeros. If so, repeating the preceding arguments we find

$$g(y) = \sum_{v=0}^{t} b_v y^{vp}, \quad tp = m;$$

hence, with $z = y^p$,

$$f(x) = g(x^p) = \sum_{v=0}^{t} b_v z^v = \sum_{v=0}^{t} b_v x^{vp^2}.$$

Ultimately there exists an integer $e \geq 1$, called the **exponent of inseparability**, such that

$$f(x) = h(x^{p^e}) = \sum_{i=0}^{n^*} d_i x^{ip^e}$$

with $\quad h(u) = \sum_{i=0}^{n^*} d_i u^i \in F[u] \subset F[x], \quad u = x^{p^e},$

such that $h(u)$ is irreducible in $F[u]$ and does not have multiple roots in any field $K \supseteq F$. Such irreducible polynomials $f(x)$ are said to be **inseparable** of **reduced degree** $n^* = n/p^e$; we call p^e the **degree of inseparability**.

This definition of reduced degree n^* does *not* imply that $(n^*, p) = 1$. (We shall consider inseparable polynomials further in §8.5.) A nonconstant irreducible polynomial $g(x)$ is said to be **separable** if $g'(x)$ is not identically zero.

Proposition 3. Suppose that F has characteristic $p > 0$ and that $f(x)$ is an irreducible polynomial with multiple zeros. Then all zeros have the same multiplicity p^e, $e \geq 1$.

Proof. By Proposition 2, $f(x) = h(x^{p^e}) = h(y)$, where $y = x^{p^e}$ and the irreducible polynomial $h(y) \in F[y]$ has zeros with multiplicity 1. Let $h(y) = \prod_{i=1}^{n^*}(y - \eta_i)$, η_i in some field $\Lambda \supseteq F$, and $f(x) = \prod_{i=1}^{n^*}(x^{p^e} - \eta_i)$. The polynomials $x^{p^e} - \eta_i$ factor completely in some field $K \supseteq \Lambda$, according to the theorem of §5.4, so that $\xi_i^{p^e} - \eta_i = 0$ in K. Consequently

$$x^{p^e} - \eta_i = x^{p^e} - \xi_i^{p^e} = (x - \xi_i)^{p^e},$$

and $f(x) = \prod_{i=1}^{n^*} (x - \xi_i)^{p^e}$ in $K[x]$.

§5.6 The Interpolation Formula of Lagrange

In §5.2 we noted that a polynomial of degree n with coefficients in an integral domain D had at most n roots in D. Now conversely, given n distinct elements $a_1, \ldots, a_n$ in D, we can find a polynomial $m(x) \in D[x]$ of degree n, whose roots are precisely the given elements a_i, $1 \le i \le n$; namely,

$$m(x) = (x - a_1)(x - a_2) \cdots (x - a_n).$$

More generally, we have the formula of Joseph Louis Lagrange (1736–1813) given below. While we shall give a straightforward proof, verification that Lagrange's formula follows from the Chinese Remainder Theorem of §2.11 is left as an exercise in §5.7.

Lagrange's Interpolation Formula. Let $a_1, \ldots, a_{n+1}$ be *distinct* elements and $b_1, \ldots, b_{n+1}$ any $n+1$ elements of the field F. Then there exists a polynomial $m(x) \in F[x]$ of degree at most n such that $m(a_i) = b_i$.

Proof. Consider $h_i(x) = \prod_{j \ne i} (x - a_j)$. Then

$$h_i(a_k) = 0 \quad \text{if } k \ne i$$
$$\ne 0 \quad \text{if } k = i.$$

Now

$$m(x) = \sum_{i=1}^{n+1} b_i h_i(x) (h_i(a_i))^{-1}$$

has degree at most n since $\deg h_i(x) = n$. Clearly $m(a_i) = b_i$, $1 \le i \le n+1$.

Alternatively we can set $h(x) = \prod_{i=1}^{n+1} (x - a_i)$; then $m(x)$ has the expression

$$m(x) = \prod_{i=1}^{n+1} b_i h(x) [h'(a_i)(x - a_i)]^{-1}.$$

For example, to determine the polynomial of least degree with rational coefficients whose graph contains the points $(1, 1)$, $(2, 4)$, $(5, -5)$ in the cartesian plane, we consider

$$h_1(x) = (x - 2)(x - 5), \quad h_1(1) = 4,$$
$$h_2(x) = (x - 1)(x - 5), \quad h_2(2) = -3,$$
$$h_3(x) = (x - 1)(x - 2), \quad h_3(5) = 12.$$

Thus
$$m(x) = (1)(x-2)(x-5)(\tfrac{1}{4}) + 4(x-1)(x-5)(-\tfrac{1}{3}) + (-5)(x-1)(x-2)(\tfrac{1}{12})$$
$$= \tfrac{1}{4}(x^2 - 7x + 10) - \tfrac{4}{3}(x^2 - 6x + 5) - \tfrac{5}{12}(x^2 - 3x + 2)$$
$$= -\tfrac{3}{2}x^2 + \tfrac{15}{2}x - 5.$$

A modification of Lagrange's Interpolation Formula yields a means of constructing idempotents in certain residue class rings.

Proposition. Let $h(x) = \prod_{i=1}^{n}(x - a_i)$, where the a_i are *distinct* elements of a field F. Any polynomial $k(x) \in F[x]$ of degree less than n satisfies

$$k(x) = \sum_{i=1}^{n} \left[\frac{k(a_i)}{h'(a_i)}\right]\left[\frac{h(x)}{(x-a_i)}\right].$$

Proof. Set
$$h_i(x) = \prod_{j=1}^{n}(x - a_j), \qquad \text{where } j \neq i,$$
and
$$q(x) = \sum_{i=1}^{n} k(a_i) h_i(x)(h_i(a_i))^{-1}.$$

Since $q(a_i) = k(a_i)$, $1 \leq i \leq n$, the polynomial $q(x) - k(x)$ of degree $n-1$ has n zeros. Therefore it must be the zero polynomial, i.e., $q(x) = k(x)$, which proves $k(x)$ has the asserted form.

For $k(x) = 1$, we have
$$1 = \sum_{i=1}^{n} h(x)[h'(a_i)(x-a_i)]^{-1} = \sum_{i=1}^{n} g_i(x),$$
so that
$$g_i(x) g_j(x) \equiv 0 \pmod{h(x)} \qquad \text{for } i \neq j,$$
and
$$g_i(x) \cdot 1 = g_i(x) \cdot \sum_{j=1}^{n} g_j(x)$$
$$\equiv g_i^2(x) + \sum_{j \neq i} g_i(x) g_j(x)$$
$$\equiv g_i^2(x) \pmod{h(x)}.$$

The residue classes of the polynomials $g_i(x)$ modulo $h(x)$ form a system of orthogonal idempotents, whose sum is [1], in the residue class ring $F[x]/(h(x))$ [cf. §§3.4 and 5.3]:
$$[g_i(x)]^2 = [g_i(x)],$$
$$[g_i(x)][g_j(x)] = [0] \qquad \text{if } i \neq j,$$
and
$$[g_1(x)] + \cdots + [g_n(x)] = [1].$$

This modified argument will be used to prove the Theorem of the Normal Basis for algebraic field extensions in §9.5.

§5.7 Polynomial Functions

We now relate the formal polynomials (infinite tuples) with which we have been working to the generalization of the polynomial functions of analysis.

Consider $f(x) \in F[x]$ and $c \in F$. The mapping $\varphi_c: F[x] \to F$, given by $\varphi_c(f(x)) = f(c)$, is a homomorphism, and $\ker \varphi_c$ is the principal ideal $(x-c)$. For the proof we have only to note that

$$f(c) = \sum_{i=0}^{n} a_i c^i \quad \text{and} \quad g(c) = \sum_{j=0}^{m} b_j c^j$$

are added and multiplied as elements of F according to the same *formal* rules used for sum and product in $F[x]$.

Example. For

$$f(x) = \sum_{i=0}^{n} a_i x^i \quad \text{and} \quad g(x) = \sum_{j=0}^{m} b_j x^j$$

we have

$$f(x)g(x) = \sum_{v=0}^{n+m} \left(\sum_{i+j=v} a_i b_j \right) x^v$$

and

$$\varphi_c(f(x)g(x)) = \sum_{v=0}^{n+m} \left(\sum_{i+j=v} a_i b_j \right) c^v = \left(\sum_{i=0}^{n} a_i c^i \right) \left(\sum_{j=0}^{m} b_j c^j \right)$$

$$= f(c)g(c) = \varphi_c(f(x)) \varphi_c(g(x)).$$

A **polynomial function** f in one variable over F is a single-valued function from F into F whose graph Γ_f in the cartesian product $F \times F$ is given by

$$\Gamma_f = \{(c, f(c)) : c \in F\},$$

with $f(c) = \sum_{i=0}^{n} a_i c^i = \varphi_c(f(x))$ for a polynomial $f(x) \in F[x]$.

We define the sum and product of polynomial functions f and g as follows:

$$f + g \quad \text{by } \Gamma_{f+g} = \{(c, f(c) + g(c)) : c \in F\}$$

and

$$fg \quad \text{by } \Gamma_{fg} = \{(c, f(c)g(c)) : c \in F\}.$$

Using the preceding remarks on the computation of $f(c) + g(c)$ and $f(c)g(c)$ for each $c \in F$, we can show that all polynomial functions form a ring $\Phi(F)$, which is an integral domain.

Note that each polynomial $f(x) = \sum_{i=0}^{n} a_i x^i \in F[x]$ determines a unique polynomial function f whose graph is

$$\Gamma_f = \left\{ \left(c, \sum_{i=0}^{n} a_i c^i \right) \right\} = \{(c, f(c)) : c \in F, \text{ where } f(c) = \varphi_c(f(x))\}.$$

The mapping $f(x) \to \varphi(f(x)) = f$ is a homomorphism of the ring $F[x]$ onto the ring of polynomial functions $\Phi(F)$. The kernel of φ is therefore the set of polynomials $k(x)$ such that $\varphi(k(x)) = 0$, i.e., $\Gamma_k = \{(c, 0) : \text{for all } c \in F\}$.

§5.7   Polynomial Functions

Proposition 1. If F is an *infinite* field, then $\Phi(F)$ and $F[x]$ are isomorphic integral domains.

Proof. We have already noted that the above mapping $\varphi: F[x] \to \Phi(F)$ is a surjective homomorphism. It remains to prove that $\ker \varphi = (0)$. Consider an arbitrary polynomial $f(x)$ in $\ker \varphi$; the graph Γ_f is the set $\{(c, 0) : c \in F\}$. In other words $f(c) = 0$ for all $c \in F$. But since F has an *infinite* number of elements and a nonzero polynomial has only a finite number of zeros, $f(x)$ must be the zero polynomial. Hence $\ker \varphi = (0)$, as asserted.

    If F is a finite field, then the collection of all distinct mappings (functions) from F to F is finite. Accordingly the ring $\Phi(F)$ of polynomial functions must be a finite set, while the ring $F[x]$ of polynomials has an infinite number of elements. Since $\varphi: F[x] \to \Phi(F)$ maps an infinite set to a finite one, $\ker \varphi \ne \{0\}$. (The explicit nature of $\ker \varphi$ is considered in Exercise 7, §8.2.)

Proposition 2 (*Wilson's Theorem*).   For every prime number p,
$$(p-1)! \equiv -1 \pmod{p}.$$

    In §2.12 we proved that the $p-1$ nonzero cosets $[a]$ in the field $\mathbf{Z}_p$ satisfy $[a]^{p-1} = [1]$. Thus they are the $p-1$ distinct zeros of $x^{p-1} - [1]$ in $\mathbf{Z}_p[x]$. Consequently
$$x^{p-1} - [1] = (x - [1]) \cdots (x - [p-1])$$
or
$$x^{p-1} - 1 \equiv (x - 1) \cdots (x - p + 1) \pmod{p}.$$

Furthermore, using the map $x \to [0]$, we obtain
$$[0]^{p-1} - [1] = ([0] - [1]) \cdots ([0] - [p-1])$$
or
$$-1 \equiv (-1) \cdots (-p+1) \pmod{p},$$
$$-1 \equiv (-1)^{p-1}(p-1)! \pmod{p}.$$

Hence for odd p, $(p-1)! \equiv -1 \pmod{p}$; for $p = 2$, then $1! \equiv -1 \equiv 1 \pmod 2$.

    This theorem, due to John Wilson (1741–1793), was published in 1770 by his teacher, Edward Waring (1734–1793).

Exercises

1. Prove that the ring R of all polynomial functions of one variable with coefficients in $\mathbf{Z}_p$ is not isomorphic to $\mathbf{Z}_p[x]$. (Show that there exists an epimorphism of $\mathbf{Z}_p[x]$ onto R.)
2. Prove that the ring R in Exercise 1 is isomorphic to the residue class ring $\mathbf{Z}_p[x]/(x^p - x)$.

3. Consider a prime number p of the form $p = 4n+1$, $n > 0$. Prove that the congruence $x^2 \equiv -1 \pmod{p}$ has a solution. (*Hint:* Investigate $x = [(p-1)/2]!$.)
4. Find a polynomial $p(x) \in \mathbf{Q}[x]$ for which:
   a. $p(1) = 2$, $p(2) = 4$, $p(3) = 8$, $p(4) = 16$
   b. $p(3) = 5$, $p(0) = 0$, $p(4) = 1$
   c. $p(8) = -3$, $p(-3) = 8$.
5. Repeat Exercise 4 when the coefficient field is $\mathbf{Z}_7$; that is, replace 2 by $[2]_7$, etc.
6. Let $f(x)$ be a polynomial function defined on $\mathbf{Z}_p$. Show that there exists a polynomial function $g(x)$ of degree less than p such that $f(a) = g(a)$ for all $a \in \mathbf{Z}_p$.
7. Derive Lagrange's Interpolation Formula of §5.6 from the Chinese Remainder Theorem, §2.11. *Hint:* Since the elements a_i of the statement of Lagrange's formula are distinct, the Chinese Remainder Theorem, applied to elements of $F[x]$, states that the n congruences $m(x) \equiv b_i$ $(\mod x - a_i)$ can be solved simultaneously.

§5.8 Primitive and Irreducible Polynomials

We consider now polynomials with integral coefficients considered as elements in both $\mathbf{Z}[x]$ and $\mathbf{Q}[x]$. The discussion begins with primitive polynomials and the important Lemma of Gauss.

A polynomial $\sum_{h=0}^{n} c_h x^h \in \mathbf{Z}[x]$ is called a **primitive polynomial** if the greatest common divisor $(c_0, c_1, \ldots, c_n)$ is 1.

Lemma of Gauss. The product of primitive polynomials is primitive.

Proof. Consider primitive polynomials

$$u(x) = \sum_{\mu=0}^{m} a_\mu x^\mu \quad \text{and} \quad v(x) = \sum_{\nu=0}^{n} b_\nu x^\nu,$$

and let

$$u(x)v(x) = \sum_{h=0}^{m+n} c_h x^h, \quad \text{with } c_h = \sum_{\mu+\nu=h} a_\mu b_\nu.$$

Suppose now that $u(x)v(x)$ is not primitive. Then there exists a prime p such that $c_h \equiv 0 \pmod{p}$, $0 \le h \le m+n$. By assumption, not all of the coefficients of $u(x)$ and $v(x)$ are divisible by p. Consequently there is a smallest subscript i such that $p \nmid a_i$ but $p \mid a_k$, $0 \le k < i$, and a smallest subscript j such that $p \nmid b_j$ but $p \mid b_k$, $0 \le k < j$. Hence, by the formula for c_h with $h = i+j$, the product

$$a_i b_j = c_{i+j} - (a_0 b_{i+j} + \cdots + a_{i-1} b_{j+1} + a_{i+1} b_{j-1} + \cdots + a_{i+j} b_0)$$

is divisible by p. This contradicts the fact that both a_i and b_j are not divisible by p. Thus the coefficients c_h of $u(x)v(x)$ must have GCD 1.

The following theorem makes use of the fact that for any $f(x) \in \mathbf{Q}[x]$ there exists a rational number α such that $\alpha f(x)$ is a primitive polynomial in $\mathbf{Z}[x]$. For the proof simply write $f(x) \in \mathbf{Q}[x]$ of degree n as

$$f(x) = \sum_{i=0}^{n} (s_i/t_i) x^i, \quad \text{with } s_i, t_i \in \mathbf{Z}, (s_i, t_i) = 1.$$

Let c be the least common multiple of $t_0, t_1, \ldots, t_n$. Then

$$cf(x) = \sum_{i=0}^{n} (cs_i/t_i) x^i \in \mathbf{Z}[x].$$

Now let d be the GCD of the integers cs_i/t_i, $0 \le i \le n$. Then the coefficients of $(cd^{-1}) f(x)$, that is, the elements $d^{-1}(cs_i/t_i)$, have GCD 1. Thus $(cd^{-1}) f(x)$ is a primitive polynomial in $\mathbf{Z}[x]$.

Theorem 1. *If a primitive nonconstant polynomial $f(x)$ in $\mathbf{Z}[x]$ is reducible in $\mathbf{Q}[x]$, then it is also reducible in $\mathbf{Z}[x]$.*

Proof. It must be proved that a factorization $f(x) = g(x) h(x)$ with nonconstant polynomials $g(x), h(x)$ in $\mathbf{Q}[x]$ leads to a factorization $f(x) = g^*(x) h^*(x)$ with polynomials $g^*(x), h^*(x)$ in $\mathbf{Z}[x]$. By the preceding argument there exist rational numbers α and β such that the polynomials $g_1^*(x) = \alpha g(x)$ and $h^*(x) = \beta h(x)$ belong to $\mathbf{Z}[x]$ and are primitive.

Next $f(x) = g(x) h(x)$ implies that $\alpha \beta f(x) = g_1^*(x) h^*(x) \in \mathbf{Z}[x]$. Since $f(x)$ is primitive the product $\alpha \beta$ must be an integer. The Lemma of Gauss states that $g_1^*(x) h^*(x)$ is primitive. Hence $\alpha \beta$ is a unit of $\mathbf{Z}$, i.e., $\alpha \beta = \pm 1$. Consequently

$$f(x) = g^*(x) h^*(x), \quad \text{where } g^*(x) = (\alpha \beta)^{-1} g_1^*(x).$$

Corollary. The preceding results remain valid if $\mathbf{Z}$ is replaced by a polynomial ring $F[t]$, F a field. The field of rational numbers $\mathbf{Q}$ is then replaced by the quotient field $Q(F[t])$, i.e., the field $F(t)$ of all rational functions $a(t)/b(t)$ with $a(t), b(t) \ne 0$ in $F[t]$ [see §5.1]. The units ± 1 of $\mathbf{Z}$ are replaced by the units of $F[t]$, i.e., the nonzero constants of F.

NOTE. The factors $g^*(x)$ and $h^*(x)$ are unique to within units of $\mathbf{Z}$ and $F[t]$, respectively.

A principal application of Theorem 1 is to show that irreducibility of a polynomial in $\mathbf{Z}[x]$ implies its irreducibility in the larger ring $\mathbf{Q}[x]$, a result due to Ferdinand Gotthold Eisenstein (1823–1852). The following theorem gives an important test for irreducibility.

Theorem 2 (*Eisenstein's Criterion*). *A polynomial $f(x) = \sum_{i=0}^{n} a_i x^i$ of degree $n \ge 1$ in $\mathbf{Z}[x]$ is irreducible in $\mathbf{Q}[x]$ if there exists a prime $p \in \mathbf{Z}$ such that*

$$a_n \not\equiv 0 \pmod{p}$$
$$a_i \equiv 0 \pmod{p} \quad \text{for } 0 \le i < n$$
$$a_0 \not\equiv 0 \pmod{p^2}.$$

Proof. We assume without loss of generality that $f(x)$ is primitive. By Theorem 1 it suffices to show that a factorization $f(x) = g(x)h(x)$ in $\mathbf{Z}[x]$, where $g(x)$ and $h(x)$ are not constants, leads to a contradiction. Let

$$g(x) = b_\alpha x^\alpha + \cdots + b_0 \quad \text{and} \quad h(x) = c_\beta x^\beta + \cdots + c_0,$$

with integral coefficients $b_\alpha, \ldots, c_0$. Note that $b_\alpha c_\beta \neq 0$. Moreover because p does not divide $a_n = b_\alpha c_\beta$, it does not divide c_β. Since p divides $a_0 = b_0 c_0$, but p^2 does not divide a_0, one, but not both of b_0, c_0 must be divisible by p. Suppose that p divides c_0 but not b_0.

Next, there is a coefficient c_i of

$$h(x) = c_\beta x^\beta + c_{\beta-1} x^{\beta-1} + \cdots + c_1 x + c_0$$

farthest to the right that is not divisible by p. Since $\deg h(x) < n$, necessarily $i \neq n$. Also, $i > 0$, since we assumed that $p \mid c_0$. Now, p cannot divide

(∗) $\qquad a_i = b_0 c_i + b_1 c_{i-1} + \cdots + b_{i-1} c_1 + b_i c_0$

because p does not divide $b_0 c_i$, but it does divide all other terms on the right-hand side of equation (∗) by the choice of i. Since the only coefficient a_i of $f(x)$ not divisible by p is a_n, and yet $p \nmid a_i$ for some $i < n$ if a proper factorization is possible, we have reached a contradiction. Hence $f(x)$ must be irreducible, as asserted.

Corollary. Theorem 2 remains valid if $\mathbf{Z}$ is replaced by $F[t]$ and $\mathbf{Q}$ by $F(t)$ and the primes $p \in \mathbf{Z}$ are replaced by irreducible polynomials $p(t) \in F[t]$ for an arbitrary field F.

Example 1. Let p be a prime; then $x^n - p$ is irreducible in $\mathbf{Q}[x]$.

Example 2. Suppose that F is a field of characteristic $p > 0$. Let K be the field of rational functions $F(t)$; then the polynomial $x^p - t$ is irreducible in $K[x]$.

Exercises

1. Prove that $f(x) = a_0 x^n + a_1 x^{n-1} + \cdots + a_n \in \mathbf{Z}[x]$ is irreducible in $\mathbf{Q}[x]$ if there exists a prime p such that $[f(x)] = [a_0] x^n + [a_1] x^{n-1} + \cdots + [a_n]$, $[a_i] \in \mathbf{Z}_p$, $0 \le i \le n$, is irreducible in $\mathbf{Z}_p[x]$.
2. Prove that the following polynomials are irreducible in $\mathbf{Q}[x]$.
   a. $3x^3 - 4x^2 + 2x - 2$
   b. $x^4 - 2x + 2$
   c. $x^4 + 3x^2 + 3$.
3. Verify that the polynomial in Exercise 2(c) is congruent to $(x-1)(x+1)(x^2-3)$ modulo 7.
4. Assume that $x - a$, $a \in \mathbf{Q}$, is a divisor of a monic polynomial $f(x) \in \mathbf{Z}[x]$. Prove that $a \in \mathbf{Z}$.

5. Prove that the polynomial $x^4 + 5px + p^2$ is irreducible in $\mathbf{Q}[x]$ for all $p > 5$. What about $p = 2, 3$?
6. Prove that $f(x) = x^5 + 7x^2 + 1 \in \mathbf{Z}[x]$ is irreducible in $\mathbf{Q}[x]$. (*Hint:* Reduce $\mathbf{Z}$ and $f(x)$ modulo 2 and examine for reducibility in $\mathbf{Z}_2[x]$.)
7. If $f(x)$ is a monic polynomial in $\mathbf{Z}[x]$ for which $f(1) = p$ and if $f(x+1) = x^s + pg(x)$ for some $g(x) \in \mathbf{Z}[x]$, prove that $f(x)$ is irreducible.
8.   a. Prove that $f(x) = x^{p-1} + \cdots + x + 1 \in \mathbf{Z}[x]$, p a prime, is irreducible in $\mathbf{Z}[x]$. [*Hint:* Examine $f(x+1)$.] Such polynomials are called *cyclotomic* [see §9.1].
    b. Let $f(x) = (x^{p^m} - 1)/(x^{p^{m-1}} - 1)$, p a prime. Noting that $f(x) = y^{p-1} + y^{p-2} + \cdots + 1$, where $y = x^{p^{m-1}}$, prove that $f(x)$ is irreducible in $\mathbf{Q}[x]$. [*Hint:* Examine $f(x+1)$.]
9. Prove that the following polynomials are irreducible in $F = \mathbf{Z}_3(t)$.
    a. $x^3 + t^5 x^2 - t$ b. $x^3 + tx - t^5$.
10. Write out Theorem 2 and its proof with the replacements cited in the corollary.
11. Prove that

$$f(x) = \sum_{i=0}^{j} a_i x^{n-i} + p \sum_{i=j+1}^{n} a_i x^{n-i} \in \mathbf{Z}[x],$$

where $0 \le j < n$, p a prime, has at least one irreducible divisor (in $\mathbf{Z}[x]$), whose degree is at least $n - j$, provided that $a_n \not\equiv 0 \pmod{p}$ and at least one $a_i \not\equiv 0 \pmod{p}$, where $0 \le i \le j$.
12. Let p be a prime, and suppose that the polynomial $f(x) \in \mathbf{Z}[x]$ has a factorization $f(x) \equiv g_0(x) h_0(x) \pmod{p}$, where $(g_0(x), h_0(x)) \equiv 1 \pmod{p}$. Prove that $f(x) \equiv g_k(x) h_k(x) \pmod{p^k}$, where $g_k(x), h_k(x)$, taken modulo p, equal $g_0(x), h_0(x)$, respectively. [*Hint:* Use induction on k, letting $g_k(x) = g_0(x) + pu_1(x) + \cdots + p^k u_k(x)$, etc.]

§5.9 Characteristic Polynomials of Matrices

This section concludes our discussion of polynomials and continues the study of linear algebra from Chapter 4. We develop the properties of the so-called characteristic polynomial of a linear transformation of an n-dimensional vector space V/F to itself, culminating in the celebrated Cayley-Hamilton Theorem.

Although we shall not do so here, a rich interplay can be developed between the factorization into powers of irreducible factors of the characteristic polynomial of a matrix A and the direct sum decomposition of a vector space into certain subspaces invariant under the linear transformation described by A. Here a subspace $W \subseteq V$ is called *invariant* under φ if $\varphi(W) \subseteq W$. (See, for example, F. R. Gantmacher, *The Theory of Matrices*, Vol. I.) Of primary importance for work in §9.4 are the concepts of trace and determinant of a linear transformation φ, both of which can be derived from the characteristic polynomial of any matrix representing φ.

Consider an $n \times n$ matrix $A = [a_{ij}] \in F_{n,n}$ [cf. §4.4]. The **characteristic polynomial** $\chi_A(x) \in F[x]$ of the matrix A is defined to be

$$\chi_A(x) = \det(x \cdot I_n - A),$$

where I_n is the $n \times n$ identity matrix.

NOTE. Here we need to generalize the concept of determinant from §4.5 slightly to include determinants of matrices with coefficients in a ring. All the properties of determinants cited in §4.5, except Property 8 which involves inverses, remain valid if the field of coefficients F is replaced by a commutative ring, such as $F[x]$. [Note that the inverse of an $n \times n$ matrix B (for example, $x \cdot I_n - A$) with polynomial coefficients will in general be a matrix with coefficients in the field $F(x)$ of rational functions in x over F.]

Using the Laplace expansion for determinants (Property 5, §4.5), we express the characteristic polynomial in terms of the powers of x as follows:

$$\chi_A(x) = x^n - \left(\sum_{i=1}^n a_{ii}\right) x^{n-1} + \cdots + (-1)^n \det(A).$$

Example 1. Consider an arbitrary 2×2 matrix $A = [a_{ij}]$. Then,

$$\chi_A(x) = \det(x \cdot I_2 - A)$$

$$= \det \begin{bmatrix} x - a_{11} & -a_{12} \\ -a_{21} & x - a_{22} \end{bmatrix}$$

$$= (x - a_{11})(x - a_{22}) - a_{12} a_{21}$$

$$= x^2 - (a_{11} + a_{22}) x + a_{11} a_{22} - a_{12} a_{21}.$$

Example 2. The characteristic polynomial of $A = \begin{bmatrix} 1 & 4 & 5 \\ 0 & -1 & 0 \\ 4 & 8 & 3 \end{bmatrix}$ is

$$\chi_A(x) = \det \begin{bmatrix} x - 1 & -4 & -5 \\ 0 & x + 1 & 0 \\ -4 & -8 & x - 3 \end{bmatrix}.$$

Expanding by minors of the first column, we obtain

$$\chi_A(x) = (x - 1) \begin{vmatrix} x + 1 & 0 \\ -8 & x - 3 \end{vmatrix} - 4 \begin{vmatrix} -4 & -5 \\ x + 1 & 0 \end{vmatrix}$$

$$= (x - 1)(x + 1)(x - 3) - 20(x + 1)$$

$$= x^3 - 3x^2 - 21x - 17.$$

The sum $\sum_{i=1}^n a_{ii}$ of the *diagonal elements* a_{ii} of the matrix A, which is -1 times the coefficient of the x^{n-1} term of the characteristic polynomial, is called the **trace** $T(A)$ of the matrix A. Proposition 1 states simple properties of the trace.

§5.9 Characteristic Polynomials of Matrices

Proposition 1. For $n \times n$ matrices $A, B \in F_{n,n}$ and $c \in F$, the trace function satisfies the following properties:

(i) $T(A+B) = T(A) + T(B)$.
(ii) $T(cA) = cT(A)$.
(iii) $T(cI_n) = nc$.
(iv) $T(AB) = T(BA)$.

The proofs follow immediately from the definition. Considering F as a one-dimensional vector space over itself, we obtain a second proposition.

Proposition 2. The trace function T is a linear transformation in $\mathrm{Hom}_F(F_{n,n}, F)$.

Suppose now that $P \in F_{n,n}$ is a nonsingular matrix. That is, $\det(P) \neq 0$ and P^{-1} exists. We then have the following proposition.

Proposition 3. For matrices $A, C \in F_{n,n}$ satisfying $C = P^{-1}AP$,
$$\chi_A(x) = \chi_C(x).$$

Proof. By the definition of the characteristic polynomial and the properties of determinants,

$$\begin{aligned}
\chi_C(x) &= \det(x \cdot I_n - C) = \det(x \cdot I_n - P^{-1}AP) \\
&= \det(x \cdot P^{-1}I_n P - P^{-1}AP) = \det(P^{-1}(x \cdot I_n - A)P) \\
&= \det(P^{-1})\det(x \cdot I_n - A)\det(P) = \det(P^{-1})\det(P)\det(x \cdot I_n - A) \\
&= \det(I_n)\det(x \cdot I_n - A) = \det(x \cdot I_n - A) = \chi_A(x).
\end{aligned}$$

Since both the trace and determinant of the matrices A and $C = P^{-1}AP$ occur as coefficients in the characteristic polynomial, Proposition 3 has the following corollary.

Corollary.
$$T(P^{-1}AP) = T(A),$$
$$\det(P^{-1}AP) = \det(A).$$

Now consider a linear transformation φ of an n-dimensional vector space V/F. Consider further two bases $\{A_1, \ldots, A_n\}$ and $\{B_1, \ldots, B_n\}$ of V/F and the corresponding matrix representations M_φ and N_φ of φ, as in §4.4. Proposition 4, §4.4, states that for some nonsingular $n \times n$ matrix P
$$N_\varphi = P^{-1}M_\varphi P.$$
By Proposition 3 the two matrices M_φ and N_φ representing the linear transformation φ have the *same* characteristic polynomials. Thus, to $\varphi \in \mathrm{End}_F(V)$ we can associate the uniquely determined polynomial
$$\chi_\varphi(x) = \chi_A(x) \in F[x],$$

called the **characteristic polynomial** of φ. (It is the characteristic polynomial of any matrix A representing φ.)

Further, the **trace** $T(\varphi)$ of $\varphi \in \text{End}_\varphi(V)$ is defined to be -1 times the coefficient of x^{n-1} of $\chi_\varphi(x)$, and $\det(\varphi)$ to be $(-1)^n$ times the constant term of $\chi_\varphi(x)$, where $n = \dim V$. In other words, $T(\varphi)$ and $\det(\varphi)$ are the trace and determinant, respectively, of some (and hence any) matrix representing φ. Both are uniquely determined by φ.

Proposition 4 is a consequence of the properties of matrix representations of linear transformations (especially Proposition 3, §4.4) and of the trace and determinant of matrices.

Proposition 4. For $c \in F$ and $\varphi, \psi \in \text{End}_F(V)$, where $n = \dim V$,
$$T(\varphi + \psi) = T(\varphi) + T(\psi),$$
$$T(c\varphi) = cT(\varphi),$$
and
$$\det(\varphi \circ \psi) = \det(\varphi)\det(\psi),$$
$$\det(c\varphi) = c^n \det(\varphi).$$

We conclude this section with a proof of the Cayley-Hamilton Theorem. For linear transformations this theorem is expressed as follows.

Theorem. Let $\chi_\varphi(x)$ be the characteristic polynomial of the linear transformation $\varphi \in \text{End}_F(V)$. Then in $\text{End}_F(V)$,
$$\chi_\varphi(\varphi) = \varphi^n - T(\varphi)\varphi^{n-1} + \cdots + (-1)^n \det(\varphi) 1_V = 0,$$
where 1_V is the identity map on V and 0 denotes the zero map.

Since $\chi_\varphi(x)$ is defined by means of any representing matrix $A = M_\varphi$, it suffices to prove the classical statement of the Cayley-Hamilton Theorem: any matrix is a root of its characteristic polynomial.

Cayley-Hamilton Theorem. Let $\chi_A(x)$ be the characteristic polynomial of the $n \times n$ matrix A. Then in $F_{n,n}$
$$\chi_A(A) = A^n - T(A)A^{n-1} + \cdots + (-1)^n \det(A) I_n = 0.$$

Proof. The elements of the matrix $x \cdot I_n - A$ are linear and constant functions of x; thus the elements of $B = \text{Adj}(x \cdot I_n - A)$ are polynomials of degree less than n in x. [See §4.5 for the definition of the adjoint of a matrix.] Therefore B has the expression:

(*) $$B = \sum_{i=0}^{n-1} B_i x^i, \quad \text{where } B_i \in F_{n,n},\ 0 \leq i < n.$$

From equation (*) of §4.5 we have the identity

(**) $$(x \cdot I_n - A) B = \chi_A(x) \cdot I_n.$$

§5.9 Characteristic Polynomials of Matrices

For simplicity let

$$\chi_A(x) = x^n + a_{n-1}x^{n-1} + \cdots + a_0,$$

with coefficients $a_i \in F$, $0 \le i < n$. Then using the expression (∗) for B, we can rewrite equation (∗∗):

$$x\left(\sum_{i=0}^{n-1} B_i x^i\right) - A\left(\sum_{i=0}^{n-1} B_i x^i\right) = -AB_0 + \sum_{i=1}^{n-1}(B_{i-1} - AB_i)x^i + B_{n-1}x^n$$

$$= \sum_{j=0}^{n} a_j x^j I_n,$$

where $a_n = 1$. Comparing coefficients of like powers of x yields

$$-AB_0 = a_0 I_n,$$
$$B_0 - AB_1 = a_1 I_n,$$

(∗∗∗)
$$\cdots \cdots \cdots$$

$$B_{n-2} - AB_{n-1} = a_{n-1} I_n,$$
$$B_{n-1} = I_n.$$

Multiplying the equalities in (∗∗∗) on the left by $I_n, A, \ldots, A^{n-1}, A^n$, respectively, and adding we obtain

$$0 = a_0 I_n + a_1 A + \cdots + a_{n-1} A^{n-1} + A^n = \chi_A(A).$$

This observation completes the proof.

Exercises

1. Prove that a matrix A, whose coefficients are polynomials over a field F, has an inverse in $F[x]$ if and only if $\det(A)$ is a nonzero constant.
2. Determine the characteristic polynomials of the following matrices.

   a. $\begin{bmatrix} 1 & 2 \\ 4 & 0 \end{bmatrix}$
   b. $\begin{bmatrix} 1 & -1 & 0 \\ 1 & 0 & 1 \\ 1 & 1 & 1 \end{bmatrix}$

   c. $\begin{bmatrix} 2 & -1 & 0 \\ 0 & 0 & 2 \\ -2 & 1 & 1 \end{bmatrix}$
   d. $\begin{bmatrix} 4 & 3 & -1 \\ -1 & 2 & 0 \\ 8 & 5 & 0 \end{bmatrix}$.

3. Verify that the matrix in Exercise 2(a) satisfies its characteristic polynomial.
4. Determine the trace and the determinant of each of the matrices in Exercise 2.
5. Verify that the matrix of Example 2 satisfies its characteristic polynomial.

6

Group Theory

A fundamental topic in abstract algebra is that of groups. Our presentation intersperses many examples with the axiomatic discussion. Thus an extensive list of examples follows the definition of a group [§6.1]. The basic properties of groups, subgroups, and factor groups [§§6.1 through 6.4] are presented with illustrations. In §6.5 worked-out examples of homomorphisms of groups show that the concept of homomorphism is a more precise tool in studying groups than is that of normal subgroup. (Distinct homomorphisms may define the same normal subgroup, but distinct normal subgroups must always correspond to distinct homomorphisms.) Nevertheless we first use normal subgroups to define the projection homomorphisms onto the corresponding factor groups. This approach parallels our initial emphasis on extensive study of residue class rings of integers before introducing the concept of homomorphism of commutative rings.

Throughout this chapter our aim is that the reader learn to use explicit computations, no matter how tedious they may appear at first glance, to isolate the essential meaning of the basic definitions and elementary theorems of group theory. Such work is necessary so the reader can progress beyond a superficial absorption of the basic concepts of group theory. Many of the problems can be solved by "the method of ingenuity" as opposed to "the method of infinite drudgery." The latter method need not be feared, as it often will suggest the better approaches and aid understanding.

Special properties of two families of groups—cyclic [§6.6] and permutation [§6.7]—follow an introduction of structure-preserving mappings (homomorphisms) of groups in §6.5; homomorphism properties are examined in detail in §§6.8 and 6.9. The chapter concludes with an optional discussion of homomorphisms of direct products of abelian groups, presented both theoretically and in terms of explicit determination of homomorphisms defined on the direct product of cyclic groups. The direct products of groups [§6.10] are prerequisite for the Fundamental Theorem of Finitely Generated Abelian Groups to be discussed in §7.1, the construction of new groups, and the analysis of given groups. The student has the choice of following this chapter with additional topics in group theory [Chapter 7] or with the study of field theory [Chapter 8], which applies the group theory already studied in developing the Galois theory and the theory for solving polynomial equations by radicals [§9.2].

§6.1 Elements of Group Theory

A **group** G is a nonempty set of elements $G = \{a, b, c, ...\}$ together with a law of composition or binary operation (that is, a single-valued mapping), associating to each ordered pair (a,b) in the cartesian product $G \times G = \{(a,b) : a, b \in G\}$ a unique element $ab = d \in G$, which has the following properties:

Associativity. $a(bc) = (ab)c$, for all $a, b, c \in G$.
Identity (*existence of an identity*). There exists an element $e \in G$ such that $eg = ge = g$, for all $g \in G$.
Inverse. For each $g \in G$, there exists a $g' \in G$, denoted g^{-1}, such that $gg^{-1} = g^{-1}g = e$.

A group G is said to be **commutative** or **abelian** (after the Norwegian mathematician Niels Henrik Abel, 1802–1829) if $ab = ba$ for all $a, b \in G$. This is a special property of some groups, and so must be treated separately.

Comparison of the defining properties of groups with those of rings [§3.1] is invited. By proofs essentially the same as those for integers [§2.1] and rings we obtain the uniqueness of the identity element and of the inverse of each $g \in G$.

Proposition 1. The identity or unit element e in a group G is unique. For each $g \in G$, its inverse g^{-1} is unique.

Proof. As is customary in uniqueness proofs, suppose that both e, e' have the properties of an identity element. Then

$$e = ee', \quad \text{as } g = ge', \quad \text{for all } g;$$
$$= e', \quad \text{as } eg = g, \quad \text{for all } g.$$

Similarly, if for a given g, $gg^* = g^*g = e$ and $gg' = g'g = e$, then
$$g^* = eg^* = (g'g)g^* = g'(gg^*) = g'e = g'.$$

Proposition 2. Equations of the form
$$xa = b, \qquad ay = b$$
for arbitrary $a, b \in G$ have unique solutions $x, y \in G$.

In each case the proof is an immediate consequence of the existence of the (unique) inverse of a.

Corollary. The law of cancellation holds in a group. That is, for elements $a, b, c \in G$,
$$ab = cb \Rightarrow a = c.$$

REMARK. Authors of algebra texts differ in the listing of axiomatic properties for the definition of a group. The most common variation is to require the "solvability" property of Proposition 2 in place of the Identity and Inverse axioms. The two sets of axioms are seen to be equivalent. Proposition 2 shows that Identity and Inverse imply Solvability. Solution of $ax = a$, $ya = a$ yields the identity e, and solution of $ax = e$, $ya = e$ yields the inverse of a.

Systems of axiomatic properties which may not be so obviously equivalent can be given. For example, if G has a left identity ($eg = g$, for all $g \in G$) and each $g \in G$ has a left inverse g' ($g'g = e$), then we can prove $ge = g$, and $gg' = e$. That is, e is the identity element and g' the inverse of g, as given in our axiomatic description of a group. No advantage in group theory, theoretic or practical, accrues to such sets of axiomatic properties.

The *general associative law*, which states that for any $n > 3$ elements $g_1, \ldots, g_n \in G$ all products of the n elements, keeping of course their given order, are equal, is not a group postulate. Rather it is a consequence of the associativity postulate that $g_1(g_2 g_3) = (g_1 g_2)g_3$ for any three elements $g_1, g_2, g_3 \in G$. The emphasis in the general associative law is on the term "any n elements $g_i \in G$." Thus the Principle of Induction in one of its forms is required for the proof.

Define successively $\prod_{i=1}^{n} g_i$ to be $(\prod_{i=1}^{n-1} g_i)g_n$. Thus $\prod_{i=1}^{1} g_i = g_1$, $\prod_{i=1}^{2} g_i = g_1 g_2$, $\prod_{i=1}^{3} g_i = (\prod_{i=1}^{2} g_i)g_3 = (g_1 g_2)g_3$, etc. Now the general associative law will be proved, if we can demonstrate that every combination of $g_1, \ldots, g_n$ put together in product form, keeping fixed the order of the elements g_i, equals $\prod_{i=1}^{n} g_i$.

The proof requires induction on n. For $n = 1, 2$ nothing is to be proved. For $n = 3$, we have simply the given associative law. Assume now that for all products of h, $1 \leq h < n$ elements, the arrangement of parentheses in products is immaterial. This means, to be more explicit in a special case,
$$\{(y_1 y_2)[(y_3 y_4)y_5]\} = [y_1(y_2 y_3)y_4] y_5 = \cdots, \qquad \text{for } h = 5.$$
Thus assume that any ordered product of $h < n$ elements $g_1, \ldots, g_h$ (always in this order) equals $\prod_{i=1}^{h} g_i$.

Consequently an arbitrarily bracketed product of n elements $g_1, \ldots, g_n$ will, by the induction hypothesis, have the form

$$g = (g_1 \cdots g_s)(g_{s+1} \cdots g_n)$$

where $1 \leq s < n$. The induction hypothesis implies that

$$g_{s+1} \cdots g_n = \prod_{i=1}^{n-s} g_{s+i}.$$

Therefore
$$g = (g_1 \cdots g_s) \prod_{i=1}^{n-s} g_{s+i}$$

$$= (g_1 \cdots g_s) \left[\left(\prod_{i=1}^{n-s-1} g_{s+i} \right) g_n \right] \quad \text{by the definition of } \Pi$$

$$= \left[(g_1 \cdots g_s) \prod_{i=1}^{n-s-1} g_{s+i} \right] g_n.$$

Using the induction hypothesis again, we have

$$(g_1 \cdots g_s) \prod_{i=1}^{n-s-1} g_{s+i} = \prod_{i=1}^{n-1} g_i,$$

so that
$$g = \left(\prod_{i=1}^{n-1} g_i \right) g_n = \prod_{i=1}^{n} g_i.$$

Numerous examples of different kinds of groups conclude this section. Some are geometrically derived and others are abstractly presented. On a first reading, the student should not attempt to master all of these examples. While studying later sections on group theory, the reader is urged to return to this section to review appropriate examples and to work out the details of group-theoretic arguments in terms of explicit computations with elements of particular groups.

Examples of Groups
1. a. The elements of any ring R with respect to the additive law of composition [cf. §3.1].
    b. The elements of the particular rings $\mathbf{Z}, \mathbf{Q}, \mathbf{R}, \mathbf{C}$, and $\mathbf{Z}_m$ with addition as the law of composition.
2. a. The units of any ring R with respect to the multiplicative law of composition.
    b. The nonzero elements of $\mathbf{Q}, \mathbf{R}$, and $\mathbf{C}$, denoted $\mathbf{Q}^*, \mathbf{R}^*$, and $\mathbf{C}^*$, respectively, with multiplication as the law of composition.
    c. The set of prime residue classes $[a]$ of $\mathbf{Z}_m$ with multiplication as the law of composition. In particular, the $p-1$ nonzero cosets of $\mathbf{Z}_p$, p a prime number.

3. **a.** The set T of all complex numbers ζ with absolute value $|\zeta| = 1$, with multiplication as the law of composition. Such complex numbers can be expressed in the form $\zeta = \cos\varphi + i\sin\varphi$, $i^2 = -1$.

   b. The set of all complex numbers of the form

   $$\cos\left(\frac{2\pi}{m}r\right) + i\sin\left(\frac{2\pi}{m}r\right), \qquad 0 \leq r < m,$$

   for a given nonzero integer m with multiplication as the law of composition.

4. **a.** The set of all nonsingular $n \times n$ matrices with coefficients in **Q**, **R**, **C**, or $\mathbf{Z}_p$ (p a prime number) with matrix multiplication for the law of composition.

   b. The set of all $n \times n$ matrices with determinant ± 1, with coefficients and the law of composition as given in part (a).

   c. The set of all 2×2 matrices with determinant 1, integral coefficients, and matrix multiplication as the law of composition.

   For the particular matrices

   $$S = \begin{bmatrix} 1 & 1 \\ 0 & 1 \end{bmatrix}, \quad T = \begin{bmatrix} 0 & 1 \\ -1 & 0 \end{bmatrix}, \quad W = TS = \begin{bmatrix} 0 & 1 \\ -1 & -1 \end{bmatrix},$$

   note that $T^4 = W^3 = \begin{bmatrix} 1 & 0 \\ 0 & 1 \end{bmatrix}$, and that $S^n = \begin{bmatrix} 1 & n \\ 0 & 1 \end{bmatrix}$,

   for every $n \in \mathbf{Z}$. Thus the element $S = T^{-1}W$ has infinite order (defined in §6.2) even though it is in the product of elements T^{-1} and W of finite order. Furthermore, $(TS)^3 = I$.

   d. The set of matrices

   $$\begin{bmatrix} a & b \\ c & d \end{bmatrix}$$

   of the group in part (c) above for which $a \equiv d \equiv \pm 1 \pmod{n}$ and $b \equiv c \equiv 0 \pmod{n}$, $n > 1$.

   e. The set of all 3×3 matrices

   $$\begin{bmatrix} 1 & u & v \\ 0 & 1 & w \\ 0 & 0 & 1 \end{bmatrix}$$

   with coefficients in **Z**, $\mathbf{Z}_m$, **Q**, **R**, or **C**, with matrix multiplication as the law of composition.

5. The cartesian product $R \times R'$ of two rings R, R' with the rule of addition defined by

   $$(a, a') + (b, b') = (a + b, a' + b') \qquad \text{for all } a, b \in R \text{ and } a', b' \in R'.$$

   a. In particular, we may choose **Z**, $\mathbf{Z}_m$, **Q**, **R**, and **C** as the rings R and R'.

   b. The cartesian product $\mathbf{Z}_m \times \mathbf{Z}_n$, with

   $$([a]_m, [b]_n) + ([c]_m, [d]_n) = ([a+c]_m, [b+d]_n).$$

   Note that this group has mn elements.

c. The set of pairs (α, β), where α belongs to $\mathbf{U}_m$, the multiplicative group of prime residues modulo m, and β belongs to $\mathbf{Z}_m$ considered as an additive group, with the law of composition defined by

$$(\alpha, \beta) * (\gamma, \delta) = (\alpha\gamma, \alpha\delta + \beta).$$

6. We now introduce "abstract" groups defined formally in terms of *generators* and *relations*.

  a. Let G be the set $\{a^m : m \in \mathbf{Z}\}$ of *all* integral powers of the symbol a, which we call the (multiplicative) group *generated* by a. The element a is then called a *generator* of G. The group operation is denoted by multiplication with the customary *law of exponents* $a^r a^s = a^{r+s}$, etc., and a^0 is the multiplicative identity.

  b. Let G be the set $\{a^m : m \in \mathbf{Z}\}$ of all integral powers of the symbol a, but now define a^i, a^j to represent the same element of the group if $i \equiv j \pmod{n}$ for some fixed $n \in \mathbf{N}$. Thus, the *distinct* elements of G are $\{a^0, a^1, \ldots, a^{n-1}\}$; any other power of a represents the same element in G as one of these.

The groups in Examples 6(a) and (b) are called **cyclic groups**. They are introduced again in §6.5 and studied in detail in §6.6. They are generated by a single element a.

In Examples 6(c) through 6(i), we consider "abstract" groups with two generators a, b. Let S be the set of *all* finite products of the symbols a, b, a^{-1}, b^{-1}, and e, where $a^0 = b^0 = e$, the multiplicative identity, and $a^r a^s = a^{r+s}$, etc. The identifications of symbols in the set S, given in Examples 6(d) through (i), are sometimes referred to as *defining relations*.

  c. The set S is a group with an infinite number of elements.

  d. From the set of symbols S we can obtain a group by making the following definitions:

   (i) ab and ba are equal (that is, they represent the same element of the group).
   (ii) For any fixed positive prime p and positive integer n, $a^i b^j$ and $a^k b^m$ are the same element of the group if $i \equiv k \pmod{p^{n-1}}$ *and* $j \equiv m \pmod{p}$.

Thus, we equate $a^{p^{n-1}}$ and b^p with the identity element $e = a^0 = b^0$. For example, when $n = p = 3$, the distinct elements of the group are

$$e, a, a^2, a^3, a^4, a^5, a^6, a^7, a^8,$$
$$b, ab, a^2 b, a^3 b, a^4 b, a^5 b, a^6 b, a^7 b, a^8 b,$$
$$b^2, ab^2, a^2 b^2, a^3 b^2, a^4 b^2, a^5 b^2, a^6 b^2, a^7 b^2, a^8 b^2.$$

Any product of powers of a, b can be reduced to one of these 27 products by the identifications cited above.

  e. From the set of symbols S, we can obtain a group by making the following definitions, for any fixed integer $n \geq 3$.

   (i) ba and $a^{-1}b$ are the same group element.
   (ii) a^i and a^j are the same group element if $i \equiv j \pmod{2^{n-1}}$.
   (iii) b^2 and $a^{2^{n-2}}$ are the same group element.

When $n = 3$, this group is called the **quaternion group**; its distinct elements are
$$a, a^2 = b^2, a^3, a^4 = e; \qquad b, ba, ba^2, ba^3.$$

For example, $ab = a^2ba = b^2ba = bb^2a = ba^3$. The *multiplication table* (that is, the array of all products of pairs of distinct elements in a group) for the quaternion group follows.

	e	a	a^2	a^3	b	ba	ba^2	ba^3
e	e	a	a^2	a^3	b	ba	ba^2	ba^3
a	a	a^2	a^3	e	ba^3	b	ba	ba^2
a^2	a^2	a^3	e	a	ba^2	ba^3	b	ba
a^3	a^3	e	a	a^2	ba	ba^2	ba^3	b
b	b	ba	ba^2	ba^3	a^2	a^3	e	a
ba	ba	ba^2	ba^3	b	a	a^2	a^3	e
ba^2	ba^2	ba^3	b	ba	e	a	a^2	a^3
ba^3	ba^3	b	ba	ba^2	a^3	e	a	a^2

When $n > 3$, such groups are called **generalized quaternion groups.**

f. From the set of symbols S, we can obtain a group by making the following definitions for any fixed integer $n \geq 3$.

(i) ba and $a^{-1}b$ are the same group element.
(ii) a^i and a^j are the same group element if $i \equiv j \pmod{n}$.
(iii) b^k and b^m are the same group element if $k \equiv m \pmod{2}$.

Such groups, denoted Γ_n, are called **dihedral**; a geometric description of them is given in Example 7 below. The group Γ_n is noncommutative, and has $2n$ (distinct) elements.

g. From the set of symbols S, we can obtain a group by making the following definitions, for any fixed integer $n \geq 4$.

(i) ba and $a^{1+2^{n-2}}b$ are the same group element.
(ii) a^i and a^j are the same group element if $i \equiv j \pmod{2^{n-1}}$.
(iii) b^k and b^m are the same group element if $k \equiv m \pmod{2}$.

h. From the set of symbols S, we can obtain a group by making the following definitions, for any fixed integer $n \geq 4$.

(i) ba and $a^{2^{n-2}-1}b$ are the same group element.
(ii) a^i and a^j are the same group element if $i \equiv j \pmod{2^{n-1}}$.
(iii) b^k and b^m are the same group element if $k \equiv m \pmod{2}$.

These groups are sometimes called **semidihedral.**

i. From the set of symbols S, we can obtain a group by making the following definitions, for any odd prime p.

(i) $a^i b^j$ and $a^k b^m$ are the same group element if $i \equiv k \pmod{p^2}$ and $j \equiv m \pmod{p}$.

§6.1 Elements of Group Theory 169

(ii) ba and $a^{1+p}b$ are the same group element.

See Example 2, §6.4, and Example 6, §6.9, for further discussion of this noncommutative group.

j. In a fashion analogous to the definition of the set of symbols S and Example 6(c), we can define "abstract" groups with any finite number of generators.

7. The following examples of groups are geometric descriptions of the dihedral groups in Example 6(f). Here let a be rotation of a regular n-gon (an n-sided polygon, all sides nonintersecting, of equal length, and inscribed in a circle) through $2\pi/n$ radians, $n \geq 3$, and let b be reflection about an arbitrary fixed axis of symmetry.

   a. The set Γ_3 of maps (rotations and reflections) of an equilateral triangle to itself is a group, with 6 elements, with composition of mappings for its group operation.

   b. The set Γ_4 of maps (rotations and reflections) of a square to itself is called the **group of the square**. Composition is the group operation; Γ_4 has 8 elements.

   c. More generally, the rotations and reflections of a regular n-gon constitute a group Γ_n of $2n$ elements.

8. For a prime p the set of all residue classes of $\mathbf{Z}_q$, $q = p^n$, whose representatives a satisfy $a \equiv 1 \pmod{p}$ where the law of composition is multiplication in $\mathbf{Z}_q$.

9. The vectors in any vector space with the law of vector addition [cf. §4.1].

10. The set L of all linear transformations of a finite dimensional vector space V to itself with zero kernel under composition of transformations [cf. §4.3].

There are several origins of the study of groups. Lagrange in the eighteenth century was concerned with the behavior of permutations of roots of polynomial equations. Hermann Weyl in the conclusion of his treatise *Symmetry* draws the lesson that "Whenever you have to do with a structure-endowed entity Σ try to determine its group of automorphisms, the group of those element-wise transformations which leave all structural relations undisturbed."[†] Geometrically-minded persons, following Leibniz, would speak of the group of symmetries, where Weyl uses the term "automorphisms." Newton and Hermann von Helmhlotz (1821–1894) preferred describing the structure of a space by the notion of congruence, wherein two parts of a space are defined to be congruent if they can be occupied by the same rigid body in two of its positions. (See H. Weyl, *Mathematische Analyse des Raumproblems*.) Such concepts underlie the mathematical analysis of symmetry in ornamental art (highly developed by the Arabians, Egyptians, and ancient Greeks) and in nature in crystallography, cellular arrangements, configurations of florets in flowers, and the like. "Symmetry is a vast subject, significant in art and nature. Mathematics lies at its root, and it would be

[†] Herman Weyl, *Symmetry*, p. 144. © 1952 by Princeton University Press, Princeton, N.J.

hard to find a better one on which to demonstrate the working of the mathematical intellect."†

Example 7 involves groups of symmetries of plane figures. In an analogous fashion we can consider groups of the five regular (Platonic) solids in 3-space.

Exercises

1. Let $g_1, \ldots, g_n$, n an arbitrary positive integer, be elements of the group G. Prove that
$$(g_1 \cdots g_n)^{-1} = g_n^{-1} \cdots g_1^{-1}.$$
(Make use of the Principle of Induction.)

2. In the definition of a group G replace the postulates about the identity e and the existence of an inverse of an element a by the following:
    (i) There exists an element $e \in G$ such that $eg = g$ for all $g \in G$.
    (ii) For every $g \in G$ there exists an element $g^* \in G$ such that $g^*g = e$.
   Prove that such a system G is a group in the sense of the definition in the text.

3. Suppose that G is a given nonempty set of elements on which an associative product is defined. Assume that all equations $xa = b$ and $ay = b$, a and b in G, have solutions x and y in G. Prove that G is a group.

4. Define the powers g^n of a group element $g \in G$ inductively: $g^0 = e$, the identity element of G, $g^2 = g \cdot g$, $g^3 = g \cdot g^2, \ldots$. Let g^{-1} denote the inverse of g, and define g^{-n}, $n \in \mathbf{N}$, to be $(g^{-1})^n$. Prove the *laws of exponents*:
$$g^u g^v = g^{u+v} \quad \text{and} \quad (g^r)^s = g^{rs}$$
for all $u, v, r, s \in \mathbf{Z}$.

5. Let G be the set of all real numbers $a, b, \ldots$ whose absolute values are less than 1. Define a law of composition $*$ by the formula
$$a * b = (a+b)(1+ab)^{-1}$$
where $a+b$ and ab denote the customary sum and product of the real numbers a and b.
    a. Prove that G, with $*$ for the law of composition, is a commutative group.
    b. What is the identity element of G and the inverse of a?

6. Let G be the set of functions $t, 1-t, t^{-1}, (1-t)^{-1}, (t-1)t^{-1}, t(t-1)^{-1}$ defined on $\mathbf{R}$.
    a. Prove that G is a group with respect to the usual composition of functions. That is, for $f, g \in G$, $(f \circ g)(t) = f(g(t))$.
    b. Which function is the identity element?
    c. Find the inverse of each of the six functions.

† *Ibid.*, p. 145.

7. a. Enumerate the elements of the group $G = \{a^m b^n : m, n \in \mathbf{Z}\}$ of all formal products of symbols a, b subject to the identifications $a^4 = b^2 = e$, $bab^{-1} = a^3$ [cf. Example 6(f)].
   b. Construct the multiplication table of this group.
   c. Identify this group with the group of the square Γ_4 in Example 7(b).
8. What can be said about the group in Example 6(h) when $n = 3$?
9. Verify that a vector space V is an abelian group with vector addition for the law of composition.
10. Let $Af(\mathbf{Z}_m)$ be the set of transformations

$$u \to \alpha u + \beta$$

where u is an indeterminate over $\mathbf{Z}_m$, α is a unit in $\mathbf{Z}_m$, and $\beta \in \mathbf{Z}_m$. Prove that $Af(\mathbf{Z}_m)$, called the **affine group** over $\mathbf{Z}_m$, is a nonabelian group with the usual composition of mappings for its multiplicative law of operation.

11. Consider the set of all formal products of the symbols $a, b, c, a^{-1}, b^{-1}, c^{-1}$, and e, where we identify a^2 with aa, ea and ae with $a, a^{-1}a$ and aa^{-1} with e, etc. For an odd positive prime p, assume that

$$a^p = b^p = c^p = e, \qquad ac = ca, \qquad bc = cb, \qquad c = aba^{-1}b^{-1}.$$

Prove that the resulting set is a group with p^3 elements.

§6.2 Subgroups and Orders of Elements

We now discuss some elementary concepts of group theory and their properties. In terms of the symmetry of geometrical objects, two concepts which arise immediately and are abstracted below are subgroups of symmetries, leaving fixed some parts of the object, and the order (sometimes called period) of a geometric transformation.

A **subgroup** H of G, denoted by $H \subseteq G$, is a nonempty subset of G such that for elements x, y in H both xy and x^{-1} belong to H. Equivalently, H is a subgroup if $xy^{-1} \in H$ for all $x, y \in H$. In other words, a subgroup is a subset which is itself a group *with respect to the same law of group composition*. A subgroup H of G is termed **nontrivial** if $\{e\} \subsetneq H \subsetneq G$. This definition is similar to that of a nontrivial subring in §3.1.

While we use the inclusion symbol $\subseteq$ for both subgroups and subsets, the meaning should be clear from the context.

The **right coset** of a with respect to H is the subset of G

$$Ha = \{xa : x \in H\}.$$

The **left coset** of a with respect to H is the subset of G

$$aH = \{ax : x \in H\}.$$

In particular, the left coset hH of $h \in H$ equals H; similarly, $Hh = H$. In fact, $aH = H \Leftrightarrow a \in H$.

As we customarily write groups multiplicatively (that is, use multiplication (or juxtaposition) to designate the group law of composition), we denote the cosets multiplicatively: aH and Ha. For additively-written groups we usually denote the cosets $a+H$ and $H+a$. An element a' of a coset aH (or $a+H$) is called a **representative** of the coset aH. In this case $a' = ah$ for some $h \in H$. Consequently,

$$aH = a'h^{-1}H = a'(h^{-1}H) = a'H,$$

since H is a subgroup. Similarly $Ha = Ha'$ if $a' = h'a$ for some $h' \in H$.

The set of right cosets of a given subgroup $H \subseteq G$ constitutes a partition of G [see §1.2]. This means in particular that two cosets Ha, Hb are either equal or disjoint. To prove it anew, note that $g \in Ha \cap Hb$ implies

$$g = ha = h'b, \quad \text{for some } h, h' \in H.$$

Consequently, $a = h^{-1}h'b \in Hb$ and hence $Ha \subseteq Hb$. Similarly $b = (h')^{-1}ha \in Ha$ and hence $Hb \subseteq Ha$. Therefore we can define an equivalence or congruence relation as follows:

$$\begin{aligned} a \equiv b \pmod{H} &\Leftrightarrow ab^{-1} \in H \\ &\Leftrightarrow Ha = Hb \\ &\Leftrightarrow a \in Hb. \end{aligned}$$

There is a one-one correspondence between the elements $h \in H$ and the elements of the coset Ha for each $a \in G$. Since cancellation holds in a group,

$$h_i a = h_j a \Leftrightarrow h_i = h_j.$$

Similarly, the set of left cosets of H constitutes a partition of G (although not in general the same as the partition by right cosets). There is again a one-one correspondence between the elements of H and of aH.

As an explicit illustration of these concepts, consider Examples 6(f) (for $n = 3$) and 7(a), the **group of the triangle** Γ_3, in §6.1. Describe the elements of Γ_3 as

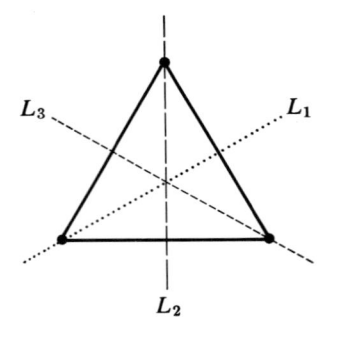

$\varepsilon = $ the identity map,
$r = $ clockwise rotation through 120°,
$r^2 = $ clockwise rotation through 240°,
$\alpha = $ reflection in the axis L_1,
$\beta = $ reflection in the axis L_2,
$\gamma = $ reflection in the axis L_3.

The multiplication table for Γ_3 follows.

§6.2 Subgroups and Orders of Elements

ε	r	r^2	α	β	γ
r	r^2	ε	γ	α	β
r^2	ε	r	β	γ	α
α	β	γ	ε	r	r^2
β	γ	α	r^2	ε	r
γ	α	β	r	r^2	ε

Check that the only subgroups of Γ_3 are $R = \{\varepsilon, r, r^2\}$, $S = \{\varepsilon, \alpha\}$, $\{\varepsilon, \beta\}$, and $\{\varepsilon, \gamma\}$. Interpreted geometrically, R is the subgroup of symmetries of the triangle which preserve the cyclical order of the vertices, whereas S and the other two subgroups leave fixed one vertex each.

The left cosets of R are

$$R = \varepsilon R = \{\varepsilon, r, r^2\}, \qquad \alpha R = \{\alpha, \beta, \gamma\}.$$

Note that $\varepsilon R = R\varepsilon$ and $\alpha R = R\alpha$. The left cosets of S are

$$S = \varepsilon S = \{\varepsilon, \alpha\}, \qquad rS = \{r, \gamma\}, \qquad r^2 S = \{r^2, \beta\}.$$

The right cosets of S are

$$S = S\varepsilon = \{\varepsilon, \alpha\}, \qquad Sr = \{r, \beta\}, \qquad Sr^2 = \{r^2, \gamma\}.$$

Schematically, we represent the subgroups of Γ_3 in a so-called *lattice diagram* ordered by inclusion.

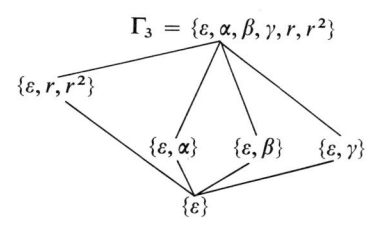

$$\Gamma_3 = \{\varepsilon, \alpha, \beta, \gamma, r, r^2\}$$

$\{\varepsilon, r, r^2\}$

$\{\varepsilon, \alpha\} \quad \{\varepsilon, \beta\} \quad \{\varepsilon, \gamma\}$

$\{\varepsilon\}$

In a group G the **order** of $g \in G$, denoted $o(g)$, is defined to be

(i) the least element in the set $\{n \in \mathbf{N} : g^n = e\}$;
(ii) ∞, if the set in (i) is empty.

In the first case, g is said to be of **finite order**, and in the second, of **infinite order**. The order of an element $g \in G$ has several general properties.

Property 1. $o(e) = 1$.

Property 2. $g \neq e \in G \Rightarrow o(g) > 1$.

Property 3. $o(g) = o(aga^{-1})$ for all $a \in G$.

Property 4. $o(g) = o(g^{-1})$.

Property 5. $o(g) = h, g^n = e \Rightarrow h | n$.

The proof of Property 5 follows from the Division Algorithm [§2.3]. Writing $n = qh + r$, where $0 \leq r < h$, we have

$$e = g^n = g^{qh+r} = (g^h)^q g^r = e^q g^r = g^r.$$

Since by definition h is the least positive integer k such that $g^k = e$, $r < h$ is necessarily zero. But that is to say $h \mid n$.

Proposition 1. For any element g of finite order in G, and any integer s,

$$o(g^s) \mid o(g).$$

Proof. Consider g of finite order n. For any integer s,

$$(g^s)^n = g^{sn} = (g^n)^s = e^s = e.$$

By Property 5 we conclude that $o(g^s) \mid n = o(g)$.

More particularly, we have the following proposition.

Proposition 2. For $g \in G$ of finite order and any integer s, $0 < s < o(g)$,

$$o(g^s) = \frac{o(g)}{(s, o(g))}.$$

Proof. Let $k = o(g^s)$, $n = o(g)$, and $d = (s, n)$. Then write $n = md$ and $s = td$. We wish to prove that $k = m$, or in other notation, $n = kd = o(g^s)(s, o(g))$. This we do by showing that $k \mid m$ and $m \mid k$.

First, $(g^s)^m = g^{tdm} = (g^n)^t = e$; so $k \mid m$. Second $e = (g^s)^k = g^{sk}$ implies that $n \mid sk$. Now write $sk = vn$. Substituting $s = td$, $n = md$, we obtain $tdk = vmd$ or $tk = vm$. Since $(m, t) = 1$ by the definition of d as the GCD (s, n), we conclude that $m \mid k$.

A corollary of Proposition 2 is that

$$o(g^s) = o(g) \Leftrightarrow (s, o(g)) = 1.$$

Proposition 3. For $g \in G$ of order mn, where $(m, n) = 1$, there exist unique elements $a, b \in G$ such that $g = ab = ba$ and $o(a) = m$, $o(b) = n$.

We leave the proof as an exercise, with appropriate hints. It is also a consequence of Proposition 2, §6.6, or the Fundamental Theorem of Finitely Generated Abelian Groups, §7.1.

REMARK. Suppose that the elements a, b of a group G commute and that they have relatively prime orders m, n. Then $o(ab) = mn$.

This remark is in general not true for noncommuting elements. For example, the group of the triangle Γ_3 is generated by the elements r and α of orders 3 and 2, respectively. However, $r\alpha = \gamma$, which is of order 2. The product $(r\alpha)\alpha = r\varepsilon = r$ of elements $r\alpha$ and α, each of order 2, has order 3.

More generally, the dihedral group Γ_n with $2n$ elements (given in Example

6(f), §6.1) is nonabelian for $n \geq 3$. The element ab satisfies

$$(ab)^2 = abab = abab^{-1} = aa^{-1} = e;$$

it is of order 2 and is the product of elements of orders n and 2. The product of ab and b, each of order 2, has order n.

The set-theoretical intersection $\bigcap_{\lambda \in \Lambda} H_\lambda = H$ of a collection of subgroups of a group G, indexed by $\lambda \in \Lambda$ for some set Λ, is called the **intersection** of the subgroups H_λ.

The intersection H is a subgroup of G, for if $x, y \in H$, then $x, y \in H_\lambda$ and hence $xy \in H_\lambda$ for all $\lambda \in \Lambda$. Thus $xy \in H$. Furthermore $x^{-1} \in H_\lambda$ for all λ, and so $x^{-1} \in H$.

Now let S be an arbitrary nonempty collection of elements in a group G, and T be the collection of all finite products $d_1 \cdots d_k$, where either $d_j \in S$ or $d_j^{-1} \in S$, $1 \leq j \leq k$. Clearly $(d_1 \cdots d_k)(d_{k+1} \cdots d_m) = d_1 \cdots d_m \in T$; also $(d_1 \cdots d_k)^{-1} = d_k^{-1} \cdots d_1^{-1} \in T$, and $e = dd^{-1} \in T$ for $d \in S$. Hence $S \subseteq T \subseteq G$, and T is a subgroup of G. The group T is called the **subgroup** of G **generated by the set** $S \subseteq G$. (This is the group-theoretic analogue of the vector space concept of span.)

Proposition 4. *Let H denote the intersection $\bigcap_\mu H_\mu$ of all subgroups $H_\mu \subseteq G$ which contain a subset $S \subseteq G$, and T the group generated by S. Then $H = T$.*

Proof. First, $S \subseteq T \subseteq G$, and hence T is one of the groups H_μ. Therefore, on the one hand, $\bigcap_\mu H_\mu \subseteq T$. On the other hand, for each μ, $S \subseteq H_\mu$ implies that H_μ contains all finite products $d_1 \cdots d_k$, where either $d_i \in S$ or $d_i^{-1} \in S$. Hence $T \subseteq H_\mu$ for all μ, so $T \subseteq \bigcap_\mu H_\mu = H$. Thus $T = H$ as asserted.

The **product** $H \cdot K$ of two subgroups H and K of a group G is the subgroup of G generated by the set of elements $H \cup K$. Thus $H \cdot K$ consists of all finite products $h_1 k_1 \cdots h_i k_i \cdots h_\nu k_\nu$ of elements $h_i \in H$ and $k_i \in K$, $1 \leq i \leq \nu$, for all $\nu \in \mathbf{N}$.

For the sake of emphasis, we state the following proposition.

Proposition 5. *The product $H \cdot K$ of two subgroups H, K of a group G is the intersection of all subgroups L of G such that $L \supseteq H$ and $L \supseteq K$.*

Corollary. *The product $H \cdot K$ of two subgroups of a group G is the least subgroup of G containing both H and K.*

"Least" is used here in the sense that *no proper* subgroup of $H \cdot K$ contains both H and K.

The definition of the product of two subgroups of a given group G extends to the product of any finite number of subgroups $H_1, \ldots, H_s$ with $H_1 \cdot H_2 \cdot \cdots \cdot H_s$ being the least subgroup of G containing all groups H_i, $i = 1, \ldots, s$. The product subgroup $H \cdot K$ must not be confused with the *complex* (set) *of products* $HK = \{hk : h \in H, k \in K\}$. For example, in the dihedral group Γ_n of $2n$ elements with generators a, b, for which $a^n = b^2 = e$, for subgroups $H = \{e, ab\}$ and $K = \{e, b\}$, then

$$H \cdot K = \Gamma_n,$$

$$HK = \{e, ab, b, abb = a\}.$$

The latter is not a subgroup of Γ_n.

From the definition it is evident that the product $K \cdot H$ denotes the same subgroup as $H \cdot K$.

Exercises

1. a. Prove the first assertion in the remark.
   b. Let a and b be elements of a group G with respective orders u and v. Prove that the order of ab divides the least common multiple $[u, v]$ if $ab = ba \neq e$. Note that the example of Γ_3 in the remark provides a counterexample when $ab \neq ba$.
   c. With reference to part (b), provide an example to show that the order of ab need not equal $[u, v]$.
2. Prove that if every element different from the identity has order 2 in a group G, then G is abelian.
3. Find all subgroups of the additive group of integers $\mathbf{Z}$.
4. Consider a subgroup H of a group G.
   a. Prove that $aH = H = Ha$ if and only if $a \in H$.
   b. Prove that there is a one-one correspondence between elements of H and those of the coset aH for any given $a \in G$.
5. a. Prove that the relation $\equiv$, defined as $a \equiv b \pmod{H}$ if and only if $ab^{-1} \in H$ for a subgroup H of the group G, is an equivalence relation in the sense of §1.2. Do the same for the relation $a \sim b \pmod{H}$ if and only if $b^{-1}a \in H$.
   b. Show that in the case of the group Γ_3 the two relations in part (a) are different (i.e., define different equivalence classes) for an appropriately chosen subgroup H.
6. Let H be a subgroup of a group G. For each $g \in G$, prove that

$$gHg^{-1} = \{ghg^{-1} : h \in H\}$$

is a subgroup of G. It is called a **conjugate subgroup** of H in G.

7. Given an abelian group G, prove that $G_t = \{g \in G : o(g) < \infty\}$ is a subgroup called the **torsion subgroup**.
8. Given a group G with subgroup H, show that the set $N_G(H) = \{g \in G : gH = Hg\}$, called the **normalizer** of H in G, is a subgroup of G.
9. With reference to Exercise 8, prove more generally that the normalizer $N_G(S) = \{g \in G : gS = Sg\}$ of any set S in a group G is a subgroup of G.

10. Let $C = \{a \in G : ag = ga,\text{ for all } g \in G\}$; it is called the **center** of G.
    a. Verify that C is a subgroup of G.
    b. What is the normalizer $N_G(C)$ of C in G?
11. a. Prove Property 3.   b. Prove Property 4.
12. Prove Proposition 5.
13. Consider finite subgroups S and T of a group G. Let ST denote the **complex** $\{st : s \in S, t \in T\}$. Prove that

$$\text{Card}(ST) = \text{Card}(S)\,\text{Card}(T)/\text{Card}(S \cap T),$$

where the **cardinality** $\text{Card}(U)$ denotes the number of distinct elements in a finite subset U of G.

14. Verify that the intersection of two subgroups of a group G is again a subgroup of G.
15. a. Prove that if a finite subset S of a group G is closed with respect to the group operation (i.e., if $a, b \in S \Rightarrow ab \in S$) and contains the identity element of G, then it is a subgroup.
    b. Give a counterexample in the case of an infinite subset.
16. Prove Proposition 3.
    a. Show that $o(g^n) = m$, $o(g^m) = n$.
    b. Let $\alpha = g^n$, $\beta = g^m$, and write $1 = vn + um$. Show that $o(\alpha^v) = o(\alpha) = m$ and $o(\beta^u) = o(\beta) = n$.
    c. Set $a = \alpha^v$, $b = \beta^u$, and prove that a, b are unique. *Hint:* Suppose a, b are not unique and raise $ab = a'b'$ to the (mu)th power. We must show that $b^{mu} = (b')^{mu}$ implies $b = b'$, but this can be accomplished by noting that $mu = 1 - vn$.
17. Generalize Proposition 3 to the case where $o(g) = m_1 \cdots m_s$, the product of s pairwise relatively prime factors.

§6.3 Coset Decompositions and the Theorem of Lagrange

We return now to the cosets of the previous section, using them as a means to enumerate the elements in a finite group.

The **order** of a group G, denoted by either $o(G)$ or $[G : e]$, is defined to be

(i) n, if G has n elements,
(ii) ∞, if G has an infinite number of elements.

The example of the group of the triangle Γ_3 shows that the order of a nonabelian group generated by two or more elements of finite order may contain factors relatively prime to the orders of the generators. In Γ_3 the elements α and $r\alpha$ of order 2 are generators, but $3 \mid [\Gamma_3 : e] = 6$.

Let G be a group (either finite or infinite) and H a subgroup (also finite or infinite). We can express

$$G = \bigcup_{g \in G} gH$$

or more particularly,
$$G = \bigcup_{i \in I} g_i H,$$
where the $g_i H$, $i \in I$, are disjoint left cosets. Similarly, we can write
$$G = \bigcup_{j \in J} H g_j,$$
where the Hg_j, $j \in J$, are disjoint right cosets. Such a representation of a group G, called decomposition into left (right) cosets modulo H, is utilized in proving the well-known theorem discovered by Lagrange in his study of the behavior of roots of polynomial equations.

Theorem of Lagrange. If G is a finite group with $o(G) = [G:e]$ elements, then for any subgroup H, $o(H) | o(G)$.

Proof. In §6.2 we noted that the elements of H are in one-one correspondence with the elements of any coset Ha, and that $G = Ha_1 \cup \cdots \cup Ha_r$, where $Ha_i \cap Ha_j = \emptyset$ for $i \neq j$. Consequently $o(G) = o(H) + \cdots + o(H) = ro(H)$, which completes the proof.

Let $r = [G:H]_r$ denote the number of distinct right cosets of H in G. If $G = b_1 H \cup \cdots \cup b_\ell H$ (decomposition of G into left cosets modulo H, $b_i H \cap b_j H = \emptyset$ for $i \neq j$), then $o(G) = o(H) + \cdots + o(H) = \ell o(H)$, where $\ell = [G:H]_\ell$ is the number of distinct left cosets of H in G. Thus $r = \ell$.

The common number $r = \ell$ of right and left cosets is called the **index** of H in G, denoted by $[G:H]$.

The Theorem of Lagrange has the equivalent statement that if H is a subgroup of a finite group G, then
$$[G:e] = [G:H][H:e].$$

Lagrange gave this theorem in 1770–71, but the first complete proof is due to Pietro Abbati (1768–1842) some thirty years later. Note, though, that the term "group" was not introduced (by Galois) until 1830.

Corollary 1. The order $o(g)$ of an element g in the finite group G divides the order of G.

For the proof note that the subgroup $H = \{g^i : i \in \mathbf{Z}\}$, generated by g, contains $o(g)$ distinct elements.

Corollary 2. If H, K are subgroups of a finite group G and $K \subseteq H$, then $[G:K] = [G:H][H:K]$.

Proof. Write, using the Theorem of Lagrange three times,
$$[G:K][K:e] = [G:e] = [G:H][H:e]$$
$$= [G:H][H:K][K:e].$$
Canceling $[K:e]$, we have
$$[G:K] = [G:H][H:K].$$

Exercises

1. Prove that any finite group G of even order contains an element of order 2. (This result is generalized in §7.5 to the statement that a prime p divides $o(G)$ if and only if G contains an element of order p.)
2. a. If the set S of Exercise 9, §6.2, consists of the single element $\{g_0\}$, prove that the index of the subgroup $N_G(\{g_0\})$ in a group G equals the number of elements g in G conjugate to g_0.

An element g is said to be **conjugate** to g_0 if there exists an element $a \in G$ such that $g = ag_0 a^{-1}$. An element g_0 is said to be **self-conjugate** if every conjugate $ag_0 a^{-1}$ is equal to g_0 itself. Thus, the center C of G is simply the set of self-conjugate elements [cf. Exercise 10, §6.2].

    b. Conclude that the number of conjugates of an element g_0 in a finite group G divides $o(G)$.

3. Let $a, b, \ldots$ be elements of a group G. Define $[a, b]$ to be the **commutator** $aba^{-1}b^{-1}$. Prove that $[ab, c] = \{a[b, c]a^{-1}\}[a, c]$ and $[a, bc] = [a, b]\{b[a, c]b^{-1}\}$.
4. Using the notation of Exercise 3, prove by induction on n that $(ab)^n = [b, a]^q a^n b^n$ where $q = (n-1)n/2$ for all $n \in \mathbf{N}$, provided that the commutator $[a, b]$ lies in the center of the group G; that is, provided $g[a, b] = [a, b]g$, for all $g \in G$.
5. Prove that a group G with p elements, p a prime number, must be abelian.
6. The **minimal exponent** of a group G is the least integer $m \in \mathbf{N}$ such that $g^m = e$ for all $g \in G$. If G is a finite abelian group with elements $a_1, \ldots, a_n$, prove that the minimal exponent of G is the LCM of $o(a_1), \ldots, o(a_n)$.
7. Noting that the prime residue classes in $\mathbf{Z}_m$ constitute a multiplicative subgroup $\mathbf{U}_m$, prove that

$$x^{\varphi(m)} \equiv 1 \pmod{m}$$

for all $x \in \mathbf{N}$ for which $(x, m) = 1$. This is known as Euler's Theorem, and, in the special case when m is a prime, as Fermat's Little Theorem [cf. §2.12]. (Here φ denotes the Euler φ-function.)
8. If $(a, m) = 1$, prove that x satisfies the congruence $ax \equiv b \pmod{m}$ if and only if $x = ba^{\varphi(m)-1} + km$, for some $k \in \mathbf{Z}$.
9. Consider a one-dimensional vector subspace U of $\mathbf{R}^2$.
   a. Verify that U is an additive abelian subgroup of $\mathbf{R}^2$ considered as a group.
   b. Describe the cosets of U.
10. Describe geometrically the cosets in $\mathbf{R}^3$ of a one-dimensional vector subspace U and of a two-dimensional vector subspace V.
11. If $[G : H]_\ell < \infty$ and $[G : K]_\ell < \infty$, prove that $[G : H \cap K]_\ell < \infty$, where $H \cap K$ is the intersection of the subgroups H and K. (This result is due to Henri Poincaré, 1854–1912.)

12. Prove that if $g \in G$ has finite order, then $H = \{g^i : i \in \mathbf{Z}\}$ has $o(g)$ *distinct* elements.

13. a. Referring to Exercise 6, §6.2, prove for all $g \in G$ that
$$[G:H]_\ell < \infty \Rightarrow [G:gHg^{-1}]_\ell = [G:H]_\ell.$$
b. Prove that $o(gHg^{-1}) = o(H)$, when $o(H) < \infty$.

§6.4 Normal Subgroups and Factor Groups

A class of subgroups of particular interest are the so-called normal subgroups, for using them we can define groups of cosets called factor or quotient groups, analogous to the residue classes modulo an ideal in ring theory.

A subgroup H of a group G is called **normal** if $Ha = aH$ for every $a \in G$, or equivalently, if $aHa^{-1} = H$ for every $a \in G$. Note that $aHa^{-1} = H$ means $\{aha^{-1} : h \in H\} = H$.

Propositions 1 through 3 are immediate consequences of this definition.

Proposition 1. If H and K are normal subgroups of a group G, then $H \cap K$ is a normal subgroup.

Proof. First recall from Exercise 14, §6.2, that $H \cap K$ is a subgroup. Consider $x \in H \cap K$; then axa^{-1} lies in H and in K. Hence $axa^{-1} \in H \cap K$, for all $a \in G$.

Proposition 2. If N is a normal subgroup and H is any subgroup of G, then the product $N \cdot H$ is equal to the set $NH = \{nh : n \in N, h \in H\}$. In fact, $NH = N \cdot H = HN$.

Proof. First for $n_i \in N$, $h_i \in H$, $i = 1, 2$, the product $n_1 h_1 n_2 h_2$ can be written as $n'h'$ with $n' \in N$, $h' \in H$:
$$n_1 h_1 n_2 h_2 = n_1 (h_1 n_2 h_1^{-1}) h_1 h_2 = n_1 n_2' h_1 h_2 = n'h',$$
by the normality of N and the fact that H and N, being subgroups, are closed with respect to multiplication. Induction on s implies that $n_1 h_1 n_2 h_2 \cdots n_s h_s$ can be written as $n'h'$. Thus any element in $N \cdot H$ can be expressed in the form $n'h' \in NH$, and $N \cdot H \subseteq NH$. Since $NH \subseteq N \cdot H$, we have $N \cdot H = NH$.

Further, as N is a normal subgroup, the sets NH and HN are equal. A product hn is equal to $n'h$ for some $n' \in N$. We simply observe that $hn = hn(h^{-1}h) = (hnh^{-1})h$ and set $n' = hnh^{-1}$. Similarly $nh = h(h^{-1}nh) = hn''$, $n'' \in N$. Consequently, when either N or H is a normal subgroup, we shall commonly write NH for the product subgroup $N \cdot H$.

Proposition 3. If N and H are both normal subgroups of a group G, then NH is a normal subgroup of G.

§6.4   **Normal Subgroups and Factor Groups**   181

Proof. For any $g \in G$,

$$NH = g^{-1}(gNg^{-1})(gHg^{-1})g$$
$$= g^{-1}NHg,$$

since $gNg^{-1} = N$ and $gHg^{-1} = H$, as both N and H are normal subgroups. Hence NH is a normal subgroup.

REMARK 1. Note carefully that the normality of N in G means $gN = Ng$, for all $g \in G$. It does *not* mean $ng = gn$. Rather, given g, we have $ng = gn'$ for some $n' \in N$; the elements $n, n' \in N$ are in general not equal. For example, in the group of the triangle Γ_3 [§6.2], R is a normal subgroup, but $r\alpha \ne \alpha r$ although $\alpha R = R\alpha$.

If N is a *normal* subgroup of a group G, then in a natural way the set of cosets $\{gN : g \in G\}$ can be given the structure of a group. This new group, called the **factor** or **quotient group** of G modulo N and denoted G/N, is analogous to the residue class rings considered in §§2.9 and 2.10.

The *product of two cosets* aN and bN is defined to be the coset of $a'b'$, namely $(a'b')N$, where a', b' represent aN, bN, respectively. As was the case for addition in $\mathbf{Z}_m$, we must prove that this rule for composition of two cosets is independent of the choice of representatives $a' \in aN$, $b' \in bN$ [cf. §2.9]. In particular, to prove

$$\left.\begin{array}{l} a'N = aN \\ b'N = bN \end{array}\right\} \Rightarrow a'b'N = abN,$$

we note that

$$a' \in aN, \ b' \in bN \Rightarrow a'b' \in abN,$$

since $a' = an$, $b' = bn'$ imply

$$a'b' = anbn' = ab(b^{-1}nb)n' = abn''n' \in abN.$$

Therefore $a'b'N \subseteq abN$. A symmetric argument yields $abN \subseteq a'b'N$, which completes the proof.

The identity or unit element in G/N is

$$eN = \{g \in G : g \in N\},$$

since $(eN)(gN) = (eg)N = gN = Ng = (gN)(eN)$, for all $g \in G$. Finally, the inverse of gN is $g^{-1}N$, the coset of g^{-1}. Associativity of multiplication of cosets gN is a consequence of the associativity of multiplication in G. We conclude that the set G/N of cosets of the normal subgroup N is itself a group.

Note that in the proof that multiplication of cosets is independent of the choice of representatives of the cosets, essential use was made of the fact that N was normal in G. The set of cosets of a nonnormal subgroup H cannot be given a corresponding structure of a group. In the case of the group of the triangle Γ_3 [§6.2] we can of course define a rule of multiplication on the three cosets of the subgroup $S = \{\varepsilon, \alpha\}$, but no such rule of multiplication will *correspond* to the product in Γ_3. The subgroup S is not a normal subgroup of Γ_3.

REMARK 2. A trivial consequence of the definition of normality is that every subgroup H of a commutative group G is a normal subgroup. Thus for a subgroup $H \subseteq G$, G commutative, we can always form the quotient group G/H.

Example 1. Consider the abelian group
$$G = \{a^m b^n : m, n \in \mathbf{Z}\}$$
where $ab = ba$ and $o(a) = 4$, $o(b) = 12$.

The element $a^2 b^3$ has order 4 and generates the subgroup $H = \{e, a^2 b^3, b^6, a^2 b^9\}$. Since $o(G) = 48$, the factor group G/H has order 12. To prove that G/H is generated by the coset abH, it suffices to prove that the order of abH in G/H is 12. Observe that
$$ab \notin H, \qquad a^2 b^2 \notin H, \qquad a^3 b^3 \notin H, \qquad a^4 b^4 \notin H, \qquad a^6 b^6 \notin H.$$
Consequently, $o(abH)$ is not 1, 2, 3, 4, or 6; thus it must be 12 since it divides $12 = o(G/H)$.

Similarly, the element $a^2 b \in G$ generates a subgroup K of order 12,
$$K = \{e, a^2 b, b^2, a^2 b^3, b^4, a^2 b^5, b^6, a^2 b^7, b^8, a^2 b^9, b^{10}, a^2 b^{11}\},$$
and G/K is a group of order 4; each element is a power of aK (and of $a^3 K$).

As a second example, consider a nonabelian group of order p^3, p a prime. (When $p = 2$, the group in Example 2 is the dihedral group Γ_4; cf. Example 6(f), §6.1.)

Example 2. Let
$$G = \{a^m b^n : m, n \in \mathbf{Z}\}$$
where $a^{p^2} = b^p = e$, $bab^{-1} = a^{1+p}$.

To show that the subgroup
$$N = \{e, a^p, a^{2p}, \ldots, a^{p(p-1)}\}$$
generated by a^p is a normal subgroup of G, observe that
$$ba^p b^{-1} = (bab^{-1})^p = (a^{1+p})^p = a^{p+p^2} = a^p e = a^p$$
implies
$$(a^m b^n) a^{kp} (a^m b^n)^{-1} = a^m (b^n a^{kp} b^{-n}) a^{-m} = a^{kp}$$
for all m, n and k.

The factor group G/N has p^2 elements and is abelian because $bab^{-1}a^{-1} = a^p \in N$. Thus
$$bN \cdot aN \cdot b^{-1} N \cdot a^{-1} N = eN$$
or
$$bN \cdot aN = aN \cdot bN,$$
and for any $m, n \in \mathbf{Z}$,
$$(aN)^m (bN)^n = (bN)^n (aN)^m.$$
(In fact, any group of order p^2, p a prime, must be abelian as shown in §7.4, Exercise 4.)

Furthermore, G/N consists of the p^2 distinct cosets $a^\alpha b^\beta N$, $0 \leq \alpha, \beta < p$.

§6.4 Normal Subgroups and Factor Groups

Exercises

1. With reference to Exercise 8 of §6.2, prove that if H is an arbitrary subgroup of a group G, then H is a normal subgroup of $N_G(H)$.
2. Prove that if a subgroup H of G is a normal subgroup of K, and $N_G(H) \subseteq K \subseteq G$, then $K = N_G(H)$.
3. Prove that the center C of any group G is a normal subgroup.
4. Prove that if the subgroup H of G has index 2, then H is a normal subgroup.
5. Consider subgroups H and K of the group G and let HK be the totality of all products hk with $h \in H$ and $k \in K$. Prove the following statements.
   a. HK is a subgroup if and only if $HK = KH$.
   b. HK is a subgroup if one of the groups H and K is a normal subgroup.
   c. HK is a normal subgroup of G if both H and K are normal subgroups.
6. Find the center C of the dihedral group Γ_7 [see Examples 6(f) and 7(c), §6.1] of order 14 [cf. Exercise 10, §6.2].
7. Let Γ_4 denote the dihedral group of order 8 [see Example 7(b), §6.1] commonly called the group of the square.
   a. Find all subgroups of Γ_4.
   b. Determine which subgroups are normal.
   c. Exhibit the multiplication tables of the corresponding factor groups.
   d. Construct the lattice diagram of subgroups, as in §6.2.
8. a. Consider the integers **Z** and the rational numbers **Q** as additive groups. Show that every element of the factor group **Q/Z** has finite order.
   b. Is the corresponding statement true for **R/Z**, where the set of real numbers **R** is considered as an additive group?
9. For a group G with a subgroup H and a normal subgroup N, prove that $H \cap N$ is a normal subgroup of H.
10. Given a group G with subgroups M, M_0, N, and N_0, where M_0 is a normal subgroup of M, and N_0 is a normal subgroup of N, prove that $M_0(M \cap N_0)$ is a normal subgroup of $M_0(M \cap N)$.
11. Prove that multiplication in the quotient group G/N is associative, where N is a normal subgroup of a group G.
12. Given a group G with subgroup H, show that the **centralizer** $C_G(H)$ of H in G is a normal subgroup of the normalizer $N_G(H)$ of H in G, where
    $$C_G(H) = \{a \in G : aha^{-1} = h \text{ for all } h \in H\}.$$
13. Show that the two equivalence relations defined in Exercise 5(a) of §6.2 are the same if and only if H is a normal subgroup of the group G.
14. Let $\alpha = (a_1, a_2)$ and $\beta = (b_1, b_2)$ be two independent vectors in $\mathbf{R}^2$, i.e., pairs of real numbers such that $a_1 b_2 - b_1 a_2 \neq 0$. Define a group structure on $\mathbf{R}^2$ by componentwise addition [see Example 5 of §6.1].
    a. Prove that $S = \{m\alpha + n\beta : m, n \in \mathbf{Z}\}$ is a subgroup of $\mathbf{R}^2$. Since $\mathbf{R}^2$ is abelian, S is a normal subgroup.
    b. Describe geometrically the quotient group $\mathbf{R}^2/S$.

15. In the cartesian plane $\mathbf{R}^2$, consider any two nonparallel translations σ, τ. Let Γ be the group generated by σ and τ.
    a. Describe the set of points
    $$U = \{(r, r') \in \mathbf{R}^2 : \gamma((0,0)) = (r, r') \text{ for some } \gamma \in \Gamma\},$$
    called the **orbit** of $(0,0)$ under Γ.
    b. Prove that U is a normal subgroup of $\mathbf{R}^2$. Such groups are special cases of transformation groups which are of particular interest in both topology and quantum mechanics.
16. If $H \subseteq G$ is the only subgroup of order n in a group G, prove that H is a normal subgroup of G. *Hint:* See Exercise 6, §6.2.
17. Let U be a subspace of a finite dimensional vector space V over a field F.
    a. Show that U can be viewed as a normal subgroup of V.
    b. Define on the set of (additive) cosets of U the structure of a vector space over F [cf. Exercise 9, §6.3]. We call this the **quotient space** of V with respect to U, denoted V/U.
    c. Prove that $\dim(V/U) = \dim V - \dim U$.
18. With reference to Example 2, prove that the cosets $a^\alpha b^\beta N$, $0 \le \alpha, \beta < p$ are distinct. *Hint:* Consider $a^\alpha b^\beta N = a^\gamma b^\delta N$ and use the commutativity of G/N to write
    $$N = b^{-\delta} a^{a-\gamma} b^\beta N = b^{-\delta} N \cdot a^{a-\gamma} N \cdot b^\beta N$$
    $$= b^{\beta-\delta} a^{a-\gamma} N.$$
19. Consider the group G of 2×2 nonsingular matrices
    $$\begin{bmatrix} a & b \\ c & d \end{bmatrix}$$
    with coefficients in a field F [cf. Example 4(a), §6.1]. Prove that the set of matrices
    $$N = \left\{ \begin{bmatrix} a & 0 \\ 0 & a \end{bmatrix} : a \in F \right\}$$
    is a normal subgroup. The factor group G/N is called the **projective linear group** $PL(2, F)$.
20. Carry out the explicit steps in the induction argument in the proof of Proposition 2.

§6.5 Group Homomorphisms

We now turn to one of the most important concepts of group theory, that of homomorphism. A homomorphism is a mapping of one group into another so that the group operations are preserved. To demonstrate the significance of such mappings we present various detailed examples which show in part that the theory of congruences is an essential tool for the study of finite groups. The examples also show that the abstract definitions are not

§6.5 Group Homomorphisms

vacuous. Further discussion of homomorphisms will be found in §§6.9 and 6.11.

Suppose that there are given two groups G and G' with respective identity elements e and e'. A single-valued function φ from G into G'

$$g \to \varphi(g) \in G' \quad \text{for } g \in G$$

is called a **(group) homomorphism** of G into G', if

$$\varphi(gg_1) = \varphi(g)\varphi(g_1).$$
$$\uparrow \qquad \qquad \uparrow$$
$$\text{product in } G \quad \text{product in } G'$$

Proposition 1. For any group homomorphism,

$$\varphi(e) = e' \quad \text{and} \quad \varphi(g^{-1}) = [\varphi(g)]^{-1}.$$

In the second equation, the exponent -1 on the left indicates the inverse of g in G; the exponent -1 on the right indicates the inverse of $\varphi(g)$ in G'.

Proof. To prove these assertions observe that $\varphi(g) = \varphi(ge) = \varphi(g)\varphi(e)$; hence by the uniqueness of the identity e' of G', $\varphi(e) = e'$. Furthermore, $\varphi(g)\varphi(g^{-1}) = \varphi(e) = e'$ implies that $\varphi(g^{-1}) = [\varphi(g)]^{-1}$ by the uniqueness of the inverse in G'.

For a normal subgroup N of a group G the homomorphism $\pi_N : G \to G/N$ is called the **canonical homomorphism** (or **natural projection**) of G onto the factor group G/N. Specifically, for $g \in G$,

$$\pi_N(g) = gN = Ng.$$

That π_N is a homomorphism follows, for g, g' in G, from

$$\pi_N(gg') = (gg')N = (gN)(g'N) = \pi_N(g)\pi_N(g').$$

For a (group) homomorphism $\varphi : G \to G'$ the complete inverse image $\varphi^{-1}(e') = \{x \in G : \varphi(x) = e'\}$ of the identity e' of G' is called the **kernel** ker φ of φ. The **image** Im φ of a homomorphism φ is defined to be the subset (actually a subgroup) of G':

$$\text{Im } \varphi = \varphi(G) = \{\varphi(g) \in G' : g \in G\}.$$

Proposition 2. The subset ker φ is a normal subgroup of G.

That ker φ is a subgroup is left as an exercise. To prove its normality consider $x \in \ker \varphi$ and $a \in G$. Then

$$\varphi(axa^{-1}) = \varphi(a)\varphi(x)\varphi(a^{-1}) = \varphi(a)e'[\varphi(a)]^{-1} = e',$$

i.e., $axa^{-1} \in \ker \varphi$.

Proposition 3. If the element $a \in G$ has finite order $h = o(a)$, then $o(\varphi(a))$ divides $o(a)$.

We have $a^h = e$, and hence $[\varphi(a)]^h = \varphi(a^h) = \varphi(e) = e'$. Then by Property 5, §6.2, $o(\varphi(a))$ divides h, as asserted.

As was the case with ring homomorphisms, group homomorphisms with special properties have particular names. We review the most important special properties below. For others the reader is referred to the analogous definitions in §3.2. A homomorphism $\varphi: G \to G'$ from one group G to another G' is said to be

onto (alternatively, **surjective** or an **epimorphism**) if for each $g' \in G'$ there exists at least one $g \in G$ such that $\varphi(g) = g'$;

one-one (alternatively, **injective** or a **monomorphism**) if $\varphi(g) = \varphi(g^*)$ implies $g = g^*$;

an **isomorphism** (alternatively, **bijective**) if it is both onto and one-one; that is, if $\varphi^{-1}(g')$ consists of one and only one element $g \in G$, for each $g' \in G'$;

an **automorphism** if it is an isomorphism and $G' = G$.

If there exists an isomorphism $\varphi: G \to G'$, we say that G and G' are isomorphic groups, denoted $G \cong G'$.

To facilitate our discussion of examples of group homomorphisms we introduce cyclic groups. A group G is said to be **cyclic** if there exists some $g^* \in G$ such that every $g \in G$ is a power of g^*. The element g^* is called a **generator** of G; it is not necessarily unique. For convenience let

$$C_n = \{c^i : 0 \le i < n\} = \langle c \rangle$$

denote a multiplicatively-written cyclic group of n elements with $1 = c^0$ as the identity element. The exponents i of c are counted modulo n, i.e., $c^{n-1}c^3 = c^2$, etc.

Example 1. Consider the two abelian groups C_9 and

$$G = \{a^m b^n : m, n \in \mathbf{Z}\}$$

where $ab = ba$ and $a^3 = b^6 = e$. (In other words, the integers m and n are to be counted modulo 3 and 6, respectively.) Define a homomorphism $\varphi: G \to C_9$ by

$$\varphi(a^m b^n) = c^{3(2m+n)}.$$

(Verify that this is indeed a homomorphism.)

To describe $\ker \varphi$ explicitly we need to determine all exponents m, n for which $c^{3(2m+n)} = 1$. This amounts to solving the congruence

$$3(2m + n) \equiv 0 \pmod 9,$$

or equivalently,

$$2m + n \equiv 0 \pmod 3.$$

(Why?)

§6.5 Group Homomorphisms

Solutions of the Stated Congruence		Corresponding Elements of ker φ
m	n	
0	0	e
0	3	b^3
1	1	ab
1	4	ab^4
2	2	a^2b^2
2	5	a^2b^5

Note that each element of ker φ is a power of ab; hence $K = $ ker φ is a cyclic subgroup of order 6 with generator ab.

The cosets
$$K = \{e, ab, a^2b^2, b^3, ab^4, a^2b^5\} = e,$$
$$aK = \{a, a^2b, b^2, ab^3, a^2b^4, b^5\} = \sigma,$$
$$a^2K = \{a^2, b, ab^2, a^2b^3, b^4, ab^5\} = \sigma^2$$

of K form a cyclic group G/K of three elements with the generator σ (also σ^2).

To describe Im $\varphi = \varphi(G) \subseteq C_9$ observe that $c^i \in$ Im φ if and only if there exist $m, n \in \mathbf{Z}$ such that

$$3(2m+n) \equiv i \pmod 9.$$

This congruence is satisfied by any solution to the following congruences.

For $m = 0$: $3n \equiv i \pmod 9$, which has a solution if and only if $3|i$. That is, $i = 0, 3, 6$.

For $m = 1$: $3n \equiv i-6 \pmod 9$, which has a solution if and only if $3|(i-6)$ or equivalently $3|i$.

For $m = 2$: $3n \equiv i-12 \pmod 9$, which has a solution if and only if $3|(i-12)$ or equivalently $3|i$.

Therefore, Im $\varphi = \{1, c^3, c^6\}$. The mapping φ is described in detail by

$a \to c^6$	$a^2 \to c^3$	$b \to c^3$
$ab \to 1$	$a^2b \to c^6$	$b^2 \to c^6$
$ab^2 \to c^3$	$a^2b^2 \to 1$	$b^3 \to 1$
$ab^3 \to c^6$	$a^2b^3 \to c^3$	$b^4 \to c^3$
$ab^4 \to 1$	$a^2b^4 \to c^6$	$b^5 \to c^6$.
$ab^5 \to c^3$	$a^2b^5 \to 1$	

Our next example shows that distinct homomorphisms may have the same kernel and image.

Example 2. Let C_{27} and C_{18} be cyclic groups of orders 27 and 18 with generators c and γ, respectively. Define distinct homomorphisms φ and ψ from C_{27} into C_{18} by

$$\varphi(c^k) = \gamma^{2k}, \qquad 0 \leq k < 27,$$
$$\psi(c^k) = \gamma^{4k}, \qquad 0 \leq k < 27.$$

To show that $\ker \varphi = \ker \psi$, observe that $c^k \in \ker \varphi$ implies that

$$2k \equiv 0 \pmod{18} \quad \text{or} \quad k \equiv 0 \pmod{9}.$$

Similarly, $c^k \in \ker \psi$ implies also that $k \equiv 0 \pmod 9$. Thus, $\ker \varphi = \ker \psi = \{c^0, c^9, c^{18}\}$.

Furthermore, the images coincide because $\varphi(c) = \gamma^2$ and $\psi(c) = \gamma^4$ generate the same subgroup

$$H = \{\gamma^0, \gamma^2, \gamma^4, \gamma^6, \ldots, \gamma^{16}\}$$

of C_{18}. This fact is most easily seen by observing that $\gamma^4 = (\gamma^2)^2$ lies in the subgroup generated by γ^2, and conversely that $\gamma^2 = (\gamma^4)^5$, since exponents are taken modulo 18, lies in that generated by γ^4.

We conclude this section with two propositions concerning the image of a subgroup $H \subseteq G$ under a homomorphism $\varphi \colon G \to G'$.

Proposition 4. *If $\varphi \colon G \to G'$ is a group homomorphism and H is a subgroup of G, then $\varphi(H)$ is a subgroup of G'.*

Proof. To prove that $H' = \varphi(H)$ is a subgroup it suffices to show that $a'b'$, a'^{-1} belong to H' for all $a', b' \in H'$. Corresponding to a', b' in H', there are elements $a, b \in H$ such that $\varphi(a) = a'$, $\varphi(b) = b'$. By the homomorphism property of φ, $\varphi(ab) = a'b'$ and $\varphi(a^{-1}) = a'^{-1}$; hence $a'b' \in H'$ and $a'^{-1} \in H'$.

Now suppose that H is a normal subgroup of G and that $\varphi \colon G \to G'$ is *surjective*. To prove that $H' = \varphi(H)$ is a normal subgroup of G', consider an arbitrary element $a' \in G'$, and (by the surjectivity of φ) a corresponding $a \in G$ such that $\varphi(a) = a'$. Because H is normal in G, we have $aH = Ha$, and so by the homomorphism property of φ

$$\varphi(a)\varphi(H) = \varphi(H)\varphi(a) \quad \text{and} \quad a'H' = H'a';$$

thus H' is a normal subgroup of G'. This proves Proposition 5.

Proposition 5. *If H is a normal subgroup of G, and the homomorphism $\varphi \colon G \to G'$ is surjective, then $\varphi(H)$ is a normal subgroup of G'.*

A converse to Propositions 4 and 5 is given in Exercise 15. Show by an example that Proposition 5 need not hold if φ is not surjective.

In §6.8 we examine the images of subgroups (normal subgroups) in the particular case that φ is the canonical projection $\pi_K \colon G \to G/K$, where K is a normal subgroup of G.

Exercises

1. Verify the following if $\varphi: G \to G'$ is a homomorphism defined on a group G.
   a. $\operatorname{Ker} \varphi$ is a subgroup of G.
   b. $\operatorname{Im} \varphi$ is a subgroup of G'.

2. Let n be an odd integer. Define, for real numbers x and y,
$$x * y = \sqrt[n]{x^n + y^n}$$
   where the real nth root is taken.
   a. Prove that the operation $*$ defines a group law on the set of real numbers $\mathbf{R}$.
   b. Prove that this group is isomorphic to the additive group of real numbers.
   c. Why does this construction of $*$ fail to produce a group if n is even and the nth root is taken to be nonnegative?

3. Let $\mathbf{Z}$ be the additive group of integers, and $\mathbf{C}$ the field of complex numbers. Suppose that ω is a *primitive* eighth *root of unity* (meaning that every eighth root of 1 is a power of ω). Let ψ be the mapping of $\mathbf{Z}$ into $\mathbf{C}$ given by
$$\psi(n) = \omega^n, \quad n \in \mathbf{Z}.$$
   a. Find the image $\psi(\mathbf{Z})$ and the kernel of ψ.
   b. Find all homomorphisms from $\mathbf{Z}$ to the subgroup of $\mathbf{C}$ generated by ω.

4. Consider the abelian group $G = \{a^m b^n : m, n \in \mathbf{Z}\}$, where $a^3 = b^3 = e$ and $ab = ba$, and the subgroup H generated by ab^2.
   a. Determine explicitly the cosets of G modulo H.
   b. Set up the multiplication table for G/H.

5. Consider the abelian group $G = \{a^m b^n : m, n \in \mathbf{Z}\}$, where $a^8 = b^2 = e$ and $ab = ba$, and the cyclic group C_3 with generator c.
   a. Show that $\varphi: G \to C_3$ given by $\varphi(a^m b^n) = c^{m+2n}$ is not a homomorphism.
   b. Show that the only homomorphism $\psi: G \to C_3$ is trivial, i.e., $\psi(G) = \{c^0\}$.

6. Let G be the multiplicative *group of quaternions*, consisting of the 8 elements $\pm 1, \pm i, \pm j, \pm k$, which satisfy
$$i^2 = j^2 = k^2 = -1, \quad -i = (-1)i,$$
$$ij = -ji = k, \quad jk = -kj = i, \quad ki = -ik = j,$$
$$(-1)^2 = 1.$$
   This is a special case ($n = 3$) of the generalized quaternion groups in Example 6(e), §6.1.
   a. Find all subgroups of G. Prove that they all are normal.
   b. Find a subgroup N of G such that G/N is isomorphic to Klein's Four-group V (named for Felix Klein, 1849–1925) of order 4 described by the 4 elements $\sigma^a \tau^b$, where a, b are taken modulo 2 and $\sigma\tau = \tau\sigma$, $\sigma^2 = 1$, $\tau^2 = 1$. Exhibit such an isomorphism.

c. How many such isomorphisms onto V are there?
   d. Find all homomorphisms of G into the group T of all complex numbers of absolute value 1.
   e. Find the kernel of each such homomorphism in part (d).

7. Conclude from Exercise 6(a) and from Exercise 7, §6.4, that the quaternion and dihedral groups of order 8 are not isomorphic.

8. Find all group homomorphisms of the following additive groups.
   a. $Z_{12} \to Z_5$ b. $Z_{12} \to Z_6$ c. $Z_6 \to Z_{28}$.

9. Let G be a group. Define for fixed $a \in G$ different from the identity the map L of G into itself by the equation $L_a(g) = ag$ for all $g \in G$.
   a. Prove that L_a is a one-one mapping, but not a homomorphism of G.
   b. Prove that the inverse mapping $(L_a)^{-1}$ exists and is equal to $L_{a^{-1}}$.

10. Define a mapping φ on a group G by $\varphi(g) = g^{-1}$. Prove that φ is an automorphism of G if and only if G is abelian.

11. a. Let G be an abelian group. Prove that the mapping $g \to g^s$, for fixed $s \in \mathbf{N}$, is a homomorphism of G into itself.
    b. If $o(G)$ is finite and relatively prime to s, prove that the above mapping is an isomorphism.

12. Give an example of a nonabelian group, using a suitable s, for which the statement of Exercise 11(a) is false.

13. Find all homomorphisms of the dihedral group Γ_4 of Exercise 7, §6.4, into the multiplicative group C^* of all nonzero complex numbers. Find the kernel of each homomorphism.

14. Let R be the integral domain defined in Exercise 3, §3.6. Now let $S = \{a + b\sqrt{-3} : a, b \in \mathbf{Z}\}$ be a similarly defined integral domain. Prove that R and S are isomorphic if considered as additive groups; prove that they are not isomorphic, however, if considered as rings.

15. a. Let $\varphi: G \to G'$ be a group homomorphism, and H' be a subgroup in G'. Prove that the complete inverse image $H = \varphi^{-1}(H')$ is a subgroup in G. Note that it is not required that $H' \subseteq \operatorname{Im} \varphi$.
    b. If H' is normal in G', prove that H is normal in G.

16. Consider groups G, G', and G''. If $\varphi: G \to G'$ and $\psi: G' \to G''$ are homomorphisms, prove that the map $\psi \circ \varphi: G \to G''$ described by $\psi \circ \varphi(g) = \psi(\varphi(g))$ is a homomorphism.

17. Suppose that $\sigma: G \to \Gamma$ and $\tau: G \to L$ are group homomorphisms, and σ is surjective. Prove that the following statements are equivalent.
    a. There exists a homomorphism $\varphi: \Gamma \to L$ such that $\tau = \varphi \circ \sigma$, i.e., $\tau(g) = \varphi((\sigma(g))$ for all $g \in G$.
    b. The kernels satisfy $\ker \sigma \subseteq \ker \tau$. Schematically,

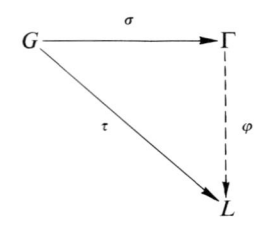

$$\begin{array}{ccc} G & \xrightarrow{\sigma} & \Gamma \\ & \tau \searrow & \downarrow \varphi \\ & & L \end{array}$$

18. Show by an example that Proposition 5 is not true in general if the homomorphism φ is not required to be surjective.
19. With reference to Example 2, find distinct integers m, n (different from 27, 18) and distinct homomorphisms φ, ψ from C_m to C_n, such that $\ker \varphi = \ker \psi$, and $\operatorname{Im} \varphi = \operatorname{Im} \psi$.

§6.6 Cyclic Groups

We undertake in this section a systematic study of cyclic groups, which we shall write multiplicatively.

Theorem. A cyclic group G is either isomorphic to the additive group $\mathbf{Z}$ or to the additive group of the residue class ring $\mathbf{Z}_m$ for some integer m.

Proof. Consider the following mapping λ of $\mathbf{Z}$ onto G, where g is a given generator of G:
$$\lambda(n) = g^n, \qquad n \in \mathbf{Z}.$$
Since $\lambda(a+b) = g^{a+b} = g^a g^b = \lambda(a)\lambda(b)$, the mapping λ is a homomorphism of $\mathbf{Z}$ onto G. Moreover $\ker \lambda$ is actually an ideal in the ring $\mathbf{Z}$ because $\lambda(-n) = g^{-n} = (g^n)^{-1} = e$ for $n \in \ker \lambda$. Thus $\ker \lambda = (0)$ or $\ker \lambda = (m)$ with $m > 0$ [cf. §2.4]. In the first case λ is an isomorphism.

In the second case, let m be the smallest positive integer such that $g^m = \lambda(m) = e$. That is, m is the order of g. As an additive group $\mathbf{Z}_m$ is generated by the coset $[1] = \{1 + sm : s \in \mathbf{Z}\}$, and $g^{1+sm} = g^1$. The m distinct elements $g^a = g^{a+sm}$, $0 \leq a < m$, are thus in one-one correspondence with the cosets $[a]$. The definition of the sum in $\mathbf{Z}_m$ then implies that the mapping $[a] \to g^a$ is an isomorphism of the additive group $\mathbf{Z}_m$ onto G.

Proposition 1. Every subgroup H of a cyclic group $G = <g>$ is itself cyclic. If $o(G) = n$, then $H = <g^k>$ for some k, such that $k \mid n$.

Proof. If H is the trivial subgroup $\{e\}$, it is generated by $g^0 = e$.

More generally, since G is cyclic and generated by g, the subgroup H is a collection of powers of g. Let
$$S = \{m \in \mathbf{N} : g^m \in H\}.$$
Because it is a nonempty set of positive integers, S has a least element k (by the Well-Ordering Principle, §2.2). To show that g^k generates H, consider an arbitrary element $g^s \in H$. By the Division Algorithm, §2.3,
$$s = qk + r, \qquad 0 \leq r < k.$$
Then
$$g^r = g^s (g^k)^{-q} \in H,$$
because g^s, g^k, and hence $(g^k)^{-q}$ belong to H. But since k is the least positive integer such that $g^k \in H$, r must be zero. Thus $k \mid s$, and $g^s = (g^k)^q$ that is, $<g^k> = H$.

If $o(G) = n$, the previous argument indicates that $k \mid n$, because $g^n = e \in H$ and $k \mid s$ for all s for which $g^s \in H$.

Conversely every divisor m of $n = o(G)$ determines a subgroup $H = \{g^{mi} : i \in \mathbf{Z}\}$ of G whose index is m. Also $o(g^m) = n/m$; thus $o(H) = n/m$. These observations prove the following proposition.

Proposition 2. If m divides the order n of a finite cyclic group G, then G has a subgroup H of index m (and of order n/m).

That such a subgroup H is unique is left as an exercise.

Proposition 3. A cyclic group G of order n has $\varphi(n)$ distinct generators.

Proof. First g^s is a generator of G if and only if $o(g^s) = n$. By Proposition 2, §6.2, $o(g^s) = o(g) = n$ if and only if $(s, o(g)) = 1$. By definition of the Euler φ-function [§2.12], $\varphi(n)$ is the number of integers s, $1 \le s < n$, relatively prime to n.

Proposition 4. The homomorphic image $G' = \psi(G)$ of a cyclic group G of order n is a cyclic group whose order divides n.

If g generates G, then clearly $\psi(g)$ generates $\psi(G)$ since $\psi(g^a) = [\psi(g)]^a$ for a homomorphism ψ. By Proposition 3, §6.5, $o(\psi(g)) = o(G')$ divides $n = o(g) = o(G)$.

In particular if H is a subgroup of a cyclic group G, then the factor group G/H is cyclic since it is the image of G under the canonical projection $\pi_H : G \to G/H$ introduced in §6.5.

Proposition 5. Let G be an abelian group of order n. If there exist for each divisor m of n at most m elements whose order divides m, then G is cyclic.

Proof. Let n have the prime factorization

$$n = \prod_{i=1}^{s} p_i^{\alpha_i}.$$

Set $m_i = p_i^{\alpha_i}$. By hypothesis, for each i, $1 \le i \le s$, at most $n_i = n/p_i < n$ elements a in G satisfy $a^{n_i} = e$. Thus at least one element $x_i \in G$ satisfies $x_i^{n_i} \ne e$. Let $y_i = x_i^{u_i}$, where $n = m_i u_i$. Then $o(y_i) | m_i$ since $y_i^{m_i} = x_i^n = e$. Moreover, because $y_i^{m_i/p_i} = x_i^{n_i} \ne e$, we have $o(y_i) = m_i$. Finally, $z = y_1 y_2 \cdots y_s$ has order n by the remark in §6.2.

Exercises

1. Find all subgroups of the cyclic group C_m of order m generated by c when

   a. $m = 8$ **b.** $m = 35$ **c.** $m = 315$.

§6.6 Cyclic Groups

2. Describe the elements of the factor group C_{18}/H, where C_{18} is the cyclic group of order 18 generated by c and H is the subgroup generated by c^6.
3. For how many distinct groups G, up to isomorphism, can a surjective homomorphism $\varphi\colon C_{54} \to G$ be determined, where C_{54} is the cyclic group of order 54? Note that G is to equal $\operatorname{Im}\varphi = \varphi(C_{54})$.
4. Prove that the subgroup H in Proposition 2 is unique.
5. Define a homomorphism $\varphi\colon C_{16} \to C_{96}$, where the cyclic groups C_{16}, C_{96} have respective generators c, γ, by $\varphi(c^m) = \gamma^{12m}$.
   a. Find the kernel of φ.   b. Find the image of φ.
   c. Find all homomorphisms ψ of C_{16} onto the subgroup G of C_{96} generated by γ^{12}. Note that if $\lambda\colon G \to G$ is an isomorphism, then $\lambda \circ \varphi$ is a surjective homomorphism of C_{16} onto G described by $\lambda \circ \varphi(c) = \lambda(\varphi(c)) = \lambda(\gamma^{12})$.
6. Consider the cyclic group C_{16} with generator c, and the abelian group $G = \{a^m b^n : m, n \in \mathbf{Z}\}$ where $a^8 = b^2 = e$, $ab = ba$, also of order 16. Define mappings φ and ψ from G to C_{16} by

   $$\varphi(a^m b^n) = c^{6m} \quad \text{and} \quad \psi(a^m b^n) = c^{4m+8n}$$

   a. Verify that φ and ψ are homomorphisms.
   b. Find the kernels of φ and ψ.
   c. Find the images of φ and ψ.
7. Suppose that N is a subgroup of the center of a group G. Prove that G is abelian if the factor group G/N is cyclic.
8. Prove that the quaternion group [Exercise 6, §6.5] of 8 elements possesses precisely three cyclic subgroups of index 2 and exactly one subgroup of order 2.
9. Prove that any group of order 4 is isomorphic to one of two abelian groups. (As a first step, exhibit two nonisomorphic groups of order 4.)
10. Let G be a noncyclic abelian group of order p^α, p a prime, and H a cyclic subgroup of G of order p^β which is not contained in any other cyclic subgroup. Observe that there exists a γ such that for all $g \in G$, $g^{p^\gamma} \in H$, but for some $g \in G$, $g^{p^{\gamma-1}} \notin H$. Prove that every $h \in H$ which is a p^γth power of an element in G is also a p^γth power of an element in H.
11. a. Consider the cyclic group C_{36} with generator c, and the homomorphism $\varphi\colon C_{36} \to C_{36}$ described by $\varphi(c) = c^s$, for some fixed $s \in \mathbf{N}$. Find the kernel of φ.
    b. Generalize the preceding to the case of $\varphi\colon C_m \to C_m$ described by $\varphi(c) = c^s$, for some fixed $s \in \mathbf{N}$, where c generates C_m. Find the kernel of φ.
12. a. Verify that the set of units $\mathbf{U}_m$ in $\mathbf{Z}_m$ for $m = 5, 6, 8$, and 15 constitute a group [cf. Example 2(c), §6.1].
    b. Is $\mathbf{U}_m$ a cyclic group for $m = 5, 6, 8$, or 15? (A general statement of the structure of $\mathbf{U}_m$ is given in Exercise 11, §6.10.)
13. Let p be a prime number and suppose that $(a, p) = 1$. Prove that the congruence $x^n \equiv a \pmod{p}$ has d solutions modulo p if $a^{(p-1)/d} \equiv 1 \pmod{p}$ and has no solution if $a^{(p-1)/d} \not\equiv 1 \pmod{p}$, where $d = (n, p-1)$.
14. Use Proposition 2 to prove Proposition 3, §6.2.

§6.7 Groups of Permutations

Let S be an arbitrary nonempty set and denote by $\Sigma(S)$ the collection $\{\alpha, \beta, \gamma, ..., \varepsilon\}$ of all **permutations** (one-one set mappings or bijections) of the set S onto itself. The set $\Sigma(S)$ is made into a group by defining a product of elements. The product $\alpha \circ \beta$, here called composition, of two mappings α, β is defined to be the bijection

$$(\alpha \circ \beta)(s) = \alpha[\beta(s)] \qquad \text{for all } s \in S.$$

The group axioms [§6.1] are readily verified for $\Sigma(S)$ with this rule for composing bijections. Note that two mappings α, α' on a set S are defined to be equal if $\alpha(s) = \alpha'(s)$ for *all* $s \in S$.

Associativity. Since for all $s \in S$

$$\begin{aligned}(\alpha \circ [\beta \circ \gamma])(s) &= \alpha[(\beta \circ \gamma)(s)] \\ &= \alpha[\beta(\gamma(s))] = [\alpha \circ \beta](\gamma(s)) \\ &= ([\alpha \circ \beta] \circ \gamma)(s),\end{aligned}$$

the composition of mappings is associative, i.e., $\alpha \circ (\beta \circ \gamma) = (\alpha \circ \beta) \circ \gamma$.

Identity. The map defined by $\varepsilon(s) = s$ for all $s \in S$ satisfies the properties of a group identity

$$\alpha \circ \varepsilon = \alpha = \varepsilon \circ \alpha$$

since for all $s \in S$

$$(\alpha \circ \varepsilon)(s) = \alpha(s) = (\varepsilon \circ \alpha)(s).$$

Inverse. An inverse α^{-1} of the bijection α is defined by

$$\alpha^{-1}(s) = s^* \qquad \text{if } \alpha(s^*) = s.$$

Since α is surjective, given $s \in S$, we can find an s^* such that $\alpha(s^*) = s$. The element s^* is unique because α is injective. The bijection α^{-1} so defined satisfies

$$(\alpha \circ \alpha^{-1})(s) = \alpha(\alpha^{-1}(s)) = \alpha(s^*) = s, \qquad \text{for all } s \in S,$$

$$(\alpha^{-1} \circ \alpha)(s^*) = \alpha^{-1}(s) = s^*, \qquad \text{for all } s^* \in S.$$

Thus $\alpha \circ \alpha^{-1} = \varepsilon = \alpha^{-1} \circ \alpha$, that is, α^{-1} is the group-theoretic inverse of α.

In particular, if $S = \{1, 2, ..., n\} = J_n$ is a finite set, $\Sigma(S) = \Sigma_n$ is called the **group of permutations of n symbols**, or the **symmetric group of n elements**. The group Σ_n has $n!$ elements as follows. For $\alpha \in \Sigma_n$, there are n possible images of 1, precisely $n-1$ possible images of 2 (since $\alpha(2) \neq \alpha(1)$), and precisely $n-2$ possible images of 3, etc. An inductive argument shows that there are $n!$ distinct bijections of J_n; that is, $[\Sigma_n : 1] = n!$

§6.7 Groups of Permutations

Notation. There are two standard methods of describing a permutation $\sigma \in \Sigma_n$. If

$$\sigma(i) = a_i \in J_n = \{1, \ldots, n\},$$

then we write

$$\begin{bmatrix} 1 & 2 & \cdots & i & \cdots & n \\ a_1 & a_2 & \cdots & a_i & \cdots & a_n \end{bmatrix}.$$

This symbol or *array* describes the map taking each element i into the element a_i below it.

If it is understood that a given permutation $\sigma \in \Sigma_n$ is defined on a set of n symbols, we often omit from the array notation symbols left unaltered by σ. For instance, the permutation

$$\begin{bmatrix} 1 & 2 & 3 & 4 & 5 \\ 2 & 5 & 3 & 4 & 1 \end{bmatrix}$$

is abbreviated

$$\begin{bmatrix} 1 & 2 & 5 \\ 2 & 5 & 1 \end{bmatrix}.$$

The alternative *cycle notation* involves writing after each element its image, thereby forming a chain or cycle, as follows:

$$[1, \sigma(1), \sigma(\sigma(1)), \sigma(\sigma(\sigma(1))), \ldots].$$

A given permutation may involve several cycles

$$[1, \sigma(1), \sigma(\sigma(1)), \ldots][a, \sigma(a), \sigma(\sigma(a)), \ldots].$$

The cycle notation, while awkward to describe in general, is, like stenography, convenient to use in specific cases.

In multiplying cycles, we read from right to left between cycles and from left to right within a cycle. That is, we write

$$[15][12] = [125] \quad \text{and} \quad [1345][243][135] = [1253].$$

This procedure is simply a convention. Some authors multiply from left to right; thus the student must be aware of the different conventions for multiplication of cycles.

For example, the permutation of J_5 mapping $1 \to 3$, $2 \to 4$, $3 \to 5$, $4 \to 2$, and $5 \to 1$ is denoted either

$$\begin{bmatrix} 1 & 2 & 3 & 4 & 5 \\ 3 & 4 & 5 & 2 & 1 \end{bmatrix}$$

or $[135][24]$. Schematically, this permutation can be displayed as

$$1 \to 3 \to 5 \to 1 \quad \text{and} \quad 2 \to 4 \to 2$$

or, literally in terms of cycles,

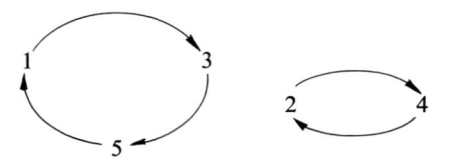

In the cycle notation, juxtaposition of two (or more) permutations denotes their product. The one to the right is performed first. In the array notation, the composition $\tau\sigma = \tau \circ \sigma$ of $\sigma, \tau \in \Sigma_n$, given by $\sigma(i) = a_i$ and $\tau(a_i) = b_i$, is

$$\begin{bmatrix} a_1 & a_2 & \cdots & a_n \\ b_1 & b_2 & \cdots & b_n \end{bmatrix} \begin{bmatrix} 1 & 2 & \cdots & n \\ a_1 & a_2 & \cdots & a_n \end{bmatrix} = \begin{bmatrix} 1 & 2 & \cdots & n \\ b_1 & b_2 & \cdots & b_n \end{bmatrix},$$

since $\tau \circ \sigma(i) = \tau(\sigma(i)) = b_i$. Hereafter, omitting the $\circ$, we shall simply denote composition of mappings by juxtaposition (see Example 1, below).

A **transposition** τ is a permutation which interchanges two elements, leaving all others fixed. In other words for a transposition τ on J_n,

$$\tau(h) = k, \qquad \tau(k) = h, \qquad \tau(i) = i$$

for all i different from any given $h, k \in J_n$. A transposition τ is denoted either by

$$\begin{bmatrix} 1 & 2 & \cdots & h & \cdots & k & \cdots & n \\ 1 & 2 & \cdots & k & \cdots & h & \cdots & n \end{bmatrix} = \begin{bmatrix} hk \\ kh \end{bmatrix}$$

or by

$$[hk][1] \cdots [n].$$

The 1-cycles $[i]$ are often omitted.

A **cycle** γ **of length** r ($r \leq n$) is a permutation such that for some r elements $i_1, \ldots, i_r$ in J_n

$$\gamma(i_1) = i_2, \qquad \gamma(i_2) = i_3, \ldots, \gamma(i_{r-1}) = i_r, \qquad \gamma(i_r) = i_1$$

and $\gamma(h) = h$ for all $h \notin \{i_1, \ldots, i_r\}$. Schematically, a cycle can be displayed as

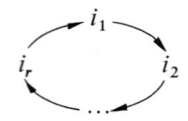

§6.7 Groups of Permutations

Thus a transposition is a cycle of length 2 or a 2-cycle:

$$i_1 \rightleftarrows i_2.$$

A cycle of length r, called an **r-cycle**,

$$[i_1, i_2, \ldots, i_r] = [i_r, i_1, i_2, \ldots, i_{r-1}]$$
$$= [i_{r-1}, i_r, i_1, i_2, \ldots, i_{r-2}], \text{ etc.}$$

has order r as an element in the group Σ_n [see §6.2]. We obtain the **inverse** of a cycle $\gamma = [i_1, i_2, \ldots, i_r]$ by writing the same elements *in reverse order*: $\gamma^{-1} = [i_r, \ldots, i_2, i_1]$. In array notation the inverse of a permutation is obtained by reading up from the second line to the first.

Example 1. The product of permutations

$$\sigma = \begin{bmatrix} 1 & 2 & 3 & 4 & 5 \\ 2 & 5 & 1 & 3 & 4 \end{bmatrix} \quad \text{and} \quad \tau = \begin{bmatrix} 1 & 2 & 3 & 4 & 5 \\ 1 & 5 & 2 & 4 & 3 \end{bmatrix}$$

is

$$\sigma\tau = \begin{bmatrix} 1 & 2 & 3 & 4 & 5 \\ 2 & 4 & 5 & 3 & 1 \end{bmatrix},$$

because

$$\sigma\tau(1) = \sigma[\tau(1)] = \sigma(1) = 2,$$
$$\sigma\tau(2) = \sigma[\tau(2)] = \sigma(5) = 4,$$
$$\sigma\tau(3) = \sigma[\tau(3)] = \sigma(2) = 5,$$
$$\sigma\tau(4) = \sigma[\tau(4)] = \sigma(4) = 3,$$
$$\sigma\tau(5) = \sigma[\tau(5)] = \sigma(3) = 1.$$

In addition,

$$\sigma^{-1} = \begin{bmatrix} 1 & 2 & 3 & 4 & 5 \\ 3 & 1 & 4 & 5 & 2 \end{bmatrix}.$$

Of particular importance in many applications of permutation groups are disjoint cycles, or cycles with no common symbols. We call two cycles $\sigma = [i_1, i_2, \ldots, i_r]$ and $\tau = [j_1, j_2, \ldots, j_s]$ **disjoint** if the set-theoretic intersection

$$\{i_1, i_2, \ldots, i_r\} \cap \{j_1, j_2, \ldots, j_s\} = \emptyset.$$

Proposition 1. Every permutation of J_n can be expressed as the (commutative) product of disjoint cycles.

The proof depends upon the following lemma.

Lemma. For any $j \in J_n$ and $\sigma \in \Sigma_n$, there exists a least integer k, such that $\sigma^k(j) = j$ and $0 \leq k \leq n$.

If $\sigma = \varepsilon$, then $k = 0$. Otherwise, consider the $n+1$ elements
$$j = \sigma^0(j), \sigma(j), \sigma^2(j), \ldots, \sigma^n(j).$$
These cannot be distinct, as J_n has only n elements. Let h be the least integer such that
$$\sigma^h(j) = \sigma^k(j)$$
for some k where $0 \leq h < k \leq n$. The proof of the lemma is completed by showing that h must equal 0. To this end, note that since σ is a permutation
$$\sigma(\sigma^{k-1}(j)) = \sigma^k(j) = \sigma^h(j) = \sigma(\sigma^{h-1}(j))$$
implies that $\sigma^{k-1}(j) = \sigma^{h-1}(j)$. Now because of the minimal choice of h, we conclude that $h-1$ is negative, that is, h must be 0.

Proof of Proposition 1. The identity map ε is the product of 1-cycles. Therefore consider a permutation σ and the cycle
$$\zeta_1 = [\sigma(1), \sigma^2(1), \ldots, \sigma^{s_1}(1) = 1],$$
where s_1 is the least positive integer k such that $\sigma^k(1) = 1$ as in the lemma, $0 \leq s_1 \leq n$. If $s_1 = n$, then $\zeta_1 = \sigma$, and the proof of Proposition 1 is complete. If $s_1 < n$, pick $j_2 \in J_n \setminus \{\sigma(1), \ldots, \sigma^{s_1}(1) = 1\}$, and consider the cycle
$$\zeta_2 = [\sigma(j_2), \sigma^2(j_2), \ldots, \sigma^{s_2}(j_2) = j_2],$$
where s_2 is the least integer k such that $\sigma^k(j_2) = j_2$. Then
$$\{\sigma(1), \sigma^2(1), \ldots\} \cap \{\sigma(j_2), \sigma^2(j_2), \ldots\} = \emptyset.$$
Repeat this process, if necessary, with
$$j_3 \in J_n \setminus \{\sigma(1), \ldots\} \cup \{\sigma(j_2), \ldots\}.$$
The process cannot be repeated indefinitely as there are only n symbols in J_n. Consequently, the permutation
$$\sigma = \zeta_1 \zeta_2 \cdots \zeta_\ell$$
where $\zeta_\nu \zeta_\mu = \zeta_\mu \zeta_\nu$, since the symbols involved in the permutations ζ_ν, ζ_μ form disjoint sets.

For our analysis of the structure of Σ_n it is useful to note that an r-cycle $\gamma = [i_1, i_2, \ldots, i_r]$ can be expressed as the product of transpositions:
$$[i_1, i_2, \ldots, i_r] = [i_1, i_2][i_2, i_3] \cdots [i_{r-1}, i_r]$$
$$= [i_1, i_r][i_1, i_{r-1}] \cdots [i_1, i_2].$$
Then we obtain the following corollary to Proposition 1.

Corollary. Every permutation of J_n can be expressed as the product of transpositions.

Example 2. The permutation
$$\sigma = \begin{bmatrix} 1 & 2 & 3 & 4 & 5 \\ 5 & 3 & 1 & 2 & 4 \end{bmatrix} \in \Sigma_5$$

is itself a cycle, written [15423] in cycle notation, since $\sigma(1) = 5$, $\sigma(5) = 4$, $\sigma(4) = 2$, etc.

Example 3. The permutation

$$\sigma = \begin{bmatrix} 1 & 2 & 3 & 4 & 5 \\ 2 & 5 & 1 & 4 & 3 \end{bmatrix} \in \Sigma_5$$

is expressed as the product $[1253][4]$ of disjoint cycles since $\sigma(1) = 2$, $\sigma(2) = 5$, $\sigma(5) = 3$, and $\sigma(3) = 1$. Also, $\sigma(4) = 4$.

Example 4. The product

$$\begin{bmatrix} 1 & 2 & 3 & 4 \\ 3 & 1 & 2 & 4 \end{bmatrix} \cdot \begin{bmatrix} 1 & 2 & 3 & 4 \\ 1 & 2 & 4 & 3 \end{bmatrix}$$

of cycles is written as a single cycle

$$\begin{bmatrix} 1 & 2 & 3 & 4 \\ 3 & 1 & 4 & 2 \end{bmatrix};$$

in cycle notation, $[132][4][1][2][34] = [2134]$; and as the product of trans-transpositions, $[2134] = [24][23][21]$.

Example 5. The permutation

$$\sigma = \begin{bmatrix} 1 & 2 & 3 & 4 & 5 & 6 & 7 & 8 & 9 \\ 4 & 5 & 6 & 2 & 1 & 3 & 8 & 9 & 7 \end{bmatrix} \in \Sigma_9$$

can be written as the product

$$\begin{bmatrix} 1 & 2 & 4 & 5 \\ 4 & 5 & 2 & 1 \end{bmatrix} \begin{bmatrix} 3 & 6 \\ 6 & 3 \end{bmatrix} \begin{bmatrix} 7 & 8 & 9 \\ 8 & 9 & 7 \end{bmatrix} = [1425][36][789]$$

of disjoint cycles, and as the product

$$[15][12][14][36][79][78]$$

of transpositions.

The following proposition is used in discussing the solvability of groups [§7.6] and of polynomial equations by algebraic means [§9.2].

Proposition 2. If $n \geq 5$, every 3-cycle $\rho = [r, k, i]$ can be written in the form $\sigma\tau\sigma^{-1}\tau^{-1}$, called the **commutator** of σ and τ, for some 3-cycles σ and τ.

Proof. Given $\rho = [r, k, i]$, consider $\sigma = [i, j, k]$ and $\tau = [k, r, s]$, where j and s are arbitrary, but *distinct* from r, k, and i, and $j \neq s$. Then, $\sigma^{-1} = [k, j, i]$ and $\tau^{-1} = [s, r, k]$. Consequently,

$$\sigma\tau\sigma^{-1}\tau^{-1} = [i, j, k][k, r, s][k, j, i][s, r, k] = [r, k, i] = \rho,$$

as asserted.

Exercises

1. Compute in the symmetric group Σ_6 the products $\alpha^2\beta$, $\gamma^{-1}\alpha^2$, $\alpha\gamma\alpha^{-1}$, where $\alpha = (143)(2)(5)(6)$, $\beta = (13)(26)(4)(5)$, and $\gamma = (1546)(2)(3)$.
2. Find the permutation ξ for which $\alpha\xi\beta = \gamma$, where α, β, γ are as given in Exercise 1.
3. Let
$$\alpha = \begin{bmatrix} 1 & 2 & 3 & 4 & 5 & 6 \\ 4 & 2 & 1 & 3 & 5 & 6 \end{bmatrix} \quad \text{and} \quad \beta = \begin{bmatrix} 1 & 2 & 3 & 4 & 5 & 6 \\ 2 & 3 & 5 & 4 & 1 & 6 \end{bmatrix}.$$
Express $\alpha\beta\alpha^{-1}$ as a product of cycles. If the result is a cycle, find its order as an element of Σ_6.
4. Prove that two disjoint cycles commute.
5. Prove that the order of a permutation σ considered as an element in the group Σ_n is the least common multiple of the lengths of its disjoint cycles.
6. Consider distinct indeterminates $x_1, \ldots, x_n$, $n > 1$, over the field $\mathbf{Q}$ and define the polynomial
$$\Delta(x_1, \ldots, x_n) = \prod_{1 \le i < j \le n} (x_j - x_i).$$
For $\sigma \in \Sigma_n$ define
$$\sigma\Delta(x_1, \ldots, x_n) = \prod_{1 \le i < j \le n} (x_{\sigma(j)} - x_{\sigma(i)}),$$
which polynomial is equal to $\text{sgn}(\sigma)\Delta(x_1, \ldots, x_n)$ where $\text{sgn}(\sigma) = \pm 1$.
   a. Prove that the mapping $\sigma \to \text{sgn}(\sigma)$ is a homomorphism from Σ_n to the multiplicative group $\{\pm 1\}$ of two elements.
   b. Prove that the mapping $\sigma \to \text{sgn}(\sigma)$ is surjective.
   c. We define A_n to be the kernel of this mapping, and call it the **alternating subgroup** of Σ_n. Prove that $[\Sigma_n : A_n] = 2$.

We shall call elements of A_n **even permutations**, and speak of A_n as the group of even permutations of n symbols. A permutation is called **odd** if it is not even.

7. Determine whether a transposition is an odd or an even permutation.
8. a. Prove that a permutation is even if and only if it can be expressed as the product of an even number of transpositions.
   b. For which values of m is an m-cycle an even permutation?
9. Write the permutation
$$\sigma = \begin{bmatrix} 1 & 2 & 3 & 4 & 5 & 6 & 7 & 8 \\ 4 & 2 & 5 & 1 & 7 & 3 & 8 & 6 \end{bmatrix}$$
as a product of disjoint cycles. What is the order of σ in the group Σ_8?
10. Prove that a finite group G of order n is isomorphic to a subgroup of the symmetric group Σ_n. (This statement is commonly known as Cayley's Theorem.) *Hint:* Prove that $g \to ag$, for fixed $a \in G$, is a permutation of the elements of G.

11. Note that $\sigma = [12][3]$ and $\tau = [123]$ generate the group Σ_3.
    a. Verify that $\sigma^2 = \tau^3 = \varepsilon$ and $\sigma\tau\sigma = \tau^2$.
    b. Prove that the mapping $\varphi: \Sigma_3 \to \Sigma_3$, defined by $\varphi(\sigma^i\tau^j) = \sigma^i$, is a homomorphism.
    c. Determine the kernel and image of the homomorphism in part (b).
    d. Prove that Σ_3 is isomorphic to the dihedral group of the triangle Γ_3 [see §6.2].
12. Using induction, prove that
$$[i_1, \ldots, i_r] = [i_1, i_2][i_2, i_3] \cdots [i_{r-1}, i_r].$$
13. Construct a nontrivial homomorphism from Σ_4 to Σ_3. *Hint:* Examine the subgroup
$$\{[1][2][3][4], [12][34], [13][24], [14][23]\},$$
which is isomorphic to Klein's Four-group.
14. Let G be the subgroup of Σ_8 generated by $\alpha = [1234][5876]$ and $\beta = [1537][2648]$. Prove that G is isomorphic to the quaternion group of 8 elements.
15. a. Prove that the dihedral group Γ_n [Example 7, §6.1] is isomorphic to a subgroup of Σ_n.
    b. Show that Γ_4 is not isomorphic to Σ_4.

§6.8 The Isomorphism Theorems of Group Theory

The isomorphism theorems furnish most important tools for more advanced results in group theory, such as the Jordan-Hölder Theorem in §7.6, and in the theory of rings and modules [§7.8]. The isomorphism theorems permit transfer of a problem which might be cumbersome in one group to another group where the solution may be easier.

Theorem 1 (*The Isomorphism Theorem*). For a homomorphism φ of G into G' with kernel K, there exists an isomorphism λ of the factor group G/K onto the image $\varphi(G) \subseteq G'$, such that $\varphi = \lambda \circ \pi$, where π is the canonical projection of G onto G/K.

In other words, there exists an isomorphism λ, such that the *diagram*

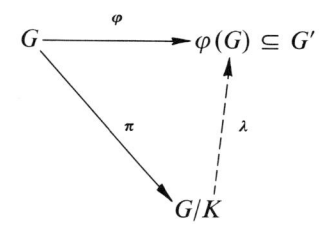

$$\begin{array}{ccc} G & \xrightarrow{\varphi} & \varphi(G) \subseteq G' \\ & \searrow_\pi \quad \uparrow_\lambda & \\ & G/K & \end{array}$$

commutes, meaning $\varphi = \lambda \circ \pi$.

Proof. (i) Define a mapping $\lambda: G/K \to G'$ by $\lambda(aK) = \varphi(a)$. First we must verify that λ is well-defined, i.e., is independent of the choice of a representing the coset aK. To this end, suppose that a and b each represent the coset $aK = bK$. Then $a^{-1}b \in K$ and $\varphi(a^{-1}b) = e'$, the identity of G', or $\varphi(a) = \varphi(b)$.

(ii) It is easy to check that λ is a group homomorphism:

$$\lambda(aK \cdot bK) = \lambda(abK) = \varphi(ab)$$
$$= \varphi(a)\varphi(b) = \lambda(aK)\lambda(bK).$$

(iii) If $\lambda(aK) = \varphi(a) = e'$, then $a \in K$ and hence $aK = K$. Thus the mapping λ is injective.

(iv) To verify the surjectivity of λ consider any $a' \in \varphi(G)$. By definition there exists an element $a \in G$ such that $\varphi(a) = a'$. Then $\lambda(aK) = \varphi(a) = a'$; hence λ is surjective. Thus λ is an isomorphism of G/K onto $\varphi(G)$ as asserted. That $\varphi = \lambda \circ \pi$ is a consequence of the definition of λ.

NOTE. Examples show that there may be many isomorphisms of G/K with $\varphi(G)$. Such isomorphisms can be constructed by following the isomorphism λ of the theorem by a nontrivial isomorphism α of the image $\varphi(G)$ onto itself, i.e., $gK \to \lambda(gK) = \varphi(g) \to \alpha(\varphi(g)) \in \varphi(G)$.

For example, suppose $\varphi(G)$ is isomorphic to the additive group $\mathbf{Z}_m$, $m > 2$. The mapping $[c] \to [a][c] = \alpha([c])$, $[c] \in \mathbf{Z}_m$, where $[a]$ is a prime residue modulo m, is an automorphism of $\mathbf{Z}_m$. Thus $\alpha \circ \lambda: G/K \to \varphi(G)$ is an isomorphism.

A simple relationship exists between the subgroups of G and those of its factor group modulo a normal subgroup K via homomorphic images. Suppose that $\pi = \pi_K$ is the canonical homomorphism of G *onto* the factor group $G^* = G/K$, where K is a normal subgroup of G. If H is a subgroup of G, then $\pi(H)$ is a subgroup of G^* by Proposition 4, §6.5. Further, if H is a normal subgroup of G, then by Proposition 5, §6.5, $\pi(H)$ is a normal subgroup of G^*.

Conversely, if H^* is a subgroup (normal subgroup) of G^*, then the complete inverse image $H = \pi^{-1}(H^*)$ of H^*—that is, the set of all $g \in G$ for which $\pi(g) \in H^*$—is a subgroup (normal subgroup) of G satisfying

$$\pi(H) = H^* \quad \text{and} \quad H \supseteq K.$$

This is a special case of the proofs requested in Exercise 15, §6.5. In summary we have the group-theoretic analogue of Proposition 3, §3.4.

Proposition. There is a one-one correspondence between the subgroups (normal subgroups) of G containing K and the subgroups (normal subgroups) of G/K. The correspondence is given by taking the projection mapping π on the subgroups $H \supseteq K$ and the inverse images of the subgroups H^* of G/K.

Theorem 2. If $\varphi: G \to G'$ and $\psi: G' \to G''$ are surjective group homomorphisms, then

$$G'/N' \cong G/N,$$

where $N' = \ker \psi$ and $N = \varphi^{-1}(N')$. Schematically,

$$G \xrightarrow{\varphi} G' \xrightarrow{\psi} G''$$
$$\cup\vert \qquad \cup\vert$$
$$\varphi^{-1}(N') = N \xrightarrow{\varphi} N' = \ker \psi.$$

Proof. By Exercise 16, §6.5, the composite $\psi \circ \varphi = \psi\varphi$ is a homomorphism of G onto G''. Further

$$\begin{aligned} N = \ker \psi\varphi &= \{g \in G : \psi(\varphi(g)) = e''\} \\ &= \{g \in G : \varphi(g) \in N' = \ker \psi\} \\ &= \varphi^{-1}(N'). \end{aligned}$$

Now by the Isomorphism Theorem (for the surjective mappings ψ and $\psi\varphi$)

$$G'' \cong G'/N' \qquad \text{and} \qquad G'' \cong G/N,$$

and hence $G'/N' \cong G/N$, as asserted.

Corollary 1. *If G is a finite group, then $[G : N] = [G' : N']$.*

Now suppose a group G has normal subgroups N and H, $H \subseteq N$. Consider the surjective homomorphisms $\varphi = \pi_H$ and ψ:

$$G \xrightarrow{\pi_H} G/H \xrightarrow{\psi} G/N,$$

where $\psi(gH) = gN$. Then, since $\ker \psi = N/H$, we have a second corollary.

Corollary 2. $G/N \cong (G/H)/(N/H)$.

Theorem 3. *If N is a normal subgroup of G and if H is a subgroup of G, then*

$$NH/N \cong H/(H \cap N).$$

Proof. First, the sets of cosets involved in the theorem are well-defined quotient groups. Propositions 1 and 2, §6.4, state that NH is a group (which contains N as a normal subgroup) and that $H \cap N$ is a normal subgroup of H.

Next the mapping $\lambda: H \to HN/N$, given by

$$\lambda(h) = hN \in HN/N \qquad \text{for } h \in H,$$

is surjective, the elements of HN being the products hn, $h \in H$, and $n \in N$ [see Proposition 2, §6.4]. Furthermore, λ is a homomorphism, $\lambda(hh') = (hh')N = (hN)(h'N) = \lambda(h)\lambda(h')$, for $h, h' \in H$. Also the kernel of λ is $N \cap H$, since $\lambda(h) = hN = N$ implies that $h \in N$, and thus $h \in N \cap H$. Therefore, by the Isomorphism Theorem, $H/(H \cap N) \cong HN/N$.

Corollary. *If $\varphi: G \to G'$ is a homomorphism with kernel K, then for any*

normal subgroup H of finite index in G,
$$[G:H] = [\varphi(G):\varphi(H)][K:H\cap K].$$

Proof. As in Theorem 2,
$$G \xrightarrow{\varphi} \varphi(G) \xrightarrow{\pi_{\varphi(H)}} G^*$$
and $G^* \cong \varphi(G)/\varphi(H)$. Since $KH = \varphi^{-1}(\varphi(H))$, we have
$$G^* \cong \varphi(G)/\varphi(H) \cong G/KH,$$
and hence $[G:KH] = [\varphi(G):\varphi(H)]$. Applied to the normal subgroup H and the subgroup K, Theorem 3 states that $[K:H\cap K] = [KH:H]$. Consequently, as asserted,
$$[G:H] = [G:KH][KH:H] = [\varphi(G):\varphi(H)][K:H\cap K].$$

Example. Consider the ideals H and K in $\mathbf{Z}$ generated by positive integers h and k, respectively, as additive subgroups of $\mathbf{Z}$. Then the LCM $[h,k] = m$ generates $H \cap K$ and the GCD $(h,k) = d$ generates $H+K$. (Because the group operation in $\mathbf{Z}$ is written additively the sum of subgroups replaces the product in the preceding discussion.)

The proof that $hk = dm$ [cf. Exercise 4, §2.7] is now a simple consequence of Theorem 3. (This is truly the hard way to prove that $hk = dm$!) By Corollary 2, §6.3,
$$k = [\mathbf{Z}:K] = [\mathbf{Z}:H+K][H+K:K] = d[H+K:K],$$
$$m = [\mathbf{Z}:H\cap K] = [\mathbf{Z}:H][H:H\cap K] = h[H:H\cap K].$$
As a consequence of Theorem 3, however, $[H+K:K] = [H:H\cap K]$. Thus $kd^{-1} = mh^{-1}$ or $kh = md$.

Exercises

1. Restate the Isomorphism Theorem (Theorem 1) for a ring homomorphism $\varphi: R \to R'$. Prove that there exists a ring isomorphism $\lambda: R/A \to \varphi(R)$ where A is the ideal $\ker\varphi$ (see the theorem, §3.2).

2. a. Do Exercise 27(c) of §3.6 by defining a ring homomorphism $\widetilde{\varphi}: \mathbf{Z} \to R_p/pR_p$. Prove that $\widetilde{\varphi}$ is surjective and conclude, using the Isomorphism Theorem for rings (Exercise 1, above), that
$$\mathbf{Z}/(p) \cong R_p/pR_p.$$
   b. Repeat part (a) with reference to Exercise 28 of §3.6.

3. With reference to the discussion of §3.5, in place of
$$\Lambda: \mathbf{Z}_m \to Z_1 + \cdots + Z_i + \cdots + Z_n$$
consider the map
$$\widetilde{\Lambda}: \mathbf{Z} \to Z_1 + \cdots + Z_i + \cdots + Z_n,$$

defined by $\widetilde{\Lambda}(a) = ([a]_1, \ldots, [a]_i, \ldots, [a]_n)$. Prove that $\ker \widetilde{\Lambda} = (m)$ and that $\widetilde{\Lambda}$ is a surjective ring homomorphism. Using the Isomorphism Theorem, conclude there exists an isomorphism

$$\Lambda: \mathbb{Z}_m \to Z_1 + \cdots + Z_n.$$

REMARK. A principal use of the Isomorphism Theorem for groups and rings, as well as analogous results for other algebraic structures, is to avoid having to define maps φ on sets of cosets in terms of representative elements, and then necessarily having to prove that such maps are well-defined, i.e., that their definition is independent of the choice of representatives of the cosets. To prove that such a φ is well-defined is essentially to repeat the proof of Theorem 1.

4. If H is a normal subgroup of a group G and is contained in a subgroup K of G, prove that $N_{G/H}(K/H) = N_G(K)/H$ [cf. Exercise 8, §6.2].
5. Consider an abelian group G in which every element different from the identity has infinite order, and let H be a subgroup with finite index in G. Define G^n and H^n to be the subgroups of nth powers of elements in G and H, respectively. Prove that $[G:H] = [G^n:H^n]$. *Hint:* G^n is the image of G for the homomorphism $\varphi: G \to G$ given by $\varphi(g) = g^n$.
6. Consider a finite dimensional vector space V over a field F as an additive group. If S and T are subgroups (i.e., subspaces), use Theorem 3 to prove that

$$\dim_F(S+T) = \dim_F S + \dim_F T - \dim_F(S \cap T)$$

[cf. Exercise 11, §4.2, and Exercise 17, §6.4].
7. Give an alternate proof of Theorem 3 by utilizing the canonical projection $\pi: G \to G/N$. Noting that $HN = \pi^{-1}(\pi(H))$, apply the Isomorphism Theorem to obtain $HN/N \cong \pi(H)$. Conclude the argument by observing that $\pi(H) \cong H/H \cap N$.

§6.9 Automorphisms, Center, Commutator Group

In this section we study automorphisms, a special class of group homomorphisms, and the concepts of center and commutator subgroup. The center is the largest subgroup of a group G whose elements commute with all $g \in G$, while the commutator subgroup is the smallest normal subgroup of G whose corresponding factor group is commutative. In investigating these concepts within the context of dihedral groups we again emphasize computations involving group elements.

A permutation (or one-one onto mapping) α of the elements of a group G is called an **automorphism** if α is *also* a homomorphism of G into itself. We denote the collection of automorphisms by $\text{Aut}(G)$. It is a subgroup of $\Sigma(G)$, the group of all permutations of G considered as a set [see §6.7]. Denote by ε the identity map on G, which is the identity element in the groups $\text{Aut}(G)$ and $\Sigma(G)$.

For a finite group G, the order of $\text{Aut}(G)$ can be appreciably smaller than that of $\Sigma(G)$. For example, if $G = C_n$, the cyclic group with n elements, then

$$[\text{Aut}(G) : \varepsilon] = \varphi(n) < n < n! = [\Sigma(G) : \varepsilon]$$

if $n \geq 3$, where $\varphi(n)$ is the Euler φ-function. This statement is verified in detail in Example 1.

Example 1. For the multiplicative cyclic group C_n of order n with generator c,

$$\text{Aut}(C_n) \cong \mathbf{U}_n,$$

the multiplicative group of units in the residue class ring $\mathbf{Z}_n$.

Since a homomorphism φ defined on a cyclic group is completely determined by the definition $\varphi(c)$, it is important to recall from Proposition 3, §6.6, that C_n has $\varphi(n)$ distinct generators c^s where $(s, n) = 1$. For any isomorphism $\alpha \in \text{Aut}(C_n)$, $\alpha(c)$ must generate C_n, because $o(\alpha(c)) = o(c) = n$. Therefore, $\alpha(c) = c^s$, for some s, $(s, n) = 1$. Conversely, the mapping $\lambda : C_n \to C_n$ given by $\lambda(c) = c^s$, for any s relatively prime to n is an automorphism. Thus, the distinct automorphisms of C_n are in one-one correspondence with the integers s, $1 \leq s \leq n$, $(s, n) = 1$.

It is convenient to describe the elements in $\text{Aut}(C_n)$ as α_s where $\alpha_s(c) = c^s$ and $1 \leq s \leq n$, $(s, n) = 1$. The mapping

$$\sigma : \text{Aut}(C_n) \to \mathbf{U}_n \subset \mathbf{Z}_n$$

given by $\sigma(\alpha_s) = [s]$ is a homomorphism, since

$$\sigma(\alpha_s \circ \alpha_{s'}) = \sigma(\alpha_{ss'}) = [ss']$$
$$= [s][s'] = \sigma(\alpha_s) \cdot \sigma(\alpha_{s'}).$$

Verification of injectivity and surjectivity is left to the reader.

Automorphisms of a group G of the form

$$g \to \sigma_x(g) = xgx^{-1} \quad \text{for all } g \in G \text{ and any fixed } x \in G$$

are called **inner automorphisms**. We denote the set of inner automorphisms of a group G by $I(G)$.

Using a Venn diagram to represent the inclusions and $\text{Hom}(G, G)$ the set of homomorphisms from a group G to itself, we have the following diagram.

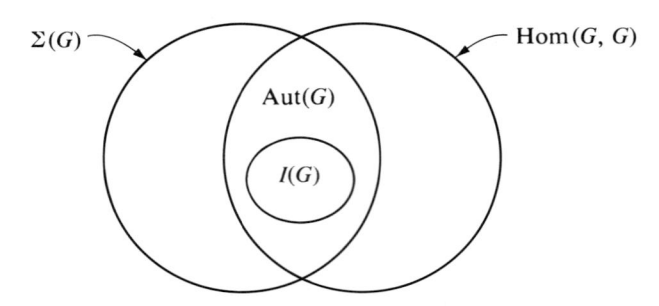

NOTE. Normal subgroups, defined in §6.4, are also called **invariant subgroups** because a subgroup H is normal if and only if $\sigma_x(H) = xHx^{-1} = H$ for all $x \in G$. In more technical terminology, we describe H as invariant (mapped to itself) under all inner automorphisms of G. In the more recent literature, greatly influenced by the French School, "invariant" is often replaced by "stable."

Proposition 1. The subset $I(G)$ of inner automorphisms of G is a normal subgroup of $\text{Aut}(G)$.

Proof. For all $g \in G$

$$(\sigma_x \circ \sigma_y)(g) = \sigma_x(ygy^{-1}) = x(ygy^{-1})x^{-1}$$
$$= (xy)g(xy)^{-1} = \sigma_{xy}(g).$$

Thus, the composite of two inner automorphisms is again an inner automorphism. Also for any $\sigma_x \in I(G)$, note that $\sigma_x \circ \sigma_{x^{-1}} = \sigma_e = \varepsilon$. Hence $\sigma_{x^{-1}} = (\sigma_x)^{-1} \in I(G)$, and $I(G)$ is a subgroup of $\text{Aut}(G)$.

Finally, for any $\alpha \in \text{Aut}(G)$,

$$(\alpha \circ \sigma_x \circ \alpha^{-1})(g) = \alpha(\sigma_x(\alpha^{-1}(g))) = \alpha(x\alpha^{-1}(g)x^{-1})$$
$$= \alpha(x)g(\alpha(x))^{-1} = \sigma_{\alpha(x)}(g),$$

for all $g \in G$, which proves the normality of $I(G)$ in $\text{Aut}(G)$.

For the group homomorphism $\Phi: G \to I(G)$ given by $\Phi(x) = \sigma_x$,

$$\ker \Phi = \{x \in G : \sigma_x(g) = \varepsilon(g) = g, \text{ for all } g \in G\}$$
$$= \{x \in G : xgx^{-1} = g, \text{ for all } g \in G\} = C,$$

called the **center** of G. Thus the Isomorphism Theorem [§6.8] implies $G/C \cong I(G)$. This fact is summarized in the following proposition.

Proposition 2. Let C be the center of a group G. Then $I(G) \cong G/C$.

For elements x, y in G, the product $xyx^{-1}y^{-1}$ is called the **commutator** of x and y. The group G^1 *generated* by all elements in G of the form $xyx^{-1}y^{-1}$, for $x, y \in G$, is called the **commutator subgroup** of G. The term *generated* means that G^1 is the group consisting of all finite products of all commutators of elements in G. Note that the product of two such finite products is a finite product of commutators, and that the inverse of a commutator is itself a commutator.

The group G^1 is also often called the **first derived group**, based on classical considerations from the Lie theory of "continuous groups," in which a chain of derived groups is developed (Sophus Lie, 1842–1899).

Proposition 3. A normal subgroup N of G contains G^1 if and only if G/N is abelian.

The proof is given by the following equivalences:

G/N is abelian $\Leftrightarrow (aN)(bN) = (bN)(aN) \quad$ for all $a, b \in G$

$\Leftrightarrow abN = baN$

$\Leftrightarrow a^{-1}b^{-1}abN = N$

$\Leftrightarrow a^{-1}b^{-1}ab \in N \quad$ for all $a, b \in G$

$\Leftrightarrow G^1 \subseteq N.$

The last equivalence is a consequence of the definition of G as the group generated by the set of all commutators.

The commutator subgroup G^1 is normal in G because for all $x, y, g \in G$,

$$\sigma_g(xyx^{-1}y^{-1}) = g(xyx^{-1}y^{-1})g^{-1}$$
$$= (gxg^{-1})(gyg^{-1})(gxg^{-1})^{-1}(gyg^{-1})^{-1},$$

and any element in G^1 is the product of commutators. Thus setting $N = G^1$ in Proposition 3 yields the following corollary.

Corollary. The factor group G/G^1 is abelian.

Since G^1 is a normal subgroup of G and is contained in any normal subgroup $N \subseteq G$ for which G/N is abelian, we can describe (or define) G^1 as the "smallest" normal subgroup of G whose corresponding factor group is abelian.

Example 2. The automorphism group $\text{Aut}(V)$ of Klein's Four-group V is isomorphic to the symmetric group Σ_3 of three symbols [cf. §6.7]. The group V consists of the four elements

$$\{a, b, ab = ba, a^2 = b^2 = e\}.$$

Any $\varphi \in \text{Aut}(V)$ can be described by

$$\varphi(a) = a^{\alpha_{11}}b^{\alpha_{12}},$$

$$\varphi(b) = a^{\alpha_{21}}b^{\alpha_{22}},$$

where $\alpha_{ij} \in \mathbf{Z}_2$, since the exponents of a, b are counted modulo 2. It is convenient to associate to $\varphi \in \text{Aut}(V)$ the 2×2 matrix

$$\lambda(\varphi) = \begin{bmatrix} \alpha_{11} & \alpha_{12} \\ \alpha_{21} & \alpha_{22} \end{bmatrix}$$

with coefficients in $\mathbf{Z}_2$. Since φ is an automorphism, $\varphi(a)$ and $\varphi(b)$ must be distinct elements of order 2; thus the matrix $\lambda(\varphi)$ must have linearly independent nonzero rows. In other words, as in §4.5, $\lambda(\varphi)$ is a nonsingular matrix. Conversely, each nonsingular 2×2 matrix with coefficients in $\mathbf{Z}_2$ determines a unique automorphism of V. In fact the mapping $\lambda: \text{Aut}(V) \to GL(2, \mathbf{Z}_2)$, the (general linear) group of nonsingular 2×2 matrices with coefficients in $\mathbf{Z}_2$, is an isomorphism.

§6.9 Automorphisms, Center, Commutator Group

Now to show that $GL(2, \mathbf{Z}_2)$ is isomorphic to Σ_3, let

$$\sigma = \begin{bmatrix} [0] & [1] \\ [1] & [0] \end{bmatrix} \quad \text{and} \quad \tau = \begin{bmatrix} [1] & [1] \\ [1] & [0] \end{bmatrix}.$$

Then

$$\sigma^2 = \tau^3 \begin{bmatrix} [1] & [0] \\ [0] & [1] \end{bmatrix}$$

and $\sigma\tau\sigma = \tau^2$. The generators σ and τ of $GL(2, \mathbf{Z}_2)$ are subject to the same relations as the generators

$$\alpha = \begin{bmatrix} 1 & 2 & 3 \\ 2 & 1 & 3 \end{bmatrix} \quad \text{and} \quad \beta = \begin{bmatrix} 1 & 2 & 3 \\ 2 & 3 & 1 \end{bmatrix}$$

of Σ_3. Thus the mapping $\lambda': GL(2, \mathbf{Z}_2) \to \Sigma_3$ described by $\lambda'(\sigma) = \alpha$ and $\lambda'(\tau) = \beta$ is an isomorphism. Composing the mappings λ and λ' we obtain an isomorphism $\lambda'': \text{Aut}(V) \to \Sigma_3$.

To illustrate the concepts of center and commutator subgroup while providing experience with computations involving group elements, we examine the dihedral groups Γ_n, $n \geq 3$, in detail. The group Γ_n can be described by

$$\Gamma_n = \{a^s b^t : s, t \in \mathbf{Z}\}$$

where $a^n = b^2 = e$ and $ba = a^{-1}b$ [cf. Examples 6(f) and 7 of §6.1].

Example 3. The center of Γ_n, $n \geq 3$, is

$$\{e\} \quad \text{if } 2 \nmid n, \quad \text{and} \quad \{e, a^{n/2}\} \quad \text{if } 2 \mid n.$$

An element $a^s b^t \in \Gamma_n$ belongs to the center C of Γ_n if and only if

$$a(a^s b^t) = (a^s b^t)a,$$
$$b(a^s b^t) = (a^s b^t)b,$$

because the "general" element of Γ has the form $a^x b^y$. To describe the center more explicitly, consider separately the cases $t = 0$, $t = 1$.

The Case $t = 0$. In this case $a^s b^t = a^s$, and conditions on s alone have to be found such that

$$aa^s = a^s a \quad \text{(no condition on } s \text{ ensues)}$$

and $$ba^s = a^s b.$$

The relation $ba = a^{-1}b$ yields

$$ba^2 = (ba)a = (a^{-1}b)a = a^{-1}a^{-1}b = a^{-2}b,$$

and in general, $ba^s = a^{-s}b$. Therefore for an element a^s in the center, $a^s b = ba^s = a^{-s}b$ implies that

$$a^{2s} = e \quad \text{or} \quad n \mid 2s.$$

If $2 \nmid n$, then $n \mid s$, so that $a^s = e$, in which case no element of the form $a^s \neq e$ can belong to C. If $2 \mid n$, however, then for $s = n/2$ we have $a^{2s} = a^n = e$, and thus $a^{n/2} \in C$.

The Case $t = 1$. If there exist elements of the form $a^s b$ such that $a(a^s b) = (a^s b)a$, then
$$a^{s+1}b = a(a^s b) = a^s ba = a^s(a^{-1}b) = a^{s-1}b.$$
Consequently $a^{s+1} = a^{s-1}$, or $a^2 = e$, contradicting the assumption that $n \geq 3$. Hence no element of the form $a^s b$ lies in the center of Γ_n.

Example 4. The factor group $\Gamma_n^* = \Gamma_n/\Gamma_n^1$ of Γ_n modulo its commutator subgroup Γ_n^1 is isomorphic, for $n \geq 3$, to $\mathbf{Z}_2$ if $2 \nmid n$, and to the Klein Four-group if $2 \mid n$.

The elements of the factor group Γ_n^* can be denoted by
$$[a^x b^y], [a^x], [b^y], [e] = \Gamma_n^1,$$
where $0 \leq x < n$ and $0 \leq y < 2$, because $[a^n] = [b^2] = [e]$. Proposition 3 asserts that Γ_n^* is abelian. Thus
$$[b][a] = [a^{-1}][b] = [b][a]^{-1} \quad \text{and} \quad [a]^2 = [e].$$
Consequently $a^2 \in \Gamma_n^1$. We now distinguish two cases.

The Case $2 \nmid n$. If $2 \nmid n$, then
$$a = ae = aa^n = a^{n+1} = a^{2k} = (a^2)^k \in \Gamma_n^1.$$
Because $bab^{-1} = a^{-1} \in \langle a \rangle$, the subgroup $\langle a \rangle$ (of order n) of Γ_n generated by a is a normal subgroup of Γ_n. Since $\Gamma_n/\langle a \rangle$ has order 2 it is abelian; therefore by Proposition 3, $\langle a \rangle \supseteq \Gamma_n^1$. But $a \in \Gamma_n^1$ implies that $\langle a \rangle \subseteq \Gamma_n^1$. Thus, when $2 \nmid n$, $\Gamma_n^1 = \langle a \rangle$, and $\Gamma_n/\Gamma_n^1 \cong \mathbf{Z}_2$.

The Case $2 \mid n$. As before $a^2 \in \Gamma_n^1$ and so $\langle a^2 \rangle \subseteq \Gamma_n^1$. Moreover $\langle a^2 \rangle$ is a normal subgroup of Γ_n since $ba^2 = a^{-2}b$ (in Example 3 we had $ba^s = a^{-s}b$) implies that $ba^2b^{-1} = (a^2)^{-1} \in \langle a^2 \rangle$. The factor group $\Gamma_n/\langle a^2 \rangle$ has order 4 and so is abelian [cf. Exercise 9, §6.6]. By Proposition 3, $\langle a^2 \rangle \supseteq \Gamma_n^1$, and therefore $\Gamma_n^1 = \langle a^2 \rangle$.

The four elements of the factor group $\Gamma_n^* = \Gamma_n/\Gamma_n^1$ are
$$e' = \langle a^2 \rangle$$
$$a' = a \langle a^2 \rangle$$
$$b' = b \langle a^2 \rangle = \langle a^2 \rangle b$$
$$a'b' = ab \langle a^2 \rangle = a \langle a^2 \rangle b.$$
Furthermore Γ_n^* is isomorphic to the Klein Four-group, as each element different from e' has order 2.

Example 5. We shall prove in §7.4 (Proposition 3) that the center of a group of prime power order is necessarily nontrivial. As a consequence any group of order p^2 is abelian, p a prime.

Example 6. As a final example, consider again the nonabelian group
$$G = \{a^m b^n : m, n \in \mathbf{Z}\},$$
where $a^{p^2} = b^p = e$ and $bab^{-1} = a^{1+p}$, of order p^3, where p is an odd prime, as in Example 2, §6.4. Since
$$ba^2 b^{-1} = (bab^{-1})(bab^{-1}) = a^{1+p} \cdot a^{1+p}$$
$$= a^{2(1+p)},$$

§6.9   Automorphisms, Center, Commutator Group

we can show by induction that for all α, $1 \leq \alpha \leq p^2$,
$$b a^\alpha b^{-1} = a^{\alpha(1+p)}.$$
Furthermore,
$$\begin{aligned} b^2 a^\alpha b^{-2} &= b(b a^\alpha b^{-1}) b^{-1} \\ &= b a^{\alpha(1+p)} b^{-1} = a^{\alpha(1+p)(1+p)} \\ &= a^{\alpha(1+p)^2}, \end{aligned}$$
and a second induction argument shows that for all β, $1 \leq \beta \leq p$,
$$b^\beta a^\alpha b^{-\beta} = a^{\alpha(1+p)^\beta} \quad \text{or} \quad b^\beta a^\alpha = a^{\alpha(1+p)^\beta} b^\beta.$$

As noted in Example 2, §6.4, the subgroup N generated by a^p is a normal subgroup of G and G/N is abelian. Hence $N \supseteq G^1$. However, since $a^p = bab^{-1}a^{-1} \in G^1$, the subgroup $\langle a^p \rangle = N$ is contained in G^1. Thus $G^1 = \langle a^p \rangle$.

To determine the center C of G, we consider conditions on m, n such that $a^m b^n \in C$. First, we must have

(*) $$\begin{aligned} a^m b^n &= b(a^m b^n) b^{-1} = (b a^m b^{-1}) b^n \\ &= a^{m(1+p)} b^n \end{aligned}$$

or $a^{m(1+p)} = a^m$. Hence $p^2 | mp$ or $p | m$. Therefore elements of the center must have the form $a^{sp} b^n$. In addition,
$$a^{sp} b^n = a^{-1}(a^{sp} b^n) a = a^{sp}(a^{-1} b^n a)$$
or
$$\begin{aligned} b^n &= a^{-1} b^n a = (a^{-1} b a)^n \\ &= (a^p b)^n = a^{pn} b^n, \end{aligned}$$
since $a^p \in C$ from equation (*) with $m = p$, $n = 0$. Therefore $a^{p^n} = e$, or $p | n$. Hence $b^n = e$. Thus $C = \{a^{sp} : 0 \leq s < p\}$, the group of order p generated by a^p.

Exercises

1. a. Determine the commutator subgroup $\Gamma_7^{\,1}$ of the dihedral group Γ_7 of order 14.
    b. Find a subgroup of the multiplicative group $\mathbf{C}^*$ of the nonzero complex numbers which is isomorphic to the factor group $\Gamma_7/\Gamma_7^{\,1}$.
2. Let G be the multiplicative abelian group of order p^m in which each element $g \neq e$ has order p.
    a. Prove that $\text{Aut}(G)$ is isomorphic to the group of units (invertible elements) in the ring of $m \times m$ matrices with coefficients in the field $\mathbf{Z}_p$. Hint: Use the fact that G is isomorphic to the standard m-dimensional vector space V over $\mathbf{Z}_p$ considered as an additive group [cf. Example 2].
    b. Find the order of $\text{Aut}(G)$.
3. Prove that if H is a normal subgroup of order 2 in a group G, then H is contained in the center of G.
4. With reference to Example 6, prove that the generator a is a power of the commutator $aba^{-1}b^{-1}$.

5. Let G be the group of Example 4.
   a. Determine all homomorphisms of the factor group G/C into the multiplicative group T of all complex numbers of absolute value 1.
   b. Write b^2a^2ba in the form b^xa^y.
6. Let G be a finite group of more than two elements, but not an abelian group of order 2^s, for any $s \in \mathbf{N}$. Prove that G has at least one automorphism α distinct from the identity.
7. Consider the integers as an additive group. Prove that $\text{Aut}(\mathbf{Z})$ is cyclic of order 2.
8. Let G be the quaternion group of 8 elements [Exercise 6, §6.5].
   a. Determine the center C of G.
   b. Determine the commutator subgroup G^1.
   c. Prove that the automorphism group $\text{Aut}(G)$ is isomorphic to Σ_4.
   d. Prove that the group of inner automorphisms $I(G)$ has 4 elements.
   e. Is $I(G)$ cyclic or is it isomorphic to Klein's Four-group?
9. With reference to Example 7(b), §6.1, and Exercise 7, §6.4, consider the dihedral group Γ_4 of order 8.
   a. Find the automorphism $\text{Aut}(\Gamma_4)$.
   b. Find the group $I(\Gamma_4)$ of inner automorphisms.
   c. Prove that $o((\text{Aut}(\Gamma_4)) = 8$. (In fact, $\text{Aut}(\Gamma_4) \cong \Gamma_4$.)
   d. Prove that $o(I(\Gamma_4)) = 4$.
   e. Describe the factor group $\text{Aut}(\Gamma_4)/I(\Gamma_4)$.
10. Prove that if x is the only element of order 2 in a group G, then x must lie in the center of G.
11. Let G be a finite group in which $(ab)^n = a^n b^n$ for all $a, b \in G$ and some $n \in \mathbf{N}$. Denote by $G^{(n)}$ the set $\{a^n : a \in G\}$ and by $G_{(n)}$ the set $\{x \in G : x^n = e\}$. Prove the following:
    a. $G^{(n)}$ and $G_{(n)}$ are normal subgroups of G.
    b. $[G^{(n)} : e] = [G : G_{(n)}]$.
12. a. If a group G of order mn contains a normal cyclic subgroup N of order m whose quotient group G/N is cyclic, prove that G is generated by two elements a, b for which $a^m = e$, $b^n = a^r$, and $bab^{-1} = a^s$, where the integers r, s satisfy $r(s-1) \equiv s^n - 1 \equiv 0 \pmod{m}$. *Hint:* Let b be a representative of a generator of G/N.
    b. Give an example of such a group, for $m, n > 1$.
13. a. With reference to Exercise 12, express $b^\alpha a^\beta b^{-\alpha}$ as a power of a.
    b. Determine the commutator subgroup in the general case in Exercise 12(a) and for your example in Exercise 12(b).
14. In a group G prove that the commutator group is the intersection of all subgroups which contain all of the commutators in G.

§6.10 Direct Product

The concept of direct product, introduced by Otto Hölder (1859–1937), has applications to the construction of new groups from given ones and to the analysis of a given group as a composite of subgroups. These applications are the group-theoretic analogues of the concepts of (internal

§6.10 Direct Product

and external) direct sums (products) encountered in ring theory [§3.3] and in vector spaces [§4.1]. Recall also the explicit examples of direct sums in the rings $\mathbf{Z}_m$ in §3.5.

When the group operation is additive, or in the case of rings, we speak of *direct sums*; when the operation is multiplicative, we speak of *direct products*.

The product $H \cdot K$ of two subgroups H and K of a group G was defined at the end of §6.2 to be the least subgroup of G containing both H and K. Proposition 2, §6.4, asserted that if either H or K is a normal subgroup of G, then the product $H \cdot K$ equals the set $\{hk : h \in H, k \in K\}$. *In this section, we consider only normal subgroups of a group G.*

A group G is said to be the **internal direct product** $G = N_1 \otimes \cdots \otimes N_s$ of the normal subgroups N_i, $1 \leq i \leq s$, if

(i) $G = N_1 \cdots N_s$ (the product of the N_i),
(ii) $(N_1 \cdots N_i) \cap N_{i+1} = \{e\}$, $1 \leq i < s$.

An equivalent statement of this definition is that G is an internal direct product if

(i) $G = N_1 \cdots N_s$,
(ii)* every $g \in G$ has a *unique* expression as a product $g = g_1 \cdots g_s$ with components $g_i \in N_i$, $1 \leq i \leq s$.

Many authors use the notation $\otimes$ for the tensor product, a concept not considered in this book. Hence no confusion should result in our use of this symbol to designate the internal direct product.

To prove this equivalence, first assume (ii) and suppose that g has two expressions as a product of elements of the N_i:

$$g = g_1 \cdots g_s = h_1 \cdots h_s, \quad \text{with } g_i, h_i \in N_i.$$

Then $\quad x = h_{s-1}^{-1} \cdots h_1^{-1} g_1 \cdots g_{s-1} = h_s g_s^{-1} \in N_s.$

In addition, being the product of elements in $N_1, \ldots, N_{s-1}$, the element x belongs to $N_1 \cdots N_{s-1}$. Hence by (ii), $h_s g_s^{-1} = e$ or $g_s = h_s$. We can show successively that $g_{s-1} = h_{s-1}, \ldots, g_1 = h_1$.

Conversely, suppose that $y \in N_1 \cdots N_i \cap N_{i+1}$. Then y has two expressions as a product of elements of N_j, $1 \leq j \leq s$:

$$y = y_1 \cdots y_i e \cdots e \quad \text{and} \quad y = e \cdots eye \cdots e.$$
$\qquad\qquad\quad \uparrow \quad\ \uparrow \qquad\qquad\qquad \uparrow \quad\ \uparrow \quad\ \uparrow$
$\qquad\qquad \in N_{i+1}\ \in N_s \qquad\qquad\quad \in N_1\ \in N_{i+1}\ \in N_s$

By the uniqueness assumption (ii)*, we must have $y_j = e$, $1 \leq j \leq i$, and $y = e$. But then $(N_1 \cdots N_i) \cap N_{i+1} = \{e\}$ for all j, $1 \leq i < s$.

If $G = N_1 \otimes N_2$, then $g_1 g_2 = g_2 g_1$ for $g_1 \in N_1$, $g_2 \in N_2$. First $g_1 g_2 g_1^{-1} g_2^{-1} = g_1 (g_2 g_1^{-1} g_2^{-1}) \in N_1$ and $(g_1 g_2 g_1^{-1}) g_2^{-1} \in N_2$ by the normality of N_1 and N_2. Consequently $g_1 g_2 g_1^{-1} g_2^{-1} \in N_1 \cap N_2 = \{e\}$. An inductive argument proves a corresponding result for the product of any finite number of normal subgroups.

Proposition 1. If $G = N_1 \otimes \cdots \otimes N_s$ is the direct product of normal subgroups N_i, then for $g_i \in N_i$ and $g_j \in N_j$, $g_i g_j = g_j g_i$, where $i \neq j$.

This is *not* the same as saying that G is an abelian group.

Consider s groups G_i with respective identity elements e_i. The set $G = \{(g_1, \ldots, g_s) : g_i \in G_i\}$ with

(i) $(g_1, \ldots, g_s) = (h_1, \ldots, h_s)$ if $g_i = h_i$, $1 \leq i \leq s$,

and

(ii) $(g_1, \ldots, g_s)(h_1, \ldots, h_s) = (g_1 h_1, \ldots, g_s h_s)$

is called the **external direct product** $G_1 \dot{\times} \cdots \dot{\times} G_s$ of the groups G_i.

That the set G is a group is readily checked because the multiplication is defined componentwise. Associativity follows from the fact that the product in each of the components G_i is associative. Furthermore, $(e_1, \ldots, e_s) = e$ is the unit for multiplication, and $(g_1^{-1}, \ldots, g_s^{-1})$ is the inverse of $(g_1, \ldots, g_s)$.

Proposition 2. The (external) direct product $G = G_1 \dot{\times} \cdots \dot{\times} G_s$ of groups G_i, $1 \leq i \leq s$, contains subgroups N_i isomorphic to G_i, $1 \leq i \leq s$, and $G = N_1 \otimes \cdots \otimes N_s$.

Proof. Define mappings $\varphi_i : G_i \to G$ by $\varphi(g_i) = (e_1, \ldots, g_i, \ldots, e_s)$ for each i, $1 \leq i \leq s$. The image subgroup $N_i = \varphi_i(G_i) \subset G$ is isomorphic to G_i, since $\ker \varphi_i = \{e_i\}$. Further, N_i is a normal subgroup of G because

$$(h_1, \ldots, h_s) \varphi_i(g_i)(h_1, \ldots, h_s)^{-1}$$
$$= (h_1 e_1 h_1^{-1}, \ldots, h_i g_i h_i^{-1}, \ldots, h_s e_s h_s^{-1})$$
$$= (e_1, \ldots, h_i g_i h_i^{-1}, \ldots, e_s)$$
$$= \varphi_i(h_i g_i h_i^{-1}) \in N_i,$$

where $h_j \in G_j$ for $1 \leq j \leq s$, and $h_i g_i h_i^{-1} \in G_i$. By the definition of equality in G, $(g_1, \ldots, g_i, \ldots, g_s)$ equals the product $\prod_{i=1}^{s}(e_1, \ldots, g_i, \ldots, e_s)$ of unique elements $(e_1, \ldots, g_i, \ldots, e_s) \in N_i$. Therefore G is the (internal) direct product of the groups N_i as asserted.

Proposition 3. Let G be a cyclic group of order $n = n_1 \cdots n_s$, where $(n_i, n_j) = 1$ for $i \neq j$. Then G is the internal direct product of s subgroups G_i, where $o(G_i) = n_i$.

This proposition is the group-theoretic analogue of the (internal) direct sum decomposition of $\mathbf{Z}_n$ in §3.5. (Recall the theorem in §6.6, which states $G \cong \mathbf{Z}_n$.)

To prove Proposition 3 without reference to §3.5, consider a generator g of G, and set $m_i = n/n_i$, $1 \leq i \leq s$. Then, as in Proposition 2, §6.6, the

§6.10 **Direct Product** 215

element g^{m_i} generates a subgroup $<g^{m_i}> = G_i$ of order n_i. To show that $G = G_1 \otimes \cdots \otimes G_s$, note first that $(m_1, m_2, \ldots, m_s) = 1$ and that there exist integers h_i such that $1 = m_1 h_1 + \cdots + m_s h_s$. For an arbitrary element $a \in G$,

$$a = a^1 = \prod_{i=1}^{s} (g^{m_i})^{h_i} \in G_1 \cdots G_s;$$

hence $G = G_1 \cdots G_s$.

To prove that this product is direct, note that $x \in (G_1 \cdots G_i) \cap G_{i+1}$ implies $o(x) | n_{i+1}$ (since $x \in G_{i+1}$) and $o(x) | n_1 \cdots n_i$ (since x is the product of elements x_i such that $o(x_i) | n_i$). Consequently $x = e$, because $(n_1 \cdots n_i, n_{i+1}) = 1$ and hence $o(x) | 1$. Thus, G equals the internal direct product $G_1 \otimes \cdots \otimes G_s$, as asserted.

Proposition 4. Suppose that G is the (internal) direct product $G_1 \otimes G_2$, and that H_1 and H_2 are normal subgroups of G_1 and G_2, respectively. Then the factor group $(G_1 \otimes G_2)/(H_1 \otimes H_2)$ is isomorphic to the (external) direct product $G_1/H_1 \dot\times G_2/H_2$.

Proof. Recall from Proposition 1 that $G = G_1 \otimes G_2$ means that $g \in G$ can be expressed uniquely as $g = g_1 g_2$, for $g_1 \in G_1$, $g_2 \in G_2$, and moreover that $g_1 g_2 = g_2 g_1$. We use these facts to construct the mapping

$$\psi: G \to G_1/H_1 \dot\times G_2/H_2$$

given by $\psi(g) = (g_1 H_1, g_2 H_2)$ for $g = g_1 g_2$. The uniqueness of the expression of $g \in G$ as the product of elements in G_1 and G_2 implies that ψ is well-defined. We now prove that ψ is a surjective homomorphism whose kernel is $H_1 \otimes H_2$.

First, ψ is a homomorphism. For $g = g_1 g_2$, $g' = g_1' g_2'$ in G,

$$\begin{aligned}\psi(gg') &= \psi(g_1 g_2 g_1' g_2') \\ &= \psi(g_1 g_1' g_2 g_2') \\ &= (g_1 g_1' H_1, g_2 g_2' H_2) \\ &= (g_1 H_1, g_2 H_2)(g_1' H_1, g_2' H_2) = \psi(g)\psi(g').\end{aligned}$$

If $g = g_1 g_2 \in \ker \psi$, then

$$\psi(g) = (g_1 H_1, g_2 H_2) = (eH_1, eH_2)$$

implies that $g_1 \in H_1$, $g_2 \in H_2$. Hence $\ker \psi \subseteq H_1 \otimes H_2$. Conversely $\psi(h_1 h_2) = (eH_1, eH_2)$, for all $h_i \in H_i$, $i = 1, 2$, and so $\ker \psi = H_1 \otimes H_2$. Next, the homomorphism ψ is surjective because the general element in $G_1/H_1 \dot\times G_2/H_2$ is $(g_1 H_1, g_2 H_2)$, the image of $g_1 g_2 \in G$ under ψ.

Since ψ is a surjective homomorphism with $\ker \psi = H_1 \otimes H_2$, the Isomorphism Theorem of §6.8 assures the existence of an isomorphism

$$\varphi: G/(H_1 \otimes H_2) \to G_1/H_1 \dot\times G_2/H_2.$$

Explicitly, the map φ is given by

$$\varphi[g(H_1 \otimes H_2)] = (g_1 H_1, g_2 H_2),$$

where $g = g_1 g_2$.

Exercises

1. Let G be the external direct sum of $\mathbf{Z}_n$ and $\mathbf{Z}_m$, considered as additive groups, where $(m,n) = 1$.
   a. Prove that G is isomorphic to $\mathbf{Z}_{mn}$.
   b. Find the image of the subgroup $\mathbf{Z}_m + \{[0]_n\}$ of G in the group $\mathbf{Z}_{mn}$.
2. Let G be the external direct sum of the additive groups $\mathbf{Z}_8$ and $\mathbf{Z}_6$.
   a. Determine the addition table of the factor group G/H where H is the subgroup generated by $x = ([4]_8, [3]_6)$, $[4]_8 \in \mathbf{Z}_8$ and $[3]_6 \in \mathbf{Z}_6$.
   b. Determine the addition table of the factor group G/K where $K = [4]_8 \mathbf{Z}_8 + [3]_6 \mathbf{Z}_6$.
3. Find all homomorphisms of the direct sum $G = \mathbf{Z}_3 + \mathbf{Z}_3$ into $\mathbf{Z}_6$.
4. Find all homomorphisms of the direct sum $G = \mathbf{Z}_2 + \mathbf{Z}_2$ into the direct sum $\Gamma = \mathbf{Z}_2 + \mathbf{Z}_2 + \mathbf{Z}_6$.
5. Consider a group $G = HN$ where H is a subgroup of G and N is a normal subgroup of G such that $H \cap N = \{e\}$. Prove that the factor group G/N is isomorphic to H.
6. If G is the (external) direct product of subgroups H, K, and if L is a subgroup of G containing $H \stackrel{*}{\times} \{e_K\}$ prove that $L = H \stackrel{*}{\times} (L \cap (\{e_H\} \times K))$.
7. Consider a finite abelian group G for which $o(g) = p$ for all $g \neq e$ in G. Prove that $o(G) = p^m$, for some $m \in \mathbf{N}$.
8. a. Prove Proposition 1.
   b. Conclude that elements of N_i commute with those in $N_1 \otimes \cdots \otimes N_{i-1} \otimes N_{i+1} \otimes \cdots \otimes N_s$.
9. For distinct groups G and G', prove that
   $$G \stackrel{*}{\times} G' \cong G' \stackrel{*}{\times} G,$$
   but that these two (external) direct products are not equal.
10. If a group G is the product NK of normal subgroups N, K and if $M = N \cap K$, prove that
    $$G/M = N/M \otimes K/M.$$
11. The following resolution of the structure of the multiplicative group of units $\mathbf{U}_n$ (group of prime residue classes) in the residue class ring $\mathbf{Z}_n$ is due to Gauss (1801). Prove each statement.
    a. If p is an odd prime, then $\mathbf{U}_{p^m}$ and $\mathbf{U}_{2p^m}$ are cyclic groups of order $(p-1)p^{m-1}$ for all $m \in \mathbf{N}$.
       (i) Prove by induction on m the existence of an element w of order $p-1$ in $\mathbf{U}_{p^m}$.
       (ii) Prove by induction on m the existence of an element z of order p^{m-1} in $\mathbf{U}_{p^m}$.
       (iii) Note that w and z generate $\mathbf{U}_{p^m}$, and since $(o(w), o(z)) = 1$, that $o(wz) = (p-1)p^{m-1}$.
       (iv) Observe that $\mathbf{U}_{2p^m} \cong \mathbf{U}_2 \stackrel{*}{\times} \mathbf{U}_{p^m} \cong \mathbf{U}_{p^m}$.
    b. If $p = 2$, then
       (i) $\mathbf{U}_2$ is cyclic of order $2^0 = 1$.
       (ii) $\mathbf{U}_{2^2}$ is cyclic of order $2^1 = 2$.

(iii) U_{2^m}, for $m \geq 3$, is the direct product of a subgroup of order 2, generated by $[-1]_{2^m}$ and a cyclic subgroup of order 2^{m-2}, generated by $[5]_{2^m}$.

 c. If the modulus n is not of the form p^m, $2p^m$ (p an odd prime), 2, or 2^2, then U_n is not a cyclic group.

12. Suppose that G is a finite abelian group of **minimal exponent** n (the least positive integer such that $g^n = e$ for all $g \in G$). Assume n has a proper factorization $n = mq$ where $(m, q) = 1$. Let S, T be the subsets of elements s, t, respectively, in G such that $s^m = t^q = e$.
 a. Prove that S and T are subgroups.
 b. Prove that the mapping $\varphi: S \overset{*}{\times} T \to G$ given by $f((s, t)) = st$ is an isomorphism.
 c. Conclude that $G = S \otimes T$.

§6.11 Homomorphisms of Abelian Groups

This section develops the group-theoretic analogue of the argument in §4.3 that the set $\operatorname{Hom}_F(U, V)$ of linear transformations from one vector space to another (over the same field F) is itself a vector space over F. Explicit examples show how the definitions lead from detailed questions on homomorphisms to simple problems in the theory of integers.

Let G and Γ be two multiplicatively written *abelian* groups. Consider the totality $\operatorname{Hom}(G, \Gamma)$ of homomorphisms of G into Γ. The set $\operatorname{Hom}(G, \Gamma)$ is itself an abelian group when the product in $\operatorname{Hom}(G, \Gamma)$ of two homomorphisms φ, ψ is defined to be the mapping $G \to \Gamma$, described for all $g \in G$ by

$$g \to (\varphi\psi)(g) = \varphi(g)\psi(g).$$

Since $\varphi(g)$ and $\psi(g)$ are elements in the *abelian* group Γ, $\varphi\psi = \psi\varphi$. More importantly, $\varphi\psi$ is again a homomorphism: for all $g, h \in G$,

$$(\varphi\psi)(gh) = \varphi(gh)\psi(gh)$$
$$= \varphi(g)\varphi(h)\psi(g)\psi(h)$$
$$= \varphi(g)\psi(g)\varphi(h)\psi(h) = (\varphi\psi)(g) \cdot (\varphi\psi)(h).$$

The homomorphism $\varepsilon: G \to \Gamma$, defined by $\varepsilon(g) = e$, the identity element in Γ, serves as an identity element in $\operatorname{Hom}(G, \Gamma)$ with respect to the product just defined, since

$$(\varphi\varepsilon)(g) = \varphi(g)\varepsilon(g) = \varphi(g)e = \varphi(g) \qquad \text{for all } g \in G.$$

The proof of the associativity, $(\varphi\psi)\lambda = \varphi(\psi\lambda)$, of multiplication in $\operatorname{Hom}(G, \Gamma)$ is left as an exercise.

Finally, for a given $\varphi \in \operatorname{Hom}(G, \Gamma)$ the mapping $\varphi^{-1}: G \to \Gamma$ defined by $\varphi^{-1}(g) = [\varphi(g)]^{-1}$ is the inverse of φ (in the sense of the law of composition for $\operatorname{Hom}(G, \Gamma)$) because $\varphi\varphi^{-1} = \varphi^{-1}\varphi = \varepsilon$. (Check that φ^{-1} is itself a homomorphism from G to Γ.) Thus $\text{Hom}(G, \Gamma)$ is an abelian group, as asserted.

Propositions 1 and 2 develop simple properties of the group $\text{Hom}(G, \Gamma)$ with respect to (internal and external) direct products.

Proposition 1. If $G = G_1 \otimes G_2$, then
$$\text{Hom}(G_1 \otimes G_2, \Gamma) \cong \text{Hom}(G_1, \Gamma) \mathbin{\dot\otimes} \text{Hom}(G_2, \Gamma).$$

To begin the proof, define for each $\varphi \in \text{Hom}(G, \Gamma)$ the "restriction" mappings $\rho_i(\varphi): G_i \to \Gamma$, for $i = 1, 2$, by
$$\rho_i(\varphi)(g_i) = \varphi(g_i) \in \Gamma$$
for all $g_i \in G_i$. Then $\rho_i(\varphi) \in \text{Hom}(G_i, \Gamma)$ because
$$\rho_i(\varphi)(g_i g_i') = \varphi(g_i g_i')$$
$$= \varphi(g_i)\varphi(g_i') = \rho_i(\varphi)(g_i)\rho_i(\varphi)(g_i'),$$
for all $g_i, g_i' \in G_i$, $i = 1, 2$.

Next, the mapping
$$\Phi: \text{Hom}(G, \Gamma) \to \text{Hom}(G_1, \Gamma) \mathbin{\dot\otimes} \text{Hom}(G_2, \Gamma)$$
given by $\Phi(\varphi) = (\rho_1(\varphi), \rho_2(\varphi))$ is an isomorphism. First,
$$\Phi(\varphi\varphi') = \Phi(\varphi)\Phi(\varphi') \quad \text{for } \varphi, \varphi' \in \text{Hom}(G, \Gamma),$$
since
$$\rho_i(\varphi\varphi')(g_i) = (\varphi\varphi')(g_i) = \varphi(g_i)\varphi'(g_i)$$
$$= \rho_i(\varphi)(g_i)\rho_i(\varphi')(g_i)$$
for all $g_i \in G_i$, $i = 1, 2$. Since $\varphi \in \ker \Phi$ implies that $\rho_i(\varphi) = \varepsilon_i$, $i = 1, 2$, where $\varepsilon_i(g_i) = e$, for all $g_i \in G_i$, we have $\ker \Phi = \{\varepsilon\}$.

Finally, to show that Φ is surjective consider $\varphi_i \in \text{Hom}(G_i, \Gamma)$, $i = 1, 2$, and define $\varphi(g) = \varphi(g_1 g_2) = \varphi_1(g_1)\varphi_2(g_2)$, with g expressed uniquely as $g_1 g_2$, $g_i \in G_i$. Then $\varphi \in \text{Hom}(G, \Gamma)$ and $\Phi(\varphi) = (\varphi_1, \varphi_2)$. Thus Φ is an isomorphism, as asserted.

Proposition 2. If $\Gamma = \Gamma_1 \otimes \Gamma_2$, then
$$\text{Hom}(G, \Gamma_1 \otimes \Gamma_2) \cong \text{Hom}(G, \Gamma_1) \mathbin{\dot\otimes} \text{Hom}(G, \Gamma_2).$$

The proof is left as an exercise.

Example 1. We use Propositions 1 and 2 to determine $\text{Hom}(G, \Gamma)$ when $G = C_6$, the cyclic group of order 6 with generator g, and Γ is the abelian group of order 27 with generators a, b of respective orders 3, 9.

First, $G \cong \langle g^2 \rangle \otimes \langle g^3 \rangle$ and $\Gamma \cong \langle a \rangle \mathbin{\dot\otimes} \langle b \rangle$, where $\langle a \rangle$ denotes the subgroup generated by a, etc. Hence,
$$\text{Hom}(G, \Gamma) \cong \text{Hom}(\langle g^2 \rangle, \langle a \rangle) \mathbin{\dot\otimes} \text{Hom}(\langle g^2 \rangle, \langle b \rangle)$$
$$\mathbin{\dot\otimes} \text{Hom}(\langle g^3 \rangle, \langle a \rangle) \mathbin{\dot\otimes} \text{Hom}(\langle g^3 \rangle, \langle b \rangle).$$

Elementary arguments on the orders of elements and their images yield

$\text{Hom}(\langle g^2 \rangle, \langle a \rangle)$ has order 3. Simply map $g^2 \to a^r$, $r = 0, 1$, or 2.

$\text{Hom}(\langle g^2 \rangle, \langle b \rangle)$ has order 3. Simply map $g^2 \to b^{3r}$, $r = 0, 1$, or 2. These are the only possible homomorphic images of g^2 since $o(\varphi(g^2)) \mid o(g^2) = 3$. (These are the only elements in $\langle b \rangle$ whose order divides 3.)

$\text{Hom}(\langle g^3 \rangle, \langle a \rangle) = \{\varepsilon\}$, since the fact that $o(\varphi(g^3))$ is a divisor of $o(g^3) = 2$ (by Proposition 3, §6.5) and of $o(\langle a \rangle) = 3$ implies $o(\varphi(g^3)) = 1$ for any homomorphism φ.

$\text{Hom}(\langle g^3 \rangle, \langle b \rangle) = \{\varepsilon\}$, since, as above, for any homomorphism φ, $o(\varphi(g^3)) \mid 1 = (2, 9)$.

Any group of order 3 is cyclic, and isomorphic to C_3. Therefore

$$\text{Hom}(G, \Gamma) \cong C_3 \stackrel{\star}{\times} C_3.$$

The elements of $\text{Hom}(G, \Gamma)$ can be described on the generator $g \in G$ by

$$\varphi_{r,s}(g) = a^r b^{3s}.$$

Example 2. We now describe $\text{Hom}(\Gamma, G)$ for the groups G, Γ of Example 1. From Propositions 1 and 2 we have

$$\text{Hom}(\Gamma, G) \cong \text{Hom}(\langle a \rangle, \langle g^2 \rangle) \stackrel{\star}{\times} \text{Hom}(\langle a \rangle, \langle g^3 \rangle)$$
$$\stackrel{\star}{\times} \text{Hom}(\langle b \rangle, \langle g^2 \rangle) \stackrel{\star}{\times} \text{Hom}(\langle b \rangle, \langle g^3 \rangle).$$

Analogous to the discussion in Exercise 1:

$$\text{Hom}(\langle a \rangle, \langle g^2 \rangle) \cong C_3,$$
$$\text{Hom}(\langle a \rangle, \langle g^3 \rangle) = \{\varepsilon\},$$
$$\text{Hom}(\langle b \rangle, \langle g^2 \rangle) \cong C_3,$$
$$\text{Hom}(\langle b \rangle, \langle g^3 \rangle) = \{\varepsilon\}.$$

Thus $\text{Hom}(\Gamma, G) \cong C_3 \stackrel{\star}{\times} C_3$, and its elements can be described in terms of the generators a, b of Γ by

$$\psi_{\mu,\nu}(a^s b^t) = (g^2)^{\mu s}(g^2)^{\gamma t} = g^{2(\mu s + \gamma t)}.$$

Exercises

1. Verify that each of the mappings $\varphi_{r,s}$, $0 \leq r, s < 3$ defined in Example 1 is a homomorphism.
2. Prove that the multiplicative group T of all complex numbers of absolute value 1 is isomorphic to the additive group $\mathbf{R}/\mathbf{Z}$ of cosets of real numbers $\mathbf{R}$ modulo the additive group of integers $\mathbf{Z}$.
3. Let G be a cyclic group of order n. Prove the following:
   a. $\text{Hom}(G, T)$, where T is the group of Exercise 2, is a cyclic group of order n.
   b. $\text{Hom}(G, T)$ contains $\varphi(n)$ (the Euler function) distinct isomorphisms from G to T.

4. Consider **Z** as an additive group. Prove the following.
   a. $\text{Hom}(\mathbf{Z}, T)$, where T is the group of Exercise 2, is isomorphic to T.
   b. The isomorphisms in $\text{Hom}(\mathbf{Z}, T)$ can be put in one-one correspondence with the irrational numbers modulo 1.
5. a. Verify that the law of composition defined for $\text{Hom}(G, \Gamma)$ is associative.
   b. Verify that the mapping $\varphi^{-1}: G \to \Gamma$ defined by $\varphi^{-1}(g) = [\varphi(g)]^{-1}$ for $\varphi \in \text{Hom}(G, \Gamma)$ is itself a homomorphism.
6. Determine the following groups.
   a. $\text{Hom}(C_6, C_{14})$ b. $\text{Hom}(C_{12}, C_{14})$
   c. $\text{Hom}(C_2 \ast C_4, C_{14})$.
7. Prove Proposition 2. *Hint:* Show that $\varphi \in \text{Hom}(G, \Gamma_1 \otimes \Gamma_2)$ gives rise to mappings $\varphi_i \in \text{Hom}(G, \Gamma_i)$, and define
$$\Psi: \text{Hom}(G, \Gamma_1 \otimes \Gamma_2) \to \text{Hom}(G, \Gamma_1) \ast \text{Hom}(G, \Gamma_2)$$
by $\Psi(\varphi) = (\varphi_1, \varphi_2)$.
8. Extend the statement of Propositions 1 and 2 to the case of s factors.
9. a. Determine explicitly, without reference to Propositions 1 and 2, the group $\text{Hom}(G, \Gamma)$ of Example 1.
   b. Determine the kernel of each of the elements of $\text{Hom}(G, \Gamma)$.
10. a. Describe $\text{Hom}(C_m, C_n)$ for $(m, n) = 1$ [cf. Proposition 3, §6.10].
    b. Describe $\text{Hom}(C_m, C_n)$ when $(m, n) = d > 1$.
11. For an abelian group G, prove that $\text{Hom}(G, G)$ can be considered as a (not necessarily commutative) ring, called the **ring of endomorphisms** of G, when a second law of operation is defined to be composition of mappings $\varphi \circ \psi(g) = \varphi(\psi(g))$, for $g \in G$, $\varphi, \psi \in \text{Hom}(G, G)$.
12. With reference to Exercise 11, describe the ring of endomorphisms of the following:
    a. C_p, p a prime   b. C_q c. $C_2 \ast C_2$.

7

Selected Topics in Group Theory

No introductory text can develop all the important aspects of group theory. We have chosen to present here five topics of special importance in applications. They may be studied in any order. Analysis of the structure of finitely generated abelian groups [§7.1] is a prerequisite for the study of algebraic topology and is basic to the group-theoretic classification of abelian groups. Topologists and quantum physicists, among others, are concerned with characters (an analogue of linear functionals on finite dimensional vector spaces) of finite abelian groups [§7.2]. Elementary analysis of finite (non-abelian) groups is based on the theorems of Sylow [§§7.3 through 7.5]. Composition series of groups [§7.6] play an important role in the examination of finite groups and in the analysis of field extensions, where the Galois theory relates group and field-theoretic arguments. Commutative algebra and algebraic geometry place major emphasis on modules, which are viewed in §7.8 as special cases of groups with operators [§7.7].

§7.1 Finitely Generated Abelian Groups

This section is devoted to the statement and proof of the Fundamental Theorem of Finitely Generated Abelian Groups; it provides a complete description (up to isomorphism) of the structure of such groups. For notational convenience we consider additively-written abelian groups and speak of direct sums. It is a simple exercise to rewrite the definition, statements, and proofs multiplicatively.

An abelian group G is said to be **finitely generated** if there exists a finite subset of elements $\{A_1, \ldots, A_m\} \subseteq G$ such that every element $A \in G$ can be written as the sum

$$A = c_1 \cdot A_1 + \cdots + c_m \cdot A_m, \quad c_i \in \mathbf{Z}, \ 1 \le i \le m.$$

The elements A_i, $1 \le i \le m$, are called **generators** of G, and $c_i \cdot A_i$ denotes the c_i-fold sum of A_i with itself. (If $c_i < 0$, then $c_i \cdot A_i = |c_i| \cdot (-A_i)$.)

No claim is made that the integers c_i in a given representation of $A \in G$ are uniquely determined by A; this definition parallels the concept of a finitely generated vector space [cf. §4.1].

For example, in the additively-written Klein Four-group V with distinct elements $a, b, c, 0$ such that $a+a = b+b = c+c = 0$, and $a+b = c$, the element a has the following two expressions in terms of generators a, b, c:

$$a = a + 0 \cdot b + 0 \cdot c,$$
$$a = 0 \cdot a + b + c.$$

Note that any finite abelian group is finitely generated, as its own elements constitute a finite set of generators.

Fundamental Theorem of Finitely Generated Abelian Groups. Every finitely generated abelian group $G \ne \{0\}$ is the (internal) direct sum of N cyclic subgroups $G_1, \ldots, G_r, G_{r+1}, \ldots, G_N$ where

(i) $o(G_i) = e_i \ge 2, \quad 1 \le i \le r,$
(ii) $e_1 | e_2 | \cdots | e_r,$
(iii) $G_{r+1}, \ldots, G_N$ are isomorphic to $\mathbf{Z}$.

The numbers r, N and the chain of divisors $e_1, \ldots, e_r$ are uniquely determined by G. However, the groups G_i, $1 \le i \le N$, are not uniquely determined when N is greater than 1.

Fundamental Theorem of Finitely Generated Abelian Groups (*Alternate Form*). Every finitely generated (nontrivial) abelian group is isomorphic to a direct sum of cyclic groups

$$\mathbf{Z}_{q_1} + \cdots + \mathbf{Z}_{q_s} + \mathbf{Z} + \cdots + \mathbf{Z},$$

where $q_i = p_i^{\alpha_i}$ and the primes p_i are not necessarily distinct; the prime powers q_i, $1 \le i \le s$, and the number of summands $\mathbf{Z}$ are uniquely determined.

We split the proof of the Fundamental Theorem into separate existence and uniqueness proofs, beginning with the existence proof. Then we introduce some terminology and two lemmas before giving the uniqueness proof. Because of their length we number the major sections of each proof. The proof of the equivalence of the two forms of the Fundamental Theorem is left as an exercise. On a first reading, some may prefer to accept the Fundamental Theorem without proof and to gain understanding of its meaning by working out a number of the exercises.

The Existence Proof

1. Consider the collection of all finite sets of generators of G. Let n be the *minimal number* of elements in the sets of generators. The existence of such an integer n is a consequence of the Well-Ordering Principle. We have two cases.

1(a). If $n=1$, the group G is generated by a single element A. If $a \cdot A = 0$ only for $a = 0$, then G is an infinite cyclic group isomorphic to $\mathbf{Z}$. If $a \cdot A = 0$ for some $a \neq 0$, there is a least positive integer m such that $m \cdot A = 0$, and G is isomorphic to $\mathbf{Z}_m$ [cf. the theorem in §6.6]. (Note that the uniqueness statement is valid trivially.)

1(b). If $n > 1$, the proof will be by induction on n.

2. Consider *all* sets of n generators, $X_1, \ldots, X_n$, of G. Furthermore, consider *all* relations

$$x_1 \cdot X_1 + \cdots + x_n \cdot X_n = 0$$

for all sets of n generators.

First no relation has a coefficient x_i that is a unit (namely, ± 1). For, if there were such a coefficient x_i then

$$X_i = -x_i^{-1}x_1 \cdot X_1 - \cdots - x_i^{-1}x_{i-1} \cdot X_{i-1} - x_i^{-1}x_{i+1} \cdot X_{i+1} - \cdots - x_i^{-1}x_n \cdot X_n$$

could be omitted as a generator, because the set $\{X_1, \ldots, X_{i-1}, X_{i+1}, \ldots, X_n\}$ of $n-1$ elements would generate G, contrary to the minimal choice of n.

There either is or is not a set of generators $X_1, \ldots, X_n$ such that $0 \cdot X_1 + \cdots + 0 \cdot X_n = 0$ is the *only* relation between them.

2(a). In the first case $A \in G$ has a unique representation

$$A = a_1 \cdot X_1 + \cdots + a_n \cdot X_n.$$

(If also $A = b_1 X_1 + \cdots + b_n X_n$, then $0 = (a_1 - b_1) \cdot X_1 + \cdots + (a_n - b_n) \cdot X_n$, whence $a_i - b_i = 0$.) Thus A is the n-fold direct sum of the infinite cyclic groups $\langle X_i \rangle = \{z \cdot X_i : z \in \mathbf{Z}\}$, $1 \leq i \leq n$, and the existence proof is finished [see §6.10 for properties of the internal direct product].

2(b). In the second case all sets of coefficients $\{x_1, \ldots, x_n\}$ (which are not all zero) for all relations of all sets of n generators contain at least one coefficient x satisfying $|x| > 1$. Therefore there exists a system of generators $A_1, A_2, \ldots, A_n$ and a relation

(*) $\qquad a_1 \cdot A_1 + a_2 \cdot A_2 + \cdots + a_n \cdot A_n = 0,$

such that $a_1 \neq 0$ and $|a_1| \leq |x_i|$ for all nonzero coefficients x_i in all possible sets of relations.

3. .To prove that $a_1 | a_i$, $2 \le i \le n$, pick any particular i. By the Division Algorithm, $a_i = a_1 q_i + r_i$, $0 \le r_i < |a_1|$. If $r_i \ne 0$, relation (∗) can be rewritten

$$(**) \qquad a_1 \cdot (A_1 + q_i \cdot A_i) + r_i \cdot A_i + \sum_{j=2}^{n} a_j \cdot A_j = 0, \qquad \text{where } j \ne i.$$

In this case the n elements $A_1 + q_i \cdot A_i$, A_i, $A_2, \ldots, A_{i-1}, A_{i+1}, \ldots, A_n$ would also generate G, but they satisfy relation (∗∗), which has a coefficient less than a_1. This contradicts the choice of a_1. Therefore, $r_i = 0$ and $a_i = a_1 q_i$, $2 \le i \le n$.

Setting $B_1 = A_1 + \sum_{i=2}^{n} q_i \cdot A_i$, note that $a_1 \cdot B_1 = 0$. Now let $G_1 = \langle B_1 \rangle = \{z \cdot B_1 : z \in \mathbf{Z}\}$ and $G_2 = \{A_2, \ldots, A_n\}_{\mathbf{Z}}$, the group of integral linear combinations of $A_2, \ldots, A_n$. We have $G = G_1 + G_2$, because $A_1, \ldots, A_n$ generate G.

4. Next, to prove that this sum is *direct* consider $C \in G_1 \cap G_2$. Then

$$C = y_1 \cdot B_1 = \sum_{i=2}^{n} y_i \cdot A_i,$$

and consequently

$$0 = y_1 \cdot B_1 - \sum_{i=2}^{n} y_i \cdot A_i.$$

Using the Division Algorithm, write $y_1 = q_1 a_1 + z_1$, where $0 \le z_1 < |a_1|$. Then since $a_1 \cdot B_1 = 0$,

$$0 = z_1 \cdot B_1 - \sum_{i=2}^{n} y_i \cdot A_i.$$

By the *minimal* choice of $|a_1|$ the coefficient z_1 must be zero. Hence $C = z_1 \cdot B_1 = 0$.

5. Since G_2 has a set of $n-1$ generators, the induction hypothesis implies that

$$G_2 = \{A_2, \ldots, A_n\}_{\mathbf{Z}} = \langle B_2 \rangle \oplus \cdots \oplus \langle B_r \rangle \oplus \langle B_{r+1} \rangle \oplus \cdots \oplus \langle B_n \rangle,$$

where $o(B_i) = e_i \ge 2$, $2 \le i \le r$, $e_2 | \cdots | e_r$, and

$$\langle B_{r+1} \rangle \cong \langle B_{r+2} \rangle \cong \cdots \cong \langle B_n \rangle \cong \mathbf{Z}.$$

6. It remains to show that $|a_1| = e_1$ divides e_2. Suppose that $e_2 = e_1 q + s$ with $0 < s < e_1$. The elements $B_1 + q \cdot B_2, B_2, B_3, \ldots, B_n$ generate G, and

$$0 = e_1 \cdot B_1 + e_2 \cdot B_2 + \cdots + e_r \cdot B_r + 0 \cdot B_{r+1} + \cdots + 0 \cdot B_n$$
$$= e_1 \cdot (B_1 + q \cdot B_2) + s \cdot B_2 + \cdots + e_r \cdot B_r + 0 \cdot B_{r+1} + \cdots + 0 \cdot B_n.$$

This contradicts the minimal choice of a_1 as the coefficient of least magnitude of any relation. Therefore $s = 0$, and $e_1 | e_2$. This concludes the existence proof.

The uniqueness proof uses the following concepts and two lemmas.

An element in a group G is called a **torsion element** if it has finite order. The torsion elements of an abelian group form a subgroup, the so-called **torsion group** G_t of G, and G is called **torsion free** if its identity element is the only torsion element.

A group is called a *p*-**group** if its order is a power of a prime p; in particular, an abelian p-group is called a *p*-**primary group**. We also refer to the **minimal exponent** m of a finite group G, that is, the least positive integer for which, in additive notation, $m \cdot g = e$ for all $g \in G$.

Lemma 1. If G_t is the torsion subgroup of an abelian group G, then G/G_t is torsion free.

For the proof consider an element (i.e., an additive coset) $g + G_t$ of G/G_t of finite order m. Since $m \cdot (g + G_t) = (m \cdot g) + G_t$, we have $(m \cdot g) + G_t = 0$, or $m \cdot g \in G_t$. This means that $m \cdot g$ has finite order, say r, and therefore g has order at most rm. Thus $g \in G_t$ and $g + G_t = 0 \in G/G_t$.

Lemma 2. For a p-primary group G with minimal exponent p^α,

$$G = G_1 \oplus \cdots \oplus G_k,$$

where the G_i are cyclic of order p^{α_i}, $1 \leq \alpha_1 \leq \alpha_2 \leq \cdots \leq \alpha_k = \alpha$, and where the α_i are uniquely determined by G.

Note that is is not claimed that the groups G_i are unique, only their orders p^{α_i} are.

For instance the Klein Four-group V, referred to in the third paragraph of this section, can be written

$$V = \langle a \rangle \oplus \langle b \rangle = \langle a \rangle \oplus \langle c \rangle,$$

$\langle b \rangle \neq \langle c \rangle$, but $o(b) = o(c) = 2$.

Proof of Lemma 2. The existence of cyclic groups G_i is a consequence of the existence portion of the Fundamental Theorem, already proved, applied to the given group G. To prove the uniqueness of the exponents α_i, we proceed by induction on the minimal exponent p^α of G.

If $\alpha = 1$, then $o(G_i) = p$, and $G_i \cong \mathbb{Z}_p$. Because G is the direct sum of the groups G_i, its order $o(G) = \prod_{i=1}^k o(G_i) = p^k$. The number k of cyclic summands is uniquely determined by the order of G.

Now assume that $\alpha \geq 2$. As the induction hypothesis suppose that the lemma is true for p-primary groups with minimal exponent less than p^α. The group $pG = \{p \cdot g : g \in G\}$ has minimal exponent $p^{\alpha-1}$, and

$$pG = pG_1 + \cdots + pG_k = pG_{m+1} + \cdots + pG_k,$$

where we omit the trivial summands (if any) $pG_1, \ldots, pG_m$. The cyclic group

pG_i is generated by the p-fold sum of a generator of G_i. Since $o(G_i) = p^{\alpha_i}$, $o(pG_i) = p^{\alpha_i - 1}$. By the induction hypothesis the integers $\alpha_i - 1$, for $m < i \le k$, are unique, and so then are the α_i, $m < i \le k$.

The number of summands $G_1, \ldots, G_m$ isomorphic to $\mathbf{Z}_p$ is unique, because the number $k - m$ of direct summands of pG is unique by the induction hypothesis and

$$o(G) = \prod_{i=1}^{k} o(G_i) = p^k \prod_{i=1}^{k} o(pG_i) = p^k \prod_{i=1}^{k} p^{\alpha_i - 1} = p^m \prod_{i=m+1}^{m} p^{\alpha_i}.$$

Application of the Principle of Induction completes the proof of Lemma 2.

Returning to the Fundamental Theorem, we now take up the uniqueness proof.

The Uniqueness Proof. Given $G = G_1 \oplus \cdots \oplus G_r \oplus G_{r+1} \oplus \cdots \oplus G_n$, where G_i is cyclic of order $e_i \ge 2$, $1 \le i \le r$, $e_1 | e_2 | \cdots | e_r$, and $G_j \cong \mathbf{Z}$, $r < j \le n$, we have to prove the uniqueness of the integers $n - r$, r, and e_i, $1 \le i \le r$. Pick generators $B_k \in G_k$, $1 \le k \le n$.

1. To prove the uniqueness of the number $n - r$ of generators of G of infinite order, let $G^* = G/G_t$. By Lemma 1, G^* is torsion free. Letting B_k^* be the coset $B_k + G_t$ of B_k in G^*, $r < k \le n$, note that $B_{r+1}^*, \ldots, B_n^*$ generate G^* and satisfy only the trivial relation $0 \cdot B_{r+1}^* + \cdots + 0 \cdot B_n^* = 0^*$.

1(a). If there were a nontrivial (i.e., not all $s_j \ne 0$) relation

$$s_{r+1} \cdot B_{r+1}^* + \cdots + s_n \cdot B_n^* = 0^*,$$

then we would have

$$s_{r+1} \cdot B_{r+1} + \cdots + s_n \cdot B_n \in G_t.$$

Hence for some nonzero $z \in \mathbf{Z}$,

$$zs_{r+1} \cdot B_{r+1} + \cdots + zs_n \cdot B_n = 0,$$

contradicting the choice of the $B_{r+1}, \ldots, B_n$.

1(b). To complete the proof of the uniqueness of $n - r$, select any positive prime p. Since

$$G^* = \langle B_{r+1}^* \rangle \oplus \cdots \oplus \langle B_n^* \rangle$$

and

$$pG^* = \langle p \cdot B_{r+1}^* \rangle \oplus \cdots \oplus \langle p \cdot B_n^* \rangle$$

Proposition 4, §6.10, yields

$(*)$ $\qquad G^*/pG^* \cong \langle B_{r+1}^* \rangle / \langle p \cdot B_{r+1}^* \rangle \dotplus \cdots \dotplus \langle B_n^* \rangle / \langle p \cdot B_n^* \rangle.$

Because $\langle B_k^* \rangle / \langle p \cdot B_k^* \rangle$ has order p, for $r < k \le n$, and the sum in equation $(*)$ is direct, G^*/pG^* has order p^{n-r}. Now G_t, G^*, and hence G^*/pG^* are uniquely determined by G (i.e., are independent of the representation of G as the direct sum of cyclic subgroups). Therefore we conclude that $n - r$ is uniquely determined by G, since $p^{n-r} = o(G^*/pG^*)$.

2. To prove the uniqueness of $e_1, \ldots, e_r$, it suffices to consider the torsion subgroup G_t. Let $M = \prod_{i=1}^{r} e_i = o(G_t)$. Writing M as the product of

powers of distinct primes,
$$M = \prod_{j=1}^{s} p_j^{\alpha_j},$$
set $m_j = M/p_j^{\alpha_j}$, and set $H_j = \{g \in G_t : p_j \text{ divides } o(g)\}$, $1 \le j \le s$.

2(a). We now show that
$$G_t = H_1 \oplus \cdots \oplus H_s.$$
Since $(m_1, \ldots, m_s) = 1$, we know [from Exercise 12, §2.5] that there exist integers h_j such that $1 = h_1 m_1 + \cdots + h_s m_s$. Thus for $A \in G_t$,
$$A = 1 \cdot A = (h_1 m_1) \cdot A + \cdots + (h_s m_s) \cdot A.$$
Noting that for all $A \in G_t$ the order of $(h_j m_j) \cdot A$ divides $p_j^{\alpha_j}$, we have that $(h_j m_j) \cdot A \in H_j$. Hence
$$G_t = H_1 + \cdots + H_s.$$

2(b). This sum is *direct*. The order of any element
$$X \in (H_1 + \cdots + H_{j-1}) \cap H_j, \quad 1 < j \le s,$$
divides $p_j^{\alpha_j}$ (being in H_j) and divides $q = \prod_{h=1}^{j-1} p_k^{\alpha_k}$ (being in $(H_1 + \cdots + H_{j-1})$). Therefore $o(X)$ divides the GCD $(q, p_j^{\alpha_j}) = 1$, from which we conclude that $o(X) = 1$ and $X = 0$.

3. Being subgroups of G_k, the groups $H_{jk} = H_j \cap G_k$, $1 \le j \le s$, $1 \le k \le r$, are cyclic. Their respective orders are powers of p_j, as they are subgroups of H_j. Let $o(H_{jk}) = e_{jk}$. From the direct sum representations
$$G_t = G_1 \oplus \cdots \oplus G_r,$$
$$G_t = H_1 \oplus \cdots \oplus H_s,$$
we obtain

(**)
$$H_j = H_j \cap G_t = H_{j1} \oplus \cdots \oplus H_{jr},$$
$$G_k = G_k \cap G_t = H_{1k} \oplus \cdots \oplus H_{sk},$$
where some of the summands might be zero.

Since these sums are direct, $e_k = o(G_k) = \prod_{j=1}^{s} e_{jk}$. Thus to prove the uniqueness of the orders e_k, it suffices to prove the uniqueness of the integers e_{jk}. For this we shall use Lemma 2.

4. Each group H_j, $1 \le j \le s$, is p_j-primary and so, by Lemma 2, has a representation as the direct sum of cyclic subgroups, whose number and orders are unique. Since equation (**) presents H_j as the direct sum of cyclic subgroups H_{jk}, the orders e_{jk} of these subgroups must be unique. This completes the uniqueness proof.

The uniqueness proof implies that the number n, determined (in part 1 of the existence proof) to be the minimal number of generators of G, is the same as the integer N in the statement of the theorem.

REMARK. If **Z** is replaced throughout this section by $F[x]$, the absolute value used in the existence part of the theorem is replaced by the degree of a polynomial, and order of a group element is replaced by "monic polynomial $m(x)$ of least degree such that $m(x) \cdot A = 0$," the theorem is the foundation for the theory of "rational similarity of matrices" [see Exercises 20 and 21, §7.8].

Exercises

1. If G is an abelian, noncyclic group of order p^2, prove without reference to the structure theorems that
$$G \cong \mathbf{Z}_p + \mathbf{Z}_p.$$

2. Prove the equivalence of the two forms of the Fundamental Theorem of Finitely Generated Abelian Groups.

3. How many nonisomorphic abelian groups are there of each of the following orders?
   a. 35   b. 36   c. 35·36
   d. 48   e. 49   f. 48·49
   g. 24   h. 63   i. 24·63.

4. If $m = m_1 m_2$, $(m_1, m_2) = 1$, and if there are n_i nonisomorphic abelian groups of order m_i, $i = 1, 2$, how many nonisomorphic abelian groups are there of order m?

5. Generalize the result of Exercise 4 to the case where m is the product of r pairwise relatively prime factors m_i, for which there are n_i nonisomorphic abelian groups of order m_i.

6. For what integers m must all abelian groups of order m be cyclic?

7. Complete the blanks.
   a. There are _____ nonisomorphic abelian groups of order 275.
   b. An abelian group of order 882 must be isomorphic to one of the following groups: _____, _____, _____, or _____.

8. Can you determine how many subgroups of order 8 there are in:
   a. An abelian group of order 48?
   b. An abelian group of order 200?

9. Prove that for an abelian group G:
   a. If a prime $p \mid o(G)$, then G has a (cyclic) subgroup of order p.
   b. If an arbitrary integer $m \mid o(G)$, then G has a subgroup of order m.
   c. If in part (b) no square divides m, then the subgroup must be cyclic.
   d. If in part (b) a square does divide m, then G might have no cyclic subgroup of order m.

10. For arbitrary $m \in \mathbf{N}$ describe the number (up to isomorphism) of abelian groups of order m (in terms of the unique prime power factorization of m).

11. Exhibit all nonisomorphic abelian groups of order:
    a. 60   b. 360   c. $2^3 5^2 7$.

12. Let F be isomorphic to the external direct sum of n copies of the additive abelian group $\mathbf{Z}$. Write F as the internal direct sum of n groups which are isomorphic to $\mathbf{Z}$. (Such groups are called **free abelian groups of rank n**.)
13. Let $F = \{(a,b) : a, b \in \mathbf{Z}\}$ be a free abelian group of rank 2. Suppose that H is the subgroup of F generated over $\mathbf{Z}$ by the elements $(2, -4)$ and $(1, 6)$. Determine the structure of the (additively-written) factor group F/H in the sense of the Fundamental Theorem.
14. Let A be the free abelian group with generators u and v, i.e., $A = \{au + bv : a, b \in \mathbf{Z}\}$. Consider the subgroup B of A generated over $\mathbf{Z}$ by the elements $x = 2u - 3v$ and $y = 7u - v$. Determine the structure of the factor group A/B.
15. Let $F = \{(a_1, a_2, a_3) : a_1, a_2, a_3 \in \mathbf{Z}\}$ be a free abelian group of rank 3. Suppose that H is the subgroup of F generated over $\mathbf{Z}$ by the elements $(2, -4, 6)$ and $(1, 6, 7)$.
    a. Determine the structure of the (additively-written) factor group F/H.
    b. Describe the group $\text{Aut}(F)$ of automorphisms of F.
16. Let A be the free abelian group with generators u, v, and w, i.e., $A = \{au + bv + cw : a, b, c \in \mathbf{Z}\}$. Suppose that B is the subgroup generated over $\mathbf{Z}$ by the elements $x = 2u - v + 4w$, $y = 4u + 5v + 6w$, and $z = 3u + 2v + 5w$.
    a. Is the factor group A/B a finite group?
    b. Determine A/B as a sum of cyclic groups.
    c. Are the summands of A as a direct sum uniquely determined?
17. Consider $\mathbf{Z}$ as an additive group.
    a. Determine all groups G such that there is a surjective homomorphism $\varphi : \mathbf{Z} \to G$.
    b. Determine the kernels of all such homomorphisms φ.
18. Prove (by induction on n) that any subgroup of a free abelian group of rank n must itself be free.
19. a. Prove that there are as many nonisomorphic abelian groups of order p^a as there are of order q^a, where p and q are arbitrary positive prime numbers.
    b. Consider an abelian group G of order $p^a q^a m$, where p and q are distinct primes and $(p, m) = (q, m) = 1$. Show that G has exactly one subgroup of order p^a and one of order q^a. Must the number of subgroups of G of order p^β be the same as the number of those of order q^β, $0 < \beta < \alpha$?
20. With reference to Exercise 6, §6.9, prove that if G is a finitely generated abelian group of order greater than 2, then $\text{Aut}(G)$ is a nontrivial group.
21. Let G be the direct product of m groups of prime order p. Prove that G contains
$$\left[\prod_{s=0}^{a-1}(p^{m-s}-1)\right]\left[\prod_{t=1}^{a}(p^t-1)\right]^{-1}$$
distinct subgroups of order p^a, $1 \le a < m$.
22. Suppose that G is a finite abelian group. Prove that $\langle g \rangle$, the group generated by an element g of maximal order, is a direct factor of G, i.e., $G = \langle g \rangle \otimes H$, where H is some subgroup of G.

§7.2 Characters of Finite Abelian Groups

In §6.11 we studied the group of homomorphisms $\text{Hom}(G, \Gamma)$, where G, Γ were arbitrary abelian groups. We now consider the special case when Γ is the multiplicative group T of all complex numbers z, for which $|z| = 1$.

For a finite abelian group G we call $\text{Hom}(G, T)$ the **group of characters** or **dual group** G^* of G. The map $\varepsilon \colon G \to \{1\} \subset T$ is the identity element of G^*, which we refer to as the identity character. Recall from §6.11 that for $\varphi, \psi \in \text{Hom}(G, T)$ the product $\varphi\psi$ is the homomorphism on G given by $(\varphi\psi)(g) = \varphi(g)\psi(g)$.

If n is the minimal exponent of G, then for every $\varphi \in \text{Hom}(G, T)$:

$$(\varphi(g))^n = \varphi(g^n) = \varphi(e) = 1.$$

Thus the character values $\varphi(g)$, for every $\varphi \in \text{Hom}(G, T)$ and every $g \in G$, are dth roots of unity for some $d \mid n$, the minimal exponent of G. Consequently the group T can be replaced by an arbitrary cyclic group of order n.

NOTATION. We shall write the nth roots of unity in T as powers of the generator

$$\zeta_n = \cos\left(\frac{2\pi}{n}\right) + i \sin\left(\frac{2\pi}{n}\right).$$

The other generators of the cyclic group of nth roots of unity are

$$\zeta_n^t = \cos\left(\frac{2\pi}{n}t\right) + i \sin\left(\frac{2\pi}{n}t\right),$$

with $(t, n) = 1$. An nth root of unity ζ is said to be **primitive** if $\zeta^n = 1$, but $\zeta^i \neq 1$, $1 \le i < n$; in other words, if ζ is a generator of the cyclic group of nth roots of unity.

Proposition 1. *If G is a cyclic group of order m, then so is its dual G^*.*

Let g be a generator of G, and define m characters χ_k, $0 \le k < m$, on G by $\chi_k(g) = \zeta_m^k$. Since ζ_m generates the group of mth roots of unity, the m powers ζ_m^k are distinct; hence the characters χ_k are distinct elements of G^*. To prove that G^* has no other elements, consider an arbitrary character χ. Then $\chi(g)$ is an mth root of unity, and so for some j, $0 \le j < m$, $\chi(g) = \zeta_m^j = \chi_j(g)$. But therefore $\chi = \chi_j$. Finally, because the character χ_1 has order m, the dual group $G^* = \{\chi_k : 0 \le k < m$, where $\chi_k(g) = \zeta_m^k\}$ is cyclic of order m.

Note that $G \cong G^*$ as they are both cyclic of the same order. Such an isomorphism is not "natural" in the sense that it depends upon the choice of the generators of the groups G and G^*.

Proposition 2. *If G is a finite abelian group, then G^* is isomorphic to G. (This isomorphism is not natural.)*

Proof. Writing $G = G_1 \otimes \cdots \otimes G_s$ as the direct product of cyclic groups by the Fundamental Theorem of Finitely Generated Abelian Groups, §7.1, we have

$$G^* = \text{Hom}(G, T) = \text{Hom}(G_1 \otimes \cdots \otimes G_s, T)$$
$$\cong \text{Hom}(G_1, T) \dot{\times} \cdots \dot{\times} \text{Hom}(G_s, T)$$
$$= G_1^* \dot{\times} \cdots \dot{\times} G_s^* \cong G_1 \otimes \cdots \otimes G_s = G,$$

where by induction we extend Proposition 1, §6.11, to the case of s factors.

Proposition 3. If $g \neq e$ in a finite abelian group G, then there exists a character $\chi \in G^*$ such that $\chi(g) \neq 1$.

Proof. In terms of the direct product expression $G = G_1 \otimes \cdots \otimes G_s$ following Proposition 2, if $g \neq e$ in G, then $g = a_1 \cdots a_s$, where at least one factor $a_i \neq e$. For a character χ defined in terms of the s generators g_j of the cyclic factors G_j of G by

$$\chi(g_k) = 1 \quad \text{for } k \neq i, \ 1 \leq k \leq s,$$

and
$$\chi(g_i) = \zeta_{n_i},$$

where ζ_{n_i} is a generator of the group of n_ith roots of unity and $n_i = o(G_i)$, we have

$$\chi(g) = \chi(a_1) \cdots \chi(a_s) = \chi(a_i) \neq 1, \quad \text{since } a_i \neq e.$$

Proposition 4. There is a natural isomorphism between G and $G^{**} = (G^*)^*$, the **double dual** of G.

Proof. By Proposition 2, $G \cong G^* \cong G^{**}$, but now the problem is to obtain an isomorphism $\kappa : G \cong G^{**}$ which is *independent* of the choice of generators of G. To this end we define $\kappa: G \to G^{**}$ by defining $\kappa(g): G^* \to T$ as follows:

$$\kappa(g)[\chi] = \chi(g) \in T \quad \text{for every } \chi \in G^*.$$

Then $\kappa(g)$ is a single-valued function on the dual G^* with values in T. To verify that $\kappa(g)$ is a homomorphism for each $g \in G$, note that by the definition of $\kappa(g)$ and of the product in G^*,

$$\kappa(g)[\chi_1 \chi_2] = (\chi_1 \chi_2)(g) = \chi_1(g)\chi_2(g) = \kappa(g)[\chi_1]\kappa(g)[\chi_2].$$

It remains to prove that κ is an isomorphism. Since

$$\kappa(g_1 g_2)[\chi] = \chi(g_1 g_2) = \chi(g_1)\chi(g_2)$$
$$= \kappa(g_1)[\chi] \cdot \kappa(g_2)[\chi] \in T,$$

κ is a homomorphism. Furthermore, κ is injective. If $\kappa(g) = \kappa(h)$, then $\kappa(g)[\chi] = \kappa(h)[\chi]$ for *all* $\chi \in G^*$, and so $\chi(g) = \chi(h)$ or $\chi(gh^{-1}) = 1$, for *all* $\chi \in G^*$. But then by Proposition 3, gh^{-1} must be 1, or $g = h$.

The surjectivity of $\kappa: G \to G^{**}$ is a consequence of the facts that κ is a one-one mapping and that G, G^{**} have the same order [see Proposition 2].

The definition of κ makes no reference to any choice of generating elements of G. Thus, κ is called a "natural" isomorphism.

In the remainder of this section we consider properties of particular subgroups $A_G(L)$ of the dual group G^*. Letting L be a subset of G, define

$$A_G(L) = \{\chi \in G^* : \chi(\ell) = 1, \text{ for all } \ell \in L\},$$

called the **annihilator** of L. We leave as an exercise the verification of the next proposition.

Proposition 5. For any subset $L \subseteq G$, $A_G(L)$ is a subgroup of G^*.

Proposition 6. For any subgroup $H \subseteq G$, we have $A_{G*}[A_G(H)] = H$, where we identify H with $\kappa(H)$ and κ is the isomorphism of Proposition 4.

Proof. By definition $A_G(H)$ consists of all those $\chi \in G^*$ for which $\chi(h) = 1$ for all $h \in H$. Therefore $\kappa(h)(\chi) = 1$ for all $\chi \in A_G(H)$; consequently, $\kappa(h) \in A_{G*}[A_G(H)]$. Making the identification of G^{**} with G, we have $H \subseteq A_{G*}[A_G(H)] \subseteq G$.

Suppose that H is a proper subgroup of the double annihilator, i.e., there exists an element g^{**}, more precisely $\kappa(g)$, in the latter which does not lie in H. This assumption leads to a contradiction as follows. If $g^{**}H \neq H$ in the factor group G/H, then there exists a character $\varphi \in (G/H)^*$ such that $\varphi(g^{**}H) \neq 1$ by Proposition 3. Next let π be the canonical homomorphism of G onto G/H, that is, $\pi(g) = gH$. Being the composition of homomorphisms the function $\psi = \varphi \circ \pi$ from G to T belongs to G^*. Furthermore, $\psi(h) = \varphi[\pi(h)] = \varphi(H)$ implies $\psi \in A_G(H)$. Consequently by the definition of $A_{G*}[A_G(H)]$ and the assumption on $g, \psi(g) = \kappa(g)[\psi] = 1$, but $\psi(g) \neq 1$ according to the choice of ψ. Hence $H \subset A_{G*}[A_G(H)]$ is false.

Proposition 7. The mapping $H \to A_G(H)$ establishes a one-one correspondence between the subgroups of G and G^* such that the following relations hold.

(i) $H \supset K$ implies $A_G(H) \subset A_G(K)$.
(ii) $A_G(HK) = A_G(H) \cap A_G(K)$.
(iii) $A_G(H \cap K) = A_G(H) A_G(K)$.

The definition of annihilator implies that $A_G(H) \subseteq A_G(K)$ if $H \supset K$. If for $H \supset K$ we had $A_G(H) = A_G(K)$, then using Proposition 6 we would obtain

$$H = A_{G*}[A_G(H)] = A_{G*}[A_G(K)] = K,$$

a contradiction. Furthermore every subgroup S^* of G^* is given as $A_G(H)$, where $H = A_{G*}(S^*)$.

Next, since $H \cdot K = HK$ contains both H and K, the annihilator $A_G(HK)$ is contained in both $A_G(H)$ and $A_G(K)$. Thus, $A_G(HK) \subseteq A_G(H) \cap A_G(K)$. For an element φ in this intersection, $\varphi(h) = \varphi(k) = 1$ for every $h \in K$ and $k \in K$. Therefore $\varphi \in A_G(HK)$ and $A_G(HK) = A_G(H) \cap A_G(K)$, which proves (ii). [The product of groups was introduced in §6.2.]

Finally, since $H \cap K$ is contained in both H and K, its annihilator $A_G(H \cap K)$ contains both $A_G(H)$ and $A_G(K)$; hence

$$A_G(H \cap K) \supseteq A_G(H) A_G(K).$$

If this last inclusion were proper, the mapping $S^* \to A_{G^*}(S^*) \subseteq G$ would imply that

$$A_{G^*}[A_G(H \cap K)] = H \cap K \subset A_{G^*}[A_G(H) A_G(K)]$$
$$= A_{G^*}[A_G(H)] \cap A_{G^*}[A_G(K)] = H \cap K,$$

according to (i) and (ii). This is a contradiction.

Proposition 8. For a subgroup H of G,

$$A_G(H) \cong (G/H)^*,$$
and
$$G^*/A_G(H) \cong H^*.$$

Note carefully that H^* is *not* a subgroup of G^*. To prove the first assertion, describe a mapping $\sigma: A_G(H) \to (G/H)^*$ by defining $\sigma(\varphi)$ by

$$\sigma(\varphi)[gH] = \varphi(g) \quad \text{for } \varphi \in A_G(H) \text{ and } g \in G.$$

The function σ acts uniquely on the cosets gH in G/H, because if $g' = gh$ with $h \in H$, then $\varphi(g') = \varphi(gh) = \varphi(g)\varphi(h) = \varphi(g)$ since $\varphi \in A_G(H)$. It is a homomorphism because for all cosets gH:

$$\sigma(\varphi\varphi')[gH] = (\varphi\varphi')(g) = \varphi(g)\varphi'(g) = \sigma(\varphi)[gH] \cdot \sigma(\varphi')[gH].$$

The mapping σ is injective. If $\sigma(\varphi)[gH] = \varphi(g) = 1$ for all $g \in G$, then $\varphi = \varepsilon$. Finally σ is surjective. Consider $\varphi^* \in (G/H)^*$ and define $\varphi = \varphi^* \circ \pi$, where π is the canonical homomorphism of G onto G/H. That is,

$$\varphi(g) = (\varphi^* \circ \pi)(g) = \varphi^*[\pi(g)] = \varphi^*(gH).$$

Then
$$\varphi(gg') = \varphi^*(gg'H) = \varphi^*(gH \cdot g'H)$$
$$= \varphi^*(gH)\varphi^*(g'H) = \varphi(g)\varphi(g'),$$

and so $\varphi \in G^*$. Next $\varphi(h) = \varphi^*(hH) = \varphi^*(H) = 1$ for all $h \in H$, thus $\varphi \in A_G(H)$. Furthermore, $\sigma(\varphi)[gH] = \varphi(g) = \varphi^*(gH)$ for all $g \in G$ implies that $\sigma(\varphi) = \varphi^*$.

To prove the second assertion, define a mapping τ from G^* to H^* as the restriction map

$$\tau(\chi)[h] = \chi(h)$$

for all $h \in H$. Clearly $\tau(\chi) \in H^*$, and τ is a homomorphism since $\tau(\chi\chi')[h] = (\chi\chi')(h) = \chi(h)\chi'(h) = \tau(\chi)[h]\tau(\chi')[h]$ for all $h \in H$. The kernel of τ is $A_G(H)$ since $\tau(\chi)[h] = \chi(h) = 1$ for all $h \in H$ implies $\chi \in A_G(H)$. Consequently, the Isomorphism Theorem [§6.8] states the existence of an injective homomorphism $\tau^*: G^*/A_G(H) \to H^*$. It satisfies

$$\tau^*(\chi A_G(H))[h] = \chi(h).$$

To establish that $G^*/A_G(H)$ and H^* are isomorphic, it remains to show that τ^* is surjective. Using the preceding isomorphism σ, we have

$$[A_G(H):1] = [(G/H)^*:1] = [G/H:1]$$
$$= [G:H] = [G:1][H:1]^{-1}$$

and $\quad [H^*:1] = [H:1],$

where for notational convenience both the identity element and identity subgroup in each group considered are denoted by 1. Consequently,

$$[G^*/A_G(H):1] = [G^*:1][A_G(H):1]^{-1}$$
$$= [G:1][A_G(H):1]^{-1} = [H:1] = [H^*:1].$$

REMARK. For a character φ of H where $[G:H] = s$, there exist precisely s distinct characters $\varphi_1, \ldots, \varphi_s$ of G whose restrictions $\tau(\varphi_1), \ldots, \tau(\varphi_s)$ to H equal φ. The existence of at least one such character follows from the preceding index relations; the precise number s is attained because $A_G(H)$ is the kernel of the map τ.

Exercises

1. Suppose the abelian group G is the direct product of the cyclic groups $H_1 = \langle s_1 \rangle$, $H_2 = \langle s_2 \rangle$, and $H_3 = \langle s_3 \rangle$, whose respective orders are 4, 8, and 6. Find all characters $\chi \in G$ for which:
    a. $\chi(s_1^2 s_2 s_3^3) = \chi(s_1^2 s_3) = 1$
    b. $\chi(s_1^2 s_2^4 s_3) = \chi(s_1^2 s_2^2) = 1$.
2. Let $G = H_1 \otimes H_2 \otimes H_3$, where the subgroups H_i are cyclic with generators γ_i, $i = 1, 2, 3$, and $\gamma_1^2 = \gamma_2^2 = \gamma_3^3 = e$. Determine the annihilator $A_G(H)$ and the multiplication table of G/H for:
    a. $H = H_2 \otimes H_3$ b. $H = \langle \gamma_1 \gamma_2 \rangle$ c. $H = \langle \gamma_1 \gamma_3 \rangle$.
3. Let $G = C_3 \otimes C_4$, where the subgroups C_m are cyclic of order m with generators γ_m, $m = 3, 4$. Determine the annihilator $A_G(H)$ and multiplication table of G/H for:
    a. $H = C_4$ b. $H = \langle \gamma_4^2 \rangle \otimes C_3$.
4. Let $G = C_6 \otimes C_9$, where the subgroups C_m are cyclic of order m with generators γ_m, $m = 6, 9$. Consider the subgroups $H_1 = \langle \gamma_6^2 \rangle \otimes \langle \gamma_9^3 \rangle$, $H_2 = \langle \gamma_6^3 \rangle$, and $H_3 = \langle \gamma_6^3 \gamma_9^3 \rangle$.
    a. Determine $A_G(H_1 H_2)$ and verify explicitly that it equals $A_G(H_1) \cap A_G(H_2)$.
    b. Determine $A_G(H_1 \cap H_3)$ and verify explicitly that it equals $A_G(H_1) A_G(H_3)$.

5. Consider an n-dimensional vector space V over a field F. The set $V^* = \text{Hom}_F(V, F)$ of linear maps $\chi: V \to F$ is called the **dual space** of V; the maps χ are called **linear functionals** on V.
   a. Prove that V^* is an n-dimensional vector space over F.
   b. Prove that there exists a natural (i.e., independent of the choice of bases for V and V^*) isomorphism between V and $V^{**} = (V^*)^*$.
   c. Is a nonzero linear functional $\chi: V \to F$ necessarily surjective?
   d. For given $\chi \in V^*$, prove that $\ker \chi$ is a subspace of V and find its dimension.
   e. If U is an m-dimensional subspace of V, determine a basis of the annihilator $A_V(U)$.
   f. Prove that $A_{V^*}(A_V(U)) = U$.
   g. Prove that for $\chi, \chi' \in V^*$, if $\ker \chi \subseteq \ker \chi'$ then there exists an element $\alpha \in F$ such that $\chi' = \alpha\chi$.
6. Prove Proposition 5.
7. Consider a group G written as the direct product of cyclic factors $G = G_1 \otimes \cdots \otimes G_s$. For each i, $1 \leq i \leq s$, let λ_i denote the character defined in the proof of Proposition 3. Prove that $G^* = \langle \lambda_1 \rangle \times \cdots \times \langle \lambda_s \rangle$.

Finite abelian groups A, B are said to be **dually paired** if there exists a single-valued mapping of their cartesian product

$$A \times B \to T,$$

such that, for all $a, a_1, a_2 \in A$ and $b, b_1, b_2 \in B$ the following relations hold:

(i) $(a_1 a_2, b) = (a_1, b)(a_2, b)$,
    $(a, b_1 b_2) = (a, b_1)(a, b_2)$.

(ii) If $(a, b) = 1$ for all $b \in B$, then $a = 1$.
     If $(a, b) = 1$ for all $a \in A$, then $b = 1$.

8. Prove that G, G^* are dually paired groups.
9. If A, B are dually paired groups, prove that $A \cong B^*$ and $A^* \cong B$.

§7.3 Bijections of Sets

This section is the first of two which are preparatory to the Sylow theory in §7.5. Bijections (permutations) of groups are used to prove the Sylow theorems, which are analytical tools for discussing both finite groups and galois field extensions [see §8.6]. In these sections we use the term *bijection* interchangeably with *permutation* to denote a one-one, surjective mapping of sets. As such it should not be confused with a bijective homomorphism of groups [cf. §6.5].

The group $\Sigma(S)$ of all permutations (bijections) of a set S onto itself was introduced in §6.7. In this section the maps $\alpha, \beta \in \Sigma(S)$ shall be written to the left of the elements on which they operate; that is, for $a \in S$, $\alpha(a)$ is replaced by αa, and $(\alpha\beta)a$ means $\alpha(\beta(a))$. (Note that some authors write $a\alpha$ for $\alpha(a)$.)

Now let K be a subgroup of $\Sigma(S)$ for a given set S. Define $a \approx b$ (mod K) for a, b in S, if there exists an element $\gamma \in K$ for which $b = \gamma a$. In this event we say that b is **conjugate** to a **with respect to the group** K. This concept of being conjugate with respect to K is an equivalence relation; that is, for $a, b, c \in S$ the following relations hold:

(i) $a \approx a \pmod{K}$
(ii) $a \approx b \pmod{K}$ implies $b \approx a \pmod{K}$
(iii) $a \approx b \pmod{K}$ and $b \approx c \pmod{K}$ imply $a \approx c \pmod{K}$.

The equivalence class of a single element $s \in S$ with respect to K is called the **orbit** of s under action by the elements of K. Denoted

$$\text{Orb}_K(s) = \{\gamma s : \gamma \in K\},$$

it consists of all elements in S conjugate to s with respect to K.

To prove that the number of elements in S conjugate to $s \in S$ (with respect to K) divides the order of the group K, we introduce the **stabilizer subgroup**

$$H_K(s) = \{\gamma \in K : \gamma s = s\} \subseteq K.$$

Definition (*from General Set Theory*). The **cardinality** of a finite set is the number of elements in the set.

Note that we consider here only finite sets.

Proposition. The cardinality of the orbit $\text{Orb}_K(s)$ of an element $s \in S$ with respect to the group $K \subseteq \Sigma(S)$ equals the index $[K : H_K(s)]$ of the subgroup $H_K(s)$ in K.

The proof follows from the observation that for a given $s \in S$ there is a one-one correspondence between elements of $\text{Orb}_K(s)$ and (left) cosets of $H = H_K(s)$ in K:

$$\gamma_i s = \gamma_j s \Leftrightarrow \gamma_j^{-1} \gamma_i s = s \Leftrightarrow \gamma_j^{-1} \gamma_i \in H$$
$$\Leftrightarrow \gamma_i H = \gamma_j H.$$

Since $H_K(s)$ is a subgroup of K, we have immediately by Lagrange's Theorem, §6.3, an important corollary.

Corollary. The cardinality of $\text{Orb}_K(s)$ divides the order of K.

Some of the arguments in the Sylow theory [§7.5] utilize a generalization of the preceding discussion in which the elements $s \in S$ are replaced by subsets $M \subseteq S$. Analogous to the definition of the stabilizer subgroup

$H_K(s) \subseteq K$ of an element $s \in S$, the **stabilizer subgroup** in K of a subset $M \subseteq S$ is

$$H_K(M) = \{\gamma \in K : \gamma M = M\}.$$

The **orbit** of M with respect to K, denoted $\text{Orb}_K(M)$, is the set $\{\gamma M : \gamma \in K\}$ of subsets of S conjugate to M with respect to K. As in the proposition, the number of elements in $\text{Orb}_K(M)$, for any subset $M \subseteq S$, equals the index $[K : H_K(M)]$ of the stabilizer subgroup of M in K and so divides the order of K.

Exercises

1. Prove that the conjugacy relation $\approx$ defined with respect to $K \subseteq \Sigma(S)$ is an equivalence relation.
2. Prove that the sets $H_K(s)$ and $H_K(M)$ are subgroups of K.
3. Verify directly from the definition of $\approx$ that the orbits of elements $s, s' \in S$ are either equal or disjoint subsets of S.
4. A group G (with identity e) is said to **act** (or operate) on a set S if to each element (g, s) in the cartesian product $G \times S$ there is associated a unique $s' = gs \in S$, and if $es = s$ and $(g_1 g_2)s = g_1(g_2 s)$ for all $s \in S$ and $g_1, g_2 \in G$.
   a. Prove that the stabilizer $H_G(s)$ is a subgroup of G.
   b. Prove that the relation defined on S by $t \sim s$ if $t \in \text{Orb}_G(s)$ is an equivalence relation on S.
   c. Suppose that G is a finite p-group and that S is a finite set, whose cardinality $\text{Card } S$ is relatively prime to p. Prove that there exists at least one element $s \in S$ satisfying $gs = s$ for all $g \in G$. (Such an element is called a **fixed point** of G on S.)
   d. Assume again that G and S are finite. Prove that for each $s \in S$,

   $$[G : 1] = [H_G(s) : 1] \cdot \text{Card Orb}_G(s).$$

5. Let G be a p group that acts on a finite set S. Let F denote the set of fixed points of S under the action of G [see Exercise 4(c)]. Prove that $\text{Card } S \equiv \text{Card } F \pmod{p}$.

§7.4 The Class Equation and Normalizers

Continuing the preparation for Sylow theory [§7.5], we now study normalizers and conjugacy relations in groups. Of interest in their own right, these concepts, originally introduced in the Exercises of §§6.2 and 6.3, provide the proof (by means of decomposing a group into conjugacy classes) that the center of any group of prime power order is nontrivial. The class equation provides a useful means of enumerating elements in this and other group-theoretic arguments.

Whereas §7.3 addressed bijections on an arbitrary (finite) set S, we now specialize to the case where the set is a finite group G. Further, we pick the subgroup $I(G) \subset \Sigma(G)$ of inner automorphisms [defined in §6.9] for the group of actions on G; that is, we replace S and K in §7.3 by G and $I(G)$, respectively.

Carrying over the definitions and results of §7.3, we have the following definitions. Subsets S, S' of G are said to be **conjugate with respect to** $I(G)$, or more simply **conjugate**, if there exists an element $x \in G$ such that $S' = \sigma_x(S) = xSx^{-1}$. A subset S is called **self-conjugate** if $\sigma_x(S) = S$ for all $x \in G$.

In particular, we refer to self-conjugate elements (which are the elements of the center of G) and self-conjugate (or normal) subgroups [cf. Exercises 6, §6.2, and 2, §6.3]. The **conjugacy class**

$$\langle\!\langle g \rangle\!\rangle = \{g' \in G : g' \text{ conjugate to } g\}$$
$$= \{\sigma_x(g) : x \in G\} = \text{Orb}_{I(G)}(g)$$

of an element $g \in G$ is the set of *all* images of g under the group $I(G)$ of inner automorphisms.

Since conjugacy is an equivalence relation, the conjugacy classes $\langle\!\langle g \rangle\!\rangle$ provide a partition of G [see §1.2 and Exercise 3, §7.3]. Thus G is the union of the distinct conjugacy classes:

$$G = \bigcup_v \langle\!\langle g_v \rangle\!\rangle.$$

From now on we consider only finite groups G.

We utilize the partition of a group G into its conjugacy classes to state the **class equation**

(*) $$[G:1] = [C:1] + \sum_v h_v$$

of G. Here C is the center of G and h_v is the number of elements in the conjugacy class $\langle\!\langle g_v \rangle\!\rangle$; the sum is taken over all distinct classes $\langle\!\langle g_v \rangle\!\rangle$ having more than one element. Again we follow the convention introduced in §7.2 that 1 indicates both the identity element and identity subgroup of the group(s) under discussion.

The class equation (*) follows immediately from the facts that the conjugacy classes constitute a partition of G and that the center C of G is the set of self-conjugate elements in G [see Exercises 9 and 10, §6.2 and 2, §6.3].

Next we develop some propositions which are useful in discussing structural properties of groups.

Proposition 1. The number h of conjugates of $g \in G$ divides $[G:1]$.

Proof. From the proposition of §7.3, the number of conjugates of a typical element $g \in G$ equals the index $h = [I(G) : H_{I(G)}(g)]$ of the stabilizer

subgroup of $\{g\}$ in $I(G)$. Hence $h \mid [I(G) : 1]$. By Proposition 2, §6.9,
$$[G : 1] = [I(G) : 1][C : 1],$$
and so $h \mid [G : 1]$.

Proposition 2. The number of subgroups in G conjugate to a given subgroup $M \subseteq G$ is $[I(G) : H_{I(G)}(M)]$ and divides $[G : 1]$.

In the proof of the Sylow theorems, §7.5, we shall use the following consequence of Proposition 2 above and the discussion at the end of §7.3.

Corollary. For a subgroup T of $I(G)$ and a nonempty subset (subgroup) M of G the number of sets (subgroups) in G conjugate to M with respect to T is $[T : H_T(M)]$ and divides $[G : 1]$.

Proposition 3. The center C of a p-group G consists of more than one element.

Proof. The class equation for G is
$$p^m = [G : 1] = [C : 1] + \sum_v h_v.$$
By Proposition 1, $h_v \mid [G : 1] = p^m$. Since by convention $h_v > 1$, we conclude that $p \mid h_v$, for each v. Hence $p \mid \sum_v h_v$, and so must divide $[C : 1]$.

We conclude this section with some observations on stabilizers (used here and in §7.3) and normalizers (introduced in Exercises 8, §6.2, and 1 and 2, §6.4). For a subset $S \subseteq G$, the stabilizer $H_{I(G)}(S)$ is a subgroup of the group $I(G)$ of inner automorphisms of G; the normalizer
$$N_G(S) = \{g \in G : gSg^{-1} = S\}$$
is a subgroup of G.

Proposition 4. For any nonempty subset $S \subseteq G$,
$$[G : N_G(S)] = [I(G) : H_{I(G)}(S)].$$

Proof. Consider the group homomorphism $\Phi : G \to I(G)$ given by $\Phi(x) = \sigma_x$, as in §6.10. The kernel of Φ equals the center of G and is a subgroup of the normalizer $N_G(S)$. The inverse image of the stabilizer subgroup $H_{I(G)}(S) \subseteq I(G)$ consists of all $y \in G$ satisfying $\sigma_y \in H_{I(G)}(S)$. That is,
$$\Phi^{-1}[H_{I(G)}(S)] = \{y \in G : ySy^{-1} = S\} = N_G(S).$$
The mapping $\Phi : N_G(S) \to H_{I(G)}(S)$ is a surjective homomorphism with kernel C, and so by the Isomorphism Theorem in §6.8, $H_{I(G)}(S) \cong N_G(S)/C$. Hence

(**) $[N_G(S) : 1] = [H_{I(G)}(S) : 1][C : 1].$

Multiplying
$$[I(G) : 1] = [I(G) : H_{I(G)}(S)][H_{I(G)}(S) : 1]$$

by $[C:1]$, and combining the result with
$$[G:1] = [I(G):1][C:1],$$
$$[G:1] = [G:N_G(S)][N_G(S):1],$$
and the equality (**), we obtain
$$[G:N_G(S)][N_G(S):1] = [I(G):H_{I(G)}(S)][N_G(S):1].$$
Canceling $[N_G(S):1]$ yields the desired equality.

The class equation (*) can now be written in the common form
$$[G:1] = [C:1] + \sum_v [G:N_v],$$
where N_v denotes the normalizer $N_G(g_v)$ of a typical, but not self-conjugate, element $g_v \in G$. This is an immediate corollary to Proposition 4, if we take for S the subset $\{g_v\} \subset G$.

Exercises

1. Prove that any group can be written as the union of mutually disjoint conjugacy classes.
2. Find all classes of conjugate elements $《g》$ for the elements g of the symmetric group Σ_3. Determine the number of elements in each class.
3. Find all conjugacy classes in each of the following groups, and determine the number of elements in each class.
   a. The nonabelian group of order 27 of Example 6(i), §6.1, with $p = 3$.
   b. The group of the square (dihedral group of order 8).
   c. The quaternion group [Example 6(e), §6.1].
4. Prove that a group of order p^2, p a prime, is necessarily abelian.
5. Complete the blanks.
   a. A group of order 215 can have at most _____ conjugacy classes.
   b. A group of order 25 must have _____ conjugacy classes.
6. Prove that a subgroup is normal only if it consists of the union of conjugacy classes. Is the union of conjugacy classes necessarily a normal subgroup?
7. For a subgroup H of a finite group G prove that $N_G(H)$ is the product of all subgroups $K \subseteq G$ in which H is a normal subgroup.
8. Prove that the normalizer $N_G(H)$ of a proper subgroup H of the p-group G properly contains H.
9. a. Suppose that G has p^2 elements, p a prime. Prove that all proper normal subgroups of G lie in the center of G.
   b. Generalize part (a) to the following: If H is a proper normal subgroup of order p in a p-group G, prove that H lies in the center of G.
10. Prove the existence of two nonisomorphic nonabelian groups of order p^3, p a prime.

§7.5 The Elementary Theorems on Sylow Subgroups

Sylow theory (named for Ludwig Sylow, 1832–1918) addresses subgroups of prime power order and thereby the structure of arbitrary finite (nonabelian) groups. As we do not assume the Fundamental Theorem of Finitely Generated Abelian Groups [§7.1], there is some repetition of argument concerning abelian groups prior to the Sylow theorems.

Theorem 1 (*Cauchy's Theorem*). If a prime number p divides the order of a finite abelian group A, then A contains an element of order p.

This is a consequence of either the Fundamental Theorem of Finitely Generated Abelian Groups or the following lemma.

Lemma. The order of a finite abelian group A divides a power of the minimal exponent of A.

If A is a cyclic group, the proof consists of noting that the exponent and order of A are equal.

For noncyclic groups the proof proceeds by induction on the order of A. To begin the inductive argument, note that the cyclic case establishes the result for a group of order 2. Suppose now that the result is valid for all groups of order less than n. Consider a group A of order n, and pick in A an element b different from the identity. Then $B = \langle b \rangle$ is a proper subgroup, and the factor group A/B (which exists since A is abelian) has order less than n.

Let e and e^* be the minimal exponents of A and A/B, respectively. By the induction hypothesis, for some $s \geq 1$,

$$[A/B : 1] \mid (e^*)^s.$$

Since the eth power of any coset aB in A/B is the identity coset, $e^* \mid e$. Therefore

$$[A : B] = [A/B : 1] \mid e^s;$$

since $o(b) = [B : 1] \mid e$, we obtain

$$[A : 1] = [A : B][B : 1] \mid e^s \cdot e = e^{s+1}.$$

Applying the Principle of Induction, we conclude that for all finite abelian groups A, the order divides some power of the exponent.

Proof of Cauchy's Theorem. Since $p \mid [A : 1]$ and $[A : 1] \mid e^n$, for some $n \in \mathbf{N}$, the prime p divides e. The minimal exponent e of A is the least common multiple of *all* the orders $o(a)$ of elements $a \in A$. Therefore $p \mid o(a)$ for some element $a \in A$ [cf. Exercise 18, §2.7]. Hence $o(a) = ph$, and consequently $o(a^h) = p$.

A **p-Sylow subgroup** of a finite group G is a p-subgroup whose order p^m is the highest power of the prime p that divides the order of G. Equivalently, S_p is a p-Sylow subgroup of G if and only if its order is a power of p and its index in G is relatively prime to p.

Theorem 2. For a given prime p, any group G for which $p \mid [G:1]$ has at least one p-Sylow subgroup S_p.

Proof. We use induction on the order of G. If $[G:1] = p$, the theorem is trivial. Now assume that the theorem is proved for all groups whose orders are less than n. Consider an arbitrary group G of order n, and suppose $p^m \mid n$ but $p^{m+1} \nmid n$, $m \geq 1$. Two cases will be distinguished.

Case 1. The group G contains a *proper* subgroup H whose index is relatively prime to p. Consequently p^m divides $[H:1]$. Since $[H:1] < [G:1]$, the induction hypothesis implies that H has a Sylow subgroup S_p of order p^m. This subgroup S_p is a p-Sylow subgroup of G.

Case 2. The index of every proper subgroup is divisible by p. If G is *non-abelian*, the class equation [see §7.4] states that

$$[G:1] = [C:1] + \sum_v [G:N_v],$$

where C is the center of G and N_v denotes the typical normalizer of a class of conjugate elements $《a_v》$. The indices $[G:N_v]$ are greater than 1 (self-conjugate elements lie in C). Hence $G \supset N_v$, and therefore p divides $[G:N_v]$. Consequently $p \mid [G:1]$ implies that $p \mid [C:1]$. If G is abelian, then $G = C$, and $p \mid [C:1]$.

By Cauchy's Theorem the abelian subgroup C contains an element a of order p. Moreover, the subgroup $\langle a \rangle$ is a normal subgroup of G because $a \in C$. Applying the induction hypothesis to the factor group $G^* = G/\langle a \rangle$, which has order $p^{m-1}s$, where $(s,p) = 1$ and $n = p^m s$, we conclude that G^* has a p-Sylow subgroup S_p^* of order p^{m-1}. Let S_p be the complete inverse image $\pi^{-1}(S_p^*)$ of S_p^*, where π denotes the canonical homomorphism mapping $g \in G$ to its coset $g\langle a \rangle$ in G. The mapping $\pi: S_p \to S_p^*$ is a surjective homomorphism with kernel $\langle a \rangle$. Hence by the Isomorphism Theorem of §6.8, $S_p^* \cong S_p/\langle a \rangle$. Thus

$$[S_p:1] = [S_p:\langle a \rangle][\langle a \rangle:1] = [S_p^*:1][\langle a \rangle:1] = p^{m-1} \cdot p = p^m.$$

Consequently G contains a subgroup S_p of order p^m, as asserted.

As a corollary note that if $p \mid [G:1]$ then G contains an element of order p. In a p-Sylow subgroup consider an element a of order p^s, $s > 0$; $a^{p^{s-1}}$ has order p.

Proposition 1. Suppose that P is a p-Sylow subgroup and N is a *normal*

§7.5 The Elementary Theorems on Sylow Subgroups

subgroup of G. Then $N \cap P$ is a p-Sylow subgroup of N, and NP/N is a p-Sylow subgroup of G/N.

Proof. Consider the following diagram for the subset relations of selected subgroups of G.

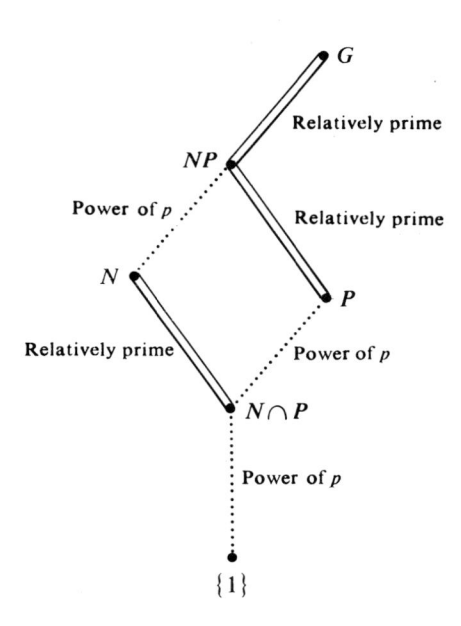

Here, "relatively prime" next to the lines from G to NP and NP to P indicates that the indices $[G:NP]$ and $[NP:P]$ are relatively prime to p. Also, "power of p" next to the lines from NP to N, P to $N \cap P$, and $N \cap P$ to $\{1\}$ indicates that the corresponding indices $[NP:N]$, etc., are powers of the prime p.

We first verify these properties of indices. Since $[G:P]$ and p are relatively prime, the equation
$$[G:P] = [G:NP][NP:P]$$
implies that $p \nmid [G:NP]$ and $p \nmid [NP:P]$. Since $[P:1] = p^m$, the indices $[P:N \cap P]$, $[N \cap P:1]$ are powers of p by Lagrange's Theorem.

Next, using Theorem 3, §6.8, with H replaced by P, we conclude that $P/(N \cap P)$ is isomorphic to NP/N; hence $[NP:N] = [P:N \cap P]$, a power of p. This equality together with the equalities
$$[NP:N \cap P] = [NP:P][P:N \cap P]$$
and
$$[NP:N \cap P] = [NP:N][N:N \cap P]$$
implies that
$$[NP:P] = [N:N \cap P].$$
Consequently $[N:N \cap P]$ and p are relatively prime because $p \nmid [NP:P]$. Hence $N \cap P$ is a p-Sylow subgroup of N since its order is a power of p.

Finally, NP/N is a p-Sylow subgroup of G/N because its order (equal to $[NP:N]$) is a power of p and its index in G/N (equal to $[G:NP]$) is relatively prime to p.

Proposition 2. If Q is a *normal* p-Sylow subgroup of G, then it is the *only* p-Sylow subgroup of G.

Proof. Suppose that P is another p-Sylow subgroup of G. Then PQ/Q is a p-Sylow subgroup of G/Q according to Proposition 1. However, since Q is a p-Sylow subgroup, $[G/Q:1]$ is relatively prime to p; consequently $PQ = Q$. Thus $P \subseteq Q$ and therefore $Q = P$ because $[Q:1] = [P:1] = p^m$.

Since any subgroup is normal in its normalizer [see §7.4 and Exercise 1, §6.4], we have the following corollary.

Corollary. A p-Sylow subgroup P of G is the only p-Sylow subgroup of $N_G(P)$.

Theorem 3. The number r of p-Sylow subgroups of G divides the order of G and satisfies $r \equiv 1 \pmod{p}$. All p-Sylow subgroups of G are conjugate.

To prove $r \equiv 1 \pmod{p}$. If there is only one p-Sylow subgroup, the theorem is trivially true. Thus, consider the case $r > 1$. (Note that *no* p-Sylow subgroup is a normal subgroup by Proposition 2.)

Suppose that $P = P_1, P_2, \ldots, P_r$ are the distinct p-Sylow subgroups of G. Borrowing the notation of §7.3, let

$$S = \{P, P_2, \ldots, P_r\},$$

$$K = \{\sigma_x \in I(G) : x \in P\},$$

the subgroup of inner automorphisms of G determined by elements $x \in P$. The set K is also a subgroup of the group $\Sigma(S)$ of bijections of S, because

$$\sigma_x P_i = x P_i x^{-1} = \{xyx^{-1} : y \in P_i\}$$

is isomorphic to P_i. Thus $\sigma_x P_i$ is a p-Sylow subgroup of G and so is an element of S. In other words, the mappings in K permute the elements of S.

Furthermore, $[K:1]$ is a power of p, a fact we shall need shortly. To verify this claim, consider the mapping $\Phi : P \to K$ which maps $x \in P$ to the inner automorphism σ_x of G [cf. §6.9]. Since Φ is surjective (by the definition of K), $K \cong P/\ker \Phi$. The order of P is p^m, and so the order of K must be a power of p.

As in §7.3, we consider the orbits in S with respect to K. Since P is a group, $\sigma_x P = P$ for all $x \in P$; hence $\{P\} \subset S$ is an orbit consisting of one element. We now show that the cardinality h_i of each of the other orbits (i.e., the number of conjugates of P_i, $i \neq 1$, with respect to K) is a *positive* power of p. First, $h_i = [K : H_K(P_i)]$ by the corollary of §7.4, where $H_K(P_i)$

is the stabilizer subgroup

$$H_K(P_i) = \{\sigma_x \in K : \sigma_x P_i = P_i\}.$$

Second, since $[K:1]$ is a power of p, we must have $h_i = p^{n_i}$ for some n_i, $0 \le n_i \le m$.

Suppose that $n_i = 0$. Then $H_K(P_i) = K$; that is,

$$xP_i x^{-1} = P_i \quad \text{for all } x \in P,$$

and consequently, $P \subseteq N_G(P_i)$, the normalizer of P_i in G. The corollary to Proposition 2 states that P_i is the *only* p-Sylow subgroup of $N_G(P_i)$. Thus $P_i = P$, a contradiction. Therefore n_i must be greater than zero for each i, $2 \le i \le r$.

Hence $r = 1 + \Sigma p^{n_i}$, $n_i > 0$, where the summation is taken over the distinct orbits. Thus $r \equiv 1 \pmod{p}$, as asserted.

To prove all p-Sylow subgroups are conjugate. Once this fact is established, it will follow immediately that r divides the order of G. The number of conjugates in G of P is

$$[I(G) : H_{I(G)}(P)] = [G : N_G(P)]$$

and divides the order of G by Propositions 2 and 4, §7.4. Thus it must be shown that, for a given (fixed) p-Sylow subgroup P and any other p-Sylow subgroup Q, an element $g \in G$ exists for which $\sigma_g P = Q$.

Let T_1 denote the class of all p-Sylow groups of G conjugate to $P_1 = P$ with respect to $I(G)$. If there is a p-Sylow group Q of G which does not lie in T_1, then the class T_2 of p-Sylow groups of G conjugate to Q with respect to $I(G)$ is distinct from T_1 since conjugacy is an equivalence relation. Using the action of $K' = \{\sigma_y \in I(G) : y \in Q\}$, where Q is a typical group in T_2, on the p-Sylow subgroups in the class T_2, we show, as in the previous part of the proof, that the number s_2 of distinct groups in the conjugacy class T_2 satisfies $s_2 \equiv 1 \pmod{p}$.

Next let the mappings in K permute the objects of the class T_2. A typical orbit within T_2 with respect to K contains $k_j = [K : H_K(Q)]$ p-Sylow groups, where $H_K(Q)$ is the set $\{\sigma_x \in K : \sigma_x Q = Q\}$. Since $H_K(Q)$ is a subgroup of K, its index k_j is a power of p. Further, k_j is a nonzero power because Q is the only p-Sylow subgroup of $N_G(Q)$. Thus s_2, the sum of the k_j over distinct orbits within T_2, is divisible by p. In other words, $s_2 \equiv 0 \pmod{p}$.

This conclusion contradicts the previously found congruence $s_2 \equiv 1 \pmod{p}$. Therefore, $T_2 = \emptyset$; there can be no p-Sylow subgroup Q that is not conjugate to P.

Theorem 4. *If G has a subgroup U whose order is a power of a prime p, then there exists a p-Sylow subgroup P such that $U \subseteq P$.*

Proof. Consider the set $S = \{P_1, \ldots, P_r\}$ of the r p-Sylow subgroups of G, and let $K = \{\sigma_u \in I(G) : u \in U\}$. As in the proof of Theorem 3, K is a subgroup of $\Sigma(S)$ and $[K:1]$ is a power of p. Also, the orbit of P_i in S contains $h_i = [K: H_K(P_i)]$ elements. The indices h_i are powers of p, but not all of them can be greater than 1, lest $p|r$. Since $r \equiv 1 \pmod{p}$ by Theorem 3, $h_i = [K: H_K(P_i)] = 1$ for *some* i, $1 \le i \le r$; that is, $K = H_K(P_i)$.

For this i, $\sigma_u P_i = u P_i u^{-1} = P_i$ for all $u \in U$. This fact implies that P_i is a normal subgroup of $U \cdot P_i$. Now by Theorem 3, §6.8,

$$U \cdot P_i / P_i \cong U/(U \cap P_i).$$

Since the order of U is a power of p, so are the orders of the factor groups $U/(U \cap P_i)$ and $U \cdot P_i / P_i$.

Finally, P_i cannot be a proper subgroup of $U \cdot P_i$, since

$$[G : P_i] = [G : U \cdot P_i][U \cdot P_i : P_i]$$

and is relatively prime to p. Thus, $U \cdot P_i = P_i$, which implies that U is contained in the p-Sylow subgroup P_i.

Exercises

1. Prove that the exponent of a cyclic group of order n is n.
2. Prove that the exponent of a finite abelian group A is the least common multiple of the orders of elements of A.
3. If B is a subgroup of the abelian group A, prove that the exponent of A/B divides the exponent of A.
4. Let G be a finite group and let H be a Sylow subgroup of G, $[H:1] = p^m$. Prove that H is the only subgroup of order p^m in $N_G(H)$.
5. With H and $N_G(H)$ as in Exercise 4, prove that $N_G[N_G(H)] = N_G(H)$.
6. Prove that any group G has a normal subgroup of order 7 if:
    a. $o(G) = 28$ b. $o(G) = 42$ c. $o(G) = 707$.
7. Let G be any group of order 105.
    a. Show that G has at least one nontrivial normal subgroup.
    b. If G has exactly one nontrivial normal subgroup N, what is $o(N)$?
8. Complete the blanks.
    a. A group of order 48 must have either _____ or _____ 2-Sylow subgroups.
    b. A group of order $595 = 5 \cdot 7 \cdot 17$ must have either _____ or _____ 7-Sylow subgroups and _____ 5-Sylow subgroups.
    c. A group of order 104 must have either _____ or _____ 2-Sylow subgroups.
    d. A group of order 122 must have either _____ or _____ 2-Sylow subgroups.
    e. An abelian group of order $255 = 3 \cdot 5 \cdot 17$ must have _____ 3-Sylow subgroups.
9. Show that every group of order $p^s m$, where $p > m > 1$, p a prime number, has a nontrivial normal subgroup.

10. Show that any group of order $595 = 5 \cdot 7 \cdot 17$ must have at least two nontrivial normal subgroups.
11. How many elements of order 5 are there in a group G of order 80 if G has no normal subgroup of order 5?
12. Suppose that the order of a group G is the product of two distinct primes p and q.
    a. If $q = 2$, prove that G has a normal subgroup of order p [cf. Exercise 4, §6.4].
    b. If $p > q$, prove that every p-Sylow subgroup of G must be normal in G.
    c. If G has normal subgroups with respective orders p and q, prove that G must be abelian and (with reference to §7.1) cyclic.
    d. If $p \not\equiv 1 \pmod{q}$ and $q \not\equiv 1 \pmod{p}$, prove that G must be cyclic.
13. Let H be a normal p-Sylow subgroup of G. Prove that $\alpha(H) = H$ for all automorphisms α of G.
14. Show that any group of order $(35)^2$ must be abelian. *Hint:* Show that such a group is the direct product of two normal abelian subgroups.
15. a. Prove that any group of order 245 must be abelian.
    b. How many nonisomorphic groups of order 245 are there?
    c. Repeat parts (a) and (b) for a group of order 85.
16. Consider primes p and q, where $p > q$ and $p \equiv 1 \pmod{q}$. Prove that, to within isomorphisms, there exists precisely one nonabelian group of order pq.
17. Consider a finite group G in which every Sylow subgroup is normal. Prove that G is the direct product of its Sylow subgroups.
18. If G is a finite group in which each element different from the identity has order p, for a fixed prime p, prove that $o(G) = p^m$, or some $m \in \mathbf{N}$.
19. Show by an example that if an integer n divides the order of an abelian group A it is not necessarily true that A contains an element of order n [cf. Cauchy's Theorem (Theorem 1)].
20. Prove that every subgroup H of G which contains the normalizer $N_G(P)$ of a p-Sylow group P is equal to its own normalizer, i.e., $N_G(H) = H$ [cf. Exercise 7, §7.4].
21. Prove that if $H \subset G$ has order p^s but is not a p-Sylow subgroup, then $H \neq N_G(H)$.
22. A subgroup $H \subset G$ is called **maximal** if there is no subgroup H' of G such that $H \subset H' \subset G$. Prove that every maximal subgroup of a p-group is normal and has index p.
23. The following steps constitute an alternate proof of Theorem 3. The method of double coset decomposition was originally used by Cauchy and later by Frobenius.
    a. Given two subgroups H, K of a group G, define the **double coset** by H and K of $g \in G$ to be
    $$HgK = \{hgk : h \in H, k \in K\}.$$
    Prove that the double cosets by H and K constitute a partition of G [see §1.2].

b. Prove that the number of elements in any double coset divides $o(G)$, when G is a finite group. *Hint:* Show that the number of right cosets Hg' in HgK equals $[K:(g^{-1}Hg) \cap K]$ and use the analogous statement for left cosets.

c. In fact, prove that the number of elements in HgK is hk/d_g, where $o(H) = h$, $o(K) = k$, and $d_g = o(gHg^{-1} \cap K)$.

d. Decomposing G by two Sylow p-groups P and Q, show that $o(gPg^{-1} \cap Q) = o(Q)$ for some g, and hence, that P and Q are conjugate subgroups of G.

24. Prove the following propositions concerning groups of prime power order by arguments similar to those used in proving the Sylow theorems. Let G have order p^m, p a prime.

a. G has at least one normal subgroup of order p^s, for each s, $0 < s < m$.

b. Every subgroup of order p^{m-1} is normal in G.

c. If $p \neq 2$, and if G has only one proper subgroup of order p^s, for each s, $0 < s < m$, then G is cyclic.

d. The number of normal subgroups of order p^s is congruent to 1 modulo p.

e. The number of subgroups of order p^s is congruent to 1 modulo p.

25. Suppose that S_1 and S_2 are two distinct p-Sylow subgroups of a subgroup H of the group G. Prove that they are not contained in the same p-Sylow subgroup of G.

26. Suppose that G is a p-group which has only one normal subgroup of index p. Prove that G is cyclic. *Hint:* Use induction on the order of G.

27. Let G be a nonabelian group of order p^n which contains a cyclic normal subgroup of index p. Prove that G contains an element a of order p^{n-1} and an element b of order p, which is not a power of a. Furthermore, prove that a and b generate G, that $\langle a \rangle$ is a normal subgroup, and that $bab^{-1} = a^r$, where $r \not\equiv 1 \pmod{p^{n-1}}$ and $r^p \equiv 1 \pmod{p^{n-1}}$.

§7.6 Composition Series and the Jordan-Hölder Theorem

More detailed description of the structure of a nonabelian group is given by the study of composition series, which are sequences of selected subgroups. The fundamental result in the study of composition series is the theorem, due to Camille Jordan (1838–1922) and Otto Hölder, stating that any two composition series of a given finite group are "isomorphic." Composition series are used in defining the "solvability" of groups, and as we shall find in §§8.6 and 9.2, this bears on the related question of solvability of polynomial equations by radicals. A principal application of composition series is in the concept of the length of such a series, which plays the role in group theory and algebraic geometry that dimension does in the study of vector spaces or field extensions [see §8.1].

Proposition 1 (*The "Butterfly Lemma" due to Hans Zassenhaus*). For a group G and four subgroups M, M_0, N, N_0, where M_0 is normal in M and N_0 is normal in N,

$$M_0 \cdot (M \cap N)/M_0 \cdot (M \cap N_0) \cong N_0 \cdot (N \cap M)/N_0 \cdot (N \cap M_0).$$

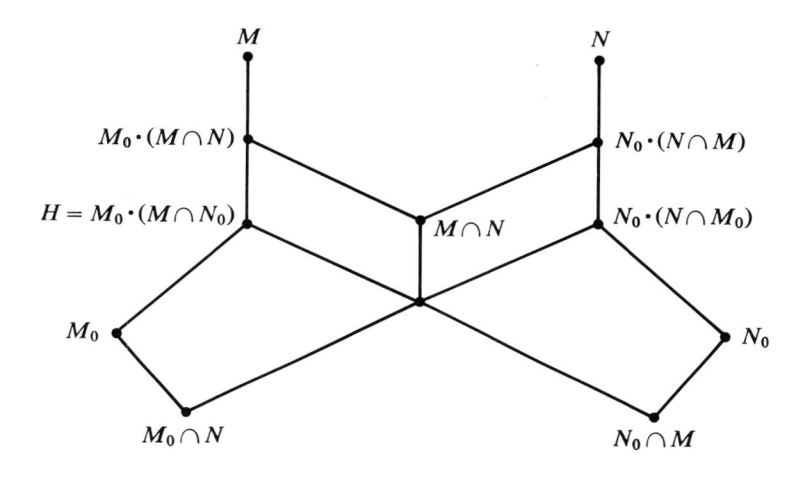

[Diagram showing lattice with nodes: M, N at top; $M_0 \cdot (M \cap N)$, $N_0 \cdot (N \cap M)$; $H = M_0 \cdot (M \cap N_0)$, $N_0 \cdot (N \cap M_0)$; $M \cap N$ in center; M_0, N_0; $M_0 \cap N$, $N_0 \cap M$ at bottom.]

The proof has several parts. First we should check that the two factor groups are indeed well-defined. That is, verify that $M_0 \cdot (M \cap N_0)$ and $N_0 \cdot (N \cap M_0)$ are normal subgroups of $M_0 \cdot (M \cap N)$ and $N_0 \cdot (N \cap M)$, respectively [cf. Exercise 10, §6.4].

Next we adopt the following notation:

$$K = M \cap N,$$
$$H = M_0 \cdot (M \cap N_0),$$
$$D = H \cap K.$$

We leave to the reader the proofs of two technical lemmas.

Lemma 1. $H \cdot K = M_0 \cdot (M \cap N_0) \cdot (M \cap N) = M_0 \cdot (M \cap N).$

Lemma 2. $D = H \cap K = (M_0 \cap N) \cdot (M \cap N_0).$

Since H is a normal subgroup of $H \cdot K = M_0 \cdot (M \cap N)$, we obtain from Theorem 3, §6.8, an isomorphism $H \cdot K/H \cong K/(H \cap K)$, or

$$M_0 \cdot (M \cap N)/M_0 \cdot (M \cap N_0) \cong (M \cap N)/(M_0 \cap N) \cdot (M \cap N_0).$$

The terms in the factor group on the right-hand side remain unaltered if M and N, M_0 and N_0 are interchanged; thus also

$$N_0 \cdot (N \cap M)/N_0 \cdot (N \cap M_0) \cong (M \cap N)/(M_0 \cap N) \cdot (M \cap N_0).$$

Consequently, combining the isomorphisms, we complete the proof of the Zassenhaus Lemma (Proposition 1).

We now consider chains or sequences of subgroups of a given group G and utilize the Zassenhaus Lemma to prove the Jordan-Hölder Theorem. A chain of subgroups of a group G

$$G = G_0 \supseteq G_1 \supseteq \cdots \supseteq G_{i-1} \supseteq G_i \supseteq \cdots \supseteq G_r = \{e\}$$

is called a **normal chain** if G_i is normal in G_{i-1}, $1 \leq i \leq r$. A normal chain

$$G = G_0 = K_{1,0} \supseteq K_{1,1} \supseteq \cdots \supseteq K_{1,s_1} = G_1$$
$$= K_{2,0} \supseteq \cdots \supseteq K_{2,s_2} = G_2 \supseteq \cdots \supseteq G_{i-1}$$
$$= K_{i,0} \supseteq \cdots \supseteq K_{i,s_i} = G_i \supseteq \cdots \supseteq G_r = \{e\}$$

is called a **refinement** of the normal chain

$$G = G_0 \supseteq G_1 \supseteq \cdots \supseteq G_r = \{e\}$$

if $K_{i,j}$ is normal in $K_{i,j-1}$, for $1 \leq i \leq r$ and $1 \leq j \leq s_i$.

A **composition series** of a group G is a normal chain

(*) $$G = G_0 \supseteq G_1 \supseteq \cdots \supseteq G_r = \{e\}$$

which admits no *proper* refinements and has no coincidences $G_{i-1} = G_i$. The number r in the composition series (*) is called the **length** of the composition series.

Equivalently, a composition series is a normal chain such that the factor groups G_{i-1}/G_i are **simple groups** (that is, have no nontrivial normal subgroups) or such that the normal subgroups G_i of G_{i-1} are *maximal normal proper* subgroups of G_{i-1}, respectively.

Two normal chains

$$G = G_0 \supseteq G_1 \supseteq \cdots \supseteq G_i \supseteq \cdots \supseteq G_r = \{e\}$$

and $$G = H_0 \supseteq H_1 \supseteq \cdots \supseteq H_j \supseteq \cdots \supseteq H_s = \{e\}$$

are called **isomorphic**, if $r = s$ and if there is a permutation τ of $\{1, \ldots, r\}$ such that $G_{i-1}/G_i \cong H_{\tau(i)-1}/H_{\tau(i)}$ for each i, $1 \leq i \leq r$.

Proposition 2. Any two normal chains of a group G have isomorphic refinements.

Proof. Consider two normal chains of subgroups of G:

$$G = G_0 \supseteq G_1 \supseteq \cdots \supseteq G_{i-1} \supseteq G_i \supseteq \cdots \supseteq G_r = \{e\}$$

and $$G = H_0 \supseteq H_1 \supseteq \cdots \supseteq H_{j-1} \supseteq H_j \supseteq \cdots \supseteq H_s = \{e\}.$$

Now define two sets of subgroups for $1 \leq i \leq r$, $1 \leq j \leq s$:

$$G_{i,j} = G_i(G_{i-1} \cap H_j), \qquad H_{j,i} = H_j(H_{j-1} \cap G_i).$$

Also set $G_{i,0} = G_{i-1}$ and $H_{j,0} = H_{j-1}$. Observe that

$$G_{i,s} = G_i, \qquad H_{j,r} = H_j,$$
$$G_{i-1} \supseteq G_{i,j} \supseteq G_i, \qquad H_{j-1} \supseteq H_{j,i} \supseteq H_j.$$

§7.6   Composition Series and the Jordan-Hölder Theorem   251

Next apply the Zassenhaus Lemma, setting $M = G_{i-1}$, $M_0 = G_i$, and $N = H_{j-1}$, $N_0 = H_j$. Then
$$G_{i,j-1}/G_{i,j} \cong H_{j,i-1}/H_{j,i}.$$

Finally, using these isomorphisms, we conclude that
$$G = G_0 \supseteq \cdots \supseteq G_{i-1}$$
$$= G_{i,0} \supseteq \cdots \supseteq G_{i,j-1} \supseteq G_{i,j} \supseteq \cdots \supseteq G_i \supseteq \cdots \supseteq G_r = \{e\}$$
and
$$G = H_0 \supseteq \cdots \supseteq H_{j-1}$$
$$= H_{j,0} \supseteq \cdots \supseteq H_{j,i-1} \supseteq H_{j,i} \supseteq \cdots \supseteq H_j \supseteq \cdots \supseteq H_s = \{e\}$$
are isomorphic refinements of the given normal chains.

Theorem 1 (*Jordan-Hölder Theorem*). If the group G has a composition series, any two of its composition series are isomorphic.

Proof. Any two composition series are normal chains, and hence by Proposition 2 have isomorphic refinements. As compostion series they have no proper refinements however, and so must themselves be isomorphic.

Thus for a group having a composition series, the length r is a property of the group, independent of the choice of composition series. Note that the theorem requires the existence of at least one composition series. This condition is satisfied whenever the group G is finite.

Proposition 3. If G has a composition series and N is a normal subgroup of G, then there exists a composition series of G one of whose terms equals N.

For the proof, apply Proposition 2 on the existence of isomorphic refinements to the normal chain $G = G_0 \supset G_1 = N \supset G_2 = \{e\}$ and to a composition series $G = H_0 \supset \cdots \supset H_j \supset \cdots \supset H_s = \{e\}$.

An important subject in group theory is that of solvability, a concept which is defined in terms of normal chains. The term "solvable" derives from the corresponding concept of solvability of polynomial equations by radicals for which the reader should refer to the Galois theory [§§8.6 and 9.2].

A group G is called **solvable** if there is a normal chain
$$G = G_0 \supset G_1 \supset \cdots \supset G_{i-1} \supset G_i \supset \cdots \supset G_s = \{e\}$$
such that each factor group G_{i-1}/G_i is abelian, $1 \leq i \leq s$. Finite solvable groups can be defined equivalently as groups G having a normal chain
$$G = G_0 \supset G_1 \supset \cdots \supset G_{j-1} \supset G_j \supset \cdots \supset G_r = \{e\}$$
such that G_{j-1}/G_j is cyclic of prime order, $1 \leq j \leq r$. The equivalence of the definitions follows from the structure of finite abelian groups [§7.1].

Example 1. All p-groups G are solvable.

These groups contain, by Proposition 3 in §7.4, a cyclic group L of order

p which lies in the center. Then G/L has order p^{m-1}. Hence we may assume, as an induction hypothesis, that $G^* = G/L = G_0^* \supset G_1^* \supset \cdots \supset G_i^* \supset \cdots \supset G_{m-1}^* \supset \{e^*\}$ where G_{i-1}^*/G_i^* is cyclic of order p. The inverse images G_i of the groups G_i^* [see §6.8] form a chain

$$G = G_0 \supset G_1 \supset \cdots \supset G_i \supset \cdots \supset L \supset \{e\}$$

of subgroups in G such that G_{i-1}/G_i, G_{m-1}/L, and $L/\{e\}$ are cyclic of order p. Hence G is solvable.

Example 2. The dihedral groups Γ_n are solvable.

If Γ_n is given [as in Examples 6(f) and 7, §6.1] by

$$\Gamma_n = \{a^s b^t : s, t \in \mathbf{Z}, a^n = b^2 = e, bab^{-1} = a^{-1}\}$$

then $\Gamma_n \supset \langle a \rangle \supset \{e\}$ is a normal chain with abelian factors, since $[\Gamma_n : \langle a \rangle] = 2$ and $\langle a \rangle$ is cyclic.

Thus, in particular, the group Σ_3 of permutations of 3 elements is solvable.

Example 3. The symmetric group Σ_4 is solvable.

The group Σ_4 of permutations of 4 elements has order 24; the alternating subgroup A_4 of even permutations has order 12 and index 2 [see Exercise 6, §6.7]. Thus A_4 is normal in Σ_4. Furthermore A_4 has a normal subgroup V, generated by the even permutations $\alpha = [12][34]$ and $\beta = [13][24]$. Since $\alpha^2 = \beta^2 = \varepsilon = [1][2][3][4]$, V is isomorphic to Klein's Four-group.

Before proving that V is normal A_4, we observe that A_4 is generated by α, β, and $\gamma = [123] = [13][12]$. Since α, β, γ are all even permutations, A_4 contains the group B generated by them. Moreover, $o(B)$ must be either 6 or 12 since it is greater than 4 and divides $12 = o(A_4)$. The proof that $B = A_4$ can be accomplished by exhibiting seven distinct elements in B such as:

$$\gamma^2 = \gamma^{-1} = [213], \qquad \gamma^2 \alpha = [423],$$
$$\gamma\alpha = [413], \qquad \gamma^2\beta = [412],$$
$$\gamma\beta = [432], \qquad \gamma^2\alpha\beta = [314].$$
$$\gamma\alpha\beta = [214],$$

Hence

$$A_4 = \{\varepsilon, \gamma, \gamma^2, \alpha, \gamma\alpha, \gamma^2\alpha, \beta, \gamma\beta, \gamma^2\beta, \alpha\beta, \gamma\alpha\beta, \gamma^2\alpha\beta\}.$$

Since V and A_4 are generated by $\{\alpha, \beta\}$ and $\{\alpha, \beta, \gamma\}$ respectively, the proof that V is normal in A_4 is reduced to verifying that $V\gamma = \gamma V$. Check that

$$\alpha\gamma = [12][34][13][12] = [243] = \gamma\beta,$$
$$\beta\gamma = [13][24][13][12] = [214] = \gamma\alpha\beta,$$
$$\alpha\beta\gamma = \alpha\gamma\alpha\beta = \gamma\beta\alpha\beta = \gamma\alpha.$$

We then have the composition series for Σ_4

$$\Sigma_4 \supset A_4 \supset V \supset \{\varepsilon, \alpha\} \supset \{\varepsilon\},$$

whose factors are cyclic groups of respective orders 2, 3, 2, 2.

§7.6   Composition Series and the Jordan-Hölder Theorem

To prove that the symmetric group Σ_n, $n \geq 5$, is not solvable we need the following lemma.

Lemma. If N is a normal subgroup of a subgroup $H \subseteq \Sigma_n$, $n \geq 5$, where H contains all 3-cycles, and if H/N is abelian, then N itself contains all 3-cycles.

Proof. Proposition 2 in §6.7 stated that, for $n \geq 5$, every 3-cycle is the commutator of two 3-cycles. Since H/N is abelian, N contains the commutator subgroup of H, by Proposition 3, §6.9. In particular, N contains all 3-cycles, as asserted.

Theorem 2. The symmetric group Σ_n, $n \geq 5$, is not solvable.

Suppose to the contrary that Σ_n were solvable, i.e., there exists a composition series

$$\Sigma_n = H_0 \supset H_1 \supset \cdots \supset H_{i-1} \supset H_i \supset \cdots \supset H_s = \{e\},$$

where H_i is normal in H_{i-1} and H_{i-1}/H_i is abelian. Then using the lemma for $\Sigma_n = H$ and $H_1 = N$, we conclude that H_1 must contain all 3-cycles. Repeating this argument s times, we conclude that $H_s = \{e\}$ must contain all 3-cycles, an absurdity. Thus no such composition series can exist.

Corollary. The alternating group A_n, $n \geq 5$, is not solvable.

If A_n were solvable, then Σ_n would be too since A_n is a normal subgroup of Σ_n and since the quotient Σ_n/A_n is an abelian group.

The structure of A_n, $n \geq 5$, is made more explicit by the following proposition, the proof of which is left as an exercise.

Proposition 4. The alternating group A_n, $n \geq 5$, is simple.

Exercises

1. Find a composition series of the quaternion group G of order 8 [see Exercise 6, §6.5].
2. Let G be a solvable group. Prove that
   a. Every subgroup of G is solvable.
   b. Every homomorphic image of G is solvable.
3. Let H be a subgroup and N a normal subgroup of the group G. Show that HN is solvable if H and N are solvable.
4. Let N be a normal subgroup of the group G. Assume that both N and its factor group G/N are solvable. Prove that G is solvable.

5. **a.** With reference to Exercise 10, §6.1, prove that the set H of (affine) maps τ on $\mathbf{Z}_7$ given by $\tau(u) = \alpha u + \beta$, where $\alpha = 1$, is a normal subgroup of the affine group $\text{Af}(\mathbf{Z}_7)$. (We call H the subgroup of *translations*.)
   b. Prove that $\text{Af}(\mathbf{Z}_7)/H \cong \mathbf{Z}_7^*$, the multiplicative group of $\mathbf{Z}_7$.
   c. Find a composition series of $\text{Af}(\mathbf{Z}_7)$.
   d. Is $\text{Af}(\mathbf{Z}_7)$ a solvable group?
   e. Find the commutator subgroup of $\text{Af}(\mathbf{Z}_7)$.
   f. Find the center C of $\text{Af}(\mathbf{Z}_7)$.
6. Prove that the affine group $\text{Af}(\mathbf{Z}_m)$ is solvable.
7. Prove that any abelian group is solvable.
8. Find all composition series of a cyclic group G of order 28.
9. Consider the following chain of subgroups $G^{(i)} \subseteq G$ where $G^{(0)} = G$, $G^{(1)} = G^1$, $G^{(2)} = [G^{(1)}]^1$, ..., $G^{(i)} = [G^{(i-1)}]^1$, ... where G^1 is the commutator subgroup of G. Prove that
   a. $G^{(i)}$ is not only a normal subgroup of $G^{(i-1)}$, $i \geq 1$, but is *also* a normal subgroup of G.
   b. G is solvable if and only if $G^{(r)} = \{e\}$ for some positive integer r.
10. Using the higher commutator subgroups $G^{(i)}$ (as in Exercise 9) of a solvable group G, prove the following statements [cf. Exercise 4].
    a. If N is a normal subgroup of G, then the factor group G/N is solvable.
    b. If H is a subgroup of G, then H is solvable.
11. Prove that every group of prime power order is solvable.
12. Prove Lemmas 1 and 2.
13. The following steps outline a proof of the simplicity of A_n, $n \geq 5$. Complete the proof of each step.
    a. The symmetric group Σ_n is generated by the transpositions $[12]$, $[13]$, ..., $[1n]$.
    b. The alternating group A_n is generated by the 3-cycles $[123]$, $[124]$, ..., $[12n]$.

    In the remaining steps assume H is a normal subgroup of A_n which contains a permutation $\sigma = \tau_1 \tau_2 \cdots \tau_j$, where the τ_i are disjoint cycles. In each case prove that $H = A_n$.

    c. Suppose $\sigma = \tau_1 = [132]$ (i.e., $j = 1$), and observe that $[12m] = [m31][132][13m]$ belongs to H for all m, $3 < m \leq n$. (This generalizes to the statement that if H contains one 3-cycle, then it contains all 3-cycles, or $H = A_n$.)
    d. Suppose $\tau_1 = [a_1, a_2, ..., a_m]$ has length $m > 3$, and observe that for $\rho = [a_1, a_2, a_3] \in A_n$ we have $\rho\sigma^{-1}\rho^{-1}\sigma = \rho\tau_1^{-1}\rho^{-1}\tau_1 = [a_1, a_m, a_3] \in H$ since disjoint cycles commute.
    e. Suppose $\tau_1 = [a_1, a_2, a_3]$ and $\tau_2 = [a_4, a_5, a_6]$ and show that H contains a permutation σ' involving a cycle of length $m > 3$ by considering $\rho = [a_2, a_3, a_4]$ and $\rho\sigma^{-1}\rho^{-1}\sigma$.
    f. Suppose τ_1 is a 3-cycle and $\tau_2, ..., \tau_j$ are transpositions. Observe that $\tau_1^2 \in H$.

g. Suppose the τ_i, $1 \leq i \leq j$, are all transpositions with $\tau_1 = [a_1, a_2]$, $\tau_2 = [a_3, a_4]$. Let $\rho = [a_2, a_3, a_4]$ and $\eta = [a_1, a_4, a_5]$. (Why must η involve an a_5?) Show that $(\rho\sigma^{-1}\rho^{-1}\sigma)\eta(\sigma^{-1}\rho\sigma\rho^{-1})\eta^{-1}$ is a 3-cycle in H.

h. Conclude that A_n is simple.

14. Prove that A_n is the only normal subgroup of Σ_4, $n \geq 5$.
15. Exhibit two nonisomorphic groups which have isomorphic composition series.
16. Consider a group G of order $p_1 p_2 p_3$, where the p_i are distinct primes. Prove that if G contains a nontrivial normal subgroup then it is solvable.
17. Suppose that H is a proper subgroup of a p-group G, so that $N_G(H) \supset H$ [cf. Exercise 7, §7.4]. Prove that H occurs as a subgroup in some composition series of G.
18. Prove that a p-group G has a composition series $G = G_0 \supset \cdots \supset G_i \supset \cdots \supset \{e\}$ such that the subgroups G_i are normal in G.
19. Let G be a group with a composition series and a normal subgroup N. Prove that the length of G is equal to the sum of the lengths of N, G/N [cf. Exercise 17(c), §6.4].

§7.7 Groups with Operators

In this section and the next we generalize two topics previously considered: vector spaces from §4.1 and bijections of sets from §7.3.

A vector space V/F [cf. §4.1] is an additive abelian group together with a field F such that each $r \in F$ determines a linear transformation

$$\omega_r : V \to V \quad \text{given by } \omega_r(v) = rv$$

for all $v \in V$. Thus in the sense of our study of bijections of sets [§7.3], the nonzero elements of F operate on or permute the elements of V.

In §7.3 we discussed bijections of sets in which a subgroup K of $\Sigma(S)$ acted on elements of a set S. Then in §7.4 we considered the group $I(G)$ of inner automorphisms acting on a group G. We now consider the more general situation of a set Ω acting on a group. Hence we have the term "group with operators." As suggested above, vector spaces are special cases of groups with operators.

Much of the group-theoretic discussion of subgroups, homomorphisms, factor groups, composition series, and the like carries over with only minor modification to the so-called groups with operators and thus to modules and vector spaces. In this section we address the more general case of groups with operators, where there are no structure requirements on the set of operators. Then in §7.8 we shall specialize our discussion to consideration of modules.

A group G is said to admit (or to have) a set Ω as a domain of operators if each $\omega \in \Omega$ determines a homomorphism of G into itself. We then refer to G as a **group with operators** Ω. It is convenient to write ωg in place of $\omega(g)$ for $\omega \in \Omega$, $g \in G$.

We begin our discussion of groups with operators with three examples.

Example 1. An abelian group A always admits $\mathbf{Z}$ as a domain of operators because each $n \in \mathbf{Z}$ determines a homomorphism $\omega_n: A \to A$ defined as follows [cf. Exercise 11, §6.5]:

$$\text{for } n \geq 0, \qquad \omega_n a = a^n;$$
$$\text{for } n < 0, \qquad \omega_n a = (a^{-1})^{-n}.$$

In each case ω_n is a homomorphism:

$$\omega_n(ab) = (ab)^n = a^n b^n = (\omega_n a)(\omega_n b)$$

or $\quad \omega_n(ab) = ((ab)^{-1})^{-n} = (a^{-1}b^{-1})^{-n} = (a^{-1})^{-n}(b^{-1})^{-n} = (\omega_n a)(\omega_n b)$.

If the law of composition for A is expressed additively, then for $n \geq 0$, we have $\omega_n a = n \cdot a$, the n-fold sum of a. For $n < 0$, $\omega_n a = |n| \cdot (-a)$. Thus in particular, $\mathbf{Z}_m$ is a group with the operators $\mathbf{Z}$.

Example 2. An arbitrary (not necessarily commutative) ring R, considered as an abelian group, admits itself as a domain of operators. For each $a \in R$ define the operator $\omega_a: R \to R$ by $\omega_a r = ar$, an additive group homomorphism.

These mappings ω_a and the mappings $\omega'_a: R \to R$ given by $\omega'_a r = ra$, which make R into a right operator group, play an important role in the study of non-commutative rings. Of course if R is commutative, then $\omega_a = \omega'_a$.

Example 3. An arbitrary group G admits the inner automorphisms $I(G)$ as a domain of operators. For each $g \in G$, the inner automorphism σ_g, given by $\sigma_g x = gxg^{-1}$ for $x \in G$, is an operator on G [cf. §7.4].

We turn now to adaptation of the concepts of subgroup, homomorphism, factor group, etc. to groups with operators. The reader is invited in §7.8 to note that these are generalizations of corresponding statements for modules.

A subgroup H of a group G with operator domain Ω is called Ω-**admissible** (or Ω-**stable**) if $\omega h \in H$ for all $h \in H$ and all $\omega \in \Omega$.

Considering the preceding examples, check that all subgroups of abelian groups are $\mathbf{Z}$-admissible. The normal subgroups of a group G are precisely the admissible subgroups with respect to the operator domain $\Omega = I(G)$ [cf. the note in §6.9]. The subspaces of a vector space V over a field F are precisely the admissible subgroups of V (considered as an abelian group) with operator domain $\Omega = \{\omega_r : r \in F\}$, where $\omega_r v = rv$, for $v \in V$.

As a note of caution, when considering $\mathbf{Q}$ and $\mathbf{Z}$ as additive (abelian) groups, observe that $\mathbf{Z}$ is a $\mathbf{Z}$-admissible subgroup of $\mathbf{Q}$, but is *not* a $\mathbf{Q}$-admissible subgroup of $\mathbf{Q}$.

Suppose that H is a normal Ω-admissible subgroup of G. The factor group G/H admits Ω as a domain of operators, if we define

$$\omega(gH) = (\omega g)H$$

for all $\omega \in \Omega$ and $g \in G$. This defines a homomorphism on G/H because

$$\omega(gH \cdot g'H) = \omega(gg'H) = (\omega(gg'))H$$
$$= (\omega g)(\omega g')H = [(\omega g)H] \cdot [(\omega g')H]$$
$$= \omega(gH)\omega(g'H).$$

If G and Γ are two groups with respective operator domains $\Omega(G)$ and $\Omega(\Gamma)$, a homomorphism $\varphi \colon G \to \Gamma$ is called an **admissible homomorphism** if there exists a single-valued mapping h of $\Omega(G)$ onto $\Omega(\Gamma)$ such that

$$\varphi[\omega(g)] = h(\omega)[\varphi(g)],$$

where $g, g' \in G$, $\omega \in \Omega(G)$, and $h(\omega) \in \Omega(\Gamma)$.

In many applications of admissible homomorphisms $\Omega(G) = \Omega(\Gamma) = \Omega$ and h is the identity map $h(\omega) = \omega$; in this case φ is called an **Ω-admissible homomorphism**.

We now carry over Proposition 2, §6.5, and the Isomorphism Theorem, §6.8, to groups with operators. The proofs follow those given earlier with allowance made now for the operators.

Lemma. The kernel K of an admissible homomorphism $\varphi \colon G \to \Gamma$ is an admissible normal subgroup of G.

Proof. Since $K = \ker \varphi$ is known to be a normal subgroup of G, it remains only to verify its admissibility with respect to the domain of operators. For any $\omega \in \Omega(G)$ and $a \in \ker \varphi$,

$$\varphi(\omega a) = h(\omega)\varphi(a) = h(\omega)e = e$$

where e is the identity of Γ. (Since $h(\omega)$ belongs to $\text{Hom}(\Gamma, \Gamma)$ it maps e to e.) Thus $\omega a \in \ker \varphi$, and $K = \ker \varphi$ is an admissible subgroup.

Theorem. If $\varphi \colon G \to \Gamma$ is an admissible homomorphism with kernel K, there exists an admissible isomorphism $\mu \colon G/K \to \varphi(G) \subseteq \Gamma$ such that the diagram

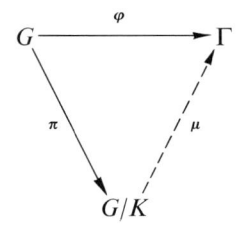

$$\begin{array}{ccc} G & \xrightarrow{\varphi} & \Gamma \\ {\scriptstyle \pi} \searrow & & \nearrow {\scriptstyle \mu} \\ & G/K & \end{array}$$

commutes (that is, $\varphi = \mu \circ \pi$), where π is the canonical projection onto the quotient.

Proof. The Isomorphism Theorem, §6.8, asserts the existence of a group isomorphism $\mu \colon G/K \to \varphi(G)$. Thus it remains only to verify the admissibility

of π and μ. The canonical projection $\pi: G \to G/K$ is Ω-admissible since
$$\pi(\omega g) = (\omega g)K = \omega(gK) = \omega\pi(g).$$
The isomorphism $\mu: G/K \to \varphi(G)$ is admissible since
$$\mu(\omega(gK)) = \mu((\omega g)K) = \varphi(\omega g)$$
$$= h(\omega)\varphi(g) = h(\omega)\mu(gK).$$

Corollary 1. The propositions in §7.6 generalize to groups with operators and operator homomorphisms and isomorphisms.

Corollary 2. The Jordan-Hölder Theorem remains valid for groups with operators if normal subgroups are replaced by admissible normal subgroups.

These corollaries are discussed in more detail in the special case of modules in §7.8.

§7.8 Modules

We now specialize the discussion of groups with operators in §7.7 to modules. Modules may also be viewed as generalizations of the concept of vector spaces V/F, because a module M over a ring R is an (additive) abelian group M and an operation (or action) of the elements in R on M. Thus this section also represents a continuation of linear algebra from Chapter 4. Modules pervade the modern study of commutative algebra, algebraic geometry, and algebraic topology.

Let R be a (commutative) ring with multiplicative identity 1. An (additive) abelian group M is called an **R-left module** if there is given an external law of composition (or mapping)
$$(r, m) \to rm \in M$$
for (r, m) in the cartesian product $R \times M$ such that
$$r(m_1 + m_2) = rm_1 + rm_2,$$
$$(r_1 r_2)m = r_1(r_2 m),$$
$$(r_1 + r_2)m = r_1 m + r_2 m,$$
$$1m = m$$
with $m, m_1, m_2 \in M$ and $1, r, r_1, r_2 \in R$.

In module-theoretic discussion we refer to the R-left module M defined above as a **unitary module** since $1m = m$ for all $m \in M$. Vector spaces V/F are examples of such modules, as are the groups in Examples 1 and 2 in §7.7.

In an analogous fashion we can define R-right modules M (R acts to the right on M), and two-sided modules. As a matter of convenience we shall refer to *R*-modules in what follows rather than to *R*-left modules. The corresponding statements for *R*-right modules are exactly equivalent. When *R* is commutative, left and right *R*-modules can be identified, but in a strict sense they are not equal. We maintain the distinction because of the carry-over of modules to the important cases of noncommutative and nonassociative rings. (If the ring *R* is noncommutative, the distinction between *R*-left and *R*-right modules is important.)

An **R-module homomorphism** $\varphi: M \to M'$ of *R*-modules M, M' is a group homomorphism which preserves (respects) the module structure; that is, for $a, b \in M$ and $r \in R$

$$\varphi(a+b) = \varphi(a) + \varphi(b),$$

$$\varphi(ra) = r\varphi(a).$$

The following statements from §7.7 need no further proof.

Lemma. The kernel K of an *R*-module homomorphism $\varphi: M \to M'$ is an *R*-submodule of M.

Proposition 1. The (additively-written) quotient group M/H is an *R*-module for any submodule $H \subseteq M$.

Theorem 1. If $\varphi: M \to M'$ is an *R*-module homomorphism with kernel K, there exists an *R*-module isomorphism $\mu: M/K \to \varphi(M) \subseteq M'$ such that the diagram

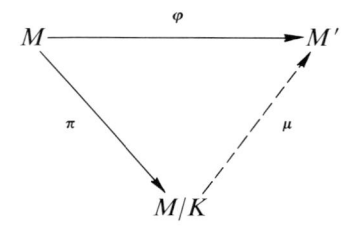

$$\begin{array}{ccc} M & \xrightarrow{\varphi} & M' \\ {\scriptstyle \pi} \searrow & & \nearrow {\scriptstyle \mu} \\ & M/K & \end{array}$$

commutes, where π is the canonical projection onto the quotient module M/K.

As asserted in Corollaries 1 and 2, §7.7, the concept of composition series and the Jordan-Hölder Theorem of §7.6 carry over to groups with operators and hence to modules. In particular, a **composition series** of a module M is a chain of submodules which admits no proper refinements. Therefore we have the following theorem.

Theorem 2 (*Jordan-Hölder Theorem*). If the module M has a composition series, any two composition series are isomorphic.

An *n*-dimensional $(n \geq 1)$ vector space V/F, considered as an *F*-module, is **simple** if and only if $\dim_F V = 1$. (Simplicity means in this

context that V has no nontrivial subspaces.) Thus a composition series of V in the sense of F-modules is a chain of subspaces
$$V = V_0 \supset V_1 \supset \cdots \supset V_{i-1} \supset V_i \supset \cdots \supset V_r = \{0\}$$
such that the factor spaces V_{i-1}/V_i (over F) are simple F-modules, i.e., $\dim_F(V_{i-1}/V_i) = 1$, $1 \le i \le r$. Consequently $\dim_F V$ is the length r of a composition series of V, because the r vectors α_i, $0 \le i < r$, form a basis of V/F where α_i is an arbitrary nonzero vector in $V_{i-1}\backslash V_i$. Thus we have proved the following proposition.

Proposition 2. The length of a composition series of a finite dimensional vector space is equal to its dimension.

As a corollary, for subspaces S, T of a finite dimensional vector space V/F considered as F-modules we have
$$\dim_F(S+T) - \dim_F S = \dim_F T - \dim_F(S \cap T),$$
as in Exercise 11, §4.2. This relation is sometimes referred to as Grassmann's relation, after Hermann Grassmann (1809–1877). It follows from the isomorphism of Theorem 3, §6.8,
$$(S+T)/S \cong T/(S \cap T),$$
and the consequence of the Jordan-Hölder Theorem [cf. Exercise 19, §7.6] that for any subspace $U \subset V$
$$\dim_F(V/U) = \dim_F V - \dim_F U.$$

Analogous to the discussion of generating sets for vector spaces V/F and of finitely generated abelian groups, an R-left module M is said to be **finitely generated** if there exists a finite set of elements $m_1, \ldots, m_s$ in M such that each element m in M can be written as a sum
$$m = r_1 m_1 + \cdots + r_s m_s,$$
with elements r_i, $1 \le i \le s$, in R. It is important to remember that the elements r_i are *not required to be uniquely determined* relative to the set of generators m_i, $1 \le i \le s$.

The remainder of this section is devoted to the study of R-module homomorphisms. Given R-modules M, M', and M'' and R-module homomorphisms $\alpha \colon M \to M'$ and $\beta \colon M' \to M''$, we call the sequence
$$M \xrightarrow{\alpha} M' \xrightarrow{\beta} M''$$
exact if $\operatorname{Im} \alpha = \ker \beta$. For example, each R-submodule $N \subseteq M$ determines an exact sequence
$$0 \longrightarrow N \xrightarrow{\iota} M \xrightarrow{\pi} M/N \longrightarrow 0,$$
where ι is the inclusion map ($\iota(n) = n$, considered as an element of M, for all $n \in N$) and π is the canonical projection onto the quotient.

§7.8 Modules

Denote by $\operatorname{Hom}_R(M', M)$ the set of R-module homomorphisms $\alpha: M' \to M$. As in §6.11, $\operatorname{Hom}_R(M', M)$ is an additive abelian group. Moreover it becomes an R-module if we define $r\alpha$ for $r \in R$, $\alpha \in \operatorname{Hom}_R(M', M)$ to be the mapping $r\alpha: M' \to M$ given by $r\alpha(m') = r[\alpha(m')]$. (Check that $r\alpha$ is an R-module homomorphism.)

Proposition 3. For R-modules M, M', M'', each $\alpha \in \operatorname{Hom}_R(M', M)$ induces an R-module homomorphism

$$\alpha^*: \operatorname{Hom}_R(M, M'') \to \operatorname{Hom}_R(M', M'')$$

and an R-module homomorphism

$$\alpha_*: \operatorname{Hom}_R(M'', M') \to \operatorname{Hom}_R(M'', M).$$

For $\beta \in \operatorname{Hom}_R(M, M'')$ define $\alpha^*(\beta) \in \operatorname{Hom}_R(M', M'')$ by

$$\alpha^*(\beta)(m') = \beta(\alpha(m')) \in M''.$$

Since α and β are R-module homomorphisms, so is their composition $\alpha^*(\beta) = \beta \circ \alpha$. Similarly, define $\alpha_*(\gamma)(m'') = \alpha(\gamma(m''))$ for $\gamma \in \operatorname{Hom}_R(M'', M')$, and of course $\alpha_*(\gamma) = \alpha \circ \gamma$ is an R-module homomorphism.

As a simple consequence we note the following more general corollary.

Corollary. Corresponding to any sequence

$$(*) \qquad M' \xrightarrow{\alpha} M \xrightarrow{\beta} M''$$

of R-modules we have the "induced" sequences

$$(**) \quad \begin{array}{c} \operatorname{Hom}_R(M', N) \xleftarrow{\alpha^*} \operatorname{Hom}_R(M, N) \xleftarrow{\beta^*} \operatorname{Hom}_R(M'', N) \\ \operatorname{Hom}_R(N, M') \xrightarrow{\alpha_*} \operatorname{Hom}_R(N, M) \xrightarrow{\beta_*} \operatorname{Hom}_R(N, M'') \end{array}$$

for any R-module N. Also, $\alpha^* \circ \beta^* = (\beta \circ \alpha)^*$, and $\alpha_* \circ \beta_* = (\alpha \circ \beta)_*$.

Leaving this verification and the proof of the fact that if the sequence $(*)$ is exact then so are the sequences $(**)$ as exercises, we conclude this discussion with the so-called Short Five Lemma. Its proof is commonly referred to as an exercise in diagram chasing.

Short Five Lemma. Consider exact sequences of R-modules with R-module homomorphisms as indicated below

$$\begin{array}{ccccccccc} 0 & \longrightarrow & M' & \xrightarrow{\varphi} & M & \xrightarrow{\psi} & M'' & \longrightarrow & 0 \\ & & \downarrow{\alpha} & & \downarrow{\beta} & & \downarrow{\gamma} & & \\ 0 & \longrightarrow & N' & \xrightarrow{\varphi'} & N & \xrightarrow{\psi'} & N'' & \longrightarrow & 0 \end{array}$$

such that the diagram commutes. If α and γ are isomorphisms, then so is β.

Exercises

1. Let $\alpha: M' \to M$ be an R-module homomorphism of R-modules M', M. For $r \in R$ define a mapping $r\alpha: M' \to M$ by $(r\alpha)(m') = r(\alpha(m'))$. Prove that $r\alpha$ is an R-module homomorphism.
2. a. Prove that $\alpha_*(\beta)$ and $\alpha^*(\gamma)$, as given in Proposition 3, are R-module homomorphisms.
   b. Verify that the mappings α^* and α_* defined in Proposition 3 are R-module homomorphisms.
3. Consider R-modules $M \supseteq M'$, $N \supseteq N'$, and an R-module homomorphism $\varphi: M \to N$ such that $\varphi(M') \subseteq N'$. Construct an R-module homomorphism $\psi: M/M' \to N/N'$ [cf. §6.8].
4. a. Verify the statements in the corollary.
   b. Prove that if the sequence (*) is exact, then so are the sequences (**).
5. a. Given R-modules N', N'', define (as in §3.3) the external direct sum $N' \dotplus N''$ and verify that it is an R-module.
   b. Define maps α, β so that the sequence
   $$0 \longrightarrow N' \xrightarrow{\alpha} N' \dotplus N'' \xrightarrow{\beta} N'' \longrightarrow 0$$
   is exact.
   c. Conclude that the sequence
   $$0 \longrightarrow \mathrm{Hom}_R(M, N') \xrightarrow{\alpha_*} \mathrm{Hom}_R(M, N' \dotplus N'')$$
   $$\xrightarrow{\beta_*} \mathrm{Hom}_R(M, N'') \longrightarrow 0$$
   is exact for any R-module M.
6. Consider R-modules M, M', N, and N' and R-module homomorphisms $\alpha: M \to M'$, $\beta: N \to N'$, and $\varphi: M' \to N$.
   a. Show that $\beta\varphi\alpha \in \mathrm{Hom}_R(M, N')$.
   b. Thus, show that the mapping f given by $f(\varphi) = \beta\varphi\alpha$ is an R-module homomorphism from $\mathrm{Hom}_R(M', N)$ to $\mathrm{Hom}_R(M, N')$. Schematically,

$$\begin{array}{ccc} M & \xrightarrow{\alpha} & M' \\ \Big\downarrow & & \Big\downarrow \varphi \\ N' & \xleftarrow{\beta} & N \end{array}$$

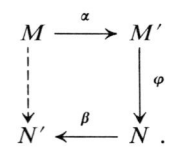

7. a. In the Short Five Lemma, prove that if α and γ are injective, so is β.
   b. Prove also that if α and γ are surjective, so is β.
   c. Hence prove the Short Five Lemma.
8. Considering the exact sequences
   $$0 \longrightarrow Z_2 \longrightarrow Z_4 \longrightarrow Z_2 \longrightarrow 0,$$
   $$0 \longrightarrow Z_2 \longrightarrow Z_2 \dotplus Z_2 \longrightarrow Z_2 \longrightarrow 0,$$
   show that in the Short Five Lemma the mapping β must be given. That is, there can exist isomorphisms α and γ, but no homomorphism β such that the diagram commutes.

9. An R-module M is called **irreducible** if it has no nontrivial submodules. Prove the lemma of Issai Schur (1875–1941) that every R-module homomorphism $\varphi: M \to M$ is either an isomorphism or the zero map ($\varphi(m) = 0$, for all $m \in M$).

10. The **group ring** $G_\mathbf{Q}$ of a multiplicative finite group $G = \{g_1, \ldots, g_n\}$ over the field of rational numbers $\mathbf{Q}$ is defined to be
$$G_\mathbf{Q} = \left\{ \sum_{i=1}^{n} a_i g_i : a_i \in \mathbf{Q} \right\};$$
addition is componentwise and multiplication is distributive over addition. Prove that
$$q = \frac{1}{n} \sum_{i=1}^{n} g_i$$
is an idempotent element (that is, $q^2 = q$).

11. Let V be the Klein Four-group and suppose that $F = \mathbf{Z}_2$. Consider, analogous to Exercise 10, the group ring V_F.
   a. Prove that $\gamma = \sum_{g \in V} g$ is nilpotent.
   b. Find all ideals $A \subseteq V_F$ for which $A^2 = \{0\}$.
   c. Prove that there is an ideal $B \subseteq V_F$ such that $A \subseteq B$ for all ideals A for which $A^2 = \{0\}$.

12. Suppose that a cyclic group $\Gamma = \langle \gamma \rangle$, generated by γ of order n, operates on the additively-written abelian group A. Let D and N be the mappings of A into A given for $a \in A$ by
$$D(a) = \gamma a - a = (\gamma - 1)a$$
and
$$N(a) = a + \gamma a + \cdots + \gamma^{n-1} a = \left(\sum_{i=0}^{n-1} \gamma^i \right) a.$$
   a. Prove that D and N belong to $\text{Hom}(A, A)$.
   b. The quotient $q(A) = [\ker D : N(A)][\ker N : D(A)]^{-1}$ is called the **Herbrand Quotient** (after Jacques Herbrand, 1908–1931) if both indices are finite. Suppose that B is a Γ-stable subgroup of A (i.e., $\gamma b \in B$ for all $b \in B$), and define an action of Γ on the factor group A/B.
   c. If any two of the quotients are finite, show that
$$q(A) = q(B) q(A/B).$$
   d. Prove that $q(A) = 1$ if A is finite.

13. Let R be a subring of a (commutative) ring S, and consider subsets M and N of S which are R-modules.
   a. Prove that the set MN of finite sums of elements mn, $m \in M$, $n \in N$, is an R-submodule of S.
   b. Prove that if M and N are finitely generated R-modules, then so is MN.

14. State and prove the analogue, for groups with operators, of:
   a. Theorem 2, §6.8.
   b. Theorem 3, §6.8.
   c. Corollary 2 to Theorem 2, §6.8.

15. **a.** Let $0 \longrightarrow M \xrightarrow{\alpha} N \xrightarrow{\beta} P$ be an exact sequence of R-module homomorphisms. Prove for every R-module A the exactness of the sequence $0 \longrightarrow \operatorname{Hom}_R(A, M) \xrightarrow{\alpha_*} \operatorname{Hom}_R(A, N) \xrightarrow{\beta_*} \operatorname{Hom}_R(A, P)$.

    b. If $M \xrightarrow{\alpha} N \xrightarrow{\beta} P \longrightarrow 0$ is exact, prove that for every R-module A the sequence
    $$0 \longrightarrow \operatorname{Hom}_R(P, A) \xrightarrow{\beta^*} \operatorname{Hom}_R(N, A) \xrightarrow{\alpha^*} \operatorname{Hom}_R(M, A)$$
    is exact.

16. Given an exact sequence of vector spaces over a field F,
    $$0 \longrightarrow W \xrightarrow{\varphi} V \xrightarrow{\psi} U \longrightarrow 0,$$
    prove the exactness of the sequence
    $$0 \longrightarrow U^* \longrightarrow V^* \longrightarrow W^* \longrightarrow 0.$$

17. Given an n-dimensional vector space V/F and a linear transformation $T: V \to V$, define the action of the polynomial $f(x) = a_n x^n + \cdots + a_1 x + a_0 \in F[x]$ on V by
    $$f(x)v = (a_n T^n + \cdots + a_1 T + a_0 I)v$$
    $$= a_n T^n(v) + \cdots + a_1 T(v) + a_0 v, \quad \text{for } v \in V.$$

    a. Prove that V/F thus can be considered as a finitely generated $F[x]$-module.

    b. Prove that the set $W(v) = \{a(x)v : a(x) \in F[x]\}$ is a subspace of V, the so-called **cyclic subspace** generated by v with respect to T.

    c. Verify that $\{v, T(v), \ldots, T^n(v)\}$ is a generating set of the subspace $W(v)$.

    d. Find a necessary and sufficient condition on $b(x) \in F[x]$ such that $W(b(x)v) = W(v)$.

18. **a.** Continuing Exercise 17, for $v \neq 0$, prove that $A(v) = \{g(x) \in F[x] : g(x)v = 0\}$ is a nontrivial ideal in $F[x]$. The monic generator $m_v(x)$ of $A(v)$ is called the **order** of v (relative to T). Thus this exercise states that each $v \neq 0$ in V is a torsion element.

    b. Prove that $m_v(x)$ divides the characteristic polynomial $\chi_T(x)$ of T.

A subspace W is said to be $p(x)$-**primary** for an irreducible polynomial $p(x)$ if $p(x)^s w = 0$ for some $s > 0$ and all $w \in W$.

c. Prove that V can be uniquely expressed (except for arrangement of summands) as the internal direct sum $V = W_1 \oplus \cdots \oplus W_h$ of $p_i(x)$-primary subspaces W_i, $1 \leq i \leq h$, where the polynomials $p_i(x)$ are the irreducible factors of the characteristic polynomial $\chi_T(x)$ of T.

d. Finally, prove that each of the $p_i(x)$-primary submodules W_i is a direct sum of $p_i(x)$-cyclic subspaces W_{ij} of orders $p_i(x)^{f_{ij}}$, $1 \leq j \leq k(i)$. Prove that
$$\prod_{i=1}^{h} \prod_{j=1}^{k(i)} p_i(x)^{f_{ij}} = \chi_T(x).$$

[Cf. the Fundamental Theorem, §7.1.]

NOTE. The discussion in these last two problems is a carry-over of the direct sum decomposition of finitely generated abelian groups. Here the degree of a polynomial replaces the absolute value of integers. Instead of positive numbers we use monic polynomials. The dimension of a vector space replaces the order of an abelian group. In general though the $F[x]$-module structure of the vector space V/F depends upon the choice of the linear transformation T.

8

Field Theory

While undergraduate courses commonly do not discuss fields beyond what we have already covered in Chapter 5, we provide, for honors courses or a second semester, the basic theory of algebraic field extensions, both separable and inseparable (although the latter case can be stripped from the arguments) so that the interplay of groups and fields might be displayed in the Fundamental Theorem of the Galois Theory [§8.7] and in subsequent explicit examples of galois groups. The existence of algebraically closed fields [§8.9] requires the Kronecker construction of roots of irreducible polynomials (from Chapter 5, §5.4) and Zorn's Lemma in the ideal-theoretic argument given. The proof of the Fundamental Theorem of Algebra (namely, that the complex numbers constitute an algebraically closed field) requires elementary real analysis and further algebraic argument; it is given in Chapter 9, §9.9. The Galois theory, especially field extensions and galois groups, might be followed by discussion of cyclotomic extensions, solution of equations by radicals, and ruler and compass constructions, which are all topics in Chapter 9.

§8.1 Algebraic Elements

This section begins our study of field theory with consideration of extensions of a given field F. A field K containing the field F is called an **extension field** (or overfield) of F, and F is called a **ground field** (base field) of K. This relationship will be denoted K/F.

§8.1 Algebraic Elements

An element $a \in K$ is called **algebraic** over F if there exists a finite dimensional vector space $V(a)/F$ in K (considered as a vector space over F) such that *all* powers a^i, $i \geq 0$, of a lie in $V(a)$. An extension K/F is called **algebraic** over F if each $a \in K/F$ is algebraic over F.

Proposition 1. If $a \in K$ is algebraic over F, then a is a zero of a (unique) monic irreducible polynomial $m_a(x) \in F[x]$. Further, $m_a(x)$ divides any polynomial $g(x)$ of which a is a root.

We call $m_a(x)$ the **minimal polynomial** of a over F, and its degree the **degree** of a over F.

Proof. If $a = 0$, then $m_a(x) = x$. Now consider $a \neq 0$ and the ring homomorphism $\varphi_a : F[x] \to K$ given by $\varphi_a(f(x)) = f(a)$. Set $M(a) = \ker \varphi_a$. By Proposition 2, §5.2, $M(a)$ is a principal ideal, and so, if it is nontrivial, it has a (unique) monic generator $m_a(x)$. Since it generates $M(a)$, $m_a(x)$ divides all polynomials $g(x) \in M(a)$.

Let $r = \dim V(a)$. Since then the elements $\{1, a, a^2, \ldots, a^r\}$ constitute a linearly dependent subset of $V(a)$, there exist elements $b_i \in F$, $0 \leq i \leq r$, not all zero, for which $\sum_{i=0}^{r} b_i a^i = 0$. Thus a is a zero of $h(x) = \sum_{i=0}^{r} b_i x^i \in F[x]$, and $M(a)$ is a nontrivial ideal.

To prove that the monic generator $m_a(x)$ of $M(a)$ is irreducible, suppose to the contrary that $m_a(x)$ has the proper factorization $m_a(x) = g(x)h(x)$. Then $0 = m_a(a) = g(a)h(a)$, a contradiction since $g(a), h(a)$ are nonzero elements of the field K.

Proposition 2 is the converse to Proposition 1.

Proposition 2. If $a \in K$ is a root of a polynomial $f(x) \in F[x]$, then a is algebraic over F.

Proof. Consider a polynomial $f(x) = \sum_{i=0}^{n} b_i x^i$, $b_n \neq 0$, of which a is a root. Then

$$a^n = -\left(\frac{b_{n-1}}{b_n}\right)a^{n-1} - \left(\frac{b_{n-2}}{b_n}\right)a^{n-2} - \cdots - \left(\frac{b_0}{b_n}\right).$$

Multiplying by a, we note that

$$a^{n+1} = -\left(\frac{b_{n-1}}{b_n}\right)a^n - \left(\frac{b_{n-2}}{b_n}\right)a^{n-1} - \cdots - \left(\frac{b_0}{b_n}\right)a$$

$$= -\left(\frac{b_{n-1}}{b_n}\right)\left[-\left(\frac{b_{n-1}}{b_n}\right)a^{n-1} - \cdots - \left(\frac{b_0}{b_n}\right)\right]$$

$$- \left(\frac{b_{n-2}}{b_n}\right)a^{n-1} - \cdots - \left(\frac{b_0}{b_n}\right)a$$

is an element of the vector space $V(a) = (1, a, \ldots, a^{n-1})_F$ generated over F

by $\{1, a, ..., a^{n-1}\}$. An inductive argument shows that a^{n+k}, $k \geq 0$, lies in $V(a)$. Hence a is algebraic over F.

We use $F[a]$ to denote the image of the homomorphism $\varphi_a: F[x] \to K$ given by $\varphi_a(f(x)) = f(a)$, for $a \in K$. The subring $F[a]$ of K is isomorphic to the residue class ring $F[x]/(m_a(x))$. (Recall the Isomorphism Theorem for rings, Exercise 1, §6.8.) Explicitly, an isomorphism $\varphi: F[x]/(m_a(x)) \to F[a]$ is described by

$$\psi([f(x)]) = \varphi_a(f(x)) = f(a) \in F[a].$$

Proposition 3. The polynomial ring $F[a]$ equals its field of quotients $F(a)$ when a is algebraic over F.

Consider a nonzero element $A = f(a)/g(a)$ in $F(a)$. Because $m_a(x)$ is irreducible, $(g(x), m_a(x)) = 1$; hence there are polynomials $k(x)$ and $h(x)$ in the ring $F[x]$ satisfying $g(x)k(x) + m_a(x)h(x) = 1$. Therefore in $F[a]$

$$1 = g(a)k(a) + m_a(a)h(a) = g(a)k(a).$$

Consequently $A = f(a)k(a) \in F[a]$.

Proposition 3 has the obvious restatement that every rational function of an algebraic element over a field F is a polynomial function over F. The dimension of the field $F(a)$, considered as a vector space over F, equals the degree of the minimal polynomial $m_a(x)$. It is denoted $\dim_F F(a)$ or $[a:F]$.

Theorem 1. If $K \supseteq F$, then all elements of K which are algebraic over F form a subfield $L \supseteq F$, the so-called **algebraic closure of F in K**.

Proof. Consider a, b in the set L of elements of K algebraic over F. Then $a^i \in V(a) = (\xi_1, ..., \xi_s)_F \subseteq K$ and $b^j \in V(b) = (\eta_1, ..., \eta_t)_F \subseteq K$ for all $i, j \geq 0$. This means $a^i = \sum_{\mu=1}^s f_{i\mu} \xi_\mu$ and $b^j = \sum_{\nu=1}^t g_{j\nu} \eta_\nu$ with coefficients $f_{i\mu}, g_{j\nu}$ in F. Hence

$$a^i b^j = \left(\sum_\mu f_{i\mu} \xi_\mu\right)\left(\sum_\nu g_{j\nu} \eta_\nu\right) = \sum_{\mu,\nu} h_{\mu\nu}^{(i,j)} \xi_\mu \eta_\nu$$

with coefficients $h_{\mu\nu}^{(i,j)} \in F$; thus $a^i b^j \in (\xi_1 \eta_1, ..., \xi_\mu \eta_\nu, ..., \xi_s \eta_t)_F = V$ for all $i, j \geq 0$. Since $\dim_F V$ is finite (at most st), ab belongs to L. Next

$$(a+b)^m = a^m + \binom{m}{1} \cdot a^{m-1} b + \cdots + \binom{m}{i} \cdot a^{m-i} b^i + \cdots + b^m$$

lies in V for all $m \in \mathbb{N}$. Consequently $a + b \in L$. A nonzero element $a \in L$ satisfies a polynomial

$$a^s + e_1 a^{s-1} + \cdots + e_s = 0,$$

with coefficients $e_i \in F$, where $e_s \neq 0$. Hence $e_s(a^{-1})^s + \cdots + e_1(a^{-1}) + 1 = 0$, which implies that $a^{-1} \in L$. Thus L is a field. Moreover, every rational

§8.1 Algebraic Elements

expression $p(a,b)/q(a,b)$ with polynomials $p(a,b)$ and $q(a,b) \neq 0$, whose coefficients lie in F, is an element of L.

An extension K/F is called a **finite algebraic extension** of the field F if K, considered as a vector space over F, has finite dimension $n = [K:F]$. This dimension is also called the **degree** of K over F.

REMARK. The elements of a finite algebraic extension K/F are algebraic over F. We can take $V(a) = (k_1, \ldots, k_n)_F$ for all $a \in K$ where $\{k_1, \ldots, k_n\}$ is a basis of K over F. The proof of Theorem 1 indicates the utility of defining the concept of algebraic element in terms of a vector space and not just as the root of a polynomial over the base field.

Theorem 2. If $F \subseteq K \subseteq E$ and $[K:F] = m$, $[E:K] = n$, then $[E:F] = mn$.

Proof. Let $k_1, \ldots, k_m$ be a basis of K over F and $e_1, \ldots, e_n$ a basis of E over K. Then every element $a \in E$ is a linear combination of the mn products $k_i e_j$ with coefficients in F. Specifically, $a = \sum_{j=1}^n b_j e_j$ with $b_j = \sum_{i=1}^m c_{ji} k_i$, where $c_{ji} \in F$; hence

$$a = \sum_{i,j} c_{ji} k_i e_j.$$

Thus $\dim_F E = [E:F]$ is at most mn. To prove that the elements $k_i e_j$ are linearly independent over F consider $\sum_{i,j} d_{ji} k_i e_j = 0$ for some coefficients $d_{ij} \in F$. Then

$$\sum_j \left(\sum_i d_{ji} k_i \right) e_j = 0;$$

by the linear independence of the elements e_j over K, $\sum_i d_{ji} k_i = 0$. Moreover $d_{ji} = 0$ since the elements k_i are linearly independent over F. Thus $[E:F] = [E:K][K:F] = mn$.

Corollary. Let K/F be an algebraic, not necessarily finite, extension. If an element a in an extension field M/K is algebraic over K, then it is necessarily algebraic over F.

For the proof let $f(x) = x^n + b_1 x^{n-1} + \cdots + b_n \in K[x]$ be the minimal polynomial of a. The field $L = F(b_1, \ldots, b_n)$ is a finite algebraic extension of F since each field $F(b_1, \ldots, b_{i-1})(b_i)$ is a finite algebraic extension of $F(b_1, \ldots, b_{i-1})$. Consequently, Theorem 2 implies that $L(a)$ is a finite extension of F. Thus a is algebraic over F.

Let F be a field contained in a field Ω. If Σ is a set of elements in Ω, the intersection $\bigcap L_\nu$ of all subfields L_ν/F in Ω that contain the set Σ is denoted by $F(\Sigma)$, and $F(\Sigma)$ is said to have been **obtained by adjoining the elements** of Σ to the field F. Hereafter we shall be concerned mainly with adjunctions $F(a_1, \ldots, a_m)$ where the elements $a_1, \ldots, a_m$ are algebraic over F.

In this case the elements of $F(\Sigma) = F(a_1, \ldots, a_m)$ are polynomials

$$\sum_{0 \le i_j < \infty} b_{i_1, \ldots, i_m} a_1^{i_1} \cdots a_m^{i_m}, \qquad 1 \le j \le m,$$

with coefficients $b_{i_1, \ldots, i_m}$ in F. To verify this fact, note that $F(a_1) = F[a_1]$ by Proposition 3 and by induction that

$$F(a_1, \ldots, a_{j-1})(a_j) = F(a_1, \ldots, a_{j-1})[a_j]$$
$$= F[a_1, \ldots, a_{j-1}][a_j]$$
$$= F[a_1, \ldots, a_j].$$

Now consider $f(x) \in F[x]$. According to the basic construction of Kronecker [see §5.4], there exist $n = \deg f(x)$ algebraic elements $a_1, \ldots, a_n$ in some extension field Ω/F such that

$$f(x) = \prod_{i=1}^{n} (x - a_i)$$

in Ω/F. A field K/F, obtained by adjoining to F the zeros $a_1, \ldots, a_n$ of a polynomial $f(x)$ of degree n in $F[x]$, which lie in some field Ω/F, is called a **splitting** (or decomposition) **field** of $f(x)$;

$$K = F(a_1, \ldots, a_n).$$

The existence of a splitting field for each $f(x) \in F[x]$ is the subject of the next proposition. Essentially it restates the theorem of §5.4. The field $K = F(a_1, \ldots, a_n)$, obtained by adjoining to F the zeros $a_1, \ldots, a_n$, is a splitting field of $f(x)$.

Proposition 4. For each polynomial $f(x) \in F[x]$ there exists a field L/F such that $f(x)$ is a product of linear factors in $L[x]$.

If $f(x)$ is irreducible in $F[x]$ the zeros of $f(x)$ in K are called **conjugates** of each other. In case the polynomial $f(x)$ is separable—that is, $f(x) \ne g(x^p)$ with $g(x) \in F[x]$—the results of §5.5 imply that $f(x)$ has $\deg f(x)$ distinct roots in any field containing a splitting field of $f(x)$. A zero of such a polynomial is called a **separable element** over F. An algebraic extension K/F is called a **separable extension** over F if each of its elements is separable over F.

Exercises

1. Suppose that $a \in K/F$, a finite algebraic extension of degree n. Prove that the degree of the irreducible monic polynomial $m_a(x) \in F[x]$ satisfied by a divides n.
2. Consider polynomials $f(x) = x^2 + x + 1$ and $g(x) = x^3 - 2$ in $\mathbf{Q}[x]$.
   a. Prove that $f(x), g(x)$ are irreducible in $\mathbf{Q}[x]$.
   b. Prove that $g(x)$ is reducible in $F[x]$, where F is the splitting field of $f(x)$.

3. **a.** Prove that $[Q(\sqrt{-2}, \sqrt{3}) : Q] = 4$.
   b. Exhibit a polynomial $f(x) \in Z[x]$ of degree 4 such that
   $$f(\sqrt{-2} + 4\sqrt{3}) = 0.$$
4. Prove that $Q(\sqrt{-2} + 4\sqrt{3}) = Q(\sqrt{-2}, \sqrt{3})$.
5. Let K_1 and K_2 be extensions of the field F in some field L/F. Assume that $[K_1 : F] = p_1$, $[K_2 : F] = p_2$ for prime numbers p_1 and p_2 (which need not be distinct). Prove that either $K_1 = K_2$ or $K_1 \cap K_2 = F$.
6. Prove that a subring R of the algebraic extension K/F, where $R \supseteq F$, is a field. Is this statement true if K contains nonalgebraic (called **transcendental**) elements over F? Give an example to substantiate your answer.
7. **a.** Find the degree of $\sqrt{7} + \sqrt{2}$ over Q.
   b. Find the degree of $\sqrt{2} + \sqrt{-2}$ over Q; prove that the field $Q(\sqrt{2} + \sqrt{-2})$ contains $\sqrt{-1} = i$.
   c. Prove that the degree of $Q(\sqrt[3]{2} + \omega)$ over Q is 6, where ω is a complex number satisfying $\omega^2 + \omega + 1 = 0$.
8. **a.** Prove that the polynomial $f(x) = x^3 - x + 1$ is irreducible in $Q[x]$.
   b. Now let $K = Q(a)$ where a is a zero of $f(x)$. Find the monic polynomial with coefficients in Q which has $b = 1 - a^2$ for a root. Furthermore express $1/(1 - a^2)$ as a polynomial in $1, a, a^2$ with rational coefficients.
9. **a.** Find the minimal polynomial $f(x) \in Q[x]$ of $2 + 3\sqrt{7}$.
   b. Find the other zero of $f(x)$.
   c. Express $(2 + 3\sqrt{7})^{-1}$ in the form $a + b\sqrt{7}$ with $a, b \in Q$.
10. Find $[Q(\sqrt{2} + \sqrt{-5}) : Q]$.
11. Let $x^p - a$ be an irreducible polynomial over a field F of characteristic $p > 0$. Prove that the polynomial $x^{p^h} - a$ is also irreducible over F.
12. If a_1, a_2, a_3 are the zeros of the polynomial $x^3 - x^2 + 1 \in Q[x]$, find cubic polynomials $f(x), g(x) \in Q[x]$ such that:
    a. $f(x)$ has roots a_1^2, a_2^2, and a_3^2.
    b. $g(x)$ has roots $1 - 1/a_1$, $1 - 1/a_2$, and $1 - 1/a_3$.
13. **a.** Find the minimal polynomial in $Q[x]$ of $\sqrt[3]{1 + \tfrac{2}{3}\sqrt{\tfrac{7}{3}}} + \sqrt[3]{1 - \tfrac{2}{3}\sqrt{\tfrac{7}{3}}}$.
    b. Determine $[Q(\sqrt[3]{1 + \tfrac{2}{3}\sqrt{\tfrac{7}{3}}}) : Q]$.

§8.2 Finite Fields

Whereas the previous section involved algebraic extensions of arbitrary fields, we now limit our consideration to finite extensions of finite fields. The discussion culminates in Proposition 6, the finite field analogue of the Fundamental Theorem of Galois Theory in §8.7. Throughout this section we consider a field F with q elements and prime field $P_p \cong Z_p$. As a vector space over P_p, F has dimension m; hence it has $p^m = q$ elements.

Proposition 1. The multiplicative group F^* of the nonzero elements of F is a cyclic group of $q - 1 = p^m - 1$ elements.

Proof. By Lagrange's Theorem [§6.3], $a^{q-1} = 1$ for all $a \in F^*$, and by Theorem 2, §5.2, the polynomial $x^d = 1$ has at most d roots. Now apply Proposition 5, §6.6, to conclude that F^* is cyclic. Since of course $0^q = 0$, all the elements a in F are zeros of the polynomial $x^q - x \in P_p[x]$. This polynomial is separable because its derivative is -1. Consequently, the elements of F are precisely the q distinct zeros of $x^q - x$.

Proposition 2. If ω generates the group F^*, then ω is a zero of an irreducible polynomial of degree m with coefficients in P_p.

Proof. Since $[F : P_p] = m$, the $m+1$ elements $\{1, \omega, \ldots, \omega^m\}$ are linearly dependent over P_p. Thus the minimal polynomial $f(x) \in P_p[x]$ of ω has degree $h \leq m$. Any nonzero $a \in F$ is a power of ω (since ω generates F^*); furthermore $a = g(\omega)$ for some polynomial $g(x) \in P_p[x]$ of degree less than h, since $F = P_p[\omega] \cong P_p[x]/f(x)$ [cf. the discussion preceding Proposition 3, §8.1]. There are p^h such polynomials (including the zero polynomial); hence F must have p^h elements. But F has p^m elements; therefore $m = h = \deg f(x)$, as asserted.

In the note at the end of §3.6 we introduced the **Frobenius automorphism**

$$\varphi : a \to a^p = \varphi(a),$$

defined on a finite field F of characteristic p. The Frobenius map φ is not simply a group automorphism of F^* (that is, $\varphi(ab) = \varphi(a)\varphi(b)$) but is an automorphism of the additive group as well: $\varphi(a+b) = \varphi(a) + \varphi(b)$. The Frobenius automorphism has the following three significant properties.

Property 1. $\varphi(a) = a$ if and only if $a \in P_p \subseteq F$.

Property 2. The order of φ is m. That is, φ^m is the identity map on F while φ^s is different from the identity map, $0 < s < m$.

Property 3. The automorphism φ generates the group G of all automorphisms of F that leave fixed the elements of P_p. Consequently G is a cyclic group of order m.

The verification of Property 1 is left as an exercise.

Verification of Property 2. The order of a generator ω of F^* is $p^m - 1$. Thus the elements $\varphi^s(\omega) = \omega^{p^s}$ are different from ω for $0 < s < m$ and equal to ω for $s = m$. Consequently φ has order m.

Verification of Property 3. The generator ω of F^* satisfies the polynomial $f(x) = x^m + a_{m-1} x^{m-1} + \cdots + a_0$ of Proposition 2. Since φ has order m, $\{\varphi^s(\omega) : 0 \leq s < m\}$ is a set of m distinct roots of $f(x)$. But as $\deg f(x) = m$, this set contains all the roots of $f(x)$.

For any automorphism $\sigma \in G$, the image $\sigma(\omega)$ must again be a root of $f(x)$, since

$$0 = \sigma(0) = \sigma(\omega^m + a_{m-1}\omega^{m-1} + \cdots + a_0)$$
$$= (\sigma(\omega))^m + a_{m-1}(\sigma(\omega))^{m-1} + \cdots + a_0.$$

Consequently $\sigma(\omega) = \varphi^s(\omega)$ for some s, $0 \leq s < m$. Because the two automorphisms σ and φ^s agree on the generator ω of the cyclic group F^*, they are equal. Thus, every $\sigma \in G$ is a power of φ, and G is a cyclic group of order m.

The third property is stated more generally for extensions of the field F as follows.

Proposition 3. Let F be a field with $q = p^m$ elements and K/F an extension of degree n. The group $G(K/F)$ of *all* automorphisms of K that leave fixed the elements of F is generated by $\psi = \varphi^m$, where φ is the Frobenius automorphism.

Proof. The multiplicative group K^* of K is cyclic of order $q^n - 1 = p^{mn} - 1$ and has a generator ω. As in the proof of Proposition 2 we show that ω is a zero of an irreducible polynomial

$$f(x) = b_0 + b_1 x + \cdots + b_{n-1} x^{n-1} + x^n, \quad b_i \in F$$

in $F[x]$ by noting that $\{1, \omega, \ldots, \omega^n\}$ are linearly independent over F. If $\deg f(x) = h$ were to be less than n, then K would have $q^h < q^n$ elements, a contradiction.

For elements $b \in F$, $\varphi^m(b) = b$, and thus $\varphi^m \in G(K/F)$. Setting $\psi = \varphi^m$, consider the n automorphisms ψ^v, $0 \leq v < n$, of K, where ψ^0 denotes the identity map. The mappings ψ^v belong to $G(K/F)$, i.e., $\psi^v(b) = b$ for all $b \in F$. Therefore, $\psi^v(\omega)$ is also a zero of $f(x)$, since $0 = \psi^v(0) = \psi^v(f(\omega)) = f(\psi^v(\omega))$. As the zeros $\psi^v(\omega) = \omega^{p^{mv}}$, $0 \leq v < n$, are distinct so are the n automorphisms ψ^v in $G(K/F)$.

Now to prove that $G(K/F)$ is a cyclic group, consider any element α. Then, $0 = \alpha(0) = \alpha(f(\omega)) = f(\alpha(\omega))$, since α leaves fixed all $b \in F$. Consequently, $\alpha(\omega)$ is a root of $f(x)$ and so must equal one of the n roots $\psi^v(\omega)$. But if $\alpha(\omega) = \psi^v(\omega)$, then $\alpha = \psi^v$, because ω is a generator of K^*. Thus, every $\alpha \in G(K/F)$ is a power of $\psi = \varphi^m$, as asserted.

Proposition 4. For every prime integer p and every positive integer m there exists a field F containing $q = p^m$ elements.

The proof follows immediately from the observation that the q zeros of the polynomial $f(x) = x^p - x \in P_p[x]$ are distinct and form a subfield of the splitting field L of $F(x)$. The details are left to the reader.

Proposition 5. All fields of p^m elements are isomorphic.

Proof. Consider two fields $F, \bar{F}$, each having $q = p^m$ elements, with respective prime fields $P_p, \bar{P}_p$. The prime fields are isomorphic since each is isomorphic to $\mathbb{Z}_p$. Extend the isomorphism $\lambda: P_p \to \bar{P}_p$, in the usual way, to an isomorphism $\lambda: P_p[x] \to \bar{P}_p[x]$.

As in Proposition 2 let ω be a generator of F^*. The minimal polynomial $f(x) \in P[x]$ of ω has degree m; it divides $h(x) = x^q - x$, since $h(\omega) = 0$. In $\bar{P}_p[\bar{x}]$ the polynomial $\lambda(f(x)) = \bar{f}(\bar{x})$ divides $\lambda(h(x)) = \bar{x}^q - \bar{x}$. In fact since $\bar{F}^*$ contains all of the roots of $\bar{x}^q - \bar{x}$, it contains the roots of $\bar{f}(\bar{x})$. Pick a root α of $\bar{f}(\bar{x})$. The m elements $1, \omega, \ldots, \omega^{m-1}$ are a basis for F/P_p, and $1, \alpha, \ldots, \alpha^{m-1}$ are a basis for $\bar{F}/\bar{P}_p$. The mapping $\Lambda: F \to \bar{F}$ described by $\Lambda(\omega^i) = \alpha^i$, $0 < i < m$, and $\Lambda(a) = \lambda(a)$ for $a \in P_p$ is the desired isomorphism.

Proposition 6. Consider a field F with $q = p^m$ elements. Let K/F be an extension of degree n and φ be the Frobenius automorphism of K. There is a one-one correspondence between the subfields L/F of K/F and the subgroups H of $G = \{\psi^v : 0 \le v < n\}$ where $\psi = \varphi^m$.

The correspondence is described by associating to a subgroup H of G the intermediate field $L = \{a \in K : \sigma(a) = a \text{ for all } \sigma \in H\}$ of elements in K invariant under H; it is called the *fixed field* of H. Conversely, to the extension L/F of degree h associate the subgroup (of order n/h) $H = \{\psi^v \in G : \psi^v(b) = b, \text{ for all } b \in L\} = G(K/L)$; it is called the *galois group* of K/L.

Proof. Consider a given subgroup $H = \{\psi^{hv} : 0 \le v < n/h\}$ of index h in G. The set $\{a \in K : \psi^h(a) = a\}$ is a subfield L of K because, for $a, b \in L$,

$$\psi^h(a \pm b) = \psi^h(a) \pm \psi^h(b) = a \pm b,$$
$$\psi^h(ab) = \psi^h(a)\psi^h(b) = ab,$$
$$\psi^h(a^{-1}) = [\psi^h(a)]^{-1}.$$

Since $\psi^h(a) = a^{q^h}$, the elements of L are precisely the q^h zeros of the polynomial $x^{q^h} - x$, and hence L has q^h elements, thus $[L : F] = h$. Consequently associated to the subgroup H of order n/h is an intermediate field L/F for which

$$[K : L] = [K : F]/[L : F] = n/h = [H : 1].$$

Finally H is exactly the subgroup, leaving fixed the elements of L. Suppose to the contrary that $\sigma(b) = b$ for all $b \in L$ for some $\sigma \in G \backslash H$. Let $\bar{H}$ be the product of the subgroups $\langle \sigma \rangle$ and H, and $\bar{L}$ be the fixed field associated to $\bar{H}$. By the preceding argument, $[K : \bar{L}] = [\bar{H} : 1] > [H : 1] = [K : L]$, and therefore $\bar{L} \subset L$. But if $\sigma(b) = b$ for all $b \in L$, then every element in L is fixed under the automorphisms in $\bar{H}$ and hence $L \subset \bar{L}$, a contradiction.

Conversely consider a given intermediate field L/F of degree h, and let H be the subset $\{\psi^v \in G : \psi^v(b) = b \text{ for all } b \in L\}$ of the cyclic group $G = G(K/F)$ of order n. By definition H is the group $G(K/L)$ of Proposition 3 and is generated by $\varphi^{hm} = \psi^h$. Hence H is a subgroup of G of order $o(\psi^h) = n/h$ and of index h in G.

Exercises

1. Prove that the polynomial $x^3 + x^2 + [1]_2$ is irreducible in $\mathbf{Z}_2[x]$. Let α be a zero of this polynomial in a splitting field K of $x^3 + x^2 + [1]_2$.
   a. Express the other zeros β and λ as polynomials in 1, α, α^2 with coefficients in $\mathbf{Z}_2$.
   b. Prove that $[\mathbf{Z}_2(\alpha) : \mathbf{Z}_2] = 3$.
   c. Show that α is a primitive 7th root of unity.
2. Prove that the polynomial $x^3 + tx^2 - t^2$ is irreducible in the field of rational functions $\mathbf{Z}_3(t)$.
3. Prove that the equation $x^2 = -[1]$ has a solution in $\mathbf{Z}_p$ if and only if $p = 4m + 3$, p a prime.
4. Define $(u, v) \cdot (x, y) = (ux + [7]vy, uy + vx)$, where $u, v, x, y, [0], [1], [7] \in \mathbf{Z}_{11}$. Prove that there exists a pair (x, y) such that $(u, v) \cdot (x, y) = ([1], [0])$ for given $(u, v) \neq ([0], [0])$.
5. a. Let p be an odd prime. Prove that every element of $\mathbf{Z}_p$ is a sum of two squares.
   b. If F is a finite field, prove that every element in F is a sum of two squares.
6. Suppose that $F = \mathbf{Z}_p(\alpha) = \{\bar{0}, \bar{1}, \ldots, \bar{a}, \bar{b}, \ldots\}$ is the field of $q = p^m$ elements where α is a primitive $(q-1)$st root of unity. Let $SL(2, F) = U$ be the group of 2×2 matrices
$$\begin{bmatrix} \bar{a} & \bar{b} \\ \bar{c} & \bar{d} \end{bmatrix}$$
with determinant $\bar{1}$. Prove that
   a. $[U : 1] = q(q^2 - 1)$.
   b. The matrices
$$\sigma = \begin{bmatrix} \bar{0} & \bar{1} \\ -\bar{1} & \bar{0} \end{bmatrix} \quad \text{and} \quad \tau_h = \begin{bmatrix} \bar{1} & \bar{0} \\ \alpha^h & \bar{1} \end{bmatrix}, \quad 0 \leq h < m,$$
generate the group U. Note that
$$\begin{bmatrix} \bar{1} & \bar{0} \\ \bar{a} & \bar{1} \end{bmatrix} \begin{bmatrix} \bar{1} & \bar{0} \\ \bar{b} & \bar{1} \end{bmatrix} = \begin{bmatrix} \bar{1} & \bar{0} \\ \bar{a}+\bar{b} & \bar{1} \end{bmatrix}.$$
7. Let F be a field of $q = p^m$ elements, p a prime. Suppose that $f(x) \in F[x]$, and define the polynomial function $\hat{f} \colon F \to F$ by $\hat{f}(a) = f(a) \in F$.
   a. Prove that the polynomial functions form an integral domain D.
   b. Show that $f \to \hat{f}$ is a homomorphism from $F[x]$ to D, whose kernel is the principal ideal generated by $x^q - x$.
8. Let F be the quadratic extension of $\mathbf{Z}_p$. Prove that all elements of $\mathbf{Z}_p$ are squares of elements of F.
9. Verify Property 1. *Hint:* Consider the polynomial $x^p - x \in F[x]$.

§8.3 The Theorem of the Primitive Element

An algebraic extension K/F is said to be **simple** if $K = F(A)$ for some $A \in K$. The element A is called a **primitive element** of K over F.

Theorem of the Primitive Element. If F is an infinite field and L/F an extension containing elements $c_1, \ldots, c_h$ algebraic over F such that $c_2, \ldots, c_h$ are separable over F, then there exists a primitive element $w \in K = F(c_1, \ldots, c_h)$.

The proof is by induction on h. First consider elements $a = c_1$ and $b = c_2$, where b is separable over F, and monic irreducible polynomials $f(x)$ and $g(x)$ in $F[x]$ such that $f(a) = g(b) = 0$. By the theorem of §5.4 there exists a field $M \supseteq K$ which contains all the zeros $a = a_1, \ldots, a_r$ and $b = b_1, \ldots, b_s$ (these are distinct!) of $f(x)$ and $g(x)$, respectively. We may assume that $b \notin F$, for otherwise $K = F(a)$ and the assertion is proved. Hence $s \geq 2$. Next each of the equations $a_i + ub_k = a_1 + ub_1$ has at most one zero in M, $1 \leq i \leq r$, $1 < k \leq s$. Since this is a finite set of equations and F is an infinite field, there exists an element $c \in F$, distinct from the solutions $u_{ik} = (a_i - a_1)/(b_1 - b_k)$, such that $a_i + cb_k \neq a_1 + cb_1$ for all i and all $k \neq 1$.

Setting $d = a_1 + cb_1 = a + cb \in F(a, b)$ yields $f(a) = f(d - cb) = 0$, where $f(d - cb)$ is a polynomial in b with coefficients in $F(d)[x]$. Let $h(x)$ be the GCD $(g(x), f(d - cx))$ in $F(d)[x]$. We seek to prove that $h(x) = x - b$, from which we may conclude that $b \in F(d)$.

Since $f(x), g(x)$ split in M, so do $f(d - cx)$ and $h(x)$. The element b is a common zero of $g(x), f(d - cx)$ and so $(x - b) | h(x)$. No higher power of $x - b$ can divide $h(x)$ however, because b is a simple zero of the (separable) polynomial $g(x)$. Any other factor of $h(x)$ must be a product of powers of the polynomials $x - b_k$, $k \neq 1$, as they are divisors of $g(x)$. This would imply that $f(d - cb_k) = 0$ and that $d - cb_k = a_1 + cb_1 - cb_k$ is one of the zeros $a_1, \ldots, a_r$ of $f(x)$. By the choice of c, this is impossible; that is, $f(d - cb_k) \neq 0$ for $k \neq 1$. Consequently $h(x) = x - b$ and $b \in F(d)$. Also, $a = d - cb \in F(d)$ and therefore $F(a, b) = F(d)$.

Finally, using induction, we have

$$F(c_1, c_2, \ldots, c_{h-1}) = F(v);$$

the preceding argument yields

$$F(c_1, c_2, \ldots, c_{h-1})(c_h) = F(v, c_h) = F(w).$$

REMARK. For a finite field F the validity of the theorem follows from the fact that $K = F(\omega)$ where ω is a primitive $(p^{mn} - 1)$st root of unity [see §8.2]. A finite field has no inseparable extensions because every element a algebraic over F such that $[F(a) : P_p] = p^{ms}$ is a zero of the separable polynomial $x^{p^{ms}} - x$.

Exercises

1. Let $F = \mathbb{Z}_p(x, y)$ be the field of rational functions of the indeterminates x, y. Suppose that ξ and η are zeros (not belonging to F) of the polynomials $t^p - x$ and $t^p - y$ in $F[t]$.

a. Prove that $[K:F] = p^2$ where $K = F(\xi, \eta)$.
   b. Show that there exists no element $\gamma \in K$ for which $K = F(\gamma)$. (In other words, the Theorem of the Primitive Element does not hold in this case.)
   c. Prove that there exist infinitely many distinct fields M for which $K \supset M \supset F$.
2. Prove that a finite separable algebraic extension K/F always can be written $K = F(A)$ for some $A \in K$.
3. Let K be a finite algebraic extension of the infinite field F. Prove that $K = F(A)$ if and only if there exist only finitely many fields $F \subseteq L \subseteq K$. (*Hint:* Take $B \in K$ such that $[F(B) : F]$ is maximal. Make an indirect proof by examining the fields $F(B+dC)$ for $C \in K \setminus F(B)$, $d \in F$.)
4. Let A and B be nonzero algebraic elements of the field K/F such that A is separable over the field F and $B^{p^e} \in F$ where char $F = p$, $e \geq 1$. Prove that:
   a. $F(A, B) = F(A+B)$
   b. $F(A, B) = F(AB)$.
5. Find a primitive element for each of the following extensions of **Q**.
   a. $\mathbf{Q}(\sqrt{2}, \sqrt{7})$
   b. $\mathbf{Q}(\sqrt{2}, \sqrt{-7})$
   c. $\mathbf{Q}(\sqrt[3]{7}, \omega)$, where $\omega^2 + \omega + 1 = 0$.

§8.4 Equivalence of Fields

If λ is an isomorphism of fields F and $\overline{F}$ and if $K, \overline{K}$ are extensions of $F, \overline{F}$, respectively, we call an isomorphism $\Lambda : K \to \overline{K}$ a **prolongation** (also an **extension** or **continuation**) of λ if the **restriction** $\Lambda \mid F$ of Λ to F is equal to the original isomorphism λ; that is, if $\Lambda(a) = \lambda(a)$ for all $a \in F$.

Two field extensions K/F, $\overline{K}/F$ are said to be **equivalent** (or **isomorphic**) over F if there exists an isomorphism $\varphi : K \to \overline{K}$ prolonging the identity map on F; in other words, an isomorphism $\varphi : K \to \overline{K}$ such that $\varphi(a) = a$ for all $a \in F$. In particular, an **automorphism of K over F** is a (field) automorphism $\varphi : K \to K$ for which $\varphi(a) = a$ for all $a \in F$.

Equivalent fields K/F and $\overline{K}/F$ contained in a common field Ω/F are called **conjugate fields**. In particular, elements $A \in K$ and $B \in \overline{K}$ are said to be **conjugate** if there is an isomorphism φ of the extensions $F(A)$ and $F(B)$ for which $\varphi(A) = B$ and $\varphi(a) = a$ for all $a \in F$. Note that these definitions extend the concept of conjugate roots of a polynomial [cf. the end of §8.1].

An isomorphism λ of fields F and $\overline{F}$ extends to a ring isomorphism of $F[x]$ and $\overline{F}[y]$, where y is an indeterminate over $\overline{F}$, as follows. Define $\lambda(x) = y$, and for $f(x) = a_n x^n + a_{n-1} x^{n-1} + \cdots + a_0$ in $F[x]$, define $\lambda(f(x))$ by

$$\lambda(f(x)) = \lambda(a_n) y^n + \lambda(a_{n-1}) y^{n-1} + \cdots + \lambda(a_1) y + \lambda(a_0).$$

That λ is an isomorphism of the polynomial rings is a consequence of its being an isomorphism of the ground fields. Further, $\lambda(f(x)) = \bar{f}(y)$ is irreducible in $\overline{F}[y]$ if and only if $f(x)$ is irreducible in $F[x]$.

The next two lemmas on prolongations are significant in the subsequent development of normal and separable extensions [§§8.5 and 8.6].

Lemma 1. If A is a root of an irreducible polynomial $f(x)$ in some extension L/F, and if B is a root of $\lambda(f(x)) = \bar{f}(y)$ in some extension $\bar{L}/\bar{F}$, then λ has a prolongation Λ to $F(A)$ such that $\Lambda(F(A)) = \bar{F}(B)$ and $\Lambda(A) = B$.

Proof. Consider the following isomorphisms:

(i) $\varphi: F[A] \to F[x]/(f(x))$, where $\varphi(g(A)) = [g(x)]$, the residue class of $g(x) \bmod f(x)$.

(ii) $\psi: \bar{F}[B] \to \bar{F}[y]/(\bar{f}(y))$, where $\psi(\bar{g}(B)) = [\bar{g}(y)]$, the residue class of $\bar{g}(y) \bmod \bar{f}(y)$.

(iii) $\lambda': F[x]/(f(x)) \to \bar{F}[y]/(\bar{f}(y))$, where λ' is the map on the residue class ring $F[x]/(f(x))$ induced by the original isomorphism $\lambda: F[x] \to \bar{F}[y]$. Specifically, $\lambda'([g(x)]_{f(x)}) = [\bar{g}(x)]_{\bar{f}(x)}$.

Schematically,

$$F(A) = F[A] \xrightarrow{\Lambda} \bar{F}[B] = \bar{F}(B)$$
$$\downarrow \varphi \qquad\qquad \downarrow \psi$$
$$F[x]/(f(x)) \xrightarrow{\lambda'} \bar{F}[y]/(\bar{f}(y)).$$

Recall that $F[A] = F(A)$ [Proposition 3, §8.1]. Composing these isomorphisms we obtain the desired prolongation $\Lambda = \psi^{-1}\lambda'\varphi: F(A) \to \bar{F}(B)$ of λ to $F(A)$.

Corollary 1. If the isomorphism λ is an automorphism of F, and if A and B are zeros of the irreducible polynomials $f(x)$ and $\bar{f}(x)$ (here x and $\lambda(x) = y$ are identified for notational simplicity) in an extension L/F, then λ can be extended to an isomorphism $\Lambda: F(A) \to F(B)$, such that $\Lambda(A) = B$.

In particular, we have another corollary, as follows.

Corollary 2. If A, B are zeros of an irreducible polynomial $f(x) \in F[x]$, then $F(A)$ and $F(B)$ are equivalent extensions of F.

REMARK. The irreducibility of $f(x)$ is essential in the hypotheses of Lemma 1 and its corollaries. For example, $[\mathbf{Q}(\sqrt{2}) : \mathbf{Q}] = 2$ and $[\mathbf{Q}(\sqrt[3]{3}) : \mathbf{Q}] = 3$. Hence $\mathbf{Q}(\sqrt{2})$ and $\mathbf{Q}(\sqrt[3]{3})$ are not isomorphic (i.e., not equivalent) extensions of $\mathbf{Q}$. Yet, both $\sqrt{2}$ and $\sqrt[3]{3}$ are roots of $h(x) = (x^2 - 2)(x^3 - 3) \in \mathbf{Q}[x]$.

Lemma 2. Let λ be an isomorphism from F to $\bar{F}$ extended as in Lemma 1 to the polynomial rings $F[x]$ and $\bar{F}[y]$. Consider a (not necessarily irreducible) polynomial $f(x) \in F[x]$ of degree n with a splitting field $L = F(A_1, \ldots, A_n)$; correspondingly, let $\bar{L} = \bar{F}(B_1, \ldots, B_n)$ be a splitting field of $\bar{f}(y)$. There exists a prolongation Λ of λ to $F(A_1, \ldots, A_n)$ such that

§8.4 Equivalence of Fields

$\Lambda(F(A_1, ..., A_n)) = \bar{F}(B_1, ..., B_n)$ where the zeros B_i are arranged so that $\Lambda(A_i) = B_i$, $1 \leq i \leq n$.

Proof. We proceed by induction on the degree n of $f(x)$, starting with A_1. This element is a zero of an *irreducible* factor $g(x)$ of $f(x)$ in $F[x]$. The corresponding polynomial $\bar{g}(y) \in \bar{F}[y]$ has a zero in $\bar{F}(B_1, ..., B_n)$. Now label the B_j's so that B_1 is a zero of $\bar{g}(y)$, and apply Lemma 1 to obtain a prolongation Λ_1 of λ to $F(A_1)$ such that $\Lambda_1(A_1) = B_1$ and $\Lambda_1(F(A_1)) = \bar{F}(B_1)$.

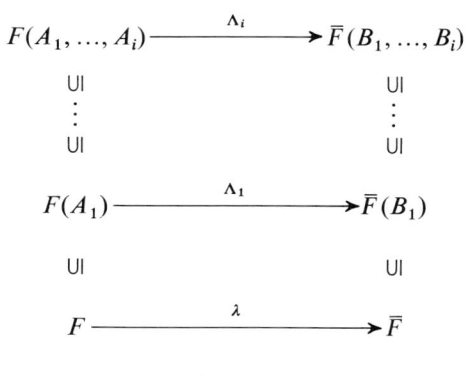

$$\begin{array}{ccc}
F(A_1, ..., A_i) & \xrightarrow{\Lambda_i} & \bar{F}(B_1, ..., B_i) \\
\cup| & & \cup| \\
\vdots & & \vdots \\
\cup| & & \cup| \\
F(A_1) & \xrightarrow{\Lambda_1} & \bar{F}(B_1) \\
\cup| & & \cup| \\
F & \xrightarrow{\lambda} & \bar{F}
\end{array}$$

Figure 8.1

Now assume that λ has been extended to an isomorphism

$$\Lambda_i : L_i = F(A_1, ..., A_i) \to \bar{L}_i = \bar{F}(B_1, ..., B_i)$$

such that $\Lambda_i(A_j) = B_j$, $1 \leq j \leq i$, as in Figure 8.1. Next, factor $f(x)$ in $L_i[x]$, obtaining

$$f(x) = \left[\prod_{j=1}^{i}(x - A_j)\right] h_{i+1}(x) \cdots h_s(x),$$

where $h_{i+1}(x), ..., h_s(x)$ are irreducible (but not necessarily distinct) polynomials in $L_i[x]$. There is a corresponding factorization in $\bar{L}_i[y]$,

$$\bar{f}(y) = \left[\prod_{j=1}^{i}(y - B_j)\right] \bar{h}_{i+1}(y) \cdots \bar{h}_s(y).$$

In the splitting fields L and $\bar{L}$ the polynomials $h_v(x)$ and $\bar{h}_v(y)$, $i < v \leq s$, split into linear factors $(x - A_k)$ and $(y - B_k)$, $i < k \leq n$. Now let A_{i+1} be a zero of the polynomial $h_{i+1}(x)$ and relabel the zeros B_k so that B_{i+1} satisfies $\bar{h}_{i+1}(y) = \Lambda_i(h_{i+1}(x))$.

Either $L_i(A_{i+1}) = L_i$, which implies that $h_{i+1}(x)$ is linear and correspondingly $\bar{L}_i(B_{i+1}) = \bar{L}_i$, or $h_{i+1}(x)$ has degree greater than one. In the first case, set $\Lambda_{i+1} = \Lambda_i$; and in the second apply Lemma 1 to obtain a prolongation Λ_{i+1} of Λ_i (and hence of λ) to $L_i(A_{i+1})$ such that

$$\Lambda_{i+1}(L_i(A_{i+1})) = \bar{L}_i(\Lambda_{i+1}(A_{i+1})) = \bar{L}_i(B_{i+1}).$$

Thus, by induction, the lemma is true for all n.

Application of Lemma 2 to splitting fields $F(A_1, \ldots, A_n)$ and $F(B_1, \ldots, B_n)$ of a polynomial $f(x) \in F[x]$ of degree n proves the following theorem.

Theorem. Any two splitting fields of a polynomial $f(x) \in F[x]$ are isomorphic over F.

Exercises

1. Let n be an odd positive integer, p a prime number. Prove that the field $\mathbf{Q}(\sqrt[n]{p})$ admits only the identity automorphism over $\mathbf{Q}$.
2. Prove that the only automorphism of $\mathbf{Q}$ is the identity map.
3. Prove that the subfield $\mathbf{Q}(\sqrt[3]{2})$ of $\mathbf{C}$ is equivalent to two distinct subfields in $\mathbf{C}/\mathbf{Q}$, other than itself.
4. a. Prove that $\mathbf{Q}(i, \sqrt{3})/\mathbf{Q}$ has degree 4 over $\mathbf{Q}$.
   b. What are the conjugate fields $\alpha(K)$ (over $\mathbf{Q}$) of $K = \mathbf{Q}(i, \sqrt{3})$ in $\mathbf{C}$?
   c. Prove that the mappings α form a group isomorphic to the direct product of two cyclic groups of order 2. (Observe that $\alpha(K) = K$.)
5. Consider $A = \sqrt{2} - \sqrt{3} \in \mathbf{C}$. Find
   a. $[\mathbf{Q}(A) : \mathbf{Q}]$ and $[\mathbf{Q}(A, \sqrt{2}) : \mathbf{Q}(\sqrt{2})]$.
   b. The monic irreducible polynomials in $\mathbf{Q}[x]$ and $\mathbf{Q}(\sqrt{2})[x]$ which have A for a zero.
   c. The conjugates of A over $\mathbf{Q}$ and $\mathbf{Q}(\sqrt{2})$.
6. If $\omega^2 + \omega + 1 = 0$, prove that the fields $\mathbf{Q}(\sqrt[3]{p})$ and $\mathbf{Q}(\omega\sqrt[3]{p})$, p a prime number, are isomorphic over $\mathbf{Q}$.

§8.5 Counting of Isomorphisms and Separability

We examine now the number of fields equivalent to a given field K/F.

Lemma 1. Let K/F and $\bar{K}/\bar{F}$ be given extensions, and the isomorphism $\Lambda: K \to \bar{K}$, a prolongation of $\lambda: F \to \bar{F}$. If $a \in K$ is a root of the (not necessarily irreducible) polynomial $f(x) \in F[x]$, then $\Lambda(a) \in \bar{K}$ is a root of $\lambda(f(x)) = \bar{f}(x) \in \bar{F}[x]$.

Writing $f(x) = \sum_{i=0}^{n} \alpha_i x^i$, $\alpha_i \in F$, we have

$$0 = f(a) = \sum_{i=0}^{n} \alpha_i a^i$$

and

$$0 = \Lambda(0) = \sum_{i=0}^{n} \lambda(\alpha_i)(\Lambda(a))^i = \bar{f}(\Lambda(a)).$$

In particular, if σ is an *automorphism* of K over F and if $a \in K$ is a root of $f(x) \in F[x]$, then $\sigma(a)$ is also a root of $f(x)$. From this observation and Corollary 2, §8.4, we have the following proposition.

Proposition. Suppose K/F is the splitting field of an irreducible polynomial $f(x) \in F[x]$. For any two roots α, β of $f(x)$, there exists an isomorphism of K/F mapping α to β. Furthermore any isomorphism of K/F permutes the roots of $f(x)$.

Lemma 2. Let a be an algebraic element of K/F and n^* the reduced degree of $m_a(x)$, the minimal polynomial of a over F. Assume that K/F contains a splitting field of $m_a(x)$. Then $F(a)$ has precisely n^* conjugate fields in K/F.

In other words, precisely n^* distinct isomorphisms defined on $F(a)$ extend the identity map of F. If a is a separable element, the reduced degree n^* coincides with the degree n of $m_a(x)$; otherwise, $n = n^* p^e$, where $p = \operatorname{char} F$. Since K contains all of the (distinct) roots a_i, $1 \le i \le n^*$, of $m_a(x)$, K/F contains precisely n^* conjugate fields $F(a_i)/F$. These n^* conjugate fields may very well coincide; what counts is that the n^* isomorphisms over F are distinct.

For example, $\mathbf{Q}(\sqrt{2})$ and $\mathbf{Q}(-\sqrt{2})$ are coincident conjugate fields over $\mathbf{Q}$ in $\mathbf{C}$, but $\sqrt{2}$ and $-\sqrt{2}$ are distinct roots of $x^2 - 2 \in \mathbf{Q}[x]$. The automorphisms given by

$$\sigma(a + b\sqrt{2}) = a + b\sqrt{2},$$
$$\tau(a + b\sqrt{2}) = a - b\sqrt{2}, \qquad a, b \in \mathbf{Q},$$

are distinct.

Proof of Lemma 2. Let $\{a = a_1, a_2, \ldots, a_{n^*}\} \subset K$ be the distinct roots of $m_a(x)$. By Corollary 1, §8.4, there exist isomorphisms

$$\lambda_i \colon F(a) \to F(a_i), \qquad 1 \le i \le n^*,$$

such that $\lambda_i(a) = a_i$, and $\lambda_i | F$ is the identity map. Thus at least n^* distinct isomorphisms are defined on $F(a)$. By Lemma 1 above, if λ is an isomorphism (extending the identity map on F) from $F(a)$ to some other extension L/F, $L \subseteq K$, $\lambda(a)$ must be a root of $m_a(x)$; that is, $\lambda(a) = a_i$ for some i. Consequently $L \supseteq F(a_i) \supseteq F$. Since both L and $F(a_i)$ are isomorphic to $F(a)$ and $L \supseteq F(a_i)$, they are equal and $\lambda = \lambda_i$. Therefore there are precisely n^* distinct isomorphisms defined on $F(a)$ as asserted.

Lemma 2 gives the key to the proof of the theorem below extending the enumeration of distinct isomorphisms to arbitrary finite algebraic extensions K/F, which in the case of characteristic $p > 0$ need not be simple. The precise statement needed in the theorem is the following generalization of Lemma 2.

Corollary. With notation as in Lemma 2, there are precisely n^* distinct prolongations of an isomorphism λ defined on F to isomorphisms defined on $F(a)$.

Let K/F be a finite extension of degree n with the vector space basis $\{B_1, \ldots, B_n\}$. (If $n = 1$, then $K = F$.) There exists a subset $\{A_1, \ldots, A_k\}$ of

$\{B_1, \ldots, B_n\}$ such that $K = F(A_1, \ldots, A_k)$ and
$$K_i = F(A_1, \ldots, A_i) = F(A_1, \ldots, A_{i-1})(A_i)$$
$$= K_{i-1}(A_i), \quad 1 \le i \le k,$$
where the fields K_i have degrees $n_i > 1$ over K_{i-1} [see §8.1]. Note for characteristic $p > 0$ that $n_i = n_i^* p^{e_i}$, where n_i^* is the reduced degree of K_i over K_{i-1} with the corresponding monic irreducible polynomial $g_i(y) \in K_{i-1}[y]$. Let $f_i(x) = g_i(x^{p^{e_i}}) \in K_{i-1}[x]$ be the minimal polynomial of A_i [see §5.5]. If char $F = 0$, then $n_i = n_i^*$.

The situation in Lemma 2 has the following graphic representation:

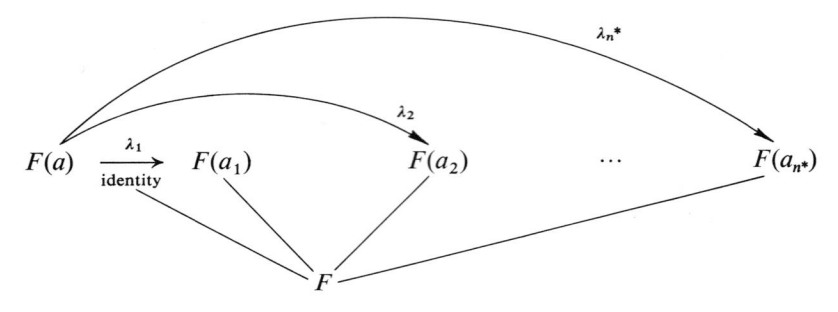

[Diagram: $F(a) \xrightarrow[\text{identity}]{\lambda_1} F(a_1)$, $F(a_2)$ via λ_2, $\ldots$, $F(a_{n^*})$ via λ_{n^*}, all over F.]

whereas that in the corollary is depicted as follows:

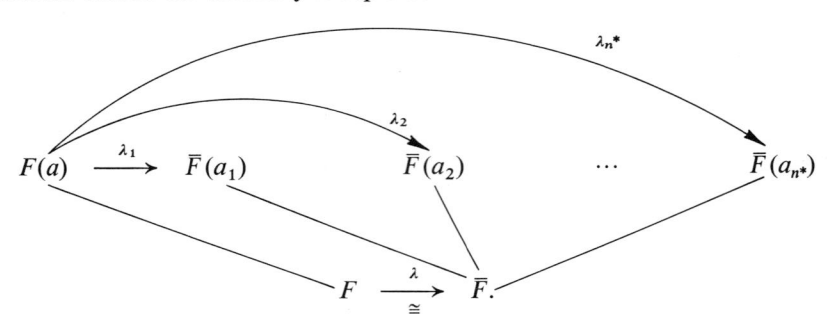

[Diagram: $F(a) \xrightarrow{\lambda_1} \bar{F}(a_1)$, $\bar{F}(a_2)$ via λ_2, $\ldots$, $\bar{F}(a_{n^*})$ via λ_{n^*}; $F \xrightarrow[\cong]{\lambda} \bar{F}$.]

With this preparation, we have the following theorem.

Theorem. Consider a finite extension $K = F(A_1, \ldots, A_k)$ of F, and let Ω/F be any extension containing all zeros of the minimal polynomials $f_i(x)$ of A_i, whose reduced degrees are n_i^*, $1 \le i \le k$. Then Ω/F contains precisely $n^* = n_1^* \cdots n_k^*$ intermediate fields conjugate to K/F.

Proof. First, consider an isomorphism σ defined on K/F (into Ω/F). Define an isomorphism σ_i on K_i/F by restriction, $1 \le i \le k$; namely, for $c_i \in K_i$,
$$\sigma_i(c_i) = \sigma(c_i).$$
Since $K_{i-1} \subset K_i$, note that $\sigma_{i-1} = \sigma | K_{i-1} = \sigma_i | K_{i-1}$. Thus, an isomorphism of K/F determines a chain of isomorphisms $\{\sigma_1, \ldots, \sigma_k\}$.

§8.5   Counting of Isomorphisms and Separability   283

If σ and τ are distinct isomorphisms on K/F, then there exists a subscript i, $1 \leq i \leq k$, such that for the *given* tower of fields

$$\sigma | K_{i-1} = \sigma_{i-1} = \tau_{i-1} = \tau | K_{i-1},$$
$$\sigma | K_i = \sigma_i \neq \tau_i = \tau | K_i.$$

That is, σ and τ restrict to distinct prolongations on K_i of $\sigma_{i-1} = \tau_{i-1}$.

Conversely, if $\{\sigma_1, \sigma_2, \ldots, \sigma_k\}$ is a chain of isomorphisms on the tower of fields $F = K_0 \subseteq K_1 \subseteq \cdots \subseteq K_k = K$ such that $\sigma_i | K_{i-1} = \sigma_{i-1}$, $2 \leq i \leq k$, the isomorphism $\sigma = \sigma_k$ of K/F satisfies $\sigma | K_i = \sigma_i$.

Now consider a typical piece K_i/K_{i-1} of the tower

$$F = K_0 \subset K_1 \subset \cdots \subset K_k = K$$

used to describe K/F. An isomorphism λ defined on K_{i-1} has, by the above corollary, precisely n_i^* distinct prolongations to K_i. It follows that the identity map on F has exactly $\prod_{i=1}^{k} n_i^*$ distinct prolongations to K/F. The total number n^* of isomorphisms of K/F (from the corollary) is independent of the tower of fields used in describing the chain of isomorphisms $\{\sigma_1, \ldots, \sigma_k\}$ associated to an isomorphism σ. Hence $n^* = \prod_{i=1}^{k} n_i^*$.

The integer n^*, denoted also $[K:F]_s$, is called the **reduced degree** of K/F or the **degree of separability**. The quotient $n/n^* = p^e = \prod_{i=1}^{k} p^{e_i}$ is called the **degree of inseparability**, denoted $[K:F]_i$, of K/F.

We conclude this section with a series of seven corollaries derived from the preceding theorem. Some of the proofs will be left to the reader.

Corollary 1. If the elements A_i of the theorem are separable over the fields K_{i-1}, then K/F has precisely $n = [K:F]$ isomorphisms which restrict to the identity on F.

A more general statement is given in Corollary 2. There "sufficiently large" means any field containing all zeros of the polynomials $\bar{f}_i(x) = \lambda(f_i(x))$ in the proof of the theorem.

Corollary 2. Each isomorphism λ from F to $\bar{F} = \lambda(F)$ has precisely $[K:F]$ prolongations to isomorphisms of K into a sufficiently large field $\bar{\Omega}/\bar{F}$.

Corollary 3. If all the elements A_i are separable over K_{i-1}, $1 \leq i \leq k$, then $K = F(A)$ where A is a separable element over F; $[A:F] = [K:F] = n$.

The Theorem of the Primitive Element [§8.3] implies that $K = F(A)$. Since K/F has, by Corollary 1, exactly $[K:F]$ distinct conjugate fields, the element A must have $[K:F]$ distinct conjugates, the zeros of the minimal $m_A(x) \in F[x]$ of A. If A were inseparable, then $m_A(x)$ would have fewer than $[K:F]$ distinct zeros [see §5.5]. Hence K/F would have fewer than $[K:F]$ conjugate fields.

Corollary 4. If $K = F(A)$ with A separable over F, then every element B in $F(A)$ is separable over F.

Proof. Consider the tower of fields $F \subseteq F(B) \subseteq F(B)(A) = K$. If B were inseparable, then $F(B)$ would have fewer than $[F(B) : F]$ isomorphisms over F. Consequently K would have fewer than $[K : F(B)][F(B) : F] = [K : F]$ isomorphisms over F, contrary to the assumption of the separability of A over F, since A is separable over $F(B)$.

This corollary together with Corollary 3 implies that a field $K = F(A_1, \ldots, A_k)$, obtained by adjoining elements A_i separable over F, is a separable extension. (An extension K/F was defined in §8.1 to be separable if each $a \in K\backslash F$ is separable over F.)

Corollary 5. A finite extension field K/F is separable if and only if K has precisely $[K : F]$ isomorphisms over F.

Proof. If K/F is separable, then the number of isomorphisms of K is $[K : F]$ by Corollary 1. Conversely, if K/F is inseparable, consider an element $A_1 \in K/F$ inseparable over F. Using the notation of the theorem, write $K = F(A_1, A_2, \ldots, A_k)$ and note that $n_1^* < n_1$. Hence the number of isomorphisms $n^* = n_1^* \cdots n_k^*$ is less than $n_1 \cdots n_k = n = [K : F]$, a contradiction.

Corollary 6. If A and B are separable algebraic elements in a field Ω/F, every rational function $h(A, B)$ of A and B with coefficients in F is separable over F.

For the proof apply Corollary 4 to the extensions

$$F \subseteq F(h(A, B)) \subseteq F(A, B).$$

The field $F(A, B)$ is separable over F. Hence the element $h(A, B)$ is separable over F.

Corollary 7. If the finite extension K/F is separable and the element A is separable over K, then A is separable over F.

Exercises

1. Prove the corollary to Lemma 2.
2. Prove Corollaries 1 and 2.
3. Prove Corollary 7.

§8.6 Prelude to Galois Theory

With the preliminaries of §§8.1, 8.4, and 8.5 complete, we now begin consideration of Galois theory proper, culminating in the next section with the statement and proof of the Fundamental Theorem of Galois Theory. In this section we focus our attention on two special types of algebraic field extensions.

A field extension K/F is called **normal** if

(i) K is algebraic over F,

(ii) any irreducible polynomial $g(x) \in F[x]$, which has *a zero* in K, has *all* of its *zeros* in K.

Theorem 1. The splitting field K/F of a (not necessarily irreducible) polynomial $f(x) \in F[x]$ is a normal extension of F.

Proof. Let $A_1, ..., A_m$ be the distinct zeros of $f(x) \in F[x]$, and write $K = F(A_1, ..., A_m)$. Suppose now that the irreducible polynomial $g(x)$ has a zero B in K.

Let Ω/F be a field containing K and all zeros $B, B', ...$ of $g(x)$. Consider an arbitrary zero $B' \in \Omega$ of $g(x)$. By Corollary 1, §8.4, there is an isomorphism λ over F from $F(B)$ to $F(B')$ such that $\lambda(B) = B'$.

Lemma 2, §8.4, asserts the existence of a prolongation Λ of λ defined on $F(B)(A_1, ..., A_m)$, such that

$$\Lambda(F(B)(A_1, ..., A_m)) = F(B')(\Lambda(A_1), ..., \Lambda(A_m)).$$

To complete the argument, observe that by hypothesis an element $B \in K \subset \Omega$ is a polynomial $B = q(A_1, ..., A_m)$ in the elements $A_1, ..., A_m$ with coefficients in F. Application of Λ to B yields

$$B' = \Lambda(B) = q(\Lambda(A_1), ..., \Lambda(A_m)) \in K.$$

The element $q(\Lambda(A_1), ..., \Lambda(A_m))$ lies in K because Λ maps the set $\{A_1, ..., A_m\}$ to itself since $\Lambda(A_i)$, $1 \leq i \leq m$, must be a root of $f(x)$ (by Lemma 1, §8.5). Hence $B' \in K$, and thus $g(x)$ splits in K, as asserted.

The definition of normal extension *does not exclude* inseparable extensions. For example, let $F = \mathbf{Z}_p(t)$ where t is an indeterminate over $\mathbf{Z}_p$. The polynomial $x^p - t \in F[x]$ is irreducible. (See Eisenstein's Criterion, §5.8, and use the prime ideal (t) in $\mathbf{Z}_p[t]$.) The coset $A = x + (x^p - t)F[x]$ in $K = F[x]/(x^p - t)$ satisfies $A^p - t = 0$. Hence $y^p - t = (y - A)^p$ in $K[y]$, y an indeterminate over K. Consequently, $x^p - t$ is an irreducible inseparable polynomial whose splitting field $K = F(A)$ is normal according to Theorem 1.

Hereafter only separable polynomials and finite separable extensions will be considered unless otherwise mentioned.

Recall that an isomorphism σ defined on $K = F(A)$ is an *automorphism of K/F* if and only if $\sigma(A) \in K$ and $\sigma|F$ is the identity map. If $[K:F] = n$, there are n or fewer automorphisms of K/F [see Lemma 2, §8.5].

A separable normal extension K/F of finite degree is called a **galois extension**. It is now easy to summarize the preceding discussion on splitting fields of polynomials and separable extensions in proving the following important theorem.

Theorem 2. A galois extension K/F of degree n has precisely n automorphisms over F. These automorphisms form a group $G(K/F) = G$, the so-called **galois group** of K over F.

Proof. The separability of K/F implies that K/F has exactly $n = [K:F]$ conjugate fields over F by Corollary 5, §8.5. By the Theorem of the Primitive Element, §8.3, $K = F(A)$ for some $A \in K$. Let $m(x) = \prod_{i=1}^{n}(x - A_i) \in F[x]$ be the minimal polynomial of the primitive element $A = A_1$. By the normality of K/F, $A_i \in K$, $2 \leq i \leq n$, and so $F(A) = F(A_2) = \cdots = F(A_n)$. Thus each of the conjugates $F(A_i)$ of $F(A)$ coincides with $F(A)$, and the n mappings $\sigma_i: F(A) \to F(A_i)$ are automorphisms of $F(A)/F$.

The set $G = \{\sigma_1 = \varepsilon, \sigma_2, \ldots, \sigma_n\}$ is a group under composition of mappings. That is, $\sigma\tau(a) = \sigma(\tau(a))$ *for all* $a \in K$. It follows immediately from the definition of $\sigma\tau$ that $(\sigma\tau)\rho = \sigma(\tau\rho)$. Every automorphism σ has an inverse σ^{-1} in G since σ is an isomorphism of K onto itself. Hence $G(K/F) = G$ is a group, and the proof is complete.

Historically the automorphisms of $\sigma \in G$ were viewed as permutations of the ordered set of zeros $\{A_1, \ldots, A_i, \ldots, A_n\}$ of $f(x)$. For $\sigma(A_i) = A_{\sigma(i)}$ where $\sigma(i)$ is uniquely determined by σ for $1 \leq i \leq n$, the map

$$\sigma \to \begin{bmatrix} 1 & \cdots & i & \cdots & n \\ \sigma(1) & \cdots & \sigma(i) & \cdots & \sigma(n) \end{bmatrix} = \pi(\sigma)$$

determines a unique element $\pi(\sigma)$ of the symmetric group Σ_n, such that $\pi(\sigma\tau) = \pi(\sigma)\pi(\tau)$. (The mapping π can be shown to be an injective homomorphism of G into Σ_n.) In the older literature the galois group $G(K/F)$ is viewed as a permutation group of the zeros of the "defining polynomial $f(x)$" of K/F.

We must guard against the misconception that every element of Σ_n originates from or gives rise to an automorphism of K/F. A permutation of Σ_n *does not necessarily* give rise to an automorphism of a normal separable extension, although there are, for each $n \geq 2$, extensions K of the field of rational numbers $\mathbf{Q}$ whose galois groups are isomorphic to Σ_n.

Example 1. The splitting field of $x^3 - 2 \in \mathbf{Q}[x]$ is the extension $K = \mathbf{Q}(\omega, \sqrt[3]{2})$ of of degree 6 over $\mathbf{Q}$, where $\omega^2 + \omega + 1 = 0$. The automorphism group of $K/\mathbf{Q}$ is described as follows by its effect on the generating elements of K:

$$\sigma(\omega) = \omega^2, \quad \sigma(\sqrt[3]{2}) = \sqrt[3]{2},$$

and

$$\tau(\omega) = \omega, \quad \tau(\sqrt[3]{2}) = \omega \cdot \sqrt[3]{2},$$

so that $\sigma^2 = \tau^3 = 1$ and $\sigma\tau\sigma^{-1} = \tau^2$. Thus, $G(K/\mathbf{Q}) \cong \Sigma_3$.

Example 2. To illustrate that an irreducible cubic polynomial does not necessarily give rise to an extension of degree 6 whose group of automorphisms is isomorphic to Σ_3, consider

$$f(x) = x^3 - 21x + 28 \in \mathbf{Q}[x].$$

§8.6 Prelude to Galois Theory

This cubic polynomial is irreducible in $\mathbf{Z}[x]$, and hence also in $\mathbf{Q}[x]$ by the Lemma of Gauss [§5.8]. If $\zeta \in \mathbf{C}$ is a zero of $f(x)$, in particular a real zero since $f(x)$ has odd degree, the elements

$$\zeta' = -7 + \tfrac{3}{2}\zeta + \tfrac{1}{2}\zeta^2$$

and

$$\zeta'' = 7 - \tfrac{3}{2}\zeta - \tfrac{1}{2}\zeta^2$$

also are real zeros of $f(x)$, as is easily checked by direct substitution into $f(x)$. Consequently the group of automorphisms of the splitting field $\mathbf{Q}(\zeta)/\mathbf{Q}$ is cyclic of order 3.

Exercises

1. a. Determine the degree of $[\mathbf{Q}(i) : \mathbf{Q}]$.
   b. Is $\mathbf{Q}(\sqrt{i})/\mathbf{Q}$ a galois extension? If so, determine its galois group.
2. Consider the finite normal extension $K = \mathbf{Q}(\sqrt{2}, \sqrt{8}, \sqrt{7})$ of $\mathbf{Q}$. Complete the blanks.
   a. The degree $[K : \mathbf{Q}]$ is _____.
   b. The order of the galois group $G(K/\mathbf{Q})$ is _____.
   c. The order of $G(L/\mathbf{Q})$ is _____ when $L = \mathbf{Q}(\sqrt{14})$.
   d. The order of $G(L/\mathbf{Q})$ is _____ when $L = \mathbf{Q}(\sqrt{2}, \sqrt{7})$.
   e. The order of $G(L/\mathbf{Q})$ is _____ when $L = \mathbf{Q}(\sqrt{2} + \sqrt{8})$.
   f. The order of $G(K/L)$ is _____ when $L = \mathbf{Q}(\sqrt{2} + \sqrt{7})$.
3. Let K be the field of Example 1, and ω a root of $x^2 + x + 1$.
   a. Find a primitive element A of K over $\mathbf{Q}$.
   b. Determine the minimal polynomial of A over $\mathbf{Q}$ and over $\mathbf{Q}(\omega)$.
   c. Find all conjugates of A over $\mathbf{Q}$ and over $\mathbf{Q}(\sqrt[3]{2})$.
   d. Find the conjugates of $\alpha = A(1 + 2\omega - \omega^2)$ over $\mathbf{Q}$ and over $\mathbf{Q}(\omega)$.
4. Prove that the polynomial $f(x) = x^4 + 4x^2 + 2$ is irreducible in $\mathbf{Q}[x]$. Let α be a root of $f(x)$ in the splitting field $K/\mathbf{Q}$ of $f(x)$.
   a. Prove that the mapping

   $$\sum_{i=0}^{3} a_i \alpha^i \to \sum_{i=0}^{3} a_i \beta^i,$$

   with $a_i \in \mathbf{Q}$ and $\beta = \alpha^3 + 3\alpha$, is an automorphism of $\mathbf{Q}(\alpha)/\mathbf{Q}$.
   b. Find the galois group of $\mathbf{Q}(\alpha)/\mathbf{Q}$ and determine its structure.
5. Find the galois group of $\mathbf{Q}(i, \sqrt{11})$ over $\mathbf{Q}$ where $i^2 = -1$.
6. Prove that $x^4 + 1 \in \mathbf{Q}[x]$ is irreducible. Find the splitting field and determine the galois group and all the subfields of the splitting field.
7. Let $K = \mathbf{Q}(\sqrt{2}, \sqrt{-2}, \sqrt{3})$. Prove that $K/\mathbf{Q}$ is a galois extension. Determine the effect of the elements of the galois group $G(K/\mathbf{Q})$ on $\sqrt{2}, \sqrt{-2}, \sqrt{3}$.
8. Why is the splitting field of a separable polynomial a galois extension?
9. Prove that the galois group of a splitting field of $x^n - a \in \mathbf{Q}[x]$ is solvable, when a is not a dth power of an element of $\mathbf{Q}$, $d \mid n$.

10. Let A be a zero of $f(x) = x^3 - 3x + 1 \in \mathbf{Q}[x]$.
   a. Prove that $-2 + A^2$ is another zero of $f(x)$, and that $K = \mathbf{Q}(A)$ is a galois extension of $\mathbf{Q}$.
   b. What is the third zero of $f(x)$?
11. a. Prove that $x^4 - 2$ is irreducible in $\mathbf{Q}[x]$.
   b. Let K be the splitting field of $x^4 - 2$ in $\mathbf{C}$. Prove that $K = \mathbf{Q}(\alpha, i)$, where $i^2 = -1$ and α is a *real* 4th root of 2.
   c. Prove that the galois group $G(K/\mathbf{Q})$ has order 8 and contains a cyclic subgroup of order 4.
   d. Find a set of generators and relations of $G(K/\mathbf{Q})$.

§8.7 The Fundamental Theorem of Galois Theory

The following terminology is useful in providing a succinct formulation of the Fundamental Theorem of Galois Theory.

The **fixed field** $\Phi(H)$ of a subgroup H of $G = G(K/F)$ is defined to be

$$\Phi(H) = \{a \in K : \sigma(a) = a \text{ for all } \sigma \in H\}.$$

It is a subfield of K since for $a, b \in \Phi(H)$

$$\sigma(a+b) = \sigma(a) + \sigma(b) = a + b,$$
$$\sigma(ab) = \sigma(a)\sigma(b) = ab;$$

furthermore

$$\sigma(aa^{-1}) = \sigma(a)\sigma(a^{-1}) = \sigma(1) = 1$$

implies that $\sigma(a^{-1}) = (\sigma(a))^{-1}$ for $a \neq 0$. As σ is an automorphism of K/F, by definition $\sigma(c) = c$ for all $c \in F$, and thus $F \subseteq \Phi(H)$.

Now associate to an intermediate field L/F of the extension K/F (that is, L is a subfield of K) the subset

$$\Gamma(L) = \{\sigma \in G(K/F) : \sigma(l) = l \text{ for all } l \in L\}$$

of $G = G(K/F)$. That $\Gamma(L)$ is a subgroup of G is easily verified. Furthermore, $\Gamma(L) = G(K/L)$, the group of automorphisms of K whose restriction to L is the identity mapping.

Lemma. If L is an intermediate field of the galois extension K/F, then K is also a galois extension of L.

Let $g(x) \in L[x]$ be an irreducible polynomial which has a zero $a \in K$. To prove that all other zeros of $g(x)$ lie in K consider the minimal polynomial $m_a(x) \in F[x]$ of a. Since $m_a(x)$ belongs to $L[x]$, as well as to $F[x]$, and since $g(x)$ is the minimal polynomial of a with coefficients in L, we have that $g(x) | m_a(x)$. Therefore, each root of $g(x)$ is a root of $m_a(x)$. Since K/F is a normal extension and $a \in K$, all roots of $m_a(x)$ belong to K. Hence all roots of $g(x)$ belong to K, and K/L is a galois extension as asserted.

The Fundamental Theorem of the Galois Theory. Let K/F be a galois extension with the galois group $G(K/F) = G$.

  (i)   If L/F is a subfield of K, then $\Phi[\Gamma(L)] = L$.
  (ii) If H is a subgroup of G, then $\Gamma[\Phi(H)] = H$.
  (iii) $[K : L] = [\Gamma(L) : 1]$ and $[H : 1] = [K : \Phi(H)]$.
  (iv) If L/F and $\bar{L}/F$ are conjugate subfields (in the sense of §8.4) of K/F, then $\Gamma(L)$ and $\Gamma(\bar{L})$ are conjugate subgroups of G by an inner automorphism. Conversely, if H and $\bar{H}$ are conjugate subgroups of G, then $\Phi(H)$ and $\Phi(\bar{H})$ are conjugate subfields of K/F.
  (v)   An intermediate field L/F of K/F is a galois extension if and only if the corresponding subgroup $\Gamma(L)$ in G is normal. The galois group $G(L/F)$ is by restriction of G to L naturally isomorphic to the factor group $G/\Gamma(L)$.

Parts (i) and (ii) of the theorem can be visualized by a diagram:

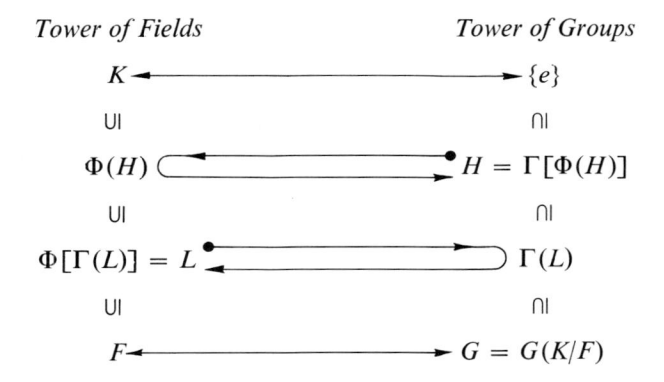

```
    Tower of Fields                    Tower of Groups

         K ←─────────────────────────────────→ {e}
         ∪|                                    ∩|
        Φ(H) ←─────────────────────────────→ H = Γ[Φ(H)]
         ∪|                                    ∩|
      Φ[Γ(L)] = L ←───────────────────────→ Γ(L)
         ∪|                                    ∩|
         F ←─────────────────────────────────→ G = G(K/F)
```

Proof of (i). If $l \in L$, then $\sigma(l) = l$, for all $\sigma \in \Gamma(L)$ by the definition of $\Gamma(L)$. Hence $L \subseteq \Phi[\Gamma(L)]$. Consider an element B of the fixed field $\Phi[\Gamma(L)]$. Then

$$K \supseteq L(B) \supseteq L \supseteq F,$$

and K is a galois extension of $L(B)$ by the preceding lemma. By the definition of $\Phi[\Gamma(L)]$ we have $\sigma(B) = B$ for all $\sigma \in \Gamma(L)$, and thus $\Gamma(L) \subseteq \Gamma(L(B))$. Since every automorphism σ of K which prolongs the identity on $L(B)$ certainly prolongs the identity on L, we also have $\Gamma(L(B)) \subseteq \Gamma(L)$. Thus, $\Gamma(L) = \Gamma(L(B))$. However, if $L(B)/L$ were to be a proper extension, then we would have

$$[\Gamma(L(B)) : 1] = [K : L(B)] < [K : L] = [\Gamma(L) : 1].$$

The equalities are taken from Theorem 2, §8.6, and the inequality from Theorem 2, §8.1. This observation contradicts the equality $\Gamma(L) = \Gamma(L(B))$. Thus, $L(B)/L$ must be an improper extension; $B \in L$; and $\Phi[\Gamma(L)] = L$.

Proof of (ii). Let $M = \Phi(H)$. For $A \in M$ we have $\sigma(A) = A$ for all $\sigma \in H$; thus $H \subseteq G(K/M) = \Gamma(M)$. Therefore, by Theorem 2, §8.6,

(*) $$[K : M] = [G(K/M) : 1] \geq [H : 1] = h.$$

By the Theorem of the Primitive Element, §8.3, $K = M(B)$ for some element $B \in K$. Now let $\varepsilon = \sigma_1, \ldots, \sigma_h$ be the elements of H and consider the polynomial

$$g(x) = \prod_{i=1}^{h}(x - \sigma_i(B)) = x^h - e_1 x^{h-1} + \cdots + (-1)^h e_h,$$

where the coefficients $e_1, \ldots, e_h$ are the **elementary symmetric functions** of $\sigma_1(B), \ldots, \sigma_h(B)$. That is,

$$e_1 = \sigma_1(B) + \cdots + \sigma_h(B) = \sum_i \sigma_i(B),$$

$$e_2 = \sigma_1(B)\sigma_2(B) + \cdots + \sigma_{h-1}(B)\sigma_h(B) = \sum_{i<j} \sigma_i(B)\sigma_j(B),$$

$$e_3 = \sum_{i<j<k} \sigma_i(B)\sigma_j(B)\sigma_k(B),$$

. . . .

$$e_h = \sigma_1(B) \cdots \sigma_h(B) = \prod_i \sigma_i(B).$$

Then $\sigma_i(e_j) = e_j$ for $1 \leq i, j \leq h$.

Consequently $e_j \in \Phi(H) = M$, and thus $B = \sigma_1(B)$ is a zero of the polynomial $g(x) \in M[x]$ which has degree h. Therefore, $[K : M] \leq h = [H : 1]$. This inequality combined with the inequality (*) implies that

$$H = \Gamma(M) = G(K/M) = \Gamma[\Phi(H)].$$

Proof of (iii). Since $\Gamma(L) = G(K/L)$, by the preceding arguments Theorem 2, §8.6, implies

$$[\Gamma(L) : 1] = [K : L].$$

Similarly, $\quad [K : \Phi(H)] = [G(K/\Phi(H)) : 1] = [H : 1].$

Proof of (iv). Suppose that L/F and $\bar{L}/F$ are conjugate subfields in the galois extension K/F. Let λ be an isomorphism $L \cong \bar{L}$ extending the identity mapping on F, and let Λ be a prolongation of λ to K/F, as in §8.4. Then $\Lambda \in G(K/F)$ since every isomorphism defined on K/F is an automorphism of K.

Let $H = G(K/L)$ and $\bar{H} = G(K/\bar{L})$. Then

$$[H : 1] = [K : L] = [K : F][L : F]^{-1} = [K : F][\bar{L} : F]^{-1}$$
$$= [K : \bar{L}] = [\bar{H} : 1],$$

since L and $\bar{L}$ are (isomorphic) conjugates over F. Finally, for an arbitrary element $\bar{l} = \lambda(l) = \Lambda(l)$ of $\bar{L}$ each automorphism $\sigma \in H$ satisfies $(\Lambda \sigma \Lambda^{-1})\bar{l} = [(\Lambda \sigma \Lambda^{-1})\Lambda] l = \Lambda(l) = \bar{l}$. Thus $\Lambda H \Lambda^{-1} \subseteq \bar{H}$, and consequently $\Lambda H \Lambda^{-1} = \bar{H}$, since $[\Lambda H \Lambda^{-1} : 1] = [H : 1] = [\bar{H} : 1]$.

Now conversely, choose $\sigma \in G$ such that $\bar{H} = \sigma H \sigma^{-1}$. Writing $\Phi(H)$ as $F(A)$ for some appropriate $A \in \Phi(H)$, consider $M = F(\sigma(A))$. Since A and $\sigma(A)$ are roots of the same irreducible polynomial in $F[x]$, the fields $F(A)$ and $M = F(\sigma(A))$ are conjugate extensions of F; hence they have the same degree over F. We show next that $M = F(\sigma(A))$ is the fixed field $\Phi(\bar{H})$ of $\bar{H}$. For any $\bar{\tau} \in \bar{H}$ and any $m \in M$, where m is a rational function of $\sigma(A)$ with coefficients in F, observe that

$$\bar{\tau}(m) = \sigma \tau \sigma^{-1}(m) = m,$$

since $\tau = \sigma^{-1}\bar{\tau}\sigma \in H$, and hence $\tau(A) = A$. Therefore, $M \subseteq \Phi(\bar{H})$. Finally, since M and $\Phi(\bar{H})$ have equal degrees over F, as follows from the equalities

$$[M : F] = [K : F][K : M]^{-1} = [K : F][K : F(A)]^{-1}$$
$$= [K : F][H : 1]^{-1} = [K : F][\bar{H} : 1]^{-1}$$
$$= [K : F][K : \Phi(\bar{H})]^{-1}$$
$$= [\Phi(\bar{H}) : F],$$

they coincide. Therefore, $\Phi(H) \cong M = \Phi(\bar{H})$.

Proof of (v). Let L/F be a galois extension of F with the galois group $G(L/F)$. As before, let $H = G(K/L) = \Gamma(L)$. The restrictions $\sigma|L$ of the automorphisms $\sigma \in G$ are automorphisms $\bar{\sigma} \in G(L/F)$, since L/F is a galois extension (i.e., every isomorphism defined on L/F is an automorphism of L). The restriction mapping $\sigma \to \sigma|L = \bar{\sigma}$ is a homomorphism $\rho: G(K/F) \to G(L/F)$. The kernel of ρ consists of those automorphisms σ of K which restricted to L are the identity map on L. That is, $\ker \rho = \Gamma(L)$, and thus $\Gamma(L) = H$ is a normal subgroup of $G(K/F)$.

Consider now $\bar{\tau} \in G(L/F)$. The lemmas on prolongations in §8.4 imply that $\bar{\tau}$ has an extension τ to K such that $\tau|L = \bar{\tau}$. Furthermore, $\tau \in G = G(K/F)$ since $\bar{\tau}|F$ is the identity map. Thus, the restriction map ρ is surjective. By the Isomorphism Theorem of §6.8,

$$G/\Gamma(L)) \cong \rho(G) = G(L/F).$$

Conversely, suppose that H is a normal subgroup of G with the fixed field L/F; $H = G(K/L)$. As in the proof of (iv), a conjugate field $\bar{L}/F$ of L/F is equal to $\sigma(L)/F$ for some $\sigma \in G$, and $G(K/\bar{L}) = \sigma G(K/L)\sigma^{-1} = \sigma H \sigma^{-1} = H = G(K/L)$. Part (i) implies that

$$\bar{L} = \Phi(G(K/\bar{L})) = \Phi(G(K/L)) = L.$$

That any conjugate field $\bar{L}$ of L must coincide with L is to say, however, that L is a galois extension field over F. Now let $L = F(B)$. The element B is a root of an irreducible polynomial $f(x) \in F[x]$. Any other root $\bar{B}$ gives rise to a conjugate extension $\bar{L} = F(\bar{B})$, which coincides with L. Thus L contains all roots of $f(x)$ and by Theorem 1, §8.6, is a normal (and hence) galois extension of F.

As a corollary observation to part (v), note that an intermediate field L is a galois extension over F if and only if L coincides with its conjugates in K.

The Fundamental Theorem of the Galois Theory provides a correspondence between intermediate fields of K/F and subgroups of the automorphism group $G(K/F)$ and between normal, separable extensions and normal subgroups. This correspondence was previously discussed in Proposition 6, §8.2, in the special case of finite fields.

Exercises

1. Consider a galois extension $K/\mathbf{Q}$ of degree p^2q, where p and q are distinct primes such that $q < p$ and $q \nmid (p^2 - 1)$.
   a. Prove the existence of intermediate fields L and M, such that $[L:\mathbf{Q}] = p^2$, $[M:\mathbf{Q}] = q$.
   b. Prove that these extensions L and M must be galois extensions of $\mathbf{Q}$.
   c. Prove that K must be an **abelian extension** of $\mathbf{Q}$ (i.e., the galois group $G(K/\mathbf{Q})$ must be abelian).
2. If K/F is a **cyclic galois extension** (i.e., has a finite cyclic galois group), prove:
   a. Every intermediate extension L is cyclic over F.
   b. For each divisor d of $[K:F]$ there exists a unique intermediate extension L_d of degree d over F.
   c. Is statement (b) necessarily true if K/F is not cyclic? Why?
3. Prove that the splitting field of $f(x) = x^4 + x^2 + 1 \in \mathbf{Q}[x]$ has degree 4 over $\mathbf{Q}$.
4. Determine the structure of the galois group of the splitting field of $f(x) = x^4 - 5x^2 + 6 \in \mathbf{Q}[x]$.
5. Find a basis of the splitting field of $f(x) = x^4 + x^3 + x^2 + x + 1$ over $\mathbf{Q}$. If $K = \mathbf{Q}(\xi)$, where ξ satisfies $f(\xi) = 0$, determine all subfields of $K/\mathbf{Q}$.
6. Find the galois group of $f(x) = x^3 - 3x^2 + 5x - 2 \in \mathbf{Q}[x]$. (*Hint:* Show that $f(x)$ is irreducible by getting rid of the "square term," let $x \to y + a$ with $a \in \mathbf{Q}$, and examine the resulting equation $y^3 + by + c \in \mathbf{Z}[x]$ modulo suitable primes $p \in \mathbf{Z}$.)
7. Determine the field of invariant elements in $F(x)$ associated with the group of automorphisms generated by the map $x \to 1/x$.
8. Let $K = \mathbf{Q}(A)$ where A is a zero of $x^4 - 10$. Find the galois group of K/F and determine all subfields of K.

§8.8 Consequences of the Fundamental Theorem

This section presents several corollaries to the Fundamental Theorem of the Galois Theory and extends the results of that theorem to the case of products of fields, analogous to the products of groups encountered in §6.2. Two significant consequences of the Galois theory are left to the next chapter:

solvability of polynomial equations by radicals, §9.2, and constructions by ruler and compass, §9.3. For the reader who wishes to study these consequences the only further prerequisite is §9.1 on cyclotomic field extensions.

Corollary 1. The following strict inclusion relations hold for the subfields $L_1/F, L_2/F$ of the galois extension K/F and for the subgroups H_1, H_2 of the galois group $G(K/F) = G$:

$$L_1 \subset L_2 \quad \text{implies} \quad \Gamma(L_1) \supset \Gamma(L_2),$$
$$H_1 \subset H_2 \quad \text{implies} \quad \Phi(H_1) \supset \Phi(H_2).$$

Suppose that $\Gamma(L_1) = \Gamma(L_2)$. Then by part (i) of the Fundamental Theorem, $\Phi[\Gamma(L_1)] = L_1 = \Phi[\Gamma(L_2)] = L_2$, a contradiction. Similarly, if $\Phi(H_1) = \Phi(H_2)$, then by part (ii) $\Gamma[\Phi(H_1)] = H_1 = \Gamma[\Phi(H_2)] = H_2$, again a contradiction.

It is convenient to let $L_1 L_2$ denote the least subfield of K containing both L_1 and L_2; we speak of the field $L_1 L_2$ as the **product** or **composite** of the fields L_1 and L_2. It is obtained by adjoining the elements of L_1 and L_2 to F (and hence consists of finite sums of products of elements in L_1 and L_2). Similarly, the product of the subgroups H_1, H_2 of G was defined at the end of §6.2 to be the least subgroup of G containing both H_1 and H_2.

Corollary 2. Consider intermediate fields L_1, L_2 of a galois extension K/F and subgroups H_1, H_2 of the galois group $G(K/F)$. Then

(i) $\quad \Gamma(L_1 L_2) = \Gamma(L_1) \cap \Gamma(L_2),$
(ii) $\quad \Phi(H_1 \cdot H_2) = \Phi(H_1) \cap \Phi(H_2),$
(iii) $\quad \Gamma(L_1 \cap L_2) = \Gamma(L_1) \cdot \Gamma(L_2),$
(iv) $\quad \Phi(H_1 \cap H_2) = \Phi(H_1) \Phi(H_2).$

The following diagrams show the inclusion of subfields and the corresponding subgroups of Corollary 2.

$$\begin{array}{ccc} K & \quad & \{\varepsilon\} \\ \cup| & & \cap| \\ L_1 L_2 & & H_1 \cap H_2 \\ \diagup \ \ \diagdown & & \diagup \ \ \diagdown \\ L_1 \qquad L_2 & H_1 \qquad H_2 \\ \diagdown \ \ \diagup & & \diagdown \ \ \diagup \\ L_1 \cap L_2 & & H_1 \cdot H_2 \\ \cup| & & \cap| \\ F & & G \end{array}$$

Proof of (i). Consider $\sigma \in \Gamma(L_1 L_2)$, $l_1 \in L_1$, and $l_2 \in L_2$. Then $l_1 l_2 \in L_1 L_2$; in particular, $\sigma(l_1) = l_1$ and $\sigma(l_2) = l_2$. Consequently $\sigma \in \Gamma(L_1)$ and $\sigma \in \Gamma(L_2)$, thus $\sigma \in \Gamma(L_1) \cap \Gamma(L_2)$. Conversely, for $\tau \in \Gamma(L_1) \cap \Gamma(L_2)$ we have $\tau(l_1) = l_1$ and $\tau(l_2) = l_2$ for all elements $l_1 \in L_1$ and $l_2 \in L_2$. Since the elements of $L_1 L_2$ are, by definition, finite sums of products of elements in L_1 and L_2, it follows that $\tau \in \Gamma(L_1 L_2)$. Hence $\Gamma(L_1 L_2) = \Gamma(L_1) \cap \Gamma(L_2)$.

Proof of (ii). We use an analogous argument. Recall from §6.2 that elements of the form $\sigma_1 \sigma_2$, where $\sigma_i \in H_i$, $i = 1, 2$, generate $H_1 \cdot H_2$. For $A \in \Phi(H_1 \cdot H_2)$ we have $\sigma_i(A) = (\sigma_i \varepsilon)(A) = A$, since $\sigma_i \varepsilon \in H_1 \cdot H_2$, $i = 1, 2$, for all $\sigma_i \in H_i$. Thus $A \in \Phi(H_i)$ and $\Phi(H_1 \cdot H_2) \subseteq \Phi(H_1) \cap \Phi(H_2)$. Conversely for $B \in \Phi(H_1) \cap \Phi(H_2)$ we have $\sigma_i(B) = B$ for all $\sigma_i \in H_i$, $i = 1, 2$. Therefore, $(\sigma_1 \sigma_2)(B) = B$ for all $\sigma_1 \sigma_2 \in H_1 \cdot H_2$ and hence for all $\tau \in H_1 \cdot H_2$, $\tau(B) = B$. Thus, $\Phi(H_1) \cap \Phi(H_2) \subseteq \Phi(H_1 \cdot H_2)$, and $\Phi(H_1 \cdot H_2) = \Phi(H_1) \cap \Phi(H_2)$.

Proof of (iii). The inclusion relations

$$L_1 \cap L_2 \subseteq L_1 \quad \text{and} \quad L_1 \cap L_2 \subseteq L_2$$

imply, as in Corollary 1, that

$$\Gamma(L_1 \cap L_2) \supseteq \Gamma(L_1) \quad \text{and} \quad \Gamma(L_1 \cap L_2) \supseteq \Gamma(L_2);$$

hence $\Gamma(L_1 \cap L_2) \supseteq \Gamma(L_1) \cdot \Gamma(L_2)$. This inclusion *cannot* be proper because, if it were, Corollary 1 and the properties of the operations Φ and Γ would imply that

$$L_1 \cap L_2 = \Phi[\Gamma(L_1 \cap L_2)] \subset \Phi[\Gamma(L_1) \cdot \Gamma(L_2)]$$
$$= \Phi[\Gamma(L_1)] \cap \Phi[\Gamma(L_2)] = L_1 \cap L_2,$$

a contradiction. Hence $\Gamma(L_1 \cap L_2) = \Gamma(L_1) \cdot \Gamma(L_2)$.

Proof of (iv). Similarly, the inclusions $H_1 \cap H_2 \subseteq H_i$, $i = 1, 2$, imply that $\Phi(H_1 \cap H_2) \supseteq \Phi(H_i)$. Hence $\Phi(H_1 \cap H_2) \supseteq \Phi(H_1)\Phi(H_2)$. This *cannot* be a proper inclusion, because

$$H_1 \cap H_2 = \Gamma[\Phi(H_1 \cap H_2)] \subset \Gamma[\Phi(H_1)\Phi(H_2)]$$
$$= \Gamma[\Phi(H_1)] \cap \Gamma[\Phi(H_2)]$$
$$= H_1 \cap H_2$$

leads to a contradiction.

Theorem 1 (*On Natural Irrationality*). Let K/F be a galois extension with the galois group $G = G(K/F)$. If L/F is an extension such that both K and L are contained in a common field Ω/F, then:

    (i) KL is a galois extension of L.
    (ii) The galois group $G(KL/L)$ is isomorphic to the galois group $G(K/(K \cap L))$.

§8.8 Consequences of the Fundamental Theorem

The inclusion relations between the various fields of this theorem may be illustrated as follows:

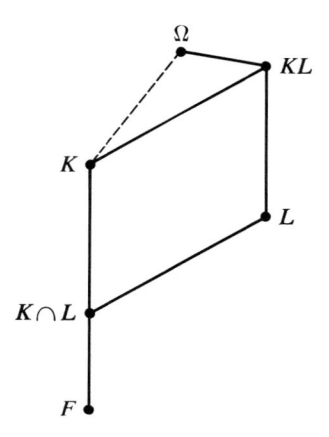

Proof of (i). Since K/F is a galois extension and hence separable, the Theorem of the Primitive Element [§8.3] implies that $K = F(A)$. Further $KL = L(A)$, as follows. First, $KL \supseteq L(A)$, since KL contains L and A. Second, by definition KL is the intersection of all fields containing both K and L. Certainly $L(A) \supseteq L$, but also $L(A) \supseteq K = F(A)$ since L contains F and A.

Part (i) follows from the fact that the zeros of the minimal polynomial $f(x) \in F[x]$ of A are polynomials in A with coefficients in F because K/F is a galois extension. Thus KL is *the* splitting field of $f(x)$ and hence a galois extension over F [see §8.6].

Proof of (ii). Consider the restriction mapping $\rho: G(KL/L) \to G(K/(K \cap L))$ given by $\rho(\sigma) = \sigma|K$ for $\sigma \in G(KL/L)$. We must verify that $\sigma|K$ belongs to $G(K/(K \cap L))$. Note that σ restricted to L (and hence to $K \cap L$ or F) is the identity map. Moreover since $\sigma(A)$ is a root of $f(x)$, it is a polynomial in A with coefficients in F. Thus $\sigma(K) \subseteq K$, and so $\sigma|K$ belongs to $G(K/F)$ and hence to $G(K/(K \cap L))$. Since $(\sigma|K)(\tau|K) = (\sigma\tau|K)$ by definition of restriction to K, the mapping ρ is a homomorphism.

Now for $\sigma \neq \tau$ in $G(KL/L)$, $\sigma(B) \neq \tau(B)$, for some element $B \in K \setminus L$, and hence $\sigma|K \neq \tau|K$. Thus, the mapping ρ is one–one. Denote Im ρ by $G(KL/L)|K$. To prove that $G(KL/L)|K = G(K/(K \cap L))$ it suffices to show that the fixed field $M \subseteq K$ of the subgroup $G(KL/L)|K \subseteq G(K/F)$ is $K \cap L$. For $b \in M$, $\sigma(b) = (\sigma|K)(b) = b$ and so $b \in L$. Therefore $M \subseteq K \cap L$. Conversely for $c \in K \cap L$, $c = \sigma(c) = (\sigma|K)(c)$ and therefore $K \cap L \subseteq M$.

As a corollary result, we have the following proposition.

Proposition. Let K/F be a galois extension and L/F a finite extension such that $K \cap L = F$. Then $[KL : F] = [K : F][L : F]$.

For the proof note that the galois groups $G(KL/L)$ and $G(K/(K \cap L)) = G(K/F)$ are isomorphic. Consequently,

$$[KL : F] = [KL : L][L : F] = [K : F][L : F].$$

As a cautionary note that this result does not hold for arbitrary finite extensions K/F and L/F, consider the extensions $K = \mathbf{Q}(\sqrt[3]{2})$ and $L = \mathbf{Q}(\omega\sqrt[3]{2})$, where $\omega^2 + \omega + 1 = 0$. Then $K \cap L = \mathbf{Q}$, $KL = \mathbf{Q}(\omega, \sqrt[3]{2})$, and $[KL : \mathbf{Q}] = 6$, but $[K : \mathbf{Q}][L : \mathbf{Q}] = 3^2$.

Theorem 2. For galois extensions K_i/F, $i = 1, 2$, contained in a field Ω/F with respective galois groups G_i:

  (i) The product field $K_1 K_2$ is a galois extension of F.
  (ii) The mapping $\varphi : G(K_1 K_2 / F) \to G_1 \dot\times G_2$ given by $\varphi(\sigma) = (\sigma | K_1, \sigma | K_2)$ is an injective homomorphism.
  (iii) The pair $(\sigma_1, \sigma_2) \in G_1 \dot\times G_2$ determines an automorphism of $K_1 K_2 / F$ if and only if
  $$\sigma_1 | (K_1 \cap K_2) = \sigma_2 | (K_1 \cap K_2).$$
  (iv) The galois group $G = G(K_1 K_2 / F)$ is isomorphic to the subgroup
  $$H = \{(\sigma_1, \sigma_2) \in G_1 \dot\times G_2 : \sigma_1 | (K_1 \cap K_2) = \sigma_2 | (K_1 \cap K_2)\}$$
  of $G_1 \dot\times G_2$.

Proof of (i). Let $K_i = F(a_i)$, $i = 1, 2$. Then the composite of the fields K_1 and K_2 can be expressed as $K_1 K_2 = K_1(a_2)$. Suppose now that $[a_2 : K_1] = k \le [a_2 : F] = [K_2 : F]$. Then the elements of $K_1 K_2$ are of the form

$$g_0 + g_1 a_2 + \cdots + g_{k-1} a_2^{k-1} \qquad \text{with } g_j \in K_1,\ 0 \le j < k.$$

Furthermore,

$$g_j = h_{j,0} + h_{j,1} a_1 + \cdots + h_{j, n_1 - 1} a_1^{n_1 - 1}$$

with coefficients $h_{j,l} \in F$. Therefore $a \in K_1 K_2$ can be written as

$$a = \sum_{j,l} h_{j,l} a_2^j a_1^l$$

where $0 \le j < k$ and $0 \le l < n_1$. For an isomorphism σ on $K_1 K_2 / F$

$$\sigma(a) = \sum_{j,l} h_{j,l} \sigma(a_2)^j \sigma(a_1)^l$$

belongs to $K_1 K_2$ since $(\sigma | K_i)(a_i) \in K_i$, $i = 1, 2$, because K_i/F is a galois extension. Thus $K_1 K_2$ is a galois extension of F.

Proof of (ii). The mapping φ is a homomorphism since for $b_i \in K_i$, $i = 1, 2$,

$$[(\sigma\tau) | K_i](b_i) = (\sigma\tau)(b_i) = \sigma(\tau(b_i))$$
$$= (\sigma | K_i)[(\tau | K_i)(b_i)].$$

§8.8 Consequences of the Fundamental Theorem

If $\sigma | K_i$ is the identity mapping on K_i/F, $i = 1, 2$, then
$$\sigma(a) = \sum_{j,l} h_{j,l} a_2{}^j a_1{}^l = a$$
for *all* $a \in K_1 K_2$. Hence σ is the identity isomorphism of $K_1 K_2/F$, and φ is an injection of G into $G_1 \dot\times G_2$.

Proof of (iii). If isomorphisms $\sigma_1 \in G_1$ and $\sigma_2 \in G_2$ have different restrictions to $K_1 \cap K_2$, there can be no automorphism σ in G such that
$$\sigma | K_1 = \sigma_1, \qquad \sigma | K_2 = \sigma_2.$$
If there were such an automorphism, then we would have the contradiction:
$$\sigma_1 | (K_1 \cap K_2) = \sigma | (K_1 \cap K_2) = \sigma_2 | (K_1 \cap K_2).$$

The converse, that an element $(\sigma_1, \sigma_2) \in G_1 \dot\times G_2$ determines an automorphism σ of $K_1 K_2/F$ such that $\sigma | K_i = \sigma_i$, $i = 1, 2$, if $\sigma_1 | (K_1 \cap K_2) = \sigma_2 | (K_1 \cap K_2)$, requires careful proof, for which we use the Theorem on Natural Irrationalities. The following diagram illustrates the relations between the various fields.

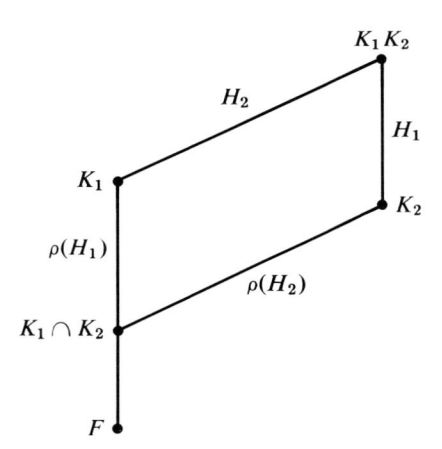

Here $\rho(H_1)$ denotes the restriction of the galois group H_1 of $K_1 K_2/K_2$ (a galois extension by Theorem 1) to the galois group of $K_1/(K_1 \cap K_2)$. Similarly $\rho(H_2)$ is the restriction of the galois group H_2 of $K_1 K_2/K_1$ to the galois group of $K_2/(K_1 \cap K_2)$.

The groups H_1 and H_2 are normal in $G(K_1 K_2/F)$ since $K_1 K_2$ is a galois extension of both K_2 and K_1 by the lemma of §8.7. Consequently the product group $H_1 \cdot H_2$ is normal in G. By Corollary 2 above the fixed field $\Phi(H_1 \cdot H_2) = K_1 \cap K_2$ and so $H_1 \cdot H_2 = G(K_1 K_2/(K_1 \cap K_2))$. Furthermore $H_1 \cap H_2 = \{e\}$ since $\Phi(H_1 \cap H_2) = K_1 K_2$. Thus the product $H_1 \cdot H_2$ is direct; by the properties of the internal direct product
$$\alpha_1 \alpha_2 = \alpha_2 \alpha_1 \qquad \text{for } \alpha_i \in H_i, \ i = 1, 2$$

[see §6.10]. Application of the Fundamental Theorem of the Galois Theory yields

(*) $\quad G((K_1 \cap K_2)/F) \cong G(K_1 K_2/F)/H_1 \otimes H_2.$

Also the automorphisms τ in $G((K_1 \cap K_2)/F)$ are restrictions $\rho(\hat{\tau})$ of automorphisms $\hat{\tau}$ in $G(K_1 K_2/F) = G$. For $\alpha_i \in H_i$, $i = 1, 2$, the restrictions $\rho(\hat{\tau}\alpha_1\alpha_2)$ and $\rho(\hat{\tau})$ are equal. By virtue of the isomorphism (*) each prolongation of $\tau \in G((K_1 \cap K_2)/F)$ to $K_1 K_2/F$ must be of the form $\hat{\tau}\alpha_1\alpha_2$ for suitable α_1, α_2.

Now finally consider mappings $\sigma_i \in G_i$, $i = 1, 2$, that are equal on $K_1 \cap K_2$. As a consequence of the isomorphism (*) $\sigma_1|(K_1 \cap K_2) = \sigma_2|(K_1 \cap K_2)$ has a prolongation $\hat{\tau}\alpha_1\alpha_2$ to $K_1 K_2$. To prove that the automorphisms $\alpha_1 \alpha_2$ can be selected so that

$$(\hat{\tau}\alpha_1\alpha_2)|K_1 = \sigma_1 \quad \text{and} \quad (\hat{\tau}\alpha_1\alpha_2)|K_2 = \sigma_2,$$

we must solve the following for α_1:

$$(\hat{\tau}\alpha_1\alpha_2)|K_1 = (\hat{\tau}|K_1) \circ (\alpha_1|K_1) \circ (\alpha_2|K_1)$$
$$= (\hat{\tau}|K_1) \circ \rho(\alpha_1) \circ 1 = \sigma_1,$$

where, according to Theorem 1, the restriction mapping ρ determines an isomorphism between $H_1 = G(K_1 K_2/K_2)$ and $G(K_1/(K_1 \cap K_2))$. Because of this isomorphism

$$\rho(\alpha_1) = (\hat{\tau}|K_1)^{-1} \circ \sigma_1,$$

and thus $\alpha_1 \in H_1$ is determined. Similarly we can determine $\alpha_2 \in H_2$, and so obtain an automorphism σ for which

$$(\sigma|K_1, \sigma|K_2) = (\sigma_1, \sigma_2) \in G_1 \dot{\times} G_2.$$

Proof of (iv). Combining parts (ii) and (iii) yields the desired isomorphism.

Exercises

1. Let σ be the map $f(x) \to f(1-x)$ on the rational function field $F(x)$. Prove that $F(x^2 - x)$ is the fixed field for $\{\varepsilon, \sigma\}$, assuming that char $F \neq 2$.
2. Assume that A and B are elements of the galois extension K/F with respective conjugates

$$A = A_1, A_2, \ldots, A_m \quad \text{and} \quad B = B_1, B_2, \ldots, B_n$$

and that $[F(A, B) : F] = mn$. Prove the existence of automorphisms $\sigma_{i,j} \in G(K/F)$ such that $\sigma_{i,j}(A) = A_i$, $\sigma_{i,j}(B) = B_j$ for $1 \leq i \leq m$, $1 \leq j \leq n$.
3. Prove that if K_1/F and K_2/F are solvable galois extensions, then so are $K_1 K_2$ and $K_1 \cap K_2$.

§8.9   Algebraic Closure

4. Denote by F the field of rational functions of x with coefficients in $\mathbf{C}$, and let ζ be a primitive nth root of unity.
   a. Prove that $\sigma(x) = \zeta x$ and $\tau(x) = x^{-1}$ determine automorphisms of F of respective orders n and 2.
   b. Prove that $\sigma\tau = \tau\sigma^{-1}$, and that the fixed field of the dihedral group generated by σ and τ is $\mathbf{C}(x^n + x^{-n})$.

5. Let $f(x) = x^3 + a_1 x^2 + a_2 x + a_3$ be an irreducible cubic in $F[x]$ whose galois group is isomorphic to Σ_3. Prove that

$$D = (A_1 - A_2)(A_1 - A_3)(A_2 - A_3)$$

   generates a subfield $L \supseteq F$ of the splitting field $K = F(A_1, A_2, A_3)$ if char $F \neq 2$.

6. If the degree of every finite algebraic extension of F is divisible by p, prove that $[K : F]$ must be a power of p, p a prime, for any K/F.

7. Let $A = \sqrt{1+4i}$, where $i^2 = -1$.
   a. Determine $[\mathbf{Q}(A) : \mathbf{Q}]$.
   b. Show that $\xi^2 - \xi + i = 0$ has a solution ξ in the least galois extension over $\mathbf{Q}$ that contains $\mathbf{Q}(A)$.
   c. Prove that $\sqrt{17}$ is contained in the galois extension of part (b).

8. Determine the galois group of the polynomial $x^4 + 2 \in \mathbf{Q}[x]$ over $\mathbf{Q}$.

9. Prove that $f(x) = x^4 + 30x^2 + 45 \in \mathbf{Q}[x]$ is irreducible. Let A be a zero of $f(x)$, and show that $\mathbf{Q}(A)/\mathbf{Q}$ is a cyclic extension of degree 4. Describe the effect of the galois group on A and exhibit the quadratic subfield $K/\mathbf{Q}$ of $\mathbf{Q}(A)/\mathbf{Q}$. (Note that A has the form $\pm\sqrt{-3(5 \pm 2\sqrt{5})}$.)

10. Suppose that K/F is a galois extension with the intermediate field L/F. Prove that the galois group $G(K/M)$ of the smallest field M for which $L \subseteq M \subseteq K$ and M/F is normal equals $\bigcap_{\sigma \in G(K/F)} \sigma G(K/L) \sigma^{-1}$.

§8.9   Algebraic Closure

In §5.4 we proved that each polynomial $f(x) \in F[x]$ splits in some finite algebraic extension of F, and that such extensions depend upon the polynomial $f(x)$. We now address the more general problem of constructing an algebraic extension of F in which every polynomial $f(x) \in F[x]$ splits. The proof requires transfinite induction or the lemma of Max A. Zorn. We prefer the latter approach, and begin with some set-theoretic preliminaries.

A set S is termed **partially ordered** if there is given for it a relation $\leq$ between certain pairs of elements x, y in S such that the following relations hold:

(i) $x \leq x$
(ii) $x \leq y$ and $y \leq x$ imply $x = y$
(iii) $x \leq y$ and $y \leq z$ imply $x \leq z$.

Note that the relation $x \leq y$ or $y \leq x$ need not hold for arbitrarily selected elements $x, y \in S$; hence the terminology *partial* order.

A subset T of a partially ordered set S is called **totally ordered** if either $x \leq y$ or $y \leq x$ for every pair of elements $x, y \in T$. An element $s \in S$ is called an **upper bound** of a nonempty subset $U \subseteq S$ if $u \leq s$ for all $u \in U$. A partially ordered set S is termed **inductively ordered** if every totally ordered subset T of S has an upper bound in S. An element m in a partially ordered set S is called **maximal** if for $x \in S$ with $m \leq x$ necessarily $m = x$.

Zorn's Lemma. If S is a partially ordered set which is inductively ordered and nonempty, then S contains at least one maximal element.

We accept this lemma as an axiom. For an interesting discussion of it, see S. Lang, *Algebraic Structures*.

Recall that an ideal M in a commutative ring R is **maximal** if $M \neq R$ and if there is *no* ideal B for which $M \subsetneq B \subsetneq R$. The residue class ring R/M of an ideal M is a field if and only if M is maximal in R [see Exercise 15, §3.6].

Lemma 1. If A is a proper ideal in a ring R, then A is contained in a maximal ideal M of R.

Proof. Let $T = \{C\}$ be a totally ordered subset of ideals C in R such that $C \supseteq A$ and $1 \notin C$. Then the set-theoretic union $K = \bigcup_{C \in T} C$ is an ideal and an upper bound for T. To verify that K is an ideal consider $x, y \in K$, $r \in R$; then $x \in C'$, $y \in C''$ for some $C', C'' \in T$. Since T is totally ordered, we may assume, without loss of generality, that $C' \subseteq C''$; therefore $x + y \in C''$, $ry \in C''$. Hence K is an ideal, and of course $K \supseteq A$. Finally $1 \notin K$, since $1 \notin C$ for all $C \in T$, and thus K is a proper ideal.

Now, applying Zorn's Lemma, we obtain a maximal element M in the set S of all proper ideals in R which contain A. Then $M \neq R$, since $1 \notin M$; furthermore M is a maximal ideal.

A field F is said to be **algebraically closed** if every polynomial $f(x) \in F[x]$ has a zero c in F, i.e., $f(x) = (x-c)g(x)$ with $g(x) \in F[x]$. That a field F is algebraically closed means that every polynomial in $F[x]$ can be written as the product of linear factors in $F[x]$.

Lemma 2. For any field F there exists an algebraically closed field Ω containing F.

Proof. To each $f(x) \in F[x]$, $\deg f(x) \geq 1$, associate a symbol X_f. Then the set S of symbols X_f is in one-one correspondence with the polynomials of degree at least 1 in $F[x]$. Next consider the polynomial ring $F[\{X_f\}]$ consisting of all polynomials in any *finite* number of the indeterminates X_f with coefficients in F. To each polynomial $f(x) \in F[x]$ associate the unique polynomial $f(X_f) \in F[\{X_f\}]$. The ideal $A = (..., f(X_f), ...)$ generated by $f(X_f)$, for all $f(x) \in F[x]$ of degree at least 1, is a proper ideal in $F[\{X_f\}]$, because if A were the whole ring then 1 would have to be a linear combination of a finite number of polynomials $f(X_f)$ with coefficients in $F[\{X_f\}]$. Let $f_1,...,f_s$ be these polynomials in the indeterminates $X_{f_i} = X_i$; then

(∗) $$1 = \sum_{i=1}^{s} g_i f_i(X_i), \qquad g_i \in F[\{X_f\}].$$

Since there are finitely many polynomials g_i, each with a finite number of indeterminates, there are only finitely many indeterminates

$$X_1,...,X_s, X_{s+1},...,X_n$$

involved in the relation (∗). We have

$$1 = \sum_{i=1}^{s} g_i(X_1,...,X_n) f_i(X_i).$$

According to the basic construction of Kronecker, §5.4, there exists a finite extension K/F in which every polynomial $f_i(X_i)$ has a zero a_i. Finally let $a_i = 0$ for $i > s$, and consider the homomorphism given by $X_i \to a_i$, for $1 \leq i \leq n$. Then in K,

$$1 = \sum_{i=1}^{s} g_i(a_1,...,a_s, 0,...,0) f_i(a_i) = 0,$$

a contradiction. Hence $1 \notin A$, and A is a proper ideal.

Lemma 1 implies that A is contained in a maximal ideal M of $F[\{X_f\}]$. Thus, the residue class ring $F[\{X_f\}]/M$ is a field K_1. This field K_1 contains a zero of every polynomial $f(x) \in F[x]$ because $f(X_f) \in A \subseteq M$; specifically, the coset X_f modulo M, denoted ξ_f, in K_1 satisfies $f(\xi_f) = 0$.

The construction of the field K_1 can be repeated recursively so that the field K_{n+1} contains a zero of every nonconstant polynomial in $K_n[x]$:

$$F \subseteq K_1 \subseteq \cdots \subseteq K_n \subseteq K_{n+1} \subseteq \cdots.$$

At each step the field K_n is identified with the residue class ring of a polynomial ring over K_{n-1}, constructed like $F[\{X_f\}]$, modulo a maximal ideal, such as M, in $F[\{X_f\}]$.

Now let $\Omega = \bigcup_n K_n$. This field is algebraically closed. First, Ω is indeed a field. If $a, b \in \Omega$, then a, b lie in K_n for some sufficiently large n. Since K_n is a field, $a \pm b$, ab, and a/b, if $b \neq 0$, are in $K_n \subseteq \Omega$. Second, if $g(x) \in \Omega[x]$, then $g(x) \in K_m[x]$, for some m, since the number of coefficients in $g(x)$ is finite. By construction, K_{m+1} contains a zero of $g(x)$; hence Ω contains a zero of $g(x)$. In other words, $\Omega[x]$ does not contain any irreducible polynomials of degree greater than 1. Therefore Ω is algebraically closed.

Theorem. Every field F has an algebraic extension K that is algebraically closed.

Such a field K is called an **algebraic closure** of F. All algebraic closures of F are isomorphic over F.

Proof. As in Lemma 2, let Ω be an algebraically closed extension of F. Define K to be the set of all elements in Ω which are *algebraic* over F. By Theorem 1, §8.1, K is a field. Furthermore it is algebraically closed because it contains the zeros of any polynomial $f(x) \in K[x]$ as follows. The coefficients of a given polynomial belong to a finite algebraic extension L/F contained in K. The zeros of $f(x)$ are algebraic over L and therefore over F. Thus they belong to K, the totality of elements in Ω that are algebraic over F.

To prove that the field K/F is unique to within isomorphisms over F, we use Zorn's Lemma. But first, if $F(a)$ is a simple extension and $f(x)$ the minimal polynomial of a, than an embedding λ of F into an algebraically closed field Ω has at most $n_0 \leq \deg f(x)$ distinct extensions to $F(a)$ [see §§8.4 and 8.5]. (Once $F(a)$ is mapped by an extension Λ into Ω, the other embeddings of $F(a)$ are obtained by mapping $\Lambda(a)$ to the other zeros of $\lambda(f(x)) \in \lambda(F) \subseteq \Omega$.)

This result generalizes to arbitrary algebraic extensions K/F. Suppose that λ embeds F in an algebraically closed field Ω which need not be algebraic over F. To prove that there exists an extension Λ of λ to K such that $\Lambda(K) \subseteq \Omega$ let S be the set of all pairs $(L/F, \sigma)$ where $L \subseteq K$ and $\sigma | F = \lambda$, σ an embedding of L in Ω/F. This is a nonempty set for $(F, \lambda) \in S$. Next introduce a partial ordering in S as follows:

$$(L/F, \sigma) \leq (L'/F, \sigma')$$

if $L \subseteq L'$ and $\sigma' | L = \sigma$.

This ordering is inductive. Let T be a nonempty, totally ordered subset of S with typical elements $(N/F, \tau)$ and $(N'/F, \tau')$. Now let $\tilde{L}/F$ be the union of all fields N occurring in T. To prove that $\tilde{L}/F$ is a field, consider $a, b \in \tilde{L}$, where $a \in N$, $b \in N'$ for some fields N, N'. Without loss of generality, we may assume because T is totally ordered that $N \subseteq N'$; hence $a, b \in N'$ and therefore ab, $a+b$, and a^{-1} (for $a \neq 0$) belong to N' and therefore to $\tilde{L}$.

Now to construct an embedding $\tilde{\sigma}: \tilde{L} \to \Omega$, define $\tilde{\sigma}(a) = \tau(a)$ for $a \in N \subseteq \tilde{L}$, where $(N/F, \tau) \in T$. This definition of $\tilde{\sigma}$ is independent of the choice of field N containing a, since if $a \in N'$ and $N \subseteq N'$, then $\tilde{\sigma}(a) = \tau'(a) = \tau(a)$ because $\tau' | N = \tau$. To verify the homomorphism properties

$$\tilde{\sigma}(a+b) = \tilde{\sigma}(a) + \tilde{\sigma}(b),$$

$$\tilde{\sigma}(ab) = \tilde{\sigma}(a)\tilde{\sigma}(b)$$

for $a \in N$, $b \in N'$, assume $N \subseteq N'$ and then use the facts that $a, b \in N'$ and that τ' is a homomorphism on N'. Therefore, since each τ occurring in T is an embedding we conclude that $\tilde{\sigma}$ is also. Thus $(\tilde{L}, \tilde{\sigma})$ is an upper bound of T.

Having verified the hypotheses of Zorn's Lemma, we apply it to obtain the existence of a maximal element $\{M, \Lambda\}$ in S, $M \subseteq K$. Necessarily $M = K$, for if there were an element $B \in K \backslash M$, we could extend Λ to an embedding Λ_0 of $M(B)$ into Ω/F, thereby contradicting the maximality of (M, Λ).

Finally assume that K/F and Ω/F are two algebraic closures of F. By the preceding argument, each can be embedded in the other. Consequently they are isomorphic over F.

§8.9   Algebraic Closure

REMARK. The separable elements of an algebraic closure K/F form a subfield K_s/F, called a **separable algebraic closure** of F [cf. §8.5]. Clearly all separable algebraic closures of F are isomorphic over F.

Exercises

1. Prove that a finite field F cannot be algebraically closed. (*Hint:* Examine $f(x) = 1 + \prod_{i=1}^{m}(x - a_i)$, where $F = \{a_i : 1 \leq i \leq m\}$. Compare with §8.2.)
2. Suppose that the elements $A_1, \ldots, A_h$ in an extension field K/F are algebraic over F. Prove that the ideal
$$\{f(x_1, \ldots, x_h) \in F[x_1, \ldots, x_h] : f(A_1, \ldots, A_h) = 0\}.$$
is a maximal ideal of $F[x_1, \ldots, x_h]$.
3. Let F be a countable field, i.e., its elements are in a one-one set correspondence with the set of positive integers $\mathbf{N}$. Prove that the polynomial ring $F[x]$ has at most a countable number of monic irreducible polynomials.
4. Let $F = \mathbf{Z}_2(x, y)$. Prove that $f(t) = t^2 + xt + y \in F[t]$ remains irreducible in $\Omega(x, y)[t]$, where Ω is an algebraic closure of $\mathbf{Z}_2$.
5. Let S be a nonempty subset of a group G, and H a subgroup of G for which $H \cap S = \emptyset$. Prove that the set of all subgroups $K \subset G$ such that $K \cap S \neq \emptyset$ and that $H \subseteq K$ contains at least one maximal subgroup of G.

9

Selected Topics in Field Theory

This chapter has a two-fold purpose: first, to develop further the Galois theory in order to lay some foundation for algebraic number theory and algebraic geometry [§§9.1, 9.4 through 9.7, and 9.11 through 9.12]; and second, to provide the solutions to problems such as "doubling the cube" and "trisecting an angle" that tantalized mathematicians from the classical Greek period through the European Renaissance. General solutions for cubic and quartic equations by Cardano, Ferrari, and Tartaglia in the sixteenth century gave rise to the query whether equations of the fifth and higher degrees could be solved by extracting radicals. These questions are treated in §§9.1 through 9.3.

In §9.9 we present a proof, derived from the work of Lagrange and Gauss, of the Fundamental Theorem of Algebra, which states that the field of complex numbers is algebraically closed. This proof is prefaced [§9.8] by a discussion of unique factorization domains, elementary symmetric functions, and the concept of integral dependence. Together with the exercises in §§3.1, 3.6, and 7.8, this provides a background for general ideal theory. We include [§9.10] E. Witt's proof of one of Wedderburn's fundamental theorems on finite dimensional algebras: that every finite division ring is commutative.

§9.1 Cyclotomic Fields

Abelian extensions of the rational number field $\mathbf{Q}$ obtained by adjunction of roots of unity are called **cyclotomic fields**. Fields of this type, specifically abelian extensions (and their subfields) over fields of finite degree over the rational numbers $\mathbf{Q}$, were significant in Kummer's attempts to prove Fermat's Last Theorem and are important in the class field theory of algebraic number fields. In this section we discuss the algebraic structure of cyclotomic fields, for which the arguments are essentially arithmetic in nature. Analytic aspects of the theory will be used in §9.10 concerning finite division rings.

Throughout this section we treat only extensions of the field $\mathbf{Q}$ of rational numbers. We denote by ζ a **primitive nth root of unity** for a fixed integer n; that is, ζ satisfies $\zeta^n = 1$, but for $1 \leq m < n$, $\zeta^m \neq 1$. (Alternatively, a primitive nth root of unity is a generator of the cyclic group of nth roots of unity.) If ζ is a primitive nth root of unity, then so is ζ^i, for each i, $1 \leq i < n$, that is relatively prime to n [cf. Proposition 2, §6.2]. For given n we define the **cyclotomic polynomial** $\Phi_n(x)$ by

$$\Phi_n(x) = \prod_{(i,n)=1} (x - \zeta^i).$$

Theorem. The cyclotomic polynomial $\Phi_n(x)$, $n \geq 1$, is irreducible in $\mathbf{Q}[x]$ and has coefficients in $\mathbf{Z}$.

Note that the degree of $\Phi_n(x)$ is the Euler φ-function $\varphi(n)$, equal to the number of generators of a cyclic group of order n [cf. Proposition 3, §6.6]. Granting that $\Phi_n(x)$ is irreducible, we know then that it is the minimal polynomial of ζ over $\mathbf{Q}$, and hence that $[\mathbf{Q}(\zeta) : \mathbf{Q}] = \varphi(n)$. The following lemma is useful in the proof of the theorem.

Lemma. For any $n \geq 1$,

$$x^n - 1 = \prod_{d \mid n} \Phi_d(x),$$

where the product is taken over all positive divisors d of n.

Proof. Because each nth root of unity is a primitive dth root of unity for one and only one divisor d of n, and because $x^n - 1$ has distinct roots (i.e., is a separable polynomial), each of the n linear factors of $x^n - 1$ in some splitting field $\Omega/\mathbf{Q}$ of $x^n - 1$ occurs precisely once as a factor of some cyclotomic polynomial $\Phi_d(x)$. Recall that $n = \sum_{d \mid n} \varphi(d)$ [cf. Exercise 11, §3.5].

Proof of the theorem. We show by induction on n that $\Phi_n(x)$ in $\mathbf{Q}[x]$ has coefficients in $\mathbf{Z}$. For $n = 1$ the statement is trivial, as $\Phi_1(x) = x - 1$. For the induction hypothesis assume that $\Phi_d(x) \in \mathbf{Z}[x]$ for all d less than n; in particular, that

$$\psi(x) = \prod_{\substack{d \mid n \\ d < n}} \Phi_d(x) \in \mathbf{Z}[x].$$

The lemma states that $\Phi_n(x)\psi(x) = x^n - 1$, and so $\Phi_n(x)$ must have integral coefficients since $x^n - 1$ is primitive [cf. Theorem 1, §5.8]. By induction we conclude that $\Phi_n(x) \in \mathbf{Z}[x]$ for all integers $n \geq 1$.

To complete the proof of the theorem we demonstrate that $\Phi_n(x)$ is the minimal polynomial of ζ over $\mathbf{Q}$ and hence is irreducible. Let $f(x) \in \mathbf{Q}[x]$ be the minimal polynomial of ζ. Then $f(x) | \Phi_n(x)$. In order to prove conversely that $\Phi_n(x) | f(x)$, consider a prime number p satisfying $(n, p) = 1$. Then ζ^p is also a primitive nth root of unity. Let $g(x)$ be its minimal polynomial. To prove that $f(x) = g(x)$, assume to the contrary that $f(x) \neq g(x)$. Then $(f(x), g(x)) = 1$ since both polynomials are irreducible. Further,

$$f(x) | (x^n - 1) \quad \text{and} \quad g(x) | (x^n - 1)$$

imply that
$$f(x)g(x) | (x^n - 1).$$

Hence $x^n - 1 = f(x)g(x)h(x)$ where $f(x)$, $g(x)$, and $h(x)$, as divisors of a primitive polynomial in $\mathbf{Z}[x]$, also lie in $\mathbf{Z}[x]$. Since ζ is a root of $f(x)$ and $g(x^p)$, these polynomials have a common factor $k(x) \in \mathbf{Z}[x]$. Let

$$f(x) = k(x)f_1(x) \quad \text{and} \quad g(x^p) = k(x)g_1(x).$$

Because these polynomials have integral coefficients, we may reduce the coefficients modulo the prime p, obtaining

$$f^*(x) = k^*(x)f_1^*(x) \quad \text{and} \quad g^*(x^p) = k^*(x)g_1^*(x)$$

in $\mathbf{Z}_p[x]$. That is, we map $\mathbf{Z}[x]$ to $\mathbf{Z}_p[x]$ homomorphically by taking coefficients modulo p, with the image of $f(x)$ denoted $f^*(x)$, etc.

Since $a^p \equiv a \pmod{p}$ for any $a \in \mathbf{Z}$ by Fermat's Little Theorem, §2.12,

$$g(x^p) \equiv (g(x))^p \pmod{p},$$

i.e., $g^*(x^p) = (g^*(x))^p$. If $q^*(x)$ is an *irreducible* factor of the common factor $k^*(x)$ of $f^*(x), g^*(x^p)$, then $q^*(x) | g^*(x)^p$ implies that $q^*(x) | g^*(x)$. Thus

$$f^*(x)g^*(x) | (x^n - [1]_p)$$

implies that
$$q^*(x)^2 | (x^n - [1]_p).$$

Consequently in some algebraic extension of $\mathbf{Z}_p$, $x^n - [1]_p$ would have a multiple root. This is impossible, however, because the derivative of $(x^n - 1) \bmod p$ is $n \cdot x^{n-1} \not\equiv 0 \pmod{p}$, according to the choice of p [see §5.5]. Hence $f(x) = g(x)$ and ζ^p is a xero of $f(x)$.

This fact implies that *all* primitive nth roots of unity are zeros of $f(x)$. Consider ζ^j, where $(j, n) = 1$, with $j = p_1 \cdots p_s$ (some of the prime factors p_i may coincide). Then $f(\zeta^{p_1}) = 0$ and

$$f(\zeta^{p_1 p_2}) = \cdots = f(\zeta^j) = 0.$$

Therefore $\Phi_n(x) | f(x)$ and consequently $\Phi_n(x) = f(x)$, the minimal polynomial of ζ.

Proposition. The galois group $G(\mathbf{Q}(\zeta)/\mathbf{Q})$ is isomorphic to the multiplicative group $\mathbf{U}_n$ of prime residue classes modulo n.

Proof. Since the roots of $x^n - 1$ are the n powers ζ^i, $1 \leq i \leq n$, of a primitive nth root of unity ζ, it follows that $\mathbf{Q}(\zeta)$ is a galois extension of degree $\varphi(n)$ over $\mathbf{Q}$ [cf. §8.6]. For each $\sigma \in G = G(\mathbf{Q}(\zeta)/\mathbf{Q})$ and for a fixed primitive nth root of unity ζ,

$$\sigma(\zeta) = \zeta^{j(\sigma)}, \qquad 0 < j(\sigma) < n,$$

is also a primitive nth root of unity, and hence $(j(\sigma), n) = 1$; that is, $j(\sigma) \in \mathbf{U}_n$. Now define a mapping $\lambda: G \to \mathbf{U}_n$ by

$$\lambda(\sigma) = [j(\sigma)]_n.$$

We seek to show that λ is an isomorphism. First,

$$\zeta^{j(\sigma\tau)} = (\sigma\tau)(\zeta) = \sigma(\tau(\zeta))$$
$$= \sigma(\zeta^{j(\tau)}) = \zeta^{j(\sigma)j(\tau)},$$

from which we conclude that λ is a homomorphism. For $\sigma \in \ker \lambda$ we have $j(\sigma) = 1$, which means that $\sigma(\zeta) = \zeta$; in other words, σ is the identity automorphism. Therefore λ is injective.

To see that λ is surjective, consider a prime residue m modulo n, and define $\sigma \in G$ by $\sigma(\zeta) = \zeta^m$. This mapping is an *automorphism* since ζ^m is a zero of the cyclotomic polynomial $\Phi_n(x)$ and hence $\mathbf{Q}(\zeta^m) = \mathbf{Q}(\zeta)$. (In more detail, certainly $\mathbf{Q}(\zeta^m) \subseteq \mathbf{Q}(\zeta)$ and $[\zeta^m : \mathbf{Q}] = [\zeta : \mathbf{Q}]$, since both have the same minimal polynomial $\Phi_n(x)$.) Finally, $\lambda(\sigma) = m$, which completes the proof.

An immediate consequence of the proposition is that the galois group of $\mathbf{Q}(\zeta)/\mathbf{Q}$ has order $\varphi(n)$.

Example. Let $n = 7$. The field $K = \mathbf{Q}(\zeta)$ of seventh roots of unity has degree $\varphi(7) = 6$ over $\mathbf{Q}$, and its galois group G is cyclic of order 6. A generator σ of G is given by $\sigma(\zeta) = \zeta^5$. Then

$$\sigma^2(\zeta) = \zeta^4, \qquad \sigma^4(\zeta) = \zeta^2, \qquad \sigma^6(\zeta) = \zeta.$$
$$\sigma^3(\zeta) = \zeta^6, \qquad \sigma^5(\zeta) = \zeta^3,$$

In particular, let ζ be the primitive seventh root of unity expressed as a complex number by

$$\zeta = \cos\frac{2\pi}{7} + i\sin\frac{2\pi}{7}.$$

Its complex conjugate is

$$\zeta^{-1} = \cos\frac{2\pi}{7} - i\sin\frac{2\pi}{7},$$

and

$$A = \zeta + \zeta^{-1} = 2\cos\frac{2\pi}{7} \in \mathbf{R}.$$

The intermediate extension $\mathbf{Q}(A)$ is associated with the subgroup $\{1, \sigma^3\}$ of order 2, since

$$\sigma^3(\zeta+\zeta^{-1}) = \sigma^3(\zeta+\zeta^6) = \zeta^6 + \zeta = \zeta + \zeta^{-1}.$$

Therefore $[\mathbf{Q}(A) : \mathbf{Q}] = [G : \{1, \sigma^3\}] = 3$. Thus $\mathbf{Q}(A)$ is the maximal real subfield of $\mathbf{Q}(\zeta)/\mathbf{Q}$.

The (irreducible) defining equation of A over $\mathbf{Q}$ can be determined as follows. Its roots are A, $\sigma(A)$, and $\sigma^2(A)$. Since

$$\sigma(A) = \sigma(\zeta+\zeta^6) = \zeta^5 + \zeta^2$$

and

$$\sigma^2(A) = \zeta^4 + (\zeta^4)^6 = \zeta^4 + \zeta^3,$$

the sum of the roots

$$A + \sigma(A) + \sigma^2(A) = -1.$$

Furthermore

$$A \cdot \sigma(A) + A \cdot \sigma^2(A) + \sigma(A) \cdot \sigma^2(A) = -2$$

and

$$A \cdot \sigma(A) \cdot \sigma^2(A) = 1.$$

Consequently,

$$x^3 + x^2 - 2x - 1 = 0$$

is the minimal polynomial of the primitive element A of the extension $\mathbf{Q}(A)/\mathbf{Q}$.

Since $\mathbf{Q}(A)/\mathbf{Q}$ is a galois extension, the zeros $\sigma(A)$ and $\sigma^2(A)$ are themselves polynomials in 1, A, and A^2 with rational coefficients. Specifically,

$$\sigma(A) = \zeta^5 + \zeta^2 = a + bA + cA^2$$
$$= a + b(\zeta+\zeta^{-1}) + c(\zeta+\zeta^{-1})^2$$
$$= a + b\zeta + b\zeta^{-1} + c\zeta^2 + 2c + c\zeta^{-2}$$
$$= (a+2c) + b\zeta + c\zeta^2 + c\zeta^5 + b\zeta^6.$$

Since $\{1, \zeta, ..., \zeta^6\}$ is a basis of $\mathbf{Q}(\zeta)/\mathbf{Q}$, we find $c = 1$, $a = -2$, and $b = 0$. Hence

$$\sigma(A) = -2 + A^2.$$

Direct computation implies that $\sigma^2(A) = 1 - A - A^2$.

Exercises

1. Find the cyclotomic polynomials $\Phi_n(x)$ over $\mathbf{Q}$ for $n = 4, 6, 8, 10$, and 12.
2. Prove that the cyclotomic polynomial $\Phi_p(x) = x^{p-1} + \cdots + 1$, p a prime, is irreducible over $\mathbf{Q}$. (*Hint:* Let $x = y+1$ and apply Eisenstein's Criterion [§5.8] to $(x^p - 1)/(x-1)$.)
3. Prove that $\Phi_{2m}(x) = \Phi_m(-x)$ for odd integers m. (*Hint:* $x^{2m} - 1 = (x^m - 1)((-x)^m - 1)$.)

4. For a given integer n, let $\zeta_1, \ldots, \zeta_n$ denote the nth roots of unity.
   a. Prove $\sum_{i=1}^{n} \zeta_i = 0$, for $n > 1$.
   b. Prove $\prod_{i=1}^{n} \zeta_i = (-1)^{n-1}$.
   c. For a given nth root, $\zeta \neq 1$, prove $\sum_{j=0}^{n-1} \zeta^j = 0$.

5. Let ζ be a primitive nth root of unity. Find the galois group of $\mathbf{Q}(\zeta)/\mathbf{Q}$, and determine the subfields of $\mathbf{Q}(\zeta)$ and their respective galois groups over $\mathbf{Q}$ for $n = 6, 8,$ and 9.

6. Let ζ be a primitive ninth root of unity. Find a primitive element for each subfield $K/\mathbf{Q}$ of $\mathbf{Q}(\zeta)$ and determine its minimal polynomial.

7. Find the subfield $K/\mathbf{Q}$ of degree 4 in the field of thirteenth roots of unity $\mathbf{Q}(\zeta)$ and describe the effect of its galois group on a suitably selected primitive element A of K.

8. Suppose that ζ is a primitive 16th root of unity.
   a. Determine the galois group of $\mathbf{Q}(\zeta)/\mathbf{Q}$ by its action on ζ.
   b. Find all subfields of $\mathbf{Q}(\zeta)$ and their associated groups.

9. Suppose that $n_1, n_2 \geq 1$; $(n_1, n_2) = 1$; and ζ_1, and ζ_2 are primitive n_1st and n_2nd roots of unity. Prove that
   a. $\mathbf{Q}(\zeta_1) \cap \mathbf{Q}(\zeta_2) = \mathbf{Q}$.
   b. $\mathbf{Q}(\zeta_1)\mathbf{Q}(\zeta_2) = \mathbf{Q}(\eta)$, where η is a primitive nth root of unity, $n = n_1 n_2$.

10. Generalize Exercise 9 to the case where $(n_1, n_2) \neq 1$. What then can be said about the fields $\mathbf{Q}(\zeta_1) \cap \mathbf{Q}(\zeta_2)$ and $\mathbf{Q}(\zeta_1)\mathbf{Q}(\zeta_2)$ as cyclotomic extensions? [Cf. Theorem 1, §8.8.]

11. Determine the following cyclotomic polynomials:
    a. $\Phi_9(x)$ in $\mathbf{Q}[x]$ and in $\mathbf{Q}(\sqrt{-3})[x]$.
    b. $\Phi_{15}(x)$ in $\mathbf{Q}[x]$ and in $\mathbf{Z}_2[x]$.

12. Let p be an odd prime and ζ a primitive pth root of unity. Prove that $\sum_{h=0}^{p-1}(x-\zeta^h)^p = p(x^p - 1)$ in $\mathbf{Q}(\zeta)[x]$.

13. Let ζ be a primitive 24th root of unity. Find $[\mathbf{Z}_5(\zeta) : \mathbf{Z}_5]$ and determine the monic polynomial $f(x) \in \mathbf{Z}_5[x]$ which has ζ for a zero.

14. Suppose that ζ is a primitive pth root of unity, p an odd prime.
    a. Prove that $-\zeta^{(p+1)/2} = -\lambda$ is a primitive $(2p)$th root of unity.
    b. Prove that $\prod_{m=1}^{2n}(\lambda^m - \lambda^{-m}) = p$.

15. Prove that A, as given in the example, is a unit in $\mathbf{Z}[A]$.

16. Consider a primitive nth root of unity ζ, $n \geq 4$. Prove that $[\mathbf{Q}(\zeta+\zeta^{-1}) : \mathbf{Q}] = \frac{1}{2}\varphi(n)$ and that $\mathbf{Q}(\zeta+\zeta^{-1}) \subseteq \mathbf{R}$. Determine explicitly the automorphism group of $\mathbf{Q}(\zeta)/\mathbf{Q}(\zeta+\zeta^{-1})$.

17. For $n > 5$ and $a = \sin(2\pi/n)$ prove that:
    a. $[\mathbf{Q}(a) : \mathbf{Q}] = \varphi(n)$ if $(n, 8) \leq 2$.
    b. $[\mathbf{Q}(a) : \mathbf{Q}] = \varphi(n)/4$ if $(n, 8) = 4$.
    c. $[\mathbf{Q}(a) : \mathbf{Q}] = \varphi(n)/2$ if $(n, 8) = 8$.

18. Find a subfield $K/\mathbf{Q}$ of $\mathbf{Q}(\zeta)$ for some nth root of unity ζ, whose galois group $G(K/\mathbf{Q})$ is:
    a. The direct product of two cyclic groups of order 2.
    b. The cyclic group of order 4.
    c. The direct product of two cyclic groups of order 4.

19. Find a primitive element and its minimal polynomial for the extensions in parts (a) and (b) of Exercise 18.
20. Prove that the derivative of $g(x) = (x^n - 1)/(x - 1) \in \mathbf{Q}[x]$ satisfies $g'(\zeta) = n/\zeta(\zeta - 1)$ for every nth root of unity $\zeta \neq 1$.
21. For an nth root of unity ζ different from 1 prove that $\sum_{h=1}^{n-1} h\zeta^h = n/(\zeta - 1)$.

§9.2 Equations Solvable by Radicals

In "olden days" students learned to solve cubic and quartic equations by means of extracting cube roots and to trisect angles by using tables of values of trigonometric functions. Thus they were familiar with the results of the Italian mathematicians Geronimo Cardano (1501–1576), Ludovico Ferrari (1522–1565), and Nicolo Tartaglia (c. 1499–1557) on solutions of equations of the form $x^3 + px = q$, $p, q \in \mathbf{Z}$, and with the reduction of a quartic to a cubic equation. For further discussion of such solutions, the reader might refer, for example, to B. L. van der Waerden, *Algebra*, Vol. I.

Toward the end of the Renaissance speculation arose as to whether there were general methods uniformly applicable to the solution of equations of degree $n \geq 5$ with rational coefficients. Would the extraction of successive mth, nth roots, etc., yield solutions of every polynomial equation? No complete answer to this question was found until the beginning of the nineteenth century. Niels Abel (at the age of 19) succeeded in completing an earlier attempt (1799) of Paolo Ruffini (1765–1822) when he proved that "in general" the quintic equation was not solvable by extracting and forming rational combinations of a finite number of nth roots (n is variable). His work is matched by that of Evariste Galois who reduced the solvability problem to one of group theory. Galois' work is related to some earlier results (c. 1770) of Lagrange on permutations of roots of polynomial functions. It is interesting to note that both Abel and Galois had their difficulties with what might be called the "scientific establishment" of their day, which scarcely designed to consider their work until some decades later.

For historical reasons and for the sake of simplicity we assume throughout the discussion below that the underlying base field F has characteristic 0.

A polynomial $f(x) \in F[x]$ is said to be **solvable by radicals** if there exists a tower of algebraic extensions

(∗) $\qquad F = E_0 \subseteq E_1 \subseteq \cdots \subseteq E_i \subseteq \cdots \subseteq E_r = E$

such that:

(i) E contains the splitting field K of $f(x)$
(ii) $E_i = E_{i-1}(A_i)$, where $A_i^{n_i} = a_i \in E_{i-1}$, $1 \leq i \leq r$.

In particular, some (or all) of the elements a_i may be 1; that is, some of the extensions E_i/E_{i-1} may be cyclotomic. Further, note that the splitting field K of $f(x)$ need not be a step in any tower (∗) with property (ii).

For example, the cyclic cubic equation
$$x^3 + x^2 - 2x - 1 = 0$$
[cf. §9.1] has $\zeta + \zeta^{-1} \in \mathbf{Q}(\zeta)$ for a root, ζ a primitive seventh root of unity. For no $n \in \mathbf{N}$ is $(\zeta + \zeta^{-1})^n$ an element of $\mathbf{Q}$; i.e., $\zeta + \zeta^{-1}$ is not a pure radical.

For the special case $r = 1$, $n_1 = n > 1$, we note the following lemma.

Lemma 1. If $K = F(A)$, where $A^n = a \in F$, and F contains all the nth roots of unity, then K/F is a cyclic extension whose degree divides n.

Proof. The minimal polynomial $g(x) \in F[x]$ of A divides $x^n - a$, and its zeros occur among the n distinct zeros $\zeta^i \cdot A$, $0 \le i < n$, of $x^n - a$, where ζ is some fixed primitive nth root of unity. Clearly the splitting field of $g(x)$ is K. Let $\sigma, \tau, \ldots$ be elements of the galois group G of K/F. Then $\sigma(A) = \zeta^{i(\sigma)} \cdot A$ for some $i(\sigma) \in \mathbf{N}$ (counted modulo n). Since $\zeta \in F$,
$$(\tau\sigma)(A) = \tau[\sigma(A)] = \tau[\zeta^{i(\sigma)} \cdot A]$$
$$= \zeta^{i(\sigma)} \cdot [\tau(A)] = \zeta^{i(\sigma)} \cdot [\zeta^{i(\tau)} \cdot A]$$
$$= \zeta^{i(\sigma) + i(\tau)} \cdot A = (\sigma\tau)(A).$$
Further $\sigma^{-1}(A) = \zeta^{-i(\sigma)} \cdot A$, because
$$A = \sigma^{-1}(\sigma(A)) = \zeta^{i(\sigma^{-1}) + i(\sigma)} \cdot A$$
implies that $i(\sigma^{-1}) \equiv -i(\sigma) \pmod{n}$. Hence G is isomorphic to a subgroup of the additive group of $\mathbf{Z}_n$. Therefore G is cyclic and $o(G)$ and $[K:F]$ divide n.

In the following theorem it will be useful to have the equivalent definition of solvability given in the following lemma.

Lemma 2. A polynomial $f(x) \in F[x]$ is solvable by radicals if and only if there exists a tower of algebraic extensions (∗) such that:

(i) E contains the splitting K of $f(x)$
(ii) $E_i = E_{i-1}(A_i)$, where $A_i^{n_i} = a_i \in E_{i-1}$, $1 \le i \le r$,
(iii) E is a galois extension of F
(iv) E_1 contains all the Nth roots of unity for $N = \prod_{i=1}^r n_i$.

Certainly these conditions imply those of the preceding definition. To prove the converse, suppose that $f(x)$ is solvable by radicals. The definition assures the existence of a tower of fields (∗) with properties (i) and (ii). To each step of the tower (∗) adjoin a primitive Nth root of unity ζ, $N = \prod_{i=1}^r n_i$. Thus
$$F = E_0 \subseteq E_0(\zeta) \subseteq \cdots \subseteq E_r(\zeta) = E(\zeta),$$
and $E_i(\zeta) = E_{i-1}(\zeta, A_i)$. This tower has properties (i) and (ii) required in the definition. For notational simplicity relabel the intermediate extensions to omit any trivial ones. Use E_i' to denote $E_i(\zeta)$ and assume that E_1' contains all Nth roots of unity (if they are not already in E_0). Thus

(∗∗) $$F = E_0 \subset E_1' \subset \cdots \subset E_s' = E'.$$

It remains only to prove that the tower can be chosen so that E'/F is a galois extension. To this end we use an inductive argument in modifying the tower (**). First, observe that E'_1 is a galois extension of F. If $\zeta \notin F$, then E'_1 is obtained by adjoining to F all roots of $x^N - 1 \in F[x]$. If $\zeta \in F$, then $E'_1 = E_0(A_1)$ and E'_1 contains *all* roots of $x^{n_1} - a_1 \in F[x]$.

For the general inductive step, suppose that the tower, $2 \le u \le s$,

$$F = E_0 \subset E'_1 \subset \cdots \subset E'_{u-1}$$

has been modified so that the field E'_{u-1} is contained in a tower

$$F = L_0 \subseteq L_1 \subseteq \cdots \subseteq L_t$$

of successive *radical* extensions such that L_1 contains all Nth roots of unity, the extensions L_i/L_{i-1} are cyclic of degree dividing N, $2 \le i \le t$, and the field L_t is a galois extension of F.

Recall that $E'_u = E'_{u-1}(A_u)$, where $A_u^{n_u} = a_u \in E'_{u-1} \subseteq L_t$. Denote by $a_{u,j}$ the (not necessarily distinct) conjugates of $a_u = a_{u,1}$, $1 \le j \le m = [L_t : F]$. All these conjugates $a_{u,j}$ belong to L_t, since L_t/F is a galois extension. The polynomial

$$g(x) = \prod_{j=1}^{m} (x^{n_u} - a_{u,j})$$

is invariant under the galois group of L_t/F, and hence $g(x) \in F[x]$. Since we consider now only fields F of characteristic 0, the extension L_t/F is separable, and hence $L_t = F(B)$ for some primitive element B. Let $h(x)$ be the minimal polynomial of B in $F[x]$, and L_{t+m} the splitting field of $g(x)h(x)$.

Since $B \in L_{t+m}$, the splitting field $L_{t+m} \supseteq L$. Since $x^{n_u} - a_u$ is a factor of $g(x)$, it splits in L_{t+m}, and thus $E_u \subseteq L_{t+m}$. Let A_j denote a zero of $x^{n_u} - a_{u,j} \in L_t[x]$, $1 \le j \le m$, where $A_u = A_{u,1}$. We have then a tower of radical extensions

$$F = L_0 \subseteq L_1 \subseteq \cdots \subseteq L_t \subseteq L_t(A_{u,1})$$
$$\subseteq L_t(A_{u,1}, A_{u,2}) \subseteq \cdots \subseteq L_t(A_{u,1}, \ldots, A_{u,m})$$
$$= L_{t+m}.$$

The degree of L_{t+j}/L_{t+j-1}, where $L_{t+j} = L_t(A_{u,1}, \ldots, A_{u,j})$, divides n_u and hence N.

Finally we conclude by induction that the tower (**) can be modified for all u so that E'_u is contained in a galois extension of F, and so that each step in the tower, after the first, is a cyclic (radical) extension [see Lemma 1]. The first step involving adjunction of Nth roots of unity is abelian, but not necessarily cyclic. In particular, E'_s is contained in a galois extension L/F. From the tower (*) we have obtained one with properties (i) through (iv).

We now turn to the crucial theorem concerning solvability.

Theorem. A polynomial $f(x) \in F[x]$ is solvable by radicals if and only if the galois group of its splitting field K/F is solvable.

Proof. Suppose that $f(x)$ is solvable. By Lemma 2 there exists a tower of fields

$$F = L_0 \subseteq L_1 \subseteq \cdots \subseteq L_u = L$$

for which L_i/L_{i-1} is cyclic, $2 \leq i \leq u$, and L_1/L_0 is abelian. The galois group $G(L/F)$ has then a composition series with cyclic factors; i.e., it is solvable. Consequently, since $K \subset L$ the group $G(K/F)$ is solvable, as it is the quotient group of a solvable group [see §§7.6 and 8.7].

Conversely, suppose that the galois group G of the splitting field K of the polynomial $f(x) \in F[x]$ is solvable. Then we have a composition series

$$G = G_0 \supset \cdots \supset G_r = \{e\}$$

with cyclic factors. Consider a primitive Nth root of unity ζ, where $N = \prod_{i=1}^r n_i$ and $n_i = [G_{i-1} : G_i]$. Let K_i be the intermediate field associated with the intermediate group G_i. Since G_{i-1}/G_i is cyclic, so is $K_i(\zeta)/K_{i-1}(\zeta)$, by Theorem 1, §8.8. That is, $K_i(\zeta)/K_{i-1}(\zeta)$ is obtained by adjoining an n_ith root of some element $a_i \in K_{i-1}$. Thus, we have a tower

$$F \subseteq F(\zeta) \subseteq K_1(\zeta) \subseteq \cdots \subseteq K_r(\zeta) = K(\zeta)$$

of radical field extensions. By definition then, $f(x)$ is solvable by radicals, as stated.

The term "solvable group," as in §7.6, is used to describe groups arising as galois groups of splitting fields of polynomial equations solvable by radicals. Polynomial equations of degree less than five are always solvable by radicals. By the remark following Theorem 2, §8.6, the galois group of a polynomial $f(x)$ of degree n is isomorphic to a subgroup (proper or not) of the symmetric group Σ_n. Since the permutation groups Σ_2, Σ_3, and Σ_4 are solvable [see §7.6], the preceding theorem implies that equations of degree at most 4 are solvable by radicals.

We should note that nonsolvability depends upon the particular ground field considered; any polynomial equation over a finite field is solvable, since any finite extension is cyclic. In special cases (cited below) the galois group of polynomials of degree n is the symmetric group Σ_n. Since the groups Σ_n, $n \geq 5$, are not solvable [§7.6], such polynomials of degree $n \geq 5$ are then not solvable by radicals.

For example let $F = \mathbf{Q}$. Oskar Perron proved that for given n the galois group of the polynomial

$$f(x) = x \prod_{i=1}^{n-1} (x - p_1 \cdots p_i k_i) + \left(\prod_{i=1}^{n-1} p_i \right) k_n$$

is the symmetric group Σ_n, where $k_i \in \mathbf{Z}$, $1 \leq i \leq n$; where the p_j, $1 \leq j < n$, are primes such that the v integers $1 \leq v \leq n-2$, $p_1 k_1, p_1 p_2 k_2, \ldots, p_1 p_2 \cdots p_v k_v$ are relatively prime to p_{v+1} and are not congruent to each other modulo p_{v+1}; and that $k_n \not\equiv 0 \pmod{\prod_{j=1}^{n-1} p_j}$ [see O. Perron, *Algebra*, Vol. II, Theorem 92].

For each degree $n \geq 5$ there exist infinitely many polynomials which are not solvable by radicals. The above statement provides an explicit construction. Quite different is an *existence* proof based on a theorem of Hilbert. If

$$f(x_1, t_1, \ldots, t_n) = x^n + t_1 x^{n-1} + \cdots + t_n$$

where $t_1, \ldots, t_n$ are algebraically independent indeterminates over $\mathbf{Q}$, then there exist infinitely many specializations $t_i \to a_i \in \mathbf{Q}$ (that is, homomorphisms of $\mathbf{Q}[t_1, \ldots, t_n]$ to $\mathbf{Q}$) such that $f(x, a_1, \ldots, a_n) \in \mathbf{Q}[x]$ is irreducible. A more arithmetic proof depending upon the theory of algebraic functions of one variable is due to Herman Ludwig Schmid (1908–1956).

Exercise

1. Consider $f(x) = x^3 - 7x + 7 \in \mathbf{Q}[x]$, and set

$$A = \sqrt[3]{\tfrac{21}{2}(-9+\sqrt{-3})} \quad \text{and} \quad B = \sqrt[3]{\tfrac{21}{2}(-9-\sqrt{-3})}.$$

Verify that

$$a_1 = \tfrac{1}{3}(\sqrt[3]{A}+\sqrt[3]{B}),$$
$$a_2 = \tfrac{1}{3}(\omega^2\sqrt[3]{A}+\omega\sqrt[3]{B}),$$

and
$$a_3 = \tfrac{1}{3}(\omega\sqrt[3]{A}+\omega^2\sqrt[3]{B}),$$

where $\omega^2 + \omega + 1 = 0$, are the zeros of $f(x)$ in $\mathbf{C}$. Prove that these three zeros are real. (This is an example of the *casus irreducibilis* of Cardano. The name is derived from the fact that the real roots cannot be expressed in terms of real radicals.)

§9.3 Constructions with Ruler and Compass

This section provides the solution to three classical problems of euclidean geometry. These solutions were obtained in the nineteenth century, more than two millenia after the problems were debated by the geometers of Plato's Academy. The central question is which coordinates in the cartesian plane can be constructed with the euclidean tools, ruler and compass. With a ruler we can draw a line between any two given points, and with a compass we can draw a circle with any given point as center and circumference passing through any second point. We can also use a compass to copy an already determined line segment onto another line.

We assume as given two perpendicular axes in a plane and a line segment of unit length. Then using a compass to lay off additional line segments

§9.3 Constructions with Ruler and Compass

of unit length and to construct perpendiculars, we can locate all points in the plane with rational components.

Geometric Lemma. If line segments of nonzero lengths a and b are given, then elementary ruler and compass constructions yield line segments of lengths $a+b$, $a-b$, ab, $1/a$, a/b, and $\sqrt{a}$.

For example, given a line segment PQ of length a, we can obtain a line segment of length $1/a$ by constructing a perpendicular at P (see Figure 9.1). On this perpendicular lay off a line segment PR of unit length, and at R construct a perpendicular to RQ to intersect at S the line PQ extended. Noting that the triangles RSP and PQR are similar, we have the proportionality $SP/RP = RP/PQ$, or length $SP = 1/a$.

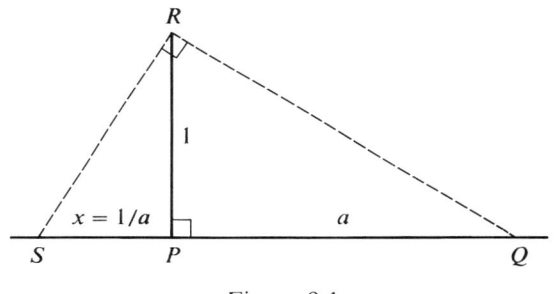

Figure 9.1

To construct a line segment of length $\sqrt{a}$ we lay off on the extension to the line segment PQ a segment UP of unit length as in Figure 9.2. Now bisect UQ at 0 and draw the semicircle centered at 0 with UQ as diameter. At P construct a perpendicular to PQ, to meet the semicircle at V. The line segments UV and VQ are perpendicular, as the angle between them is inscribed in a semicircle. Thus the triangles PUV and PVQ are similar, from which we conclude that $PU/PV = PV/PQ$ or, since the lengths $PU = 1$ and $PQ = a$, that the length $(PV)^2 = a$ and finally $PV = \sqrt{a}$.

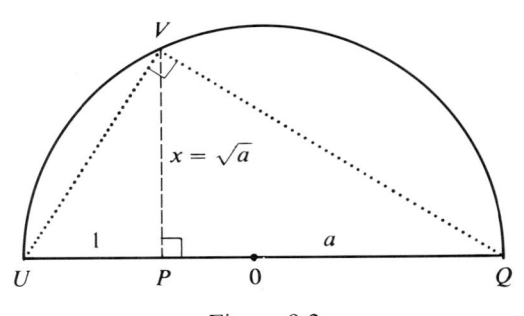

Figure 9.2

Complete verification of the geometric lemma is left as an exercise. Once a cartesian coordinate system has been established in the plane, these facts and the *geometric problems* of finding finitely often
  (a) the intersection of two lines, and
  (b) the intersection of lines and circles, or the intersection of two circles

can be translated via analytic geometry into the *algebraic problems* of finding
  (a*) the solution of a pair of linear equations, and
  (b*) the solution of a linear and a quadratic equation, or of two quadratic equations.

Conversely, the algebraic solutions of problems like (a*) and (b*) can be constructed geometrically by ruler and compass, if the coefficients of the equations are constructible.

Corollary 1. The set of real numbers (lengths of line segments) constructible by ruler and compass is a subfield of $\mathbf{R}$.

Corollary 2. If every element of a field $K \subseteq \mathbf{R}$ is constructible by ruler and compass, then so is every element in the extension $K(\sqrt{a})$ for all $a \in K$.

Corollary 3. If every element of a field $K \subseteq \mathbf{R}$ is constructible by ruler and compass, then so is every element in a galois extension L/K where $[L : K] = 2^m$.

The proof of Corollary 3 follows from the fact that $L = L_t$ contains a chain of subfields L_i, $1 \le i \le t$, such that $[L_i : L_{i-1}] = 2$ and $L_0 = K$, since the galois group $G(L/K)$ has a composition series whose factors have order 2 [see §7.6, especially Exercise 11, and §8.7]. Now $L_i = L_{i-1}(B_i)$, where $B_i^2 = b_{i-1} \in L_{i-1}$, and by Corollary 2 and an induction argument B_i can be constructed geometrically. Hence so can L_i. (In case b_{i-1} is a nonreal complex number, then B_i has constructible coordinates in the complex plane; see Exercise 2, §9.9.)

These observations can be restated as follows.

Proposition 1. If the elements of $K \subseteq \mathbf{C}$ are constructible by ruler and compass, and if L/K is a galois extension of degree 2^m, then $L = K(A)$, where A can be obtained by (a finite number of) ruler and compass constructions.

REMARK. The same result applies to nongalois extensions K/F given by a finite tower of stepwise quadratic subfields:

$$F = K_0 \subset K_1 \subset \cdots \subset K_{i-1} \subset K_i \subset \cdots \subset K_r, \qquad [K_i : K_{i-1}] = 2.$$

Conversely, suppose that $a \in L/K$ is obtained by a finite sequence of ruler and compass constructions, where L/K is a finite algebraic extension and every element of $K \subseteq \mathbf{C}$ can be constructed with ruler and compass. That is, a is obtained from the elements of K by solution of a succession of linear and quadratic equations which give rise to a tower of field extensions of degrees 1 and 2, respectively. Hence the extension $K(a)/K$ has degree 2^m, for some $m \in \mathbf{N}$. Thus, we have the following converse to Proposition 1.

Proposition 2. If the elements of $K \subseteq \mathbf{N}$ can be constructed with ruler and compass and if $a \in L/K$ is obtained from K by a succession of ruler and compass constructions, then $[K(a) : K] = 2^m$, for some $m \in \mathbf{N}$.

Behind the preceding general statements are three long-standing geometric problems, the first two of which date from the classical Greek era, Athens in the age of Pericles.

1. **Doubling of the Cube** (*Delian Problem*). Given a cube whose edges have length one, can a ruler and compass construction determine the edge x of a cube that has volume 2?
2. **Trisection of an Angle.** Given an arbitrary angle α, can a ruler and compass construction determine $\alpha/3$?
3. **Construction of Polygons.** Which regular polygons of n sides can be constructed by ruler and compass?

The answers to these problems, given via Galois theory and Proposition 2 in the nineteenth century, are determined as follows.

Problem 1. Clearly the length x of the edge must satisfy the equation $x^3 = 2$. Since $x^3 - 2 \in \mathbf{Q}[x]$ is irreducible, the real number $\sqrt[3]{2}$ has degree 3 over $\mathbf{Q}$, and so by Proposition 2 is not constructible over $\mathbf{Q}$ by ruler and compass. Consequently the cube cannot be doubled by euclidean constructions. (The oracle of Apollo at Delos is reported (c. 429 BC) to have required that a cubical altar be doubled as a condition for ending a plague. However, the plague which Phoebus Apollo sent to punish the Athenians finally abated.)

Problem 2. Given the angle $\beta = 3\alpha$, the problem is to determine the angle α. For the solution we consider the trigonometric identity $4 \cos^3 \alpha - 3 \cos \alpha = \cos 3\alpha$. It suffices to determine whether $\cos \alpha$ can be found by a ruler and compass construction. Let $\beta = 60°$; then $\cos \beta = \frac{1}{2}$, and $\cos \alpha$ is a zero of the polynomial $p(x) = 4x^3 - 3x - \frac{1}{2} \in \mathbf{Q}[x]$. Because $p(x)$ has no root in $\mathbf{Q}$, it is irreducible over $\mathbf{Q}$. Therefore, $\cos \alpha$ has degree 3 over $\mathbf{Q}$, and so by Proposition 2 is not constructible over $\mathbf{Q}$ by ruler and compass. In other words, "general" angles cannot be trisected.

Obviously this statement does not exclude trisection of certain angles, e.g., $3\alpha = \beta = 90°$. Note that it was not proved that there exist infinitely many angles which cannot be trisected. (Such a result is true, but requires the proof that $4x^3 - 3x - u \in F[u][x]$, u an indeterminate over $F = \mathbf{Q}(t_1, \ldots, t_r)$, $t_i \in \mathbf{R}$, is irreducible and possesses infinitely many consistent specializations $u \to v \in F$ such that $4x^3 - 3x - v$ is irreducible in $F[x]$.)

Problem 3. By similarity in the euclidean plane it suffices to examine regular polygons of $n = 2^\alpha \prod_{i=1}^r p_i^{\alpha_i}$ sides, inscribed in a circle of radius 1, where the

p_i are distinct odd primes. Using the formula of Abraham de Moivre (1667–1754),

$$(\cos \varphi + i \sin \varphi)^n = \cos(n\varphi) + i \sin(n\varphi),$$

we examine the primitive nth root of unity $\zeta = \cos(2\pi/n) + i \sin(2\pi/n)$, and the galois extension $\mathbf{Q}(\zeta)/\mathbf{Q}$:

$$[\mathbf{Q}(\zeta) : \mathbf{Q}] = \varphi(n) = 2^{\alpha-1} \left(\prod_{i=1}^{r} p_i^{\alpha_i - 1} \right) \prod_{i=1}^{r} (p_i - 1).$$

By Proposition 2, we must have $\varphi(n) = 2^m$, $m \geq 1$. Since $p_i \neq 2$, each α_i must equal 1, and furthermore each p_i must have the form $p_i = 1 + 2^{k_i}$, $1 \leq i \leq r$. These primes $1 + 2^{k_i}$ must further be **Fermat primes**—that is, they must have the form $1 + 2^{2^{h_i}}$—named for the pioneer of modern number theory, Pierre de Fermat [cf. Exercises 5 and 6, §2.12]. Consequently, a regular polygon of n sides can be constructed by ruler and compass if and only if $n = 2^{\alpha} \prod_{i=1}^{r} p_i$ where $p_i = 1 + 2^{2^{h_i}}$. Whether there are Fermat primes other than 3, 5, 17, 257, and 65,537 ($2^{2^m} + 1$ for $0 \leq m \leq 4$) is an open question. The integers $2^{2^m} + 1$ for $5 \leq m \leq 16$ are not Fermat primes. For example, in 1732 Euler showed that $2^{2^5} + 1 = 641 \cdot 6{,}700{,}417$.

The condition $\varphi(n) = 2^m$ for the constructibility of regular polygons of n sides implies that in the ruler and compass sense used here, it is impossible to construct, for example, the regular polygons of 7 and 9 sides.

Gauss actually constructed (c. 1796) the regular polygon of 17 sides. (For the algebraic interpretation of his solution see Exercise 2.) In this connection it should be noted that he was in possession of very considerable parts of the structure of the cyclotomic fields $\mathbf{Q}(\zeta)$ [see §9.1] and especially their connection with the quadratic law of reciprocity. These results lead ultimately to Kronecker's theorem that every quadratic field $\mathbf{Q}(\sqrt{d})$, and more generally every finite abelian extension of $\mathbf{Q}$, is a subfield of some field $\mathbf{Q}(\zeta)$ for which the complex function e^z with period $2\pi i$ is a "generating" function. The foundation of some of the most exciting research of today—the arithmetic of abelian functions, diophantine equations, rationality of certain ζ functions, and the like—were laid, we venture to say, by these early investigations.

Exercises

1. For which of the following values of n can a regular n-gon be constructed with ruler and compass?
   a. $n = 13$
   b. $n = 15$
   c. $n = 17$
   d. $n = 24$
   e. $n = 37$
   f. $n = 36$.

2. a. Find a composition series of the galois group of $K = \mathbf{Q}(\zeta)$, where ζ is a primitive seventeenth root of unity.
    b. Determine the corresponding chain of subfields of $K/\mathbf{Q}$.
3. Repeat Exercise 2, where ζ is a primitive fifteenth root of unity.
4. Given line segments of lengths a and b, find a line segment of length x by geometric construction such that:
    a. $x : a = a : 1$, or equivalently, $x = a^2$.
    b. $x : a^2 = a : 1$, or equivalently, $x = a^3$.
    c. $x : a = b : 1$, or equivalently, $x = ab$.
    Thus verify that if $t_1, \ldots, t_r$ are constructible real numbers, then every element in the field $\mathbf{Q}(t_1, \ldots, t_r)$ is geometrically constructible.
5. With the hypotheses of Proposition 2, prove that $K(a)$ is contained in a galois extension L', such that $[L' : K]$ is a power of 2 [cf. Lemma 2 and the theorem of §9.2].
6. Determine whether or not a root (in $\mathbf{C}$) of the following polynomials can be constructed by ruler and compass.
    a. $x^4 + 3x^2 + 25$ b. $x^4 + 5x^2 + 5$
    c. $x^4 - x^2 + 3$.
7. Use the result due to Perron at the end of §9.2 to exhibit a quartic polynomial in $\mathbf{Q}[x]$, none of whose roots can be constructed by ruler and compass.
8. Find the galois group of the polynomial $x^4 - 7 \in \mathbf{Q}[x]$. Prove that a primitive element of its splitting field is constructible by ruler and compass.

§9.4 Trace and Norm

This section is a prelude to the proofs of Hilbert's Theorem 90 and Noether's Equations in §9.6 and to the discussion in §9.7 of special abelian extensions of fields. That discussion relates, in turn, to the classical problem of the solution of equations by radicals, presented in §9.2.

Here we consider two mappings of a finite algebraic extension K/F into the field F. The first is the trace, a linear functional defined on K with values in F. The second is the norm, a homomorphism of the multiplicative group of K into that of F. These functions can be defined on the one hand by considering K as a (finite dimensional) vector space over F, and on the other hand by utilizing the Galois theory. For the sake of simplicity of argument in the proofs and for ease of extending the results to include inseparable extensions, we shall use both definitions interchangeably.

Let K be an extension of degree n of the field F. For fixed $a \in K$, the mapping
$$\varphi_a : K \to K \quad \text{given by } \varphi_a(c) = ac, \ c \in K$$
is a linear transformation of K considered as a vector space over F. For c_1, c_2 in K and $d \in F$,
$$\varphi_a(c_1 + c_2) = a(c_1 + c_2) = ac_1 + ac_2 = \varphi_a(c_1) + \varphi_a(c_2)$$
and
$$\varphi_a(dc) = a(dc) = d(ac) = d\varphi_a(c).$$

In fact the set of mappings $\{\varphi_a : a \in K\}$ is a subring of the ring of endomorphisms $\text{End}_F(K)$ of K/F [cf. §4.3]. We embed K in $\text{End}_F(K)$ by definining $\iota: K \to \text{End}_F(K)$ by $\iota(a) = \varphi(a)$. For all $c \in K$,

$$\varphi_{a+b}(c) = (a+b)c = ac + bc = \varphi_a(c) + \varphi_b(c)$$

and $\quad \varphi_{ab}(c) = (ab)c = a(bc) = \varphi_a(bc) = \varphi_a(\varphi_b(c)) = (\varphi_a \varphi_b)(c),$

and therefore $\varphi_{a+b} = \varphi_a + \varphi_b$, $\varphi_{ab} = \varphi_a \varphi_b$, which states that ι is a ring homomorphism. The injectivity of ι follows immediately by considering a such that $\varphi_a = 0$. Then $\varphi_a(c) = ac = 0$ for all $c \in K$, and in particular for $c = 1$. Hence $a = a \cdot 1 = 0$, and $\ker \iota = \{0\}$.

A matrix representation $M_a = [a_{ij}]$ of the linear mapping φ_a is obtained, as in §4.4, by selecting a basis $\{a_1, \ldots, a_n\}$ of K/F, and writing $aa_i = \sum_{j=1}^n a_{ji} a_j$. The **characteristic polynomial** $\chi_a(x)$ of the linear transformation φ_a (or equivalently of the matrix M_a determined by φ_a) was defined in §5.9 to be

$$\chi_a(x) = \det(x \cdot I_n - M_a)$$
$$= x^n - \left(\sum_{i=1}^n a_{ii}\right) x^{n-1} + \cdots + (-1)^n \det(M_a),$$

where I_n denotes the $n \times n$ identity matrix. While the matrix M_a corresponding to a linear transformation φ_a depends upon the choice of a basis for K/F, the characteristic polynomial $\chi_a(x)$ is independent of this choice. Thus, to each element $a \in K/F$ is associated its *unique* characteristic polynomial $\chi_a(x)$ of degree $n = [K:F]$.

We define the **trace** of $a \in K$ with respect to F, denoted $T(a)$, or more precisely $T_{K/F}(a)$, to be the coefficient of $-x^{n-1}$ in $\chi_a(x)$. Equivalently, $T(a)$ is the trace of the matrix M_a, as defined in §5.9, or the sum of the roots of $\chi_a(x)$. Similarly, the **norm** of $a \in K$ with respect to F, denoted $N(a)$, or more precisely $N_{K/F}(a)$, is the determinant of the matrix M_a. It is $(-1)^n$ times the constant term of $\chi_a(x)$, or the product of the roots of $\chi_a(x)$. Being defined in terms of the characteristic polynomial, $T(a)$ and $N(a)$ are independent of the choice of basis for K/F.

The Cayley-Hamilton Theorem of §5.9 implies the following theorem, since ι is an embedding of K in $\text{End}_F(K)$.

Theorem. An element $a \in K$ is a zero of its characteristic polynomial $\chi_a(x)$ over F.

Continuing to consider K/F, an extension of degree n with basis $\{a_1, \ldots, a_n\}$, suppose now that L/K is an extension of degree m with basis $\{b_1, \ldots, b_m\}$ over K. The mn elements $a_k b_j$ form a basis of L/F [see §8.1]. Ordering the elements of this basis as follows

$$\{a_1 b_1, \ldots, a_n b_1, a_1 b_2, \ldots, a_n b_2, \ldots, a_1 b_m, \ldots, a_n b_m\},$$

§9.4   Trace and Norm

we determine for the endomorphism φ_a of L/F (φ_a is multiplication by $a \in K$) the representing matrix

$$S(a) = \begin{bmatrix} A & \cdots & 0 \\ \vdots & \ddots & \vdots \\ 0 & \cdots & A \end{bmatrix},$$

where the matrix $A = M_a$, determined by φ_a relative to the basis $\{a_1, \ldots, a_n\}$ of K/F, appears m times on the main diagonal.

Consequently the characteristic polynomial of a (or of the matrix $S(a)$) with respect to the basis $\{a_1 b_1, \ldots, a_n b_m\}$ of L/F is the mth power of the characteristic polynomial of a (or of the matrix A) with respect to the basis $\{a_1, \ldots, a_n\}$ of K/F.

This discussion yields three propositions.

Proposition 1. Consider extensions L/K and K/F of respective degrees m and n. Then, for $a \in K$,

$$T_{L/F}(a) = m T_{K/F}(a) \quad \text{and} \quad N_{L/F}(a) = [N_{K/F}(a)]^m.$$

Proposition 2. For a tower of fields $K \supseteq F(a) \supseteq F$, with $m = [K : F(a)]$ and $[K : F] = n$, the characteristic polynomial $\chi_a(x)$ of a in K/F is the mth power of the characteristic polynomial $\mu_a(x)$ of a in the extension $F(a)/F$.

Proposition 3. The minimal polynomial $m_a(x)$ over F of $a \in K/F$ equals the characteristic polynomial of a in $F(a)/F$.

Proof. Since the characteristic polynomial $\mu_a(x)$ of a in the extension $F(a)/F$ has a for a root, $m_a(x) | \mu_a(x)$. These polynomials are monic of the same degree $[F(a) : F]$, and so they coincide.

As a consequence of this discussion, for $a \in K/F$ each root of $\chi_a(x)$ is conjugate to a because $\chi_a(x)$ is the mth power of the irreducible characteristic polynomial $\mu_a(x)$ of $a \in F(a)/F$. In fact, we have the following alternate definition of the **trace** and **norm**. For $a \in K/F$, $T(a)$ and $N(a)$ are the sum and product, respectively, of the conjugates of a, *taken with appropriate multiplicity*, in a splitting field of $\mu_a(x)$.

Proposition 4 states significant properties of the trace and norm, derived from their definition in terms of a matrix representing the element $a \in K$ [cf. Proposition 4, §5.9]. Two additional properties, significant for work in arithmetic field theory, are stated for not necessarily separable extensions in Proposition 5.

Proposition 4. For elements a, b in an extension K/F of degree n:

    (i) $T(a+b) = T(a) + T(b)$
    (ii) $N(ab) = N(a) \cdot N(b)$
    (iii) $T(ca) = cT(a)$ for $c \in F$

(iv) $\quad T(c) = nc \quad$ for $c \in F$
(v) $\quad N(ca) = c^n N(a)$ for $c \in F$
(vi) $\quad N(c) = c^n$.

Proposition 5. Let $L \supset K \supset F$ be pairwise finite extensions. Then, for $a \in L$,

(i) $\quad T_{L/F}(a) = T_{K/F}[T_{L/K}(a)]$
(ii) $\quad N_{L/F}(a) = N_{K/F}[N_{L/K}(a)]$.

One approach to the proof, which necessitates a theorem on determinants using triangular representations, involves representations of elements in L relative to bases of L/K and K/F. [See, for example, N. Jacobson, *Lectures in Abstract Algebra*, Vol. III; *Theory of Fields and Galois Theory*, pp. 66–70.] We prefer a proof based on properties and enumerations of conjugations (i.e., equivalences of L/F in a sufficiently large field Ω/F, see §8.4).

Proof. First, we recall some facts related to the fundamental lemma on prolongations. Denote by K_s the maximal separable subfield of K/F, by L_s the maximal separable subfield of L/K, and by L'_s the maximal separable subfield of L/F. Schematically,

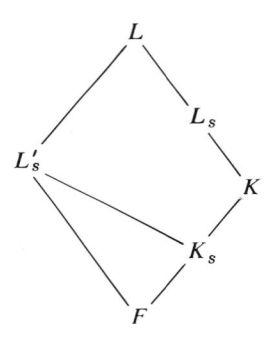

The degrees of these extensions satisfy

$$[L'_s : F] = [L_s : K][K_s : F] \quad \text{or} \quad [L : F]_s = [L : K]_s [K : F]_s$$

and $\quad [L : L'_s] = [L : L_s][K : K_s] \quad$ or $\quad [L : F]_i = [L : K]_i [K : F]_i$

as will be shown by enumerating the conjugates of L/K, L/K, and K/F in a splitting field Ω/F of some polynomial $g(x) \in F[x]$. Let $K = F(a_1, \ldots, a_s)$ and $L = K(a_{s+1}, \ldots, a_t)$, and let $g_i(x)$ be the minimal polynomial in $F[x]$ of a_i, $1 \leq i \leq t$. Then set $g(x) = \prod_{i=1}^{t} g_i(x)$ [see §§8.4 and 8.5].

There are exactly $[L : K]_s$ distinct isomorphisms σ_i of L/K in Ω/F; there are also exactly $[K : F]_s$ distinct isomorphisms τ_j of K/F in Ω/F. Furthermore, each isomorphism τ_j has exactly $[L : K]_s$ distinct prolongations to an isomorphism τ_{ij} of L/F [see the theorem in §8.5]. For each i,

$1 \le i \le [L:K]_s$, we have the prolongation $\hat{\tau}_j \sigma_i$. If $\bar{\tau}_j$ is any other prolongation of τ_j to L/F, then the restriction of $\hat{\tau}_j^{-1} \bar{\tau}_j$ to K is the identity map on K. Hence $\hat{\tau}_j^{-1} \bar{\tau}_j = \sigma_i$, for some i, and $\bar{\tau}_j = \hat{\tau}_j \sigma_i$. Thus there are precisely $[L:K]_s [K:F]_s$ distinct isomorphisms $\hat{\tau}_j \sigma_i$, $1 \le j \le [K:F]_s$ and $1 \le i \le [L:K]_s$. Note that this is the number of distinct isomorphisms of L/F (namely, $[L:F]_s$). (An isomorphism ρ of L/F induces in K/F by restriction a unique isomorphism τ_j; consequently $\rho = \hat{\tau}_j \sigma_i$ with a unique determination of σ_i after a fixed extension $\hat{\tau}_j$ of τ_j to L/K is taken.) Thus $[L:F]_s = [L:K]_s [K:F]_s$. Finally,

$$[L:F] = [L:K][K:F]$$
$$= [L:K]_s [L:K]_i [K:F]_s [K:F]_i$$
$$= [L:F]_s [L:F]_i;$$

and so $\qquad [L:F]_i = [L:K]_i [K:F]_i$.

These results will now be applied to the tower of fields

$$K \supseteq K_s \supseteq F(a) \supseteq F(a)_s \supseteq F,$$

where K_s is the maximal separable subfield of $K/F(a)$, and $F(a)_s$ is the maximal separable subfield of $F(a)/F$. Let

$$n_0 = [F(a)_s : F] = [F(a) : F]_s,$$
$$u = p^e = [F(a) : F(a)_s] = [F(a) : F]_i,$$
$$N_0 = [K_s : F(a)] = [K : F(a)]_s,$$
$$v = p^E = [K : K_s] = [K : F(a)]_i.$$

Each of the n_0 distinct isomorphisms τ_j of $F(a)/F$ extends one of the n_0 isomorphisms τ_j^* of $F(a)_s/F$. Each of the N_0 distinct isomorphisms σ_i of $K/F(a)$ extends one of the N_0 isomorphisms σ_i^* of $K_s/F(a)$. Also there are precisely $n_0 N_0$ isomorphisms of K/F given by $\hat{\tau}_j \sigma_i$, where $\hat{\tau}_j$ is chosen as above as a fixed prolongation of τ_j to K.

According to the discussion of §§8.1 and 8.4, the minimal polynomial of $a \in L$ is

$$\mu_a(x) = \prod_{j=1}^{n_0} (x - \tau_j(a))^u.$$

Next, by Propositions 2 and 3, the characteristic polynomial is

$$\chi_a(x) = [\mu_a(x)]^m,$$

where $m = [K : F(a)]$. Consequently, since $\sigma_i(a) = a$,

$$\chi_a(x) = \prod_{j=1}^{n_0} (x - \tau_j(a))^{u N_0 v}$$
$$= \prod_{i,j} (x - \hat{\tau}_j \sigma_i(a))^{uv}$$
$$= \prod_{h=1}^{n^*} (x - \rho_h(a))^{uv},$$

where the mappings ρ_h, $1 \leq h \leq n^*$, are the $n^* = [K:F]_s$ distinct isomorphisms $\hat{\tau}_j \sigma_i$ of K/F, and $uv = p^e p^E = [K:F]_i$. Therefore, by Proposition 1,

$$T_{K/F}(a) = uv \sum_{h=1}^{n^*} \rho_h(a)$$

and

$$N_{K/F}(a) = \prod_{h=1}^{n^*} \rho_h(a)^{uv}.$$

The statement of Proposition 5 follows by noting that the $[K:F]_s$ mappings ρ_h of L/F are given as $\hat{\tau}_j \sigma_k$, where the σ_k are the $[L:K]_s$ distinct isomorphisms of L/K and the $\hat{\tau}_j$ are selected prolongations of the $[K:F]_s$ distinct isomorphisms of K/F. Let $t = [L:F]_i$. For $a \in L$ we have

$$T_{L/K}(a) = [L:K]_i \cdot \sum_k \sigma_k(a);$$

hence

$$T_{K/F}(T_{L/K}(a)) = [K:F]_i \cdot [L:K]_i \cdot \sum_j \hat{\tau}_j \left(\sum_k \sigma_k(a) \right)$$

$$= t \cdot \sum_{j,k} (\hat{\tau}_j \sigma_k)(a)$$

$$= t \cdot \sum_h \rho_h(a) = T_{L/F}(a).$$

Similarly,

$$N_{L/K}(a) = \prod_k (\sigma_k(a))^{[L:K]_i};$$

consequently,

$$N_{K/F}(N_{L/K}(a)) = \left[\prod_j \hat{\tau}_j \left(\prod_k \sigma_k(a) \right)^{[L:K]_i} \right]^{[K:F]_i}$$

$$= \left[\prod_{j,k} (\hat{\tau}_j \sigma_k)(a) \right]^t$$

$$= \prod_h \rho_h(a)^t = N_{L/F}(a).$$

Exercises

1. If char $F = p > 0$, prove that a polynomial $x^p - x - a \in F[x]$ either is irreducible or factors completely in $F[x]$. (*Hint:* Observe that $(x+[1])^p = x^p + [1]^p$, $[1] \in \mathbb{Z}_p$, and examine the trace of potential irreducible factors of $x^p - x - a$.)
2. Let $x^p - a \in F[x]$, where the prime p is distinct from char F. Prove that $x^p - a$ is either irreducible in $F[x]$ or that it factors completely in $F[x]$. (*Hint:* Replace trace by norm in Exercise 1.)
3. Suppose that $F = \mathbb{Q}(i)$ where $i^2 = -1$. Find the trace of $\omega + a$ in $\mathbb{Q}$ and the norm of $\omega + a$ over $\mathbb{Q}(\omega)$, where $a \in F$ and $\omega^2 + \omega + 1 = 0$.

4. Prove Proposition 1.
5. Let $K = \mathbf{Q}(\sqrt[3]{5}, \omega)$ where $\omega^2 + \omega + 1 = 0$. Find:
   a. $T_{K/\mathbf{Q}}(\sqrt[3]{5} + \omega)$ b. $N_{K/\mathbf{Q}}(\omega)$.
6. Verify the formulas in Proposition 4.
7. Let $A \in \mathbf{Z}[\sqrt{-p}]$, p a positive prime. Prove that $N(A) = 1$ if and only if A is a unit of the ring $\mathbf{Z}[\sqrt{-p}]$.
8. Suppose that A is a zero of the irreducible polynomial $f(x) = x^3 - 2 \in \mathbf{Q}[x]$. Let $B = (1-A)^{-1}$ in $K = \mathbf{Q}(A)$.
   a. Find the minimal polynomial in $\mathbf{Q}[x]$ of B.
   b. What are the norm and trace of B?
   c. What are the norm and trace of B considered as an element of the splitting field of $f(x)$?
9. Let $K = \mathbf{Q}(\sqrt{2}, \sqrt{-5})$. Find $N_{K/\mathbf{Q}}(\sqrt{2} - \sqrt{-5})$ and $T_{K/\mathbf{Q}}(\sqrt{2})$.
10. Let ζ be a primitive pth root of unity, p a prime. Denote by T and N the trace and norm of $\mathbf{Q}(\zeta)$ over $\mathbf{Q}$. Prove:
    a. $T(\zeta) = -1$, $T(1) = p - 1$, $T(\zeta^h) = -1$, and $T(1 - \zeta^h) = p$, for $1 < h < p$.
    b. $N(1 - \zeta) = p$ and $\prod_{h=1}^{p-1}(1 - \zeta^h) = p$.
    (*Hint*: Use $x^{p-1} + \cdots + 1 = [(y+1)^p - 1]/y$ for $x = y + 1$.)
11. Let $F = \mathbf{Z}_3$. Prove that a zero A of $x^3 + [2]_3 x + [1]_3$ is a primitive 28th root of unity over F. Find:
    a. The conjugates of A as polynomials in A with coefficients in F.
    b. An element $B \in F(A)$ whose norm with respect to F is $[1]_3$.
    c. An element $C \in F(A)$ such that $B = C/\varphi(C)$, where φ is the Frobenius automorphism of $F(A)/F$ [see §8.2].
12. Let $K = \mathbf{Q}(\zeta)$, where ζ is a primitive fifth root of unity, and denote by σ a generating automorphism of $G(K/\mathbf{Q})$.
    a. Find an element $A \in K$ whose norm with respect to $\mathbf{Q}$ is 1, and determine all $B \in K$ such that $B/\sigma(B) = A$.
    b. Find a nonzero element $C \in K$ whose trace with respect to $\mathbf{Q}$ is zero, and determine all $D \in K$ such that $D - \sigma(D) = C$.

§9.5 Theorem of the Normal Basis

This section is devoted to a single theorem significant in the solution of arithmetic problems in algebraic number theory. Much of the argument involves congruences and is reminiscent of the proof of the Chinese Remainder Theorem in §2.11. The Theorem of the Normal Basis asserts that galois extensions have bases whose elements are conjugates of a single element. For the sake of simplicity and in order to avoid, for example, the representation theory of linear algebras, we shall assume that our ground field has infinitely many elements.

Theorem. Let $K = F(a)$ be a galois extension of degree n over an infinite field F with the galois group $G(K/F) = \{\sigma_1, \ldots, \sigma_n\}$. There exists an element b in $F(a)$ such that the n conjugates $\sigma_1(b), \ldots, \sigma_n(b)$ of b form a vector space basis of K over F.

Such a basis is called a **normal basis** because it consists of conjugates of a single element.

The first step in the proof is to express the unit element as the sum of orthogonal idempotents in a residue class ring of $K[x]$ [cf. §§3.4 and 5.3]. First, extend the automorphisms $\sigma_i \in G$ from K to $K[x]$ by defining $\sigma_i(x) = x$. Then let $a_i = \sigma_i(a)$, where σ_1 is the identity map, and denote by $f(x)$ the minimal (over F) polynomial of a. Then in K, $f(x) = \prod_{i=1}^{n}(x-a_i)$. By a modification of the Lagrange Interpolation Formula we obtain, as in §5.6, the equivalent expression for any polynomial $h(x)$ of degree $m < n$ in $K[x]$:

$$h(x) = \sum_{i=1}^{n} \frac{h(a_i)}{f'(a_i)} \frac{f(x)}{x-a_i}.$$

Note that for $1 \le i \le n$,

$$\sigma_i(f(x)) = f(x) \quad \text{and} \quad \sigma_i(f'(a)) = f'(a_i).$$

In particular, for $h(x) = 1$,

$$1 = \sum_{i=1}^{n} \frac{f(x)}{f'(a_i)(x-a_i)} = \sum_{i=1}^{n} \sigma_i \frac{f(x)}{f'(a)(x-a)}.$$

Setting $g(x) = f(x)/[f'(a)(x-a)]$ and $g_i(x) = \sigma_i(g(x)) \in K[x]$, we obtain

$$1 = \sum_{i=1}^{n} g_i(x).$$

Further since $(x-a_j) | g_i(x)$ for $i \ne j$,

$$g_i(x) = g_i = \sigma_i(g(x)) \equiv \xi_{ij} \pmod{x-a_j}, \quad \text{for } 1 \le j \le n.$$

As in §2.11, the polynomials $g_i(x)$ are orthogonal idempotents modulo $f(x)$; that is,

$$g_i(x) g_j(x) \equiv \delta_{ij} g_i(x) \pmod{f(x)}.$$

Since $\sigma_i(g_j(x)) = g_k(x)$ for some k and for fixed j, the trace function satisfies

$$T(g_j(x)) = \sum_{i=1}^{n} \sigma_i(g_j(x)) = \sum_{k=1}^{n} g_k(x) = 1.$$

Let $D(g_1,\ldots,g_n)$ denote the matrix $[\sigma_i(g_j)]$ with components in $K[x]$ and ${}^tD(g_1,\ldots,g_n)$ its transpose. Then

$$\begin{aligned}{}^tD(g_1,\ldots,g_n) D(g_1,\ldots,g_n) &= [\sigma_j(g_i)][\sigma_i(g_j)] \\ &= [\alpha_{ij}], \quad \text{where } \alpha_{ij} = \sum_k \sigma_k(g_i)\sigma_k(g_j) \\ &= [T(g_i g_j)].\end{aligned}$$

Consequently,

$$\Delta(x) = (\det D(g_1,\ldots,g_n))^2 \equiv 1 \pmod{f(x)}$$

§9.5 Theorem of the Normal Basis

because $g_i(x)g_j(x) \equiv \delta_{ij} g_i(x) \pmod{f(x)}$ and

$$\Delta(x) = \det[T(g_ig_j)] \equiv \det[\delta_{ij} T(g_i)]$$
$$\equiv \det[\delta_{ij}] \equiv 1 \pmod{f(x)}.$$

Therefore the discriminant $\Delta(x)$ is a nonzero polynomial in $F[x]$. Assuming now that F is an *infinite* field, we can find an element $c \in F$ for which $\Delta(c) \neq 0$ by Theorem 2, §5.2.

Finally the elements

$$b = g(c) = \sigma_1(b) \quad \text{and} \quad \sigma_i(b) = g_i(c), \quad 2 \leq i \leq n,$$

are linearly independent, because if there were a dependence relation

$$x_1 \sigma_1(b) + \cdots + x_n \sigma_n(b) = 0$$

with coefficients $x_1, \ldots, x_n$ in F, then

$$x_1 \sigma_i(\sigma_1(b)) + \cdots + x_n \sigma_i(\sigma_n(b)) = 0$$

for $1 \leq i \leq n$. The determinant of this homogeneous system of n linear equations in $x_1, \ldots, x_n$ equals $\Delta(c)$, different from zero. Consequently the system of linear equations has only the trivial solution $x_1 = \cdots = x_n = 0$ [see §4.5, following Property 9]. Thus the n conjugates $\sigma_i(b)$, $1 \leq i \leq n$, of b are linearly independent and so constitute a normal basis of K/F.

Exercises

1. Determine a normal basis for each of the following extensions of $\mathbf{Q}$.
   a. $\mathbf{Q}(\sqrt{7})$
   b. $\mathbf{Q}(\sqrt{i})$
   c. $\mathbf{Q}(\omega)$ where $\omega^2 + \omega + 1 = 0$
   d. $\mathbf{Q}(\sqrt{3}, \sqrt{7})$
   e. $\mathbf{Q}(\sqrt{3}, \sqrt{7}, \sqrt{i})$.

2. Suppose that K_1 and K_2 are distinct quadratic extensions of the field F, char $F \neq 2$. Assume that the elements A_1, A_2 of K_1, K_2, respectively, generate normal bases over F.
   a. Do the conjugates of $A_1 A_2$ determine a normal basis of $K_1 K_2$ over F?
   b. Generalize this fact to the case of n distinct quadratic extensions.

3. Let $K = \mathbf{Q}(A)$ be the cyclic cubic extension where $A = \zeta + \zeta^6$ with a primitive seventh root of unity ζ. Then $A^3 + A^2 - 2A - 1 = 0$.
   a. Show that the mapping σ defined by $\sigma(A) = -2 + A^2$ belongs to the galois group $G(K/\mathbf{Q})$.
   b. Show that $\sigma^2(A) = \zeta^4 + \zeta^3$.
   c. Find $N(A)$.
   d. Find all elements $B \in K$ for which $B/\sigma(B) = A$.

4. Verify the following steps used in proving the theorem:
   a. $\det[T(g_i g_j)] \equiv \det[T(g_i) T(g_j)] \pmod{f(x)}$
   b. $T(g_i) T(g_i) \equiv \delta_{ij} T(g_i) \pmod{f(x)}$
   c. $\sigma_i(f'(a)) = f'(\sigma_i(a))$
   d. $g_i(x) g_j(x) \equiv \delta_{ij} g_i(x) \pmod{f(x)}$.

§9.6 Hilbert's Theorem 90 and Noether's Equations

The celebrated Theorem 90 of David Hilbert (1862–1943) originally was proved about 1897 and subsequently was generalized by Amalie Emmy Noether (1882–1935). It presents some aspects of the cohomology theory prevalent in algebraic topology, in the modern version of class field theory, and in portions of the structure theory of groups. The discussion for finite fields leads ultimately to local class field theory, i.e., the study of abelian extensions of fields obtained by the completion of number fields and function fields with respect to special types of absolute values [for example, see Exercise 16, §2.7].

Suppose that $K = F(a)$ is a *separable* extension with the minimal polynomial $m_a(x) = f(x) \in F[x]$. In a splitting field $L \supseteq K$ of $f(x)$,

$$f(x) = \prod_{i=1}^{n} (x - a_i)$$

where $a = a_1, a_2, \ldots, a_n$ are the distinct conjugates of a. As in §8.5, we also denote these conjugates by $\sigma_i(a) = \dot{a}_i$; let σ_1 denote the identity automorphism of K/F.

The product $\sum_{i=2}^{n} (a - a_i) = f'(a)$ is called the **different** of the element a; and $\Delta(a) = (-1)^s N(f'(a))$ with $s = n(n-1)/2$, where N is the norm function of K/F, is called the **discriminant** of a. Since a is a primitive element of K with respect to F, necessarily $f'(a) \neq 0$ and hence $\Delta(a) \neq 0$.

NOTE. If $b \in K$ is not a primitive element of K/F, then $F(b) \subsetneq K$. Since $b \in K$ has only $[F(b) : F]$ distinct conjugates [see §8.5], the images $\sigma_i(b)$, $1 \le i \le [K : F]$, cannot be distinct. In other words, $c \in K$ has $[K : F]$ distinct conjugates in any galois extension L/F if and only if c is a primitive element of K/F.

Lemma 1. The discriminant $\Delta(a) = \prod_{1 \le i < j \le n} (a_j - a_i)^2$.

For any isomorphism σ_j, $1 \le j \le n$,

$$\sigma_j[f'(a)] = \prod_{i=2}^{n} [\sigma_j(a) - \sigma_j(a_i)].$$

Since $f'(a) \neq 0$, and σ_j is an isomorphism, then $\sigma_j(a) \neq \sigma_j(a_i)$, $2 \le i \le n$. Hence

$$N(f'(a)) = \prod_{j=1}^{n} \sigma_j[f'(a)]$$
$$= \prod_{j=1}^{n} \prod_{i=2}^{n} (\sigma_j(a) - \sigma_j(a_i))$$
$$= \prod_{i=2}^{n} \prod_{j=1}^{n} (a_j - \sigma_j(a_i))$$
$$= \prod_{i \neq j} (a_j - a_i)$$
$$= (-1)^s \prod_{i < j} (a_j - a_i)^2.$$

Then $\Delta(a) = (-1)^s N(f'(a)) = \prod_{i < j} (a_j - a_i)^2$.

Lemma 2. The discriminant $\Delta(a)$ is equal to the square of the determinant

$$D = \det \begin{bmatrix} 1 & a_1 & a_1^2 & \cdots & a_1^{n-1} \\ 1 & a_2 & a_2^2 & \cdots & a_2^{n-1} \\ \vdots & \vdots & \vdots & \cdots & \vdots \\ 1 & a_n & a_n^2 & \cdots & a_n^{n-1} \end{bmatrix} = \det[a_i^{j-1}].$$

Note that in this determinant the superscripts denote powers of the conjugates a_i of a. The assertion that $\det[a_i^{j-1}] = \prod_{i<j}(a_j - a_i)$ is a standard result in the theory of determinants due to Alexandre Theophile Vandermonde (1735–1796) for whom determinants of this type are named. We present the proof for completeness of discussion.

In D subtract a_1 times the $(n-1)$st column from the nth column so as to obtain

$$D = \det \begin{bmatrix} 1 & a_1 & \cdots & a_1^{n-2} & 0 \\ 1 & a_2 & \cdots & a_2^{n-2} & (a_2-a_1)a_2^{n-2} \\ \vdots & \vdots & \cdots & \vdots & \vdots \\ 1 & a_n & \cdots & a_n^{n-2} & (a_n-a_1)a_n^{n-2} \end{bmatrix}.$$

Next subtract a_1 times the $(n-2)$nd column from the $(n-1)$st column; consequently

$$D = \det \begin{bmatrix} 1 & a_1 & \cdots & 0 & 0 \\ 1 & a_2 & \cdots & (a_2-a_1)a_2^{n-3} & (a_2-a_1)a_n^{n-2} \\ \vdots & \vdots & \cdots & \vdots & \vdots \\ 1 & a_n & \cdots & (a_n-a_1)a_n^{n-3} & (a_n-a_1)a_n^{n-2} \end{bmatrix}.$$

Repeating these operations, we obtain

$$D = \det \begin{bmatrix} 1 & 0 & 0 & \cdots & 0 \\ 1 & a_2-a_1 & (a_2-a_1)a_2 & \cdots & (a_2-a_1)a_2^{n-2} \\ \vdots & \vdots & \vdots & \cdots & \vdots \\ 1 & a_n-a_1 & (a_n-a_1)a_n & \cdots & (a_n-a_1)a_n^{n-2} \end{bmatrix}$$

$$= \prod_{j=2}^{n}(a_j-a_1) \cdot \det \begin{bmatrix} 1 & a_2 & \cdots & a_2^{n-2} \\ \vdots & \vdots & \cdots & \vdots \\ 1 & a_n & \cdots & a_n^{n-2} \end{bmatrix}.$$

Consequently by recursion,

$$D = \prod_{i=1}^{n-1} \prod_{j=i+1}^{n} (a_j-a_i) = \prod_{1 \le i < j \le n} (a_j-a_i)$$

and

$$D^2 = \prod_{1 \le i < j \le n} (a_j-a_i)^2 = \Delta(a).$$

After these preparations it is quite easy to prove the following lemma.

Lemma 3. If $K = F(a)$ is a separable extension of F, there exists an element $d \in K$ whose trace $T(d)$ is different from 0.

Proof. Let $\{b_j : 1 \leq j \leq n\}$ be a vector space basis of K/F. Since $\{1, a, a^2, \ldots, a^{n-1}\}$ is also a basis of K/F, we may write $b_j = \sum_{k=1}^{n} c_{kj} a^{k-1}$, where the matrix of coefficients $C = [c_{kj}]$ is nonsingular and the elements c_{kj} belong to F.

Define elements b_{ij} in any galois extension L/F containing K by $b_{ij} = \sigma_i(b_j)$ for $1 \leq i \leq n$, where the σ_i are the isomorphisms on K/F. Then

$$b_{ij} = \sigma_i(b_j) = \sum_{k=1}^{n} c_{kj}(\sigma_i(a))^{k-1} = \sum_{k=1}^{n} c_{kj} a_i^{k-1}$$

and $\quad [b_{ij}] = [a_i^{j-1}] \cdot C.$

Now set

$$\delta(b_1, \ldots, b_n) = (\det[b_{ij}])^2 = D^2 \cdot (\det C)^2,$$

where $D = \prod_{1 \leq i < j \leq n}(a_j - a_i)$ as above. From matrix theory [§4.5] we recall that $\det C = \det {}'C$, where ${}'C$ is the transpose of C, and that $(\det C)^2 = \det {}'C \cdot \det C$. Thus,

$$\delta(b_1, \ldots, b_n) = \det {}'[b_{ij}] \cdot \det[b_{ij}]$$
$$= \det[T(b_i b_j)],$$

where the trace $T(b_i b_j)$ equals $\sum_{k=1}^{n} b_{ki} b_{kj}$ [cf. §9.4].

Since $D^2 \neq 0$ for the primitive element a (the conjugates of a are distinct) and since C is nonsingular, $\det[b_{ij}] \neq 0$ for each basis $b_1, \ldots, b_n$ of K over F. Consequently for each such basis there exists a pair of indices i, j such that the trace of $b_i b_j = d$ is different from zero. The existence of such an element d yields the following theorem.

Theorem (*Noether's Equations*). Let K be a galois extension of F of degree n whose galois group is $G = \{\varepsilon, \rho, \sigma, \tau, \ldots\}$. There are n nonzero elements $x_\sigma, x_\tau, \ldots$ in K satisfying the n^2 equations

$$x_\sigma \cdot \sigma(x_\tau) = x_{\sigma\tau}$$

if and only if there exists a nonzero element $y \in K$ for which

$$y/\sigma(y) = y^{1-\sigma} = x_\sigma, \qquad \sigma \in G.$$

Before proving the theorem, we note that the symbolic power notation used here and subsequently is interpreted as $a^\sigma = \sigma(a)$ for $a \in K, \sigma \in G$. Introduced in Hilbert's *Zahlbericht*, it was used most significantly in modified form in Philip Furtwängler's proof of the Principal Ideal Theorem in Class

Field Theory (1930). In particular,
$$y^{\sigma\tau} = \sigma\tau(y) = \sigma(\tau(y)) = (y^\tau)^\sigma,$$
$$(y^{1-\tau})^\sigma = \sigma[y/\tau(y)] = \sigma(y)/\sigma\tau(y) = y^{\sigma-\sigma\tau}.$$

If one of the elements x_σ satisfying $x_\sigma \cdot \sigma(x_\tau) = x_{\sigma\tau}$ is 0, then all the x_τ, $\tau \in G$, are 0. Therefore, assume in the statement of the theorem that the elements x_σ are different from 0. Also $x_\varepsilon \cdot \varepsilon(x_\tau) = x_\tau = x_\varepsilon \cdot x_\tau$ implies $x_\varepsilon = 1 \in K$ for the identity automorphism ε in G.

Proof. For $\sigma \in G$, $y^{1-\sigma} = x_\sigma$ implies that
$$y^{1-\sigma} \cdot (y^{1-\tau})^\sigma = y^{1-\sigma} \cdot y^\sigma \cdot y^{-\sigma\tau} = y^{1-\sigma\tau},$$
i.e., $x_\sigma \cdot \sigma(x_\tau) = x_{\sigma\tau}$.

For the converse, pick a primitive element $a \in K/F$ and set
$$z_i = \sum_{\sigma \in G} (\sigma(a))^i \cdot x_\sigma, \qquad 0 \le i < n.$$

At least one of the sums z_i is different from zero (i.e., this is a nonhomogeneous system of n equations), for if it were a homogeneous system, it would have only the trivial solution $x_\sigma = 0$ for all $\sigma \in G$, because $\det[\sigma_j(a^i)] = \det[a_j^i] \ne 0$ for a primitive element a according to Lemma 2.

Finally we compute as follows, letting $z = z_i$ for some i for which $z_i \ne 0$, and using Noether's equations in the simplification:
$$z/z^\sigma = \left[\sum_{\rho \in G} (\rho(a))^i \cdot x_\rho\right] \Big/ \sum_{\rho \in G} [(\sigma\rho)(a)]^i \cdot \sigma(x_\rho)$$
$$= \left[\sum_{\rho \in G} (\rho(a))^i \cdot x_\rho\right] \Big/ \left[\sum_{\rho \in G} ((\sigma\rho)(a))^i x_{\sigma\rho} x_\sigma^{-1}\right]$$
$$= x_\sigma \left[\sum_{\rho \in G} (\rho(a))^i \cdot x_\rho\right] \Big/ \left[\sum_{\rho \in G} ((\sigma\rho)(a)^i) \cdot x_{\sigma\rho}\right]$$
$$= x_\sigma,$$
i.e., $x_\sigma = z^{1-\sigma}$, as asserted.

Corollary (*Hilbert's Theorem 90*). Let K/F be a cyclic extension of degree n whose galois group is generated by σ. Then $N(x) = 1$ for $x \in K$ if and only if $x = y^{1-\sigma}$ for some $y \in K$.

Proof. Set $x_\sigma = x$ and define
$$x_{\sigma^i} = x \cdot \sigma(x) \cdots \sigma^{i-1}(x).$$

Then $N(x) = x \cdot \sigma(x) \cdots \sigma^{n-1}(x)$ implies that $x_{\sigma^i} \cdot \sigma^i(x_{\sigma^j}) = x_{\sigma^{i+j}}$ for $1 \le i, j \le n$ [see §9.4]. Note that $x_{\sigma^0} = x_{\sigma^n} = 1$ must hold since $N(x) = 1$. We have $x_\varepsilon \cdot \varepsilon(x_\sigma) = x_\sigma$, whence $x_\varepsilon = 1$, and $x_\varepsilon = x_{\sigma^n} = x \cdot \sigma(x) \cdots \sigma^{n-1}(x) = N(x)$. Therefore, by the preceding theorem $x = x_\sigma = y^{1-\sigma}$ for some $y \in K$.

For the converse note that $x = y^{1-\sigma}$ implies
$$N(x) = N(y)N(\sigma(y))^{-1} = N(y)N(y)^{-1} = 1.$$
Noether's equations have the following additive analogue.

Proposition 1. If n elements x_σ of a normal extension K/F with galois group $G = \{\sigma, \tau, \ldots\}$ satisfy the n^2 equations

(*) $$x_\sigma + \sigma(x_\tau) = x_{\sigma\tau},$$

then $x_\sigma = y - \sigma(y)$ with $y \in K$, and conversely.

Proof. The converse assertion is again easy. We have
$$(y - \sigma(y)) + \sigma[(y - \tau(y))] = y - \sigma(y) + \sigma(y) - (\sigma\tau)(y)$$
$$= y - (\sigma\tau)(y).$$

Since K/F is a separable extension, Lemma 3 assures the existence of an element $d \in K$ with nonvanishing trace. For
$$y = \left(\frac{1}{T(d)}\right) \sum_{\tau \in G} x_\tau \cdot \tau(d)$$
we have $\sigma(y) = [1/T(d)] \sum_{\tau \in G} \sigma(x_\tau) \cdot (\sigma\tau)(d)$. Then
$$x_\sigma = x_\sigma \frac{T(d)}{T(d)} = \frac{x_\sigma}{T(d)} \sum_{\tau \in G} (\sigma\tau)(d),$$
and consequently
$$\sigma(y) + x_\sigma = \frac{1}{T(d)} \sum_{\tau \in G} [\sigma(x_\tau) + x_\sigma] \cdot (\sigma\tau)(d).$$
Therefore, using the assumption in equation (*), we have
$$\sigma(y) + x_\sigma = \frac{1}{T(d)} \sum_{\tau \in G} (x_{\sigma\tau} - x_\sigma + x_\sigma) \cdot (\sigma\tau)(d)$$
$$= \frac{1}{T(d)} \sum_{\tau \in G} [x_{\sigma\tau} \cdot (\sigma\tau)(d)] = y$$
and
$$x_\sigma = y - \sigma(y).$$

Proposition 2 is the additive analogue of Hilbert's theorem. Its proof can be given by changing multiplication to addition, division to subtraction, and norm to trace in the proof of the corollary.

Proposition 2. If K/F is a cyclic extension of degree n whose galois group is generated by σ, then $T(x) = 0$ for $x \in K$ if and only if $x = y - \sigma(y)$ for some $y \in K$.

We conclude this section with discussion of two consequences of Hilbert's Theorem 90 significant in the study of "local class field theory."

Let K be a cyclic extension of the finite field F of $q = p^m$ elements, and consider the Frobenius automorphism φ of §8.2, which leaves fixed the elements of F. Then the mappings

$$x \to x^{1-\varphi} = x/\varphi(x) \quad \text{and} \quad x \to N(x)$$

are *homomorphisms* of the multiplicative group K^* of the field K into itself. Following standard notational usage, we set

$$N(K^*) = \{N(a) : a \in K^*\},$$
$$(K^*)^{1-\varphi} = \{a^{1-\varphi} : a \in K^*\};$$

and correspondingly for the additive group K,

$$T(K) = \{T(a) : a \in K\},$$
$$(1-\varphi)K = \{a - \varphi(a) : a \in K\}.$$

Then

$$[K^* : 1] = [(K^*)^{1-\varphi} : 1] \cdot [F^* : 1]$$

by the Galois theory, and

$$[K^* : 1] = [N(K^*) : 1] \cdot [(K^*)^{1-\varphi} : 1]$$

by Hilbert's Theorem 90. Consequently $[F^* : 1] = [N(K^*) : 1]$ and $F^* = N(K^*)$, since $F^* \supseteq N(K^*)$.

The additive version of Hilbert's Theorem 90 yields a corresponding result for the additive group of F. By the Galois theory,

$$[K : 0] = [(1-\varphi)K : 0] \cdot [F : 0];$$

by Proposition 2,

$$[K : 0] = [T(K) : 0] \cdot [(1-\varphi)K : 0].$$

Hence $[T(K) : 0] = [F : 0]$, and consequently, $T(K) = F$ since $F \supseteq T(K)$.

Exercises

1. Suppose that the field F has $q = p^n$ elements, p an odd prime. Prove that the equation $ax^2 + by^2 = c$, $abc \neq 0$ in F, has $q+1$ solutions if $-ab$ is not a square in F.
2. Let $K = \mathbf{Q}(\sqrt{5})$ and consider $A = 9 - 4\sqrt{5}$. Find:
   a. $N_{K/\mathbf{Q}}(A)$.
   b. An element $B \in K$ such that $A = B/\sigma(B)$ where σ is a generating automorphism of the galois group of $K/\mathbf{Q}$.

3. Let $K = \mathbf{Q}(\sqrt{3})$. Prove that $N_{K/\mathbf{Q}}(2-\sqrt{3}) = 1$ and find all elements $B \in K$ such that $B/\sigma(B) = 2 - \sqrt{3}$ where σ generates the galois group of $K/\mathbf{Q}$.
4. Let $K = \mathbf{Q}(\sqrt{-3})$. Find all elements $B \in K$ such that $5\sqrt{-3} = B - \sigma(B)$, where σ generates the galois group of $K/\mathbf{Q}$.
5. Prove Proposition 2.

§9.7 Kummer or Radical Extensions

These special abelian extensions were first systematically studied by Ernst Eduard Kummer in his investigations of Fermat's Last Theorem. Kummer extensions are significant for investigations in class field theory dealing with abelian field extensions and arithmetic properties of fields, and in algebraic geometry for problems addressing covering varieties and the resolution of singularities.

Kummer's theory is expressed multiplicatively, involving finite abelian extensions whose galois groups have exponents relatively prime to the characteristic of the base field. We present also the additive version of Kummer theory (c. 1927), due to Emil Artin (1898–1962) and Otto Schreier (1901–1929), which treats finite abelian extensions whose galois groups have exponents equal to the characteristic of the base field.

Throughout this section we shall consider abelian extensions of a field F, where F contains *all* of the nth roots of unity for a *given* $n \in \mathbf{N}$. (That is, F contains a splitting field of $x^n - 1$.) If char $F = 0$ we place no restriction on n, but if char $F = p > 0$ we require that $(n, p) = 1$.

Denote by F^* the multiplicative group of F and by ζ a primitive nth root of unity. Since the derivative of $x^n - 1$ is $nx^{n-1} \neq 0$ (because either char $F = 0$ or $(n, p) = 1$; see §5.5), the nth roots of unity are distinct. Therefore, the group $\langle \zeta \rangle = \{\zeta^i : 0 \leq i < n\}$ has order n.

Lemma 1. *If K/F is a cyclic extension of degree n with the properties described above, then there exist elements $A \in K$ and $a \in F$ such that $K = F(A)$ and A is a zero of the polynomial $x^n - a \in F[x]$.*

Proof. By Proposition 4(vi) of §9.4 for $\zeta^{-1} \in F$, the norm homomorphism $N = N_{K/F}$ satisfies
$$N(\zeta^{-1}) = \zeta^{-n} = 1.$$
Consequently, for a generator σ of the galois group $G(K/F)$ and for some $A \in K$,
$$\zeta^{-1} = A/\sigma(A)$$
by Hilbert's Theorem 90 [corollary, §9.6]. Thus $\sigma(A) = \zeta \cdot A$; hence
$$\sigma^i(A) = \zeta^i \cdot A.$$
Furthermore the equations
$$\sigma(A^n) = (\sigma(A))^n = (\zeta \cdot A)^n = A^n$$

imply that $A^n = a \in F$. The polynomial $x^n - a$ has the n distinct zeros $\sigma^i(A) = \zeta^i \cdot A$ in K. It is irreducible in $F[x]$, because the minimal (irreducible) polynomial of A divides $x^n - a$ and is satisfied by the n conjugates $\sigma^i(A) = \zeta^i \cdot A$ of A. Hence the minimal polynomial has degree n and so must equal $x^n - a$. Thus finally, $K = F(A)$ because

$$n = [K:F] = [K:F(A)][F(A):F] = [K:F(A)] \cdot n.$$

The additive version of Lemma 1 deals with cyclic extensions K/F of degree p, the characteristic of F.

Lemma 1'. *If K/F is cyclic of degree p, then $K = F(A)$ where A is a zero of an irreducible polynomial $x^p - x - a$, $a \in F$.*

Proof. By Proposition 4(iv) of §9.4, for $[-1] \in \mathbf{Z}_p \subseteq F$, the trace $T = T_{K/F}$ satisfies

$$T([-1]) = p \cdot [-1] = 0.$$

(Note that we identify the prime field of F with $\mathbf{Z}_p$; cf. §3.6.) The additive version of Hilbert's Theorem 90 [Proposition 2, §9.6] then yields the existence of $A \in K/F$ such that

$$\sigma(A) = A + [1],$$

where σ is a generator of the galois group $G(K/F)$. Furthermore $\sigma^i(A) = A + [i]$, where $[i] \in \mathbf{Z}_p \subseteq F$. Next

$$\sigma(A^p - A) = (\sigma(A))^p - \sigma(A)$$
$$= (A + [1])^p - (A + [1]) = A^p - A.$$

Thus, $A^p - A$ is invariant under the action of the galois group, and so $a = A^p - A \in F$. Consequently A satisfies the polynomial $x^p - x - a \in F[x]$. The equations

$$p = [K:F] = [K:F(A)][F(A):F] \quad \text{and} \quad [F(A):F] > 1$$

imply that $K = F(A)$ and that the *minimal* polynomial $m_A(x)$ of A over F has degree p. Therefore

$$m_A(x) = x^p - x - a,$$

which must be irreducible.

A finite abelian extension K/F is said to have **exponent** n, if $\sigma^n = \varepsilon$ for all σ in the galois group $G(K/F)$. The exponent is not unique. In fact, if K/F has exponent n then it has exponent nk, for all $k \in \mathbf{N}$. The exponent of K/F is a multiple of the minimal exponent [defined in §§7.1 and 7.2] of the galois group $G(K/F)$.

A finite abelian extension K/F with exponent n, which contains the nth roots of unity, is called a **Kummer** or **radical extension**. Again we require

$(n, p) = 1$ if char $F = p > 0$. These extensions, as we shall show below, are obtained by adjoining to F the nth roots of elements of F; hence the alternate name, radical extensions.

Lemma 2. Let K/F be a Kummer extension, and denote by $\{A\}$ the multiplicative group

$$\{A\} = \{A \in K : A^n \in F^*\}.$$

The factor group $\{A\}/F^*$ is isomorphic to the dual G^* of the galois group $G = G(K/F)$.

Proof. Since $A^n \in F^*$, for $\sigma \in G$,

$$(A/\sigma(A))^n = A^n/\sigma(A^n) = 1,$$

and therefore $A/\sigma(A)$ is an nth root of unity. Let

$$f_{A,\sigma} = A/\sigma(A) \in \langle \zeta \rangle \subseteq F^*,$$

and note that for $\sigma, \tau \in G$

$$\tau(A)/(\tau\sigma)(A) = \tau(A/\sigma(A))$$
$$= \tau(f_{A,\sigma}) = f_{A,\sigma}$$

and
$$\tau(A)/(\tau\sigma)(A) = [A/f_{A,\tau}][f_{A,\tau\sigma}/A].$$

Hence
$$f_{A,\sigma} f_{A,\tau} = f_{A,\tau\sigma} = f_{A,\sigma\tau} \in \langle \zeta \rangle.$$

(Because G is abelian $\tau\sigma = \sigma\tau$.) Thus the mapping

$$\varphi_A: G \to \langle \zeta \rangle \quad \text{given by} \quad \varphi_A(\sigma) = f_{A,\sigma}$$

is a homomorphism of G. Next, the mapping

$$\lambda: \{A\} \to G^* \quad \text{given by} \quad \lambda(A) = \varphi_A.$$

is a (group) homomorphism since for $A, B \in \{A\}$

$$(AB)/\sigma(AB) = (A/\sigma(A)) \cdot (B/\sigma(B)) = \varphi_A(\sigma) \varphi_B(\sigma) = \varphi_{AB}(\sigma).$$

To show that λ is surjective consider $\chi \in G^*$. The elements $\chi(\sigma), \chi(\tau)$ of $\langle \zeta \rangle \subseteq F^*$ satisfy

$$\chi(\sigma\tau) = \chi(\sigma)\chi(\tau) = \chi(\sigma)\sigma(\chi(\tau)),$$

and so by Noether's equations [theorem, §9.6] there exists an element $C \in K^*$ for which

$$C/\sigma(C) = \chi(\sigma)$$

for all $\sigma \in G$. Since n is the minimal exponent of G and hence of G^*, $(\chi(\sigma))^n = C^n/(\sigma(C))^n = C^n/\sigma(C^n) = 1$. Thus $C^n \in F^*$ or $C \in \{A\}$, and $\chi = \lambda(C) = \varphi_C$.

For all elements D in the kernel of λ, φ_D is the unit character of G, and consequently $D \in F^*$. Therefore the factor group $\{A\}/F^*$ is *naturally* isomorphic to the group of characters G^* according to the Isomorphism Theorem of §6.8. It follows that $\{A\}/F^*$ is isomorphic (but not naturally) to G.

In preparation for the next lemma we consider a subgroup $\{s\}$ of the multiplicative group F^* of F, such that the index $[\{s\} : F^{*n}] = m < \infty$, where F^{*n} denotes the set of nth powers of elements of F^*. Choose representatives $s_1, \ldots, s_m$ of the cosets of $\{s\}/F^{*n}$, and let K be the splitting field extension of F obtained by adjoining the roots of the polynomial

$$\prod_{i=1}^{m} (x^n - s_i).$$

Then K is a galois extension and $K = F(\ldots, \zeta^h S_i, \ldots)$ where $S_i^n = s_i$ and $1 \leq h \leq n$. Clearly K does not depend upon the choice of the m representatives $s_1, \ldots, s_m$, since for $a \in F^*$

$$F(\sqrt[n]{s_i a^n}) = F(\sqrt[n]{s_i}) = F(S_i).$$

Writing

$$\{S\} = \{s\}^{1/n} = \{\sqrt[n]{s} : s \in \{s\}\},$$

we set $K = F(\{S\}) = F(\{s\}^{1/n})$.

Furthermore, since raising elements of a (commutative) group to their nth powers is a group homomorphism, the corollary to Theorem 3, §6.8, implies that

(∗) $[\{S\} : F^*] = [\{S^n\} : F^{*n}][\langle \zeta \rangle : \langle \zeta \rangle] = [\{S^n\} : F^{*n}] = [\{s\} : F^{*n}],$

since the powers of ζ are the only elements in K whose nth powers are equal to 1, and since $\zeta \in F^*$. Now consider the homomorphism

$$\Lambda: \{S\}/F^* \to \{S^n\} F^{*n}/F^{*n} = \{s\}/F^{*n}$$

given by $\Lambda(SF^*) = S_0^n F^{*n}$; it is independent of the choice of the representative S_0 of the coset SF^*. Because of the index equalities (∗) Λ is an isomorphism.

Lemma 3. The galois group G of $K = F(\{S\})$ is abelian and is isomorphic to the dual group of $\{S\}/F^*$.

Proof. Consider $\sigma \in G$. Then $S/\sigma(S)$ is an nth root of unity because $S^n \in F^*$ implies $(S/\sigma(S))^n = S^n/\sigma(S^n) = 1$. The equation

$$\chi_\sigma(SF^*) = S/\sigma(S) \in \langle \zeta \rangle$$

defines a homomorphism $\chi_\sigma : \{S\}/F^* \to \langle \zeta \rangle$ since, for $S_1, S_2 \in \{S\}$ and $a \in F^*$,

$$(S_1/\sigma(S_1)) \cdot (S_2/\sigma(S_2)) = (S_1 S_2)/\sigma(S_1 S_2)$$

and

$$a/\sigma(a) = 1.$$

The map $\sigma \to \chi_\sigma$ is a homomorphism of the galois group G into the dual group of $\{S\}/F^*$, because

$$(S/\sigma(S))(S/\tau(S)) = (S/\sigma(S)) \cdot \sigma(S/\tau(S))$$
$$= (S/\sigma(S)) \cdot (\sigma(S)/(\sigma\tau)(S))$$
$$= S/(\sigma\tau)(S)$$

since $S/\tau(S)$ is an nth root of unity and lies in F^*. Thus $\chi_\sigma \chi_\tau = \chi_{\sigma\tau}$.

Furthermore, if $\chi_\sigma = 1$ then $\chi_\sigma(SF^*) = S/\sigma(S) = 1$ for all $S \in \{S\}$. Since $K = F(\{S\})$, the Galois theory implies that σ is the identity map in G. Thus, $\sigma \to \chi_\sigma$ is one-one and

$$[G : 1] \leq [\{S\}/F^* : 1].$$

Now conversely, to prove that $\{S\}/F^*$ is isomorphic to a subgroup of G, consider, as in the argument for Lemma 2, the mapping for a fixed $S \in \{S\}$

$$\varphi_S : G \to \langle \zeta \rangle \quad \text{given by } \varphi_S(\sigma) = S/\sigma(S).$$

The mapping φ_S is a homomorphism and, as in the first part of the proof,

$$\lambda : \{S\}/F^* \to G^* \quad \text{given by } \lambda(S) = \varphi_S$$

is a homomorphism. Furthermore if φ_s is the unit map, then

$$S/\sigma(S) = 1 \quad \text{for all } \sigma \in G.$$

Hence $S \in F^*$ and SF^* is the identity coset in $\{S\}/F^*$. Thus λ is an isomorphism from $\{S\}/F^*$ into G^*. Consequently

$$[\{S\}/F^* : 1] \leq [G^* : 1] = [G : 1].$$

Thus finally

$$[G : 1] = [\{S\}/F^* : 1]$$
$$= [\{S\} : F^*] = [\{s\} : F^{*n}],$$

and G is naturally isomorphic to the dual group of $\{S\}/F^*$. Hence, by duality, G^* is naturally isomorphic to $\{S\}/F^* \cong \{s\}/F^{*n}$

We use the preceding notation to state the principal theorem of the Kummer theory. Corollaries 1 and 2 give important consequences of this theorem.

Theorem 1. The Kummer extensions K of exponent n over a field F are in one-one correspondence with the subgroups $\{s\}$ of the multiplicative group F^* such that $[\{s\} : F^{*n}] < \infty$. Further, if $A^n \in F^*$,

$$G = G(K/F) \cong \{s\}/F^{*n} \cong \{S\}/F^* = \{A\}/F^*.$$

Proof. The first statement is a consequence of Lemmas 2 and 3. For the identity of $\{A\}$ and $\{S\}$, let $[\{s\} : F^{*n}]$ be finite. Then the galois group

$G = G(K/F)$ of $K = F(\{S\})$ is isomorphic to $\{S\}/F^* \cong \{s\}/F^{*n}$ by Lemma 3. Lemma 2 implies that $\{A\}/F^* \cong G^* \cong G$. Hence $\{A\} = \{S\}$.

Corollary 1. The galois group G of an abelian extension K/F of exponent n is determined by the effect of its elements on the group $\{S\}$ of elements in K^* whose nth powers lie in F.

Proof. Let $S_1, \ldots, S_r$ be representatives of the cosets of $\{S\}/F^*$ such that the cosets $S_i F^*$ form a *basis* of $\{S\}/F$. Let n_i be the order of $S_i F^*$—i.e., the least positive integer k such that $S_i^k = s_i \in F^*$—and note that $n_i | n$. Furthermore, let ζ_i be a fixed n_ith primitive root of unity. Also $K_i \cap K_j = F$ for $i \neq j$ where $K_i = F(S_i)$, $1 \leq i \leq r$. Then the field K is a cyclic extension of degree n_i over the product of fields $\hat{K}_i = K_1 \cdots K_{i-1} K_{i+1} \cdots K_r$, and $G_i = G(K/\hat{K}_i)$, where $K = \hat{K}_i(S_i)$, is generated by the automorphism σ_i for which $\sigma_i(S_i) = \zeta_i S_i$. Then K_i is the fixed field for the group $H_i = G_1 \cdots G_{i-1} G_{i+1} \cdots G_r$, and the restriction $\sigma_i | K_i$ generates the galois group of K_i/F.

Corollary 2. Consider multiplicative subgroups $\{s_i\}$ of F containing F^{*n} such that $[\{s_i\} : F^{*n}] < \infty$, $i = 1, 2$. Let $\{S_i\}$, $i = 1, 2$, be the set of nth roots of elements of $\{s_i\}$, and $K_i = F(\{S_i\})$ be the corresponding radical extensions of F.

(i) If $\{s_1\} \supseteq \{s_2\} \supseteq F^{*n}$, then $G(K_1/K_2) \cong \{s_1\}/\{s_2\}$.
(ii) If $\{s\} = \{s_1\} \cap \{s_2\}$, then $K = F(\{s\}^{1/n})$ equals $K_1 \cap K_2$.
(iii) If $\{s\} = \{s_1\}\{s_2\}$, then $K = F(\{s\}^{1/n})$ equals $K_1 K_2$.

These properties follow from the duality theory of finite abelian groups [§7.2] and the Galois theory [§§8.7 and 8.8]. For the proofs it is useful to recall that $\{S\} \supset \{S_i\}$ implies that the annihilator $A_G(\{S_i\})$ of $\{S_i\}$ in the group $G = G(K/F)$ is a subgroup of G, and that the fixed field of $A_G(\{S_i\})$ is a proper subfield of $K = F(\{S\})$. Consequently we can demonstrate a correspondence between

$$\{S_1\} \cap \{S_2\} \quad \text{and} \quad F(\{S_1\} \cap \{S_2\});$$
$$\{S_1\}\{S_2\} \quad \text{and} \quad F(\{S_1\}\{S_2\}).$$

The details of the proof are left as an exercise.

These results have an *additive version* for finite abelian extensions K of exponent p over a field F of characteristic $p > 0$. Replace the multiplicative group of nth roots of unity in the preceding discussion by the additive group $\mathbf{Z}_p$ and apply the additive version of Hilbert's Theorem 90.

Set $\wp(x) = x^p - x$. Then

$$\wp(x+y) = (x+y)^p - (x+y) = (x^p - x) + (y^p - y) = \wp(x) + \wp(y),$$

and $\wp(z) = 0$ if and only if $z \in \mathbf{Z}_p$, where we identify the prime field of F with $\mathbf{Z}_p$. Let B be a root of $x^p - x - b \in F[x]$. Since $\wp(x+[i]) = \wp(x) + [0] =$

$\wp(x)$ for $[i] \in \mathbf{Z}_p$, the other roots of $x^p - x - b$ are $B + [i]$, $[i] \in \mathbf{Z}_p$. Here we write $B = \wp^{-1}(b)$, whereas in the preceding multiplicative case we wrote $S = s^{1/n}$.

Theorem 2 is obtained by making the following replacements in the discussion of Lemmas 2 and 3, and in Theorem 1.

1. A Kummer extension K/F of exponent n, where char $F = p$, $(p, n) = 1$

1'. An abelian extension K/F of exponent p, $p =$ char F

2. Consider F^* as a multiplicative group

2'. Consider F as an additive group

3. The multiplicative group $\{A\}$ of all $A \in K$ such that $A^n \in F^*$

3'. The additive group $\{A\}$ of all $A \in K$ such that $\wp(A) \in F$

4. $f_{A,\sigma} = A/\sigma(A)$ for $\sigma \in G = G(K/F)$

4'. $f_{A,\sigma} = A - \sigma(A)$ for $\sigma \in G = G(K/F)$

5. $\varphi_A(\sigma) = f_{A,\sigma} \in \langle \zeta \rangle$

5'. $\varphi_A(\sigma) = f_{A,\sigma} \in \mathbf{Z}_p$

6. $f_{A,\sigma} f_{A,\tau} = f_{A,\sigma\tau}$ and thus $\varphi_A \in G^*$

6'. $f_{A,\sigma} + f_{A,\tau} = f_{A,\sigma\tau}$ and thus $\varphi_A \in \text{Hom}(G, \mathbf{Z}_p)$

7. $\varphi_{AB}(\sigma) = \varphi_A(\sigma) \varphi_B(\sigma)$ and thus $\lambda: \{A\} \to G^*$ given by $\lambda(A) = \varphi_A$ is a homomorphism

7'. $\varphi_{A+B}(\sigma) = \varphi_A(\sigma) + \varphi_B(\sigma)$ and thus $\lambda: \{A\} \to \text{Hom}(G, \mathbf{Z}_p)$ given by $\lambda(A) = \varphi_A$ is a homomorphism

8. $\{A\} \cong G^*$

8'. $\{A\} \cong \text{Hom}(G, \mathbf{Z}_p)$

9. $\{s\}$ a subgroup of F^* such that $[\{s\} : F^{*n}] < \infty$

9'. $\{s\}$ a subgroup of F such that $[\{s\} : \wp(F)] < \infty$, where $\wp(F) = \{\wp(a) : a \in F\}$

10. $K = F(\ldots, \zeta^h S_i, \ldots)$ where $S_i^n = s_i$, and ζ is a primitive nth root of unity

10'. $K = F(\ldots, S_i + [h], \ldots)$ where $\wp(S_i) = s_i$ and $[h] \in \mathbf{Z}_p$

11. $[\{S\} : F^*] = [\{s\} : F^{*n}]$

11'. $[\{S\} : F] = [\{s\} : \wp(F)]$

12. $\chi_\sigma(SF^*) = S/\sigma(S)$

12'. $\chi_\sigma(S + F) = S - \sigma(S)$

13. $[S/\sigma(S)] \cdot [S/\tau(S)] = S/(\sigma\tau)(S)$

13'. $[S - \sigma(S)] + [S - \tau(S)] = S - (\sigma\tau)(S)$.

Theorem 2. Finite abelian extensions K/F of exponent p, where p is the characteristic of F, are in one-one correspondence with the subgroups $\{s\}$ of the additive group F such that $[\{s\} : \wp(F)] < \infty$ and $G = G(K/F) \cong \{s\}/\wp(F) \cong \{S\}/F = \{A\}/F$.

REMARK. In the case that K/F has exponent p^m but not exponent p, where $p =$ char F, we can use the so-called Witt vectors [see N. Jacobson, *loc. cit.*, pp. 124–139] or an inductive argument due to A. A. Albert (1905–1972) [see *Bull. Amer. Math. Soc.* **40**(1934), pp. 625–631].

Exercises

1. Suppose that $x^p - a$ and $x^p - b$ are distinct irreducible polynomials of $\mathbf{Q}[x]$ such that $[\mathbf{Q}(A, B) : \mathbf{Q}] = p^2$ for two zeros A, B of $x^p - a$, $x^p - b$, respectively, in an extension $L/\mathbf{Q}$. Prove that $[\mathbf{Q}(A+B) : \mathbf{Q}] = p^2$.
2. Suppose that F is a finite field, char $F = p$, and x, t are indeterminates over F.
   a. Is $t^p - t - x$ irreducible in $F(x)[t]$?
   b. Show that $t^p - t - x^{-1}$ is irreducible in $F(x)[t]$.
3. Assume that $f(x) = x^p - x - a$ is an irreducible polynomial in the polynomial ring $F[x]$, char $F = p > 0$. For $\alpha \in F$ such that $f(\alpha) \neq 0$, prove that the polynomial $x^p - x - a\alpha^{p-1}$ is irreducible.
4. Let A_1, A_2 be zeros of the irreducible polynomials $x^p - x - a_1$, $x^p - x - a_2 \in F[x]$, respectively, where $p = \text{char } F$. Prove that $F(A_1) = F(A_2)$ if and only if $A_2 = [c]A_1 + b$, where $[c] \neq 0$ in $\mathbf{Z}^p$ and $b \in F$.
5. If char $F = p > 0$, prove that $x^p - x - a \in F[x]$ either factors completely or is irreducible.
6. Prove that for an odd prime p the polynomial $x^{p^n} - a \in F[x]$ is irreducible for all $n \in \mathbf{N}$ if a is not a pth power in F.
7. Prove that the galois group of the splitting field K/F of $x^n - a \in F[x]$ is isomorphic to a subgroup of the affine group of $\mathbf{Z}_n$ if $(n, \text{char } F) = 1$. (If char $F = 0$, no restriction is placed on n.)
8. Suppose that the field F contains all nth roots of unity where $(n, \text{char } F) = 1$. Prove that $F(A) = F(B)$, with $A^n = a$, $B^n = b$, and $[F(A) : F] = n$, implies that $b = a^s c^n$ where $c \in F$ and $(s, n) = 1$.
9. Prove Corollary 2 to Theorem 1 in detail.
10. In the parallel argument preceding Theorem 2, verify statement $8'$ in detail.
11. Let $K = \mathbf{Q}(\omega, \sqrt[3]{7}, \sqrt[4]{1+\sqrt[3]{7}})$, where $\omega^2 + \omega + 1 = 0$.
    a. Find the least normal extension of $K/\mathbf{Q}$ and determine its galois group G.
    b. Determine a composition series of G.
12. Let $K = \mathbf{Q}(i, \sqrt[6]{3}, \xi)$, where ξ is a primitive sixth root of unity. Find the galois group G of the least normal extension of $K/\mathbf{Q}(i)$, and determine a composition series of G.
13. For relatively prime integers m and n prove that $x^{mn} - a \in F[x]$ is irreducible if and only if both $x^m - a$ and $x^n - a$ are irreducible in $F[x]$.

§9.8 Unique Factorization Domains and Elementary Symmetric Functions

This section, essentially arithmetic in nature, provides the preparation for Gauss' proof of the Fundamental Theorem of Algebra [see §9.9]. Emphasis is placed on the concept of integral dependence. Thus we obtain a proof of the Fundamental Theorem of Elementary Symmetric Functions somewhat different from the usual one, and at the same time prepare for important concepts in ring theory: unique factorization domains and integral dependence.

Suppose now that R is an integral domain with F for its quotient field and that K/F is an algebraic (field) extension. Since we consider only elements in a field K, left and right R-modules in K can be considered to be equal. Thus in this section we shall not distinguish between left and right modules.

In §§2.6 and 2.7 we considered the factorization of integers and properties of certain "irreducible" integers, called primes, which could not be factored nontrivially. Similarly in §5.2 we considered the questions of divisibility and factorization within the integral domain of polynomials over a field. We now extend the discussion of factorization to more general integral domains.

A nonzero element a in an integral domain R is called **irreducible** if

(i) it is not a unit of R, and
(ii) whenever $a = bc$ with $b, c \in R$, then one of the elements b and c is a unit of R.

A nonzero element $a \in R$, which is not a unit, is said to have a **unique factorization into irreducible elements** of R, if

(i) $a = u \prod_{i=1}^{r} p_i$ with a unit $u \in R$ and irreducible elements p_i, $1 \le i \le r$, in R, and
(ii) if $a = v \prod_{j=1}^{s} q_j$ is another such factorization, then $r = s$ and $p_i = u_i q_{\sigma(i)}$ with units $u_i \in R$, $1 \le i \le r$, where σ is a permutation of $\{1, \ldots, r\}$.

We call an integral domain R a **unique factorization domain** (commonly abbreviated UFD) if every nonzero element has a unique factorization into irreducible elements.

A nonzero element $a \in R$ is termed a **divisor** of $b \in R$, if $ac = b$ with $c \in R$; symbolically, we write $a|b$. An element $d \in R$ is called a **greatest common divisor** of a and b if $d|a$ and $d|b$, and if every nonzero element $g \in R$ that divides a and b also divides d. Elements a, a_1 of a unique factorization domain R are said to be **associates**, $a \approx a_1$, if $a_1 = \varepsilon a$ for some unit ε of R. The relation $\approx$ of being associates is an equivalence relation.

Every element $a \in F = Q(R)$ can be written as a quotient u/v where u and v lie in R and have no common irreducible (nonunit) factors. The factorization property implies that, for each irreducible element $p \in R$,

$$a = p^\alpha a^* \ne 0 \quad \text{for a unique } \alpha \in \mathbf{Z},$$

with $a^* \in F$, where neither numerator or denominator of a^* has p as a factor. We now define an *order function* v_p on $F = Q(R)$ for a given irreducible element $p \in R$ [cf. Exercise 14, §2.7] by setting

$$v_p(a) = \alpha \in \mathbf{Z}.$$

Note that $v_{\varepsilon p}(a) = \alpha$ for every unit ε of R. For $a = 0$, we set $v_p(0) = \infty$. The

§9.8 Unique Factorization Domains and Elementary Symmetric Functions

unique factorization of elements implies that, for any pair of nonzero elements a, b of F,
$$v_p(ab) = v_p(a) + v_p(b).$$
The order function v_p extends to the polynomials
$$f = f(x) = a_n x^n + a_{n-1} x^{n-1} + \cdots + a_0$$
in $F[x]$ as follows:
$$v_p(f) = \operatorname{Min}\{v_p(a_i) : 0 \leq i \leq n\}.$$

A *p-content* $C_p(f)$ of the polynomial f is an element εp^γ, where $\gamma = v_p(f)$ and ε is a unit of R. The **content** $C(f)$ is defined to be $\prod_p C_p(f) = \eta \prod_p p^\gamma$, where η is any unit of R, and where the product is taken over a set of representatives of the distinct classes of associated irreducible elements $p \in R$. (The product is well-defined because all but a finite number of the $C_p(f)$ are equal to 1.) This definition implies immediately that
$$C(af) = aC(f)$$
for every nonzero element a of the field of coefficients F. Consequently
$$f(x) = cf_1(x)$$
where $c = C(f)$ and the content of $f_1(x) \in R[x]$ is 1. More precisely, the coefficients of $f_1(x)$ lie in R and their GCD is 1.

The arguments of the proof of Gauss' Lemma [§5.8] generalize to yield the proof of Lemma 1. In the proof, which we leave to the reader, it suffices to consider polynomials f, g for which $C(f) = C(g) = 1$ and to prove that $C(fg) = 1$—in other words, that $v_p(fg) = 0$ for all irreducible elements $p \in R$—because in the general case we write $f = af_1$, $g = bg_1$, where $C(f_1) = C(g_1) = 1$, so that $C(fg) = abC(f_1 g_1)$.

Lemma 1. If R is a unique factorization domain with quotient field $F = Q(R)$, then for $f, g \in F[x]$
$$C(fg) = C(f)C(g).$$

An immediate consequence of Lemma 1 is that if $h(x) \in R[x]$ has the factorization $h(x) = f(x)g(x)$ in $F[x]$, then
$$h(x) = C(f)C(g)f_1(x)g_1(x),$$
where $C(g)C(h) \in R$ and $f_1(x)$ and $g_1(x)$ are polynomials in $R[x]$ with content 1. A generalization of this argument proves the following lemma [cf. the unique factorization of integers in §2.7 and of polynomials, §5.2].

Lemma 2. If R is a unique factorization domain, then so is the polynomial ring $R[x]$. The irreducible elements of $R[x]$ are either irreducible elements of R or polynomials of $R[x]$ which are irreducible in $F[x]$, $F = Q(R)$, and have content 1.

The preceding lemmas together with induction on the number of indeterminates yield the proof of the following theorem.

Theorem 1. If $x_1, \ldots, x_n$ are indeterminates over a field F, then the polynomial ring $F[x_1, \ldots, x_n]$ is a unique factorization domain.

Before introducing the second objective of this section, elementary symmetric functions, we define the concept of integral dependence. The following discussion of integral elements and integral closure is parallel to that in Chapter 8 of algebraic elements and algebraic closure.

An element $a \in K$ is said to be **integral over** R, if there exists a finitely generated R-module $M_a \subseteq K$ such that

$$aM_a = \{am : m \in M_a\} \subseteq M_a = (a_1, \ldots, a_s)_R.$$

We use a series of four lemmas to develop properties of elements in K which are integral over R.

Lemma 3. The set of elements in K integral over R is a subring of K.

Proof. Consider elements $a, b \in K$, integral over R. Express M_a in terms of generators $a_1, \ldots, a_s$ as

$$M_a = Ra_1 + \cdots + Ra_s,$$

and M_b in terms of generators $b_1, \ldots, b_t$ as

$$M_b = Rb_1 + \cdots + Rb_t.$$

Now define a finitely generated R-module M by

$$M = Ra_1 b_1 + \cdots + Ra_i b_j + \cdots + Ra_s b_t,$$

$1 \leq i \leq s$, $1 \leq j \leq t$. Since $aM \subseteq M$ and $bM \subseteq M$, we have $(a+b)M \subseteq M$ and $abM \subseteq M$. Thus, $a+b$ and ab are integral over R.

Consequently the set $S = \{a \in K : a \text{ integral over } R\}$ is a subring of K; it is called the **integral closure** of R in K.

Lemma 4. An element $a \in K$ is integral over R if and only if it is a zero of a monic polynomial in $R[x]$.

Proof. First, consider $a \in K$, integral over R;

$$aa_i = \sum_{j=1}^{s} d_{ji} a_j, \quad 1 \leq i \leq s,$$

with coefficients $d_{ji} \in R$. Hence a is a zero of the characteristic polynomial $\chi(x)$ of the matrix $[d_{ij}]$ [see §5.9].

Conversely, suppose $a^k + c_{k-1} a^{k-1} + \cdots + c_0 = 0$. Set $M_a = 1 \cdot R + aR + \cdots + a^{k-1} R$. Then, by induction on $h \geq k$, we can show that $a^h \in M_a$, and hence that $aM_a \subseteq M_a$, since $a^k = -c_0 - \cdots - c_{k-1} a^{k-1}$ [cf. §8.1].

§9.8   Unique Factorization Domains and Elementary Symmetric Functions

Lemma 5.   An element $c \in K$, integral over the integral closure S of R in K, is integral over R.

Proof.   By Lemma 4, the element c satisfies a monic polynomial $x^k + c_{k-1} x^{k-1} + \cdots + c_0$ in $S[x]$. Now let $M = M_1 \cdots M_k$, the product module generated by the set of elements $\{m_1 \cdots m_k : m_i \in M_i, \ 1 \leq i \leq k\}$, where M_i is a finitely generated R-module in K for which $c_i M_i \subseteq M_i$, $1 \leq i \leq k$. Then $c_i M \subseteq M$; and M is finitely generated. Finally $\overline{M} = M + cM + \cdots + c^{k-1} M$ is a finitely generated R-module, and $c\overline{M} \subseteq \overline{M}$.

A subring of a field F is said to be **integrally closed** in F if it contains all elements of F which are integral over it. Lemma 5 states then that the integral closure S of R in F is itself integrally closed in F.

Lemma 6.   A unique factorization domain R is integrally closed in its quotient field $F = Q(R)$.

Proof.   Suppose that $a/b \in F$ is integral over R, and $a/b \notin R$ with $a, b \in R$ having no common irreducible factor. Let p be an irreducible element that divides b. Then a relation of integral dependence of the quotient a/b over R,

$$(a/b)^m + c_{m-1}(a/b)^{m-1} + \cdots + c_0 = 0,$$

with coefficients $c_i \in R$ implies that

$$a^m + c_{m-1} b a^{m-1} + \cdots + c_0 b^m = 0.$$

The assumption that $p | b$ implies then that $p | a^m$ and consequently that $p | a$, contrary to the assumption on a. Hence a/b cannot be integral over R, and R is integrally closed.

We now apply Theorem 1 to polynomial rings. Let F be a field and $R = F[x_1, \ldots, x_n]$. Then $K = Q(R) = F(x_1, \ldots, x_n)$ is the field of rational functions of $x_1, \ldots, x_n$ with coefficients in F. Suppose that

$$\sigma = \begin{bmatrix} 1 & \cdots & n \\ i_1 & \cdots & i_n \end{bmatrix}$$

is an element of the symmetric group Σ_n on n elements [see §6.7]. The mappings γ_σ, for $\sigma \in \Sigma_n$, defined by

$$\gamma_\sigma [f(x_1, \ldots, x_n)] = f(x_{i_1}, \ldots, x_{i_n})$$

are distinct automorphisms of K/F. Let L be the fixed field of $G = \{\gamma_\sigma : \sigma \in \Sigma_n\}$, as in §8.6. Then K is a galois extension over L of degree $n!$ and $K = L(x_1, \ldots, x_n)$.

The polynomial

$$g(t) = \prod_{i=1}^{n} (t - x_i) = t^n + \sum_{j=1}^{n} (-1)^j e_j t^{n-j} \quad \text{in } K[t],$$

where $e_j(x_1, \ldots, x_n) = \sum_{k_1 < k_2 < \cdots < k_j} x_{k_1} x_{k_2} \cdots x_{k_j}$ for $1 \leq j \leq n$, can be rewritten as

$$g(t) = \prod_{v=1}^{n} (t - x_{i(v)})$$

for each $\sigma \in \Sigma_n$. Consequently the coefficients of $g(t)$ remain unaltered by application of any $\sigma \in \Sigma_n$. The polynomials $e_1, \ldots, e_n$, called the **elementary symmetric functions**, therefore lie in the fixed field L.

Hence the rational function field $E = F(e_1, \ldots, e_n)$ is a subfield of the fixed field L of the automorphism group G of K. Actually E equals L. Since $K = E(x_1)$ contains the n distinct zeros $x_1, \ldots, x_n$ of the polynomial $g(t) \in E[t]$, it is a galois extension of E, and consequently the galois group $G(K/E)$ is isomorphic to a subgroup of Σ_n [see §8.6]. Therefore $[K : L] = n! = [\Sigma_n : 1]$ and $L \supseteq E$ imply $L = E$ as asserted.

To summarize, we have proved the following lemma.

Lemma 7. The field $K = F(x_1, \ldots, x_n)$ is a galois extension of the field $E = F(e_1, \ldots, e_n)$ of the elementary symmetric functions $e_1, \ldots, e_n$ with coefficients in F. The galois group of K/E is isomorphic to the symmetric group Σ_n.

Lemma 8. The elementary symmetric functions $e_1, \ldots, e_n$ of the indeterminates $x_1, \ldots, x_n$ over F are algebraically independent over F.

Proof. We shall use an indirect recursive argument. Suppose that $e_1, \ldots, e_n$ are algebraically dependent, that is,

(∗) $$g(e_1, \ldots, e_n) = 0$$

for some nonzero polynomial $g(u_1, \ldots, u_n)$ in the polynomial ring $F[u_1, \ldots, u_n]$ in the indeterminates $u_1, \ldots, u_n$. Let $f(u_1, \ldots, u_n)$ be a polynomial of minimal degree h in u_n among all polynomials satisfying relation (∗). It can be written in the form

$$f = f_0(u_1, \ldots, u_{n-1}) + \cdots + f_h(u_1, \ldots, u_{n-1}) u_n^h,$$

where necessarily $f_0(u_1, \ldots, u_{n-1}) \neq 0$, because of the minimal choice of h. Hence

$$0 = f_0(e_1, \ldots, e_{n-1}) + \cdots + f_h(e_1, \ldots, e_{n-1}) e_n^h.$$

This is a relation in $F[x_1, \ldots, x_n]$ and remains valid if the indeterminate x_n is replaced by 0 (a homomorphism of $F[x_1, \ldots, x_n]$ to $F[x_1, \ldots, x_{n-1}]$, since $e_n = \prod_{i=1}^{n} x_i$). Hence

$$f_0(e_{1,0}, \ldots, e_{n-1,0}) = 0,$$

where the terms $e'_v = e_{v,0}$ are seen by inspection to be the elementary symmetric functions of $x_1, \ldots, x_{n-1}$. That is, the elementary symmetric functions $e'_1, \ldots, e'_{n-1}$ are then algebraically dependent.

§9.8 Unique Factorization Domains and Elementary Symmetric Functions

Since $e_1 = x_1$ for $n = 1$, the preceding reduction from n to $n-1$ implies that $e_1, \ldots, e_n$ are algebraically independent.

We conclude this section with the important concept of symmetric polynomials; a basic property of such polynomials (Theorem 2) is sometimes referred to as the Fundamental Theorem of Elementary Symmetric Functions. A polynomial $f = f(x_1, \ldots, x_n) \in F[x_1, \ldots, x_n]$ is **symmetric** if, for each $\sigma \in \Sigma_n$,

$$\gamma_\sigma(f) = f.$$

Theorem 2. A symmetric polynomial f in the indeterminates $x_1, \ldots, x_n$ is equal to a polynomial $h(e_1, \ldots, e_n) \in F[e_1, \ldots, e_n]$, where the e_i are the elementary symmetric functions of the x_i, $1 \leq i \leq n$.

Proof. Since $\gamma_\sigma(f) = f$ for all $\sigma \in \Sigma_n$, the polynomial f lies in the fixed field $E = F(e_1, \ldots, e_n)$. Because the elements $x_1, \ldots, x_n$ are the zeros of the polynomial $g(t) \in F[e_1, \ldots, e_n]$ they are integrally dependent over $F[e_1, \ldots, e_n]$ [see the discussion prior to Lemma 7]. Consequently f is integrally dependent on the unique factorization domain $F[e_1, \ldots, e_n]$ by Lemma 4. Finally $F[e_1, \ldots, e_n]$ is integrally closed according to Lemma 6 together with Lemma 8 and Theorem 1. Hence $f \in F[e_1, \ldots, e_n]$, as asserted.

Exercises

1. a. Express $x_1^3 + x_2^3 + x_3^3$ as a polynomial in $\mathbf{Q}[e_1, e_2, e_3]$, where $e_1, e_2,$ and e_3 are the elementary symmetric functions of $x_1, x_2,$ and x_3.
    b. Express $\sum_{i=1}^n x_i^2$ and $\sum_{i=1}^n x_i^3$ as polynomials of the elementary functions $e_1, \ldots, e_n$ of $x_1, \ldots, x_n$.
2. A ring is called a **principal ideal ring** if all of its ideals are principal. Prove that such a ring is a unique factorization domain. (Is the converse true? Examine $\mathbf{Q}[x, y]$.)
3. Let A_1, A_2, A_3 be the zeros of the irreducible polynomial $x^3 + px + q \in \mathbf{Q}[x]$. Express $[(A_3 - A_1)(A_3 - A_2)(A_2 - A_1)]^2$ in terms of p and q.
4. Let $K = \mathbf{Q}(\omega)$, where $\omega^2 + \omega + 1 = 0$.
    a. Prove that the subring $R = \{1, m\omega\}_\mathbf{Z} = \{a + bm\omega : a, b \in \mathbf{Z}\}$ is not integrally closed in K; $m \neq 0, \pm 1$.
    b. Find the integral closure of such a ring R in K.
5. Let $K = \mathbf{Q}(\sqrt{m})$ where $[K:\mathbf{Q}] = 2$ and $m \in \mathbf{Z}$ is a product of distinct primes. Prove the following statements:
    a. If $m \equiv 2$ or $m \equiv 3 \pmod{4}$, the integral closure of $\mathbf{Z}$ in K equals $\{a + b\sqrt{m} : a, b \in \mathbf{Z}\}$.
    b. If $m \equiv 1 \pmod{4}$, the integral closure of $\mathbf{Z}$ in K equals $R = \{a + \delta b : a, b \in \mathbf{Z}\}$ where $\delta = (1 + \sqrt{m})/2$. Show that
    $$R = \{(x + y\sqrt{m})/2 : x, y \in \mathbf{Z}\}.$$

6. Let R be an integrally closed integral domain $[R \subseteq Q(R)]$. Suppose that M is a subset of R such that $0 \notin M$ and $s_1 s_2 \in M$ for all $s_1, s_2 \in M$. Generalize the construction of the quotient field [§3.6], thus defining $M^{-1}R$, and prove that $M^{-1}R$ is integrally closed.

7. Consider a ring R with subring S having the same multiplicative unit as R. Suppose that A and P are prime ideals of S and R, respectively, for which $P \cap S = A$. Prove that there exists an embedding of the residue class ring S/A into R/P. Furthermore, determine whether or not R/P consists of integral elements over the image S/A if R consists of integral elements over S.

8. a. Let R be a unique factorization domain with the quotient field $Q(R) = F$. Prove that a zero $A \in F$ of $x^n + a_{n-1}x^{n-1} + \cdots + a_0 \in R[x]$, where $a_0 \neq 0$, is necessarily an element in R which divides a_0 [cf. the Integral Root Theorem, Exercise 9, §5.2].
   b. State and prove an analogue of the Rational Root Theorem, Exercise 8, §5.2.

9. Suppose that $f(x, y) \in F[x, y]$ satisfies $k(x) f(x, y) = g(x, y) h(x, y)$ for some $k(x) \in F[x]$, and $g(x, y), h(x, y) \in F[x, y]$, and that the coefficients of powers of y in the polynomial $g(x, y)$ have GCD 1. Prove that $k(x)$ divides each of the coefficients of the powers of y in the polynomial $h(x, y)$.

10. Suppose that the polynomial $x^n + a_1 x^{n-1} + \cdots + a_n \in F[x]$ has the zeros $A_1, \ldots, A$ in some field $K \supseteq F$. Let $s_k = A_1^k + A_2^k + \cdots + A_n^k$. Prove Newton's formulas:
    a. $s_k + a_1 s_{k-1} + a_2 s_{k-2} + \cdots + a_{k-1} s_1 + k a_k = 0,$ $1 \leq k \leq n$.
    b. $s_k + a_1 s_{k-1} + \cdots + a_n s_{k-n} = 0$ for $k > n$.

11. Prove that the only units of the integral closure of $\mathbf{Z}$ in $\mathbf{Q}(\sqrt{d})$, for $d = -2$ or $d < -3$, are ± 1.

12. Prove Lemma 1.

§9.9 The Fundamental Theorem of Algebra

The Fundamental Theorem of Algebra states that the field of complex numbers $\mathbf{C} = \mathbf{R}(i)$, $i^2 = -1$, is algebraically closed. An alternate statement is that any polynomial with complex coefficients has a root in $\mathbf{C}$, and hence can be written as the product of linear factors in $\mathbf{C}[x]$.

This theorem is attributed by the mathematical historian, David E. Smith, to Peter Roth (c. 1608) and Albert Girard (1590–1633) in 1629. François Viète (1540–1603) was cognizant of the relations between the zeros of a polynomial with rational coefficients and its coefficients. In the middle of the eighteenth century Jean le Rond d'Alembert (1717–1783) made several attempts, all unsuccessful, to prove Girard's conjecture, and because of his efforts his name is commonly given to the theorem, especially in France. While Euler (in 1749) and Lagrange also attempted proofs, the first rigorous proof of the Fundamental Theorem of Algebra was given by Carl Friedrich Gauss (1777–1855) in his doctoral dissertation at the University of Helmstedt

§9.9 The Fundamental Theorem of Algebra

in 1798, published a year later. In this paper Gauss also demonstrated the inadequacy of earlier proofs, including those of Euler and Lagrange.

Complex numbers, which we write $a+bi$ with $a, b \in \mathbf{R}$, were sometimes viewed as cosets in $\mathbf{R}[x]/(x^2+1)$, for the geometrical description of the complex plane was new. In 1798 Caspar Wessel (1745–1818) published his graphical representation of complex numbers in the transactions of the Danish Academy of Sciences, and Jean Robert Argand (1768–1822) introduced the complex plane in 1806. Gauss subsequently strove, albeit without success, for a strictly algebraic proof of the theorem, presenting two new proofs in 1816 and one in 1850.

There are a number of proofs of this theorem, all of them analytical (or topological) by nature, a quality inherently due to the definition of the field $\mathbf{C}$ as a quadratic extension of the field of real numbers. The proof we present appears to be due in part to Lagrange. It uses a minimal amount of real analysis such as the Intermediate Value Theorem whose proof in turn employs not more than the existence of least upper bounds, a characteristic property of the real numbers, and a simple inequality.

Gauss reduced the problem of finding $\xi = \alpha + \beta i$ such that $\xi^2 = a+bi$ to one of finding the intersection of two hyperbolas $\alpha^2 - \beta^2 = a$ and $\alpha\beta = b/2$. For purely imaginary numbers, $a = 0$, and the hyperbola $\alpha^2 - \beta^2 = 0$ is degenerate. Thus the problem is one of finding the intersection of two lines $\alpha = \pm \beta$ and the hyperbola $\alpha\beta = b/2$.

As a preparatory lemma, we note that a real polynomial

$$f(x) = x^n + a_{n-1}x^{n-1} + \cdots + a_0 \in \mathbf{R}[x]$$

of *odd* degree has at least one real root. First, there exists a positive real number ρ such that

$$f(x) > 0 \quad \text{for } x > \rho$$

and

$$f(x) < 0 \quad \text{for } x < -\rho.$$

We have the following estimate for $a > 1$:

$$f(a) \geq a^n - (|a_{n-1}|a^{n-1} + \cdots + |a_0|)$$
$$\geq a^n - a^{n-1}(|a_{n-1}| + \cdots + |a_0|)$$
$$= a^{n-1}(a - [|a_{n-1}| + \cdots + |a_0|]) > 0$$

if $a > \text{Max}\{1, |a_0| + \cdots + |a_{n-1}|\}$. Furthermore

$$f(-a) \leq -a^n + |a_{n-1}|a^{n-1} + \cdots + |a_0|$$
$$\leq -a^n + a^{n-1}(|a_{n-1}| + \cdots + |a_0|)$$
$$= -a^{n-1}(a - [|a_{n-1}| + \cdots + |a_0|]) < 0.$$

Hence let $\rho \geq \text{Max}\{1, |a_0| + \cdots + |a_{n-1}|\} + 1$. The Intermediate Value Theorem, applied to the interval $-\rho \leq x \leq \rho$, implies that $f(x)$ has at least one real zero.

Proof of the Fundamental Theorem of Algebra. It must be shown that every polynomial $g(x) = x^n + c_{n-1}x^{n-1} + \cdots + c_0$ with complex coefficients c_v has a zero in the complex field **C**. Set
$$g^*(x) = x^n + c^*_{n-1}x^{n-1} + \cdots + c^*_0,$$
where c^*_v is the conjugate complex number of c_v, $1 \le v \le n$. Then
$$f(x) = g(x)g^*(x) = (g(x)g^*(x))^*$$
is a polynomial with *real* coefficients since it equals its own conjugate. Therefore the complex zeros of $f(x)$ occur as pairs of conjugate numbers. Consequently the proof is reduced to showing that real polynomials split into linear factors in $\mathbf{C}[x]$. Thus consider a real polynomial $f(x)$ of degree $d = 2^m q$, where $(q, 2) = 1$. We proceed by induction on m.

If $m = 0$, then d is odd and hence has a real zero according to the preliminary discussion. Suppose now that $m \ge 1$. The theory of splitting fields [see §8.1] implies the existence of an extension K of **C** such that
$$f(x) = \prod_{i=1}^{d} (x - a_i) \in K[x].$$
Now pick an element $u \in \mathbf{R}$, and define $d(d+1)/2$ elements
$$b_{ij} = a_i + a_j + ua_i a_j \in K, \qquad 1 \le i \le j \le d.$$
The coefficients of the polynomial $h(x) = \prod_{i \le j}(x - b_{ij})$ are symmetric polynomials of $a_1, \ldots, a_d$ with coefficients in **R**. Consequently they are polynomials (with real coefficients) of the elementary symmetric functions of $a_1, \ldots, a_d$, according to Theorem 2, §9.8. (Substitute indeterminates for $a_1, \ldots, a_d$, apply the theorem, and ultimately replace the indeterminates by the elements a_i.) Consequently $h(x)$ is a real polynomial of degree $d(d+1)/2 = 2^{m-1}q'$, where $q' = q(d+1)$ is odd since both q and $d+1$ are.

Now $h(x)$ has a zero z_u in **C** by the induction hypothesis because $(q', 2) = 1$. Consequently one of the elements b_{ij} must be equal to z_u; denote this element by
$$b_{i(u), j(u)} = a_{i(u)} + a_{j(u)} + ua_{i(u)} a_{j(u)}.$$
Since **R** has infinitely many elements and since the number of pairs (i, j) with $1 \le i \le j \le d$ is finite, there must exist two *distinct* real numbers u and v for which $i(u) = i(v) = r$ and $j(u) = j(v) = s$. Consequently, both
$$a_r + a_s + ua_r a_s \qquad \text{and} \qquad a_r + a_s + va_r a_s$$
belong to **C**, as do
$$v(a_r + a_s + ua_r a_s) \qquad \text{and} \qquad u(a_r + a_s + va_r a_s).$$
Thus $a_r + a_s \in \mathbf{C}$, and therefore $a_r a_s \in \mathbf{C}$.

This fact implies that a_r and a_s satisfy a quadratic equation $y^2 + 2\alpha y + \beta = 0$ with coefficients α, β in **C**. The roots of this equation are
$$-\alpha \pm \sqrt{\alpha^2 - \beta},$$

where $\alpha^2 - \beta = a + bi$ with a, b in $\mathbf{R}$. Then
$$\pm\sqrt{a+bi} = \sqrt{\tfrac{1}{2}\left(\sqrt{a^2+b^2}+a\right)} \pm i\sqrt{\tfrac{1}{2}\left(\sqrt{a^2+b^2}-a\right)}.$$
Consequently, a_r and a_s lie in $\mathbf{C}$, and $f(x)$ has zeros a_r, a_s in $\mathbf{C}$, as asserted.

Exercises

1. For $u_1, u_2 \in \mathbf{C}$, prove the *Triangle Inequality* $|u_1 + u_2| \le |u_1| + |u_2|$. (*Hint:* Write the u's in the form $u = a + bi$ with $a, b \in \mathbf{R}$.)
2. Let a and b be real numbers. Find real numbers α, β such that $a + bi = (\alpha + \beta i)^2$.
3. Find the conjugates of $A = \sqrt{3+\sqrt{10}} + \sqrt{3-\sqrt{10}} \in \mathbf{C}$ over $\mathbf{Q}(i)$, $i^2 = -1$.
4. Assume that A is a zero in the splitting field K of $f(x) = x^4 + 6x^3 + 10x^2 + 3x + 1 \in \mathbf{Q}[x]$. Prove that $B = A + A^2$ has degree 2 over $\mathbf{Q}$. Is B a real number? Find the other zeros of $f(x)$.

§9.10 Finite Division Rings

One of the famous theorems of Joseph Henry Macglagen Wedderburn (1882–1948) asserts that any finite ring with a unit for multiplication whose nonzero elements form a multiplicative group is necessarily a field. The proof we present is due to E. Witt (see *Abhandlungen des Mathematischen Seminars der Universität Hamburg*, 1931). It should be noted that although we are dealing with a strictly algebraic question, the clinching argument depends upon an analytic statement concerning the zeros of a cyclotomic polynomial. For a completely algebraic proof see I. N. Herstein, *Noncommutative Rings*.

A (not necessarily commutative) ring D with unit for multiplication whose nonzero elements form a multiplicative group D^* is called a **division ring**.

Lemma. The center F of a division ring D is a field.

By definition, the center is the set
$$F = \{x \in D : ax = xa \text{ for all } a \in D\}.$$
For $x, y \in F$, and for all $a \in D$,
$$a(x+y) = ax + ay = xa + ya = (x+y)a,$$
$$a(xy) = (xa)y = x(ay) = (xy)a,$$
$$ax^{-1} = (xa^{-1})^{-1} = (a^{-1}x)^{-1} = x^{-1}a, \quad \text{if } x \ne 0,$$
Thus $x + y$, $xy = yx$, and x^{-1} all belong to F, so that F is a field.

Theorem (*Wedderburn*). A finite division ring D is a field.

Consider D as a vector space over its center F. Since D is finite, the dimension $\dim_F D = n$ is finite. The object of the proof is to show that $n > 1$ leads to a contradiction, and thus that D must equal F. First note that if F has q elements, then D has q^n elements. Next, for $a \in D$ let

$$F_a = \{u \in D : au = ua\}$$

denote the normalizer of a [see §7.4]. Since for $u, v \in F_a$

$$a(u+v) = au + av = ua + va = (u+v)a$$

and
$$a(uv) = (ua)v = (uv)a,$$

$u+v, uv \in F_a$. The fact that D is finite implies that the set $\{1, u, u^2, \ldots\}$ of powers of any nonzero element u is finite. Consequently for some h, k with $h > k$, we have $u^h = u^k$. Thus, $u \cdot u^{h-k-1} = 1$ so that $u^{h-k-1} \in F_a$ is the inverse of u. Therefore F_a is a division ring.

The division ring F_a may be considered as a d-dimensional vector space over F; then F_a has q^d elements. Furthermore, D can be considered as a left module over F_a, where the module multiplication of $x \in D$ by $u \in F_a$ is the product ux, so that for $y \in D$ and $u, v \in F_a$ the equations

$$1x = x, \qquad u(x+y) = ux + uy,$$
$$(u+v)x = ux + vx, \qquad (uv)x = u(vx)$$

verify the axioms of a left module [see §7.8]. Just as in the theory of vector spaces, it follows that D contains a basis $x_1, \ldots, x_s$ over F_a [see §4.2]. The number of elements of D, namely q^n, must equal $(q^d)^s$; hence $d | n$.

Next we apply the class equation of §7.4 to the group D^*. First note that the number of distinct conjugates (with respect to the group of inner automorphisms of D) of a nonzero element $a \in D$ is $(q^n - 1)/(q^d - 1)$. Then

(*)
$$[D^* : 1] = [F^* : 1] + \sum \left[\frac{q^n - 1}{q^d - 1}\right]$$
$$= (q-1) + \sum \left[\frac{q^n - 1}{q^d - 1}\right],$$

where the summation is taken over the distinct conjugacy classes.

Finally, let $\Phi_n(x) = \prod (x - \zeta_i) \in \mathbf{Z}[x]$ be the nth cyclotomic polynomial where the elements ζ_i denote the $\varphi(n)$ primitive nth roots of unity [see §9.1]. If d is a proper divisor of n,

$$\Phi_n(x) \left| \left[\frac{x^n - 1}{x^d - 1}\right] \right..$$

Since all polynomials in this relation are monic, it follows as before [§9.1] that

$$\frac{x^n - 1}{x^d - 1} = \Phi_n(x) g(x),$$

with $g(x) \in \mathbf{Z}[x]$. Furthermore, $\Phi_n(x)|(x^n-1)$ implies for the same reason that
$$(x^n - 1) = \Phi_n(x) h(x),$$
with $h(x) \in \mathbf{Z}[x]$. Consequently these factorizations, which are identities in $\mathbf{Z}[x]$, yield for the homomorphism $\mathbf{Z}[x] \to \mathbf{Z}$ given by $x \to q$ that
$$\Phi_n(q) \left| \left[\frac{q^n-1}{q^d-1} \right] \right. \quad \text{and} \quad \Phi_n(q)|(q^n-1).$$

The class equation (*) implies that

(**) $$\Phi_n(q)|(q-1).$$

This relation cannot hold if $n > 1$, because for every primitive nth root of unity $\zeta = \zeta_i$, $1 \le i \le \varphi(n)$, the inequality $|q - \zeta_i| > q - 1$ holds, as is shown below. Thus
$$|\Phi_n(q)| = \left| \prod_i (q - \zeta_i) \right| > (q-1)^{\varphi(n)}$$
contradicts expression (**).

Note that $|\cos \zeta| = |\operatorname{Re} \zeta|$. Since the real part of the complex root of unity ζ is less than 1,
$$|2 \operatorname{Re} \zeta| = |\zeta + \bar\zeta| < 2,$$
$$q^2 - q|\zeta + \bar\zeta| > q^2 - 2q,$$
$$q^2 - (|\zeta|q + |\bar\zeta|q) + \zeta\bar\zeta > q^2 - 2q + 1,$$
or
$$(q - |\zeta|)(q - |\bar\zeta|) > (q-1)^2;$$
hence
$$|q - \zeta| > q - 1.$$

Exercises

1. If F is an algebraically closed field and D is a division ring whose center contains F, such that the elements of D are algebraic elements over F, prove that $D = F$.
2. Let $R = ([1], i, j, k)_{\mathbf{Z}_p}$, $p \ne 2$, where
$$i^2 = j^2 = k^2 = -[1], \quad ij = -ji, \quad jk = -jk, \quad ki = -ik.$$
Prove that R is isomorphic to the ring of 2×2 matrices with coefficients in $\mathbf{Z}_p$.
3. In Exercise 2, let $p = 2$. Does R then contain a two-sided ideal $A \ne 0$ for which $A^2 = 0$? If so, what is the structure of the residue class ring R/A?

§9.11 Simple Transcendental Extensions

An apparently easy problem in the structure theory of simple transcendental extensions—that is, fields of rational functions, $F(x_1, ..., x_n)$—is the question whether subfields properly containing the field of coefficients F are themselves fields of rational functions. A solution for $F(x)$, F an arbitrary field, is provided by the theorem of Jacob Lüroth (1844–1910). Its proof, as will be seen below, depends upon arithmetic properties of polynomials of one and two indeterminates. If $n = 2$, then algebraic-geometric arguments are needed [see the work of Guido Castelnuovo and Oscar Zariski cited in the bibliography]. In the cases of $n \geq 3$, only special results are known, most of which are negative. A paper of Frederigo Enriques (1871–1946) indicates that a general theorem is far from being realized. Recently Y. Manin and V. Iskovskibh have given a set of counterexamples in the case $n = 3$. Here is one of the outstanding challenges of algebra which will certainly require considerable advances in arithmetic algebraic geometry and analysis.

An element ξ in a proper extension E/F is called **transcendental**, or (by abuse of language) an **indeterminate** over F, if it is not algebraic over F. That is, $\xi \in E/F$ is transcendental if it is not a zero of some polynomial $f(X) \in F[X]$.

We define the **degree** of an element $\eta = f(\xi)/g(\xi)$ in the quotient field $Q(F[\xi])$ of $F(\xi)$, ξ a transcendental element, where without loss of generality we assume that $(f(\xi), g(\xi)) = 1$, to be the larger of $\deg_\xi f(\xi)$ and $\deg_\xi g(\xi)$.

Theorem 1. If the element $\eta \in F(\xi)$ does not lie in F, then η is transcendental over F and $[F(\xi) : F(\eta)] = n$, where n is the degree of η.

Proof. Consider $\eta = f(\xi)/g(\xi)$ with relatively prime polynomials $f(\xi)$ and $g(\xi)$ in $F[\xi]$. Then ξ satisfies the equation $g(\xi)\eta - f(\xi) = 0$ with coefficients in $F[\eta]$. In detail, if

$$f(\xi) = a_0 + a_1 \xi + \cdots \quad \text{and} \quad g(\xi) = b_0 + b_1 \xi + \cdots,$$

then
$$(b_0 \eta - a_0) + (b_1 \eta - a_1)\xi + \cdots + (b_n \eta - a_n)\xi^n = 0.$$

The leading coefficient $b_n \eta - a_n$ is not zero since at most one of a_n, b_n can be zero, since η has degree n and $\eta \notin F$. Consequently ξ is algebraic over the field $F(\eta)$. Hence η must be *transcendental* over the field of coefficients F, for otherwise ξ would be algebraic over F, contrary to the hypothesis that ξ is transcendental [see the corollary to Theorem 2, §8.1].

Now let X be an indeterminate over the field $F(\eta)$. Then ξ is a zero of

$$h(X) = g(X)\eta - f(X) \in F[\eta][X].$$

This polynomial has degree n with respect to X according to our initial assumption on $f(\xi)$ and $g(\xi)$. Furthermore, using the Lemma of Gauss, §5.8, it suffices to prove that $h(X)$ cannot be factored in $F[\eta][X] = F[\eta, X]$.

For the indirect proof of irreducibility assume that $h(X)$ can be factored in $F[\eta, X]$. The polynomial $h(X)$ is *linear* with respect to η, and

§9.11 Simple Transcendental Extensions

consequently reducibility implies that one of its factors must be a polynomial $k(X) \in F[X]$. Write $h(X) = k(X)S(X, \eta)$ in $F[\eta, X]$. Applying the homomorphism $F[\eta, X] \to F[X]$ obtained by setting $\eta = 0$, we find that $k(X) | f(X)$. Furthermore, $k(X)$ then divides $h(X) - f(X) = \eta g(X)$, and therefore $k(X) | g(X)$. This contradicts the assumption that $f(\xi), g(\xi)$ are relatively prime. Hence ξ is a zero of an *irreducible* polynomial of degree n with coefficients in $F(\eta)$, and as asserted, $[F(\xi) : F(\eta)] = n$.

Note the following, immediate consequence.

Corollary. The polynomial $g(X)\eta - f(X) \in F[\eta, X]$ has no factor in $F[X]\backslash F$.

Moreover, if η is replaced by its representation $f(\xi)/g(\xi)$, it follows that $g(X)f(\xi) - f(X)g(\xi)$ has no factors in $F[X]\backslash F$.

Theorem 2. The group of automorphisms (over F) of a simple transcendental extension $F(\xi)$ of F is isomorphic to the projective linear group $PL(2, F)$.

Proof. Suppose that σ is an automorphism of $F(\xi)/F$. Then $\sigma(v(\xi)/w(\xi)) = v(\sigma(\xi))/w(\sigma(\xi))$ for $v(\xi), w(\xi) \in F[\xi]$. Hence σ is completely determined by the image $\eta = \sigma(\xi)$ of ξ. Let $\eta = f(\xi)/g(\xi)$ where $(f(\xi), g(\xi)) = 1$ in $F[\xi]$. By Theorem 1

$$[F(\xi) : F(\eta)] = n,$$

where $n = \text{Max}[\deg_\xi f(\xi), \deg_\xi g(\xi)]$.

For an automorphism σ, the fields $F(\xi)$ and $F(\eta)$ must coincide. Hence the degrees of the polynomials $f(\xi), g(\xi)$ cannot exceed 1. Furthermore at least one of these degrees must be 1, for otherwise $\eta \in F$ contrary to our hypothesis. Consequently

$$\sigma(\xi) = \frac{a\xi + b}{c\xi + d}$$

where a, b, c, d lie in F, and at least one of the coefficients a, c is distinct from zero. Since $(a\xi + b, c\xi + d) = 1$,

$$\det \begin{bmatrix} a & b \\ c & d \end{bmatrix} \neq 0.$$

(If the determinant were 0, then $\eta = \sigma(\xi)$ would lie in F, since the rows of

$$\begin{bmatrix} a & b \\ c & d \end{bmatrix}$$

would be proportional.) Conversely, each nonsingular matrix

$$\begin{bmatrix} a & b \\ c & d \end{bmatrix}$$

with coefficients in F determines by $\xi \to (a\xi+b)/(c\xi+d) = \eta$ an automorphism of $F(\xi)$ for $[F(\xi) : F(\eta)] = 1$ according to Theorem 1.

Finally, we note by direct verification that the mapping

$$\begin{bmatrix} a & b \\ c & d \end{bmatrix} \to \frac{a\xi+b}{c\xi+d}$$

of the group of 2×2 matrices with coefficients in F is a homomorphism onto the automorphism group A of $F(\xi)/F$. The kernel of this map consists of the diagonal matrices

$$\begin{bmatrix} a & 0 \\ 0 & a \end{bmatrix},$$

$a \neq 0$. Hence A is isomorphic to the factor group

$$\left\{ \begin{bmatrix} a & b \\ c & d \end{bmatrix} : \det \begin{bmatrix} a & b \\ c & d \end{bmatrix} \neq 0 \right\} \Big/ \left\{ \begin{bmatrix} a & 0 \\ 0 & a \end{bmatrix} : a \neq 0 \right\};$$

that is, the two-dimensional **projective linear group** $PL(2, F)$ of the field F.

Theorem 3 (*Lüroth*). Let the proper extension K/F be a subfield of the rational function field $F(x)$. Then K is also a rational function field $F(y)$.

Proof. For any $z \in K \backslash F$, $F(x)$ is a finite algebraic extension of $F(z)$ by Theorem 1. Hence $F(x)$ is also algebraic over K; thus $F(x) = K(x)$, where $[F(x) : K] = n$. Now let

$$f(X) = X^n + a_{n-1} X^{n-1} + \cdots + a_0 \in K[X],$$

X an indeterminate over K, be *the* minimal polynomial of x. Note that $f(X) \notin F[X]$, because otherwise $a_i \in F$ and x would be algebraic over F, a contradiction. Thus, at least one of the coefficients $a_i \in K \backslash F$.

As elements of $K \subseteq F(x)$, the coefficients a_i are rational functions of x. Then $f(X)$ can be multiplied by an appropriately chosen polynomial in $F[x]$ so that the resulting product

$$f^*(x, X) = b_n(x) X^n + b_{n-1}(x) X^{n-1} + \cdots + b_0(x)$$

in $F[x, X]$ is irreducible and primitive [see §5.8]. The uniqueness of $f(X)$ implies that

$$a_i = \frac{b_i(x)}{b_n(x)}, \quad 0 \leq i < n.$$

Since $f(X) \notin F[X]$, at least one of these coefficients, say a_j, actually involves x because $f^*(x, X)$ is primitive. Now write $y = a_j = g(x)/h(x)$, where $(g(x), h(x)) = 1$. The degrees of $g(x)$ and $h(x)$ with respect to x are at most m, the degree of $f^*(x, X)$ with respect to x. Since $y \neq 0$, we have

$$P(x, X) = g(X) - yh(X) = g(X) - \left(\frac{g(x)}{h(x)}\right) h(X) \neq 0.$$

Applying the homomorphism $X \to x$ from $F[x, X]$ to $F[x]$, we obtain $P(x, x) = 0$, which implies that x satisfies $P(x, X) \in K[X]$. Therefore $P(x, X) \equiv 0 \pmod{f(X)}$ in $K[X]$ by the definition of $f(X)$.

Next the Lemma of Gauss relates divisibility in $K(x)[X]$ to that in $K[x][X]$. The preceding congruence implies that $P(x, X) = q(X)f(X)$, where $q(X) \in K[X]$. All the coefficients of the powers of X in this relation are quotients of polynomials in $F[x]$. Hence multiplication by a polynomial in $F[x]$ results in an equation in $F[x, X]$ of the form

$$s(x)f(X)q(X) = s(x)P(x, X)$$
$$= s(x)[g(X)h(x) - g(x)h(X)] = f^*(x, X)t(x, X).$$

Since $f^*(x, X)$ is a primitive polynomial, $s(x)$ must be a factor of $t(x, X)$; hence we may write

$$A(x, X) = g(X)h(x) - g(x)h(X) = f^*(x, X)u(x, X),$$

with $u(x, X)$, $A(x, A) \in F[x, X]$. By the definition of $A(x, X)$ we have $\deg_x A(x, X) \le m$, whereas $\deg_x f^*(x, X) = m$. Hence $u(x, X) \in F[X]$. Furthermore, referring to the corollary to Theorem 1, we note that $u(x, X) \in F$. Thus we may write, ignoring factors in F,

$$A(x, X) = g(X)h(x) - g(x)h(X) = cf^*(x, X),$$

$0 \ne c \in F$. Now observe that $\deg_X f^*(x, X) = n$. Furthermore $A(x, X) = -A(X, x)$ implies that $m = n$.

Finally, $F(x) \supseteq K \supseteq F(y)$—thus $[F(x) : F(y)] \le m$ by Theorem 1—and $[F(x) : K] = n$ imply that $K = F(y)$, as asserted.

Exercises

1. a. Prove that the six mappings of $f(x) \in \mathbf{Q}(x)$, given by $f(x) \to f(x)$, $f(1-x)$, $f(1/x)$, $f(1-1/x)$, $f(1/[1-x])$, and $f(x/[x-1])$, are automorphisms of $\mathbf{Q}(x)$ and form a group. Which group of order 6 is this?
   b. Show that $\mathbf{Q}(x)$ is a galois extension of degree 6 over the field of rational functions $\mathbf{Q}(r(x))$ of

   $$r(x) = \frac{(x^2 - x + 1)^2}{x^2(x-1)^2}.$$

2. Prove that $K = \mathbf{Q}(x)$ is a quadratic extension of the fields $F_1 = \mathbf{Q}(x^2)$ and $F_2 = \mathbf{Q}(x(x+1))$. Determine the automorphisms of K/F_1 and K/F_2 in terms of x. Show that K is not a finite algebraic extension of the intersection $F_1 \cap F_2$.

3. Prove that $F(x)$ is a finite algebraic extension of any subfield E that is a proper extension of F.

§9.12 Perfect Fields

In this final section we discuss the so-called perfect fields: fields which, as far as irreducible polynomials are concerned, have the same properties as does the field of rational numbers. All algebraic extensions of such fields are separable [cf. §8.5]; all irreducible polynomials have distinct roots. The existence of an algebraic extension of a given field in which all irreducible polynomials are separable will be proved. We also present Robert Gilmer's argument on the existence of an algebraically closed extension, which avoids the double induction approach used in §8.9 [see *Amer. Math. Monthly* **75**(1968), pp. 1101–1102].

A field F is called **perfect** if every irreducible polynomial in $F[x]$ is separable.

In §5.5 we proved that every field of characteristic 0 is perfect. Every finite field is also perfect. For the proof see §8.2 and note that the field of p^n elements, determined by a zero of an irreducible polynomial of degree n with coefficients in the prime field $\mathbf{Z}_p$, is a galois extension of $\mathbf{Z}_p$ [see §8.6].

An equivalent statement of the definition is given for fields of nonzero characteristic in the following lemma.

Lemma 1. A field F of nonzero characteristic p is perfect if and only if every element $a \in F$ has a pth root A in F; that is, if and only if the polynomial $x^p - a$ is reducible in $F[x]$.

Proof. Assume F is perfect. The reducibility of $x^p - a$ in $F[x]$ implies $x^p - a = g(x)h(x)$ with a proper factor $g(x) \in F[x]$ of degree $s < p$. In a splitting field K of $F[x]$, we have $x^p - a = (x - A)^p$ [see §5.5]. Hence $g(x) = (x - A)^s$. Now $(s, p) = 1$, so that $st + pq = 1$ for some $t, q \in \mathbf{Z}$. Consequently $(x - A)^1 = g(x)^t \cdot (x^p - a)^q \in F(x)$. Hence $A \in F$, and therefore $(x - A)^p = x^p - a$ in $F[x]$.

Next, if some element $a \in F$ is not the pth power of an element in F, then $x^p - a$ must be irreducible in $F[x]$ by the preceding argument. Thus F has the inseparable extension $F(A)$, $A = \sqrt[p]{a} \notin F$, which contradicts the assumption that F is perfect. Consequently $A \in F$, and a has a pth root in F.

Conversely, assume that every element of F is a pth root of an element in F. If there were an irreducible inseparable polynomial $f(x)$ in $F[x]$, then

$$f(x) = g(x^{p^e}) = y^n + a_1 y^{n-1} + \cdots + a_n$$

with $y = x^{p^e}$, where e is the maximal such integer, as in §5.5. Now $a_i = A_i^{p^e}$ with $A_i \in F$, by assumption. Consequently

$$f(x) = (x^n + A_1 x^{n-1} + \cdots + A_n)^{p^e},$$

and $f(x)$ is reducible. Hence e must be 0, which means that $f(x)$ is separable.

The proof of the existence of an algebraic closure K of F in §8.9 relied upon a double induction argument in the construction of the fields

K_i. We now indicate that such an argument is unnecessary if simple facts concerning inseparable extensions and perfect fields are used. However it should be noted that the proof in §8.9 does not require the concepts of normality, separability, and the Theorem of the Primitive Element.

Let K_1 be the field $F[\{X_f\}]/M$ of §8.9 and $\hat{F}$ be the totality of all elements $A \in K_1$ which are **purely inseparable** over F; that is, for which $A^{p^u} \in F$ for some $u \geq 0$. Then $\hat{F}$ is a field. For $A^{p^u}, B^{p^v} \in F$, the elements $(A+B)^r, (AB)^r$ belong to F where $r = \text{Max}(p^u, p^v)$.

Lemma 2. The field $\hat{F}/F$ is perfect.

Proof. Consider $C \in \hat{F}$; then $C^{p^e} \in F$ by the definition of $\hat{F}$. The polynomial $x^{p^{e+1}} - C^{p^e}$ of $F[x]$ has a zero D in K_1 by the construction of K_1 in §8.9. Hence $D^{p^{e+1}} \in F$ and $D \in \hat{F}$. Consequently $(D^p - C)^{p^e} = 0$ in $\hat{F}$, and thus $D^p = C$. Finally, Lemma 1 implies that $\hat{F}$ is a perfect field.

Lemma 3. The field K_1/F contains an algebraic closure of $\hat{F}$.

Proof. The proof of Lemma 3 consists of showing that an irreducible polynomial $g(x)$ in $\hat{F}[x]$ factors completely in $K_1[x]$. Let $g(x)$ have roots $B_1, \ldots, B_n$ and consider a splitting field $S = \hat{F}(B_1, \ldots, B_n)$ of $g(x)$. Since $g(x)$ is separable by Lemma 2, the Theorem of the Primitive Element [§8.3] yields the existence of $w \in S$, such that $S = \hat{F}(w)$. Denote by $f(x)$ the minimal polynomial of w in $\hat{F}[x]$.

The p^eth powers of the coefficients of $f(x)$ lie in F for some $e \geq 0$ by the definition of the field $\hat{F}$. Hence $(f(x))^{p^e}$ lies in $F[x]$ and has a zero A in K_1 according to the construction of K_1. In other words, $f(x) \in \hat{F}[x]$ has a root $A \in K_1$. Since A and w are roots of the same irreducible polynomial $f(x) \in \hat{F}[x]$, there is an isomorphism over $\hat{F}$

$$\lambda: \hat{F}(w) \to \hat{F}(A), \quad \hat{F}(A) \subseteq K_1.$$

Since the zeros B_i of $g(x)$ are linear combinations (over $\hat{F}$) of powers of w, they correspond to linear combinations $\lambda(B_i)$ (over $\hat{F}$) of powers of A. The elements $\lambda(B_i) \in \hat{F}(A)$ are roots of $\lambda(g(x)) = g(x)$. In other words, $g(x) = \prod_{i=1}^n (x - \lambda(B_i))$, and the polynomial $g(x)$ splits completely in $\hat{F}(A)[x] \subseteq K_1[x]$. Consequently the field K_1 contains an algebraic closure of $\hat{F}$ as asserted.

Proposition. The field K_1/F contains an algebraic closure of F.

The proof is obtained by combining Lemmas 2 and 3. As our final statement we prove the following theorem.

Theorem. For each field F of nonzero characteristic p there exists a field $F^{p^{-\infty}}/F$ which is perfect and algebraic over F. Further, none of its proper subfields containing F is perfect. The field $F^{p^{-\infty}}$ is uniquely determined to within isomorphisms over F.

Proof. The field $\hat{F}$ of Lemma 2 is perfect and, by construction, algebraic over F. Suppose that L were a proper perfect subfield of $\hat{F}$ with $L \supseteq F$. Then there would exist an element $a \in \hat{F}\setminus L$ for which $a^{p^e} = \alpha \in F$. Observe that the equation $x^p - \alpha = 0$ has a solution A_1 in L, because L is perfect by assumption. Next $x^{p^2} - \alpha = (x^p - A_1)^p$ has a root $A_2 \in L$ for the same reason. Hence, by induction, $x^{p^e} - \alpha = (x^p - A_{e-1})^{p^{e-1}}$ has a root A_e in L. Consequently,

$$A_e^{p^e} = \alpha = a^{p^e}$$

implies that
$$(A_e - a)^{p^e} = 0$$

and that $a = A_e \in L$, a contradiction.

Finally let $\hat{F}_1/F$ be another perfect extension satisfying the hypotheses of the theorem. Suppose that $\hat{F}_1$ is embedded in an algebraic closure $\bar{F}_1/F$ and that $\hat{F}$ is embedded in an algebraic closure $\bar{F}/F$. The fields $\bar{F}_1$ and $\bar{F}$ are isomorphic over F [from §8.9], and this isomorphism maps $\hat{F}_1$ onto $\hat{F}$, as both fields have the property that some power of each of their elements lies in F.

Q.E.D.

Exercises

1. Prove that $F(x)$ is not a perfect field if char $F = p > 0$.
2. Determine the smallest perfect field containing $\mathbb{Z}_p(x)$.
3. Assume that the field F, char $F = p > 0$, is not perfect. Prove that the sets
$$F^{p^{-s}} = \{a^{p^{-s}} : a \in F\}, \quad s = 1, 2, \ldots$$
are distinct fields, where $a^{p^{-s}}$ denotes a zero of $x^{p^s} - a \in F[x]$.
4. If the degree of the algebraic closure $\bar{F}$ over F of the field F is a prime π, prove that F is a perfect field.
5. Making the same assumption as in Exercise 4, show that char $F \neq \pi$.
6. Construct the smallest perfect algebraic extension of $K = F(x)$, called the **perfect closure** of K, where char $F = p > 0$. (*Hint:* Examine the fields $F^{p^{-s}}$ of Exercise 3. Denote $\bigcup_{s \in \mathbb{N}} F^{p^{-s}}$ by $F^{p^{-\infty}}$.)
7. Prove that an algebraic extension K of a perfect field F is perfect, where char $F = p > 0$.
8. Subfields K_1, K_2 of a field M are said to be **linearly disjoint** over a common subfield F if every finite set of linearly independent (over F) elements in K_1 remain linearly independent when considered over K_2 as the field of coefficients.

    Let K/F be a subfield of an algebraic closure Ω/F. Prove that K/F is separable if and only if K and $F^{p^{-1}}$ are linearly disjoint.
9. a. Prove that $\mathbb{Q}(\sqrt[3]{2})$ and $\mathbb{Q}(\sqrt[3]{5})$ are linearly disjoint fields over $\mathbb{Q}(i)$, $i^2 = -1$.
    b. Are the fields $\mathbb{Q}(\sqrt[3]{5})$ and $\mathbb{Q}(\omega\sqrt[3]{5})$ linearly disjoint over $\mathbb{Q}$, where $\omega^2 + \omega + 1 = 0$?

10. Let a be an element in a field F of characteristic $p > 0$, which is not a pth power of an element in F. Prove that the polynomials $x^{p^e} - a \in F[x]$, $a \in \mathbf{N}$, are irreducible.

11. Prove that the definition in Exercise 8 is symmetric in K_1 and K_2. *Hint:* Show that the definition is equivalent to the statement that if $x_1, \ldots, x_n \in K_1$ are linearly independent over F and if $y_1, \ldots, y_m \in K_2$ are linear independent over F, then the products $x_i y_j$, $1 \leq i \leq n$, $1 \leq j \leq m$, are linearly independent over F.

12. Let $\{a_1, \ldots, a_m\}$ be a set of elements in an algebraic extension Ω/F, and suppose $F \subseteq L \subseteq \Omega$. Prove that:
    a. $[L(a_1, \ldots, a_m) : L] \leq [F(a_1, \ldots, a_m) : F]$.
    b. Equality of the degrees in part (a) holds if and only if the fields L/F and $F(a_1, \ldots, a_m)/F$ are linearly disjoint with respect to F.

13. Prove that an element A in an algebraic extension K/F belongs to F if it is simultaneously purely inseparable and separable over F.

14. Prove that a field F of nonzero characteristic p is perfect if and only if $F^p = F$.

15. Let $\tilde{F}$ be a perfect closure of a field F, where char $F = p > 0$. Show that either $\tilde{F} = F$ or $\tilde{F}$ does not have finite degree over F.

16. Let Ω be an algebraic closure of a field F, where char $F = p > 0$. Prove that an element $A \in \Omega$ is left unaltered by any automorphism of Ω/F if and only if $A^{p^m} \in F$ for some $m \in \mathbf{N}$.

17. Prove that an algebraic element A over a field F, char $F = p > 0$, is separable over F if and only if $F(A) = F(A^p)$.

18. Prove that when char $F = p > 0$:
    a. If K/F is a separable (algebraic) extension, then $F(K^p) = K$.
    b. If $F(K^p) = K$ and $[K : F] < \infty$, then K/F is separable, where $K^p = \{a^p : a \in K\}$.

19. Let K/F be a separable algebraic extension of characteristic $p > 0$. Prove that $\{A_\lambda^p : \lambda \in \Lambda\}$ is a basis of K/F if $\{A_\lambda : \lambda \in \Lambda\}$ is a basis of K/F.

Bibliography

The following bibliography lists books of interest in the history of mathematics related to algebra, additional readings on material covered in the text, and suggested readings for students wanting to pursue subsequent algebraic topics.

ALBERT, A.A., *Structure of Algebras*, Amer. Math. Soc. Coll. Publ. XXIV, 1939.

ARTIN, E., *Galois Theory*, Notre Dame Mathematical Lectures, No. 2, University of Notre Dame, 1948.

⎯⎯⎯, *Theory of Algebraic Numbers*, Göttingen, Germany, 1959.

⎯⎯⎯, NESBITT, C.J., THRALL, R.M., *Rings with Minimum Condition*, University of Michigan Press, Ann Arbor, Mich., 1944.

BELL, E.T., *The Development of Mathematics*, second edition, McGraw-Hill Book Co., New York, 1945.

⎯⎯⎯, *Men of Mathematics*, Simon and Schuster, New York, 1937.

BIRKHOFF, G., *see* MACLANE, S.

BOYER, C., *A History of Mathematics*, John Wiley & Sons, New York, 1968.

CASTELNUOVO, G., *Sulla razionalità delle involuzioni piane*, Math. Ann. **44** (1894), 125–155.

CURTIS, C., *Linear Algebra: An Introductory Approach*, third edition, Allyn and Bacon, Boston, 1974.

DICKSON, L.E., *Modern Elementary Theory of Numbers*, The University of Chicago Press, Chicago, 1939.

ENRIQUES, F., *Sopra una involuzione non razionale dello spazio*, Atti Accad. naz. Lincei, Rend. V.s. **21**[1] (1912), 81–83.

EVES, H., *Introduction to the History of Mathematics*, third edition, Holt, Rinehart and Winston, New York, 1969.

FULTON, W., *Algebraic Curves*, W. A. Benjamin, New York, 1969.

GREUB, W.H., *Linear Algebra*, Springer-Verlag, New York, 1967.

⎯⎯⎯, *Multilinear Algebra*, Springer-Verlag, New York, 1967.

GROSSWALD, E., *Topics from the Theory of Numbers*, The Macmillan Co., New York, 1966. (For a succinct discussion of the Fermat Conjecture, see pp. 159–182.)

HALL, M., *The Theory of Groups*, The Macmillan Co., New York, 1959.

HALL, TORD, *Carl Friedrich Gauss: A Bibliography* (translated by Albert Froderberg), MIT Press, Cambridge, Mass., 1970.

HALMOS, P., *Naive Set Theory*, D. Van Nostrand Co., Princeton, N.J., 1960.

HERSTEIN, I.N., *Noncommutative Rings*, Carus Mathematical Monographs, Number 15, The Mathematical Association of America, Washington, D.C., 1968.

HILBERT, D., *Collected Works, Vol. I, Number Theory*, Verlag von Julius Springer, Berlin, 1932.

———, *Ueber die Irreducibilität ganzer rationaler Funktionen mit ganzzahligen Coefficienten*, Jour. f.d.r.u.a. Math. **110** (1892), 104–129.

HOCHSCHILD, G., *The Structure of Lie Groups*, Holden-Day, San Francisco, 1965.

HOFFMAN, K.M., and KUNZE, R.A., *Linear Algebra*, Prentice-Hall, Englewood Cliffs, N.J., 1971.

JACOBSON, N., *Lectures in Abstract Algebra*, Vols. 1–3, D. Van Nostrand Co., Princeton, N.J., 1951, 1953, 1964.

KLEIN, F., *Lectures on the Icosahedron and the Solution of Equations of the Fifth Degree*, 1884, second and revised edition, Dover Publications, New York, 1956.

KUNZE, R.A., *see* HOFFMAN, K.M.

KUROSH, A.G., *The Theory of Groups* (translated from the Russian and edited by K.A. Hirsch), second English edition, 2 vols., Chelsea Publishing Co., New York, 1960.

LANDAU, E., *Grundlagen der Analysis*, Chelsea Publishing Co., New York, 1946.

LANG, S., *Introduction to Algebraic Geometry*, Interscience, New York, 1958.

———, *Algebraic Numbers*, Addison-Wesley Publishing Co., Reading, Mass., 1964.

———, *Algebra*, Addison-Wesley Publishing Co., Reading, Mass., 1965.

———, *Algebraic Number Theory*, Addison-Wesley Publishing Co., Reading, Mass., 1970.

LEDERMANN, W., *Introduction to the Theory of Finite Groups*, Oliver and Boyd, London, 1961.

MACLANE, S., and BIRKHOFF, G., *Algebra*, The Macmillan Co., New York, 1967.

MAXFIELD, J.E., and MAXFIELD, M.W., *Abstract Algebra and Solution by Radicals*, W.B. Saunders Co., Philadelphia, 1971.

NESBITT, J.E., *see* ARTIN, E.

NEWMAN, J.R., Editor, *The World of Mathematics*, Vols. 1–4, Simon and Schuster, New York, 1956–1960.

NORTHCOTT, D.G., *An Introduction to Homological Algebra*, Cambridge University Press, London, 1960.

O'MEARA, O.T., *Introduction to Quadratic Forms*, Springer-Verlag, Berlin, 1963.

ORE, O., *Number Theory and its History*, first edition, McGraw-Hill Book Co., New York, 1948.

PERRON, O., *Algebra*, Vol. II, de Gruyter and Co., Berlin, 1927.

RIBENBOIM, P., *Algebraic Numbers*, John Wiley & Sons, New York, 1972.

ROTMAN, J., *The Theory of Groups: An Introduction*, second edition, Allyn and Bacon, Boston, 1973.

SAMUEL, P., *Algebraic Theory of Numbers* (translated by Allan J. Silberger), Houghton-Mifflin, Boston, 1970.

SEIDENBERG, A., *Elements of the Theory of Algebraic Curves*, Addison-Wesley Publishing Co., Reading, Mass., 1968.

SERRE, J.P., *Corps locaux*, Publications de l'Institut de Mathématique de l'Université de Nancago, VIII, Actualités scientifiques et industrielles, 1296, 1962.

STEINITZ, E., *Algebraische Theorie der Körper*, Jour. f.d.r.u.a. Math. **137** (1910), 167–309.

STRUIK, D.J., Editor, *A Source Book in Mathematics, 1200–1800*, Harvard University Press, Cambridge, Mass., 1969.

SUN TSU, *Suan-Ching* (trans. *Arithmetic*), Abh.-Gesch. Math. Wiss. **30** (1912), 32.

THRALL, R.M., see ARTIN, E.

VAN DER WAERDEN, B.L., *Modern Algebra*, revised edition, 2 vols., Frederick Ungar Publishing Co., New York, 1969.

VANDIVER, H.S., *Fermat's last theorem, its history and the nature of the known results concerning it*, Amer. Math. Monthly **53** (1946), 555–578.

WARING, E., *Meditationes Algebraicae*, Cambridge, England, 1770.

WEIL, A., *Foundations of Algebraic Geometry*, Amer. Math. Soc. Coll. Publ. XXIX, 1946.

WEISS, E., *Algebraic Number Theory*, McGraw-Hill Book Co., New York, 1963.

WEYL, H., *Mathematische Analyse des Raumproblems*, Verlag von Julius Springer, Berlin, 1923.

―――, *Symmetry*, Princeton University Press, Princeton, N.J., 1952.

WITT, E., *Über die Kommutativität endlicher Schiefkörper*, Abh. Math. Semin., Hamburg, **8** (1931), 413.

ZARISKI, O., *On Castelnuovo's criterion of rationality $p_a = P_2 = 0$ of an algebraic surface*, Ill. J. Math. **2** (1958), 303–315.

―――, and SAMUEL, P., *Commutative Algebra*, Vols. I and II, D. Van Nostrand Co., Princeton, N.J., 1959.

ZASSENHAUS, H., *On the fundamental theorem of algebra*, Amer. Math. Monthly **74** (1967), 485–497.

Index of Mathematicians

A

Abbati, Pietro	178
Abel, Niels Henrik	163, 310
Albert, Adrian A.	340
al-Haitam, Ibn	51(Ex. 2c)
al-Kashi	21
Apian, Peter	21
Archimedes of Syracuse	19
Argand, Jean Robert	349
Artin, Emil	334

B

Boyer, Carl B.	42n
Brahmegupta	51(Ex. 2a)
Cardano, Geronimo	304, 310, 314
Castelnuovo, Guido	354
Cauchy, Augustin-Louis	121, 144, 241, 247(Ex. 23a)

C

Cayley, Arthur	110, 124, 157, 160, 200(Ex. 10)
Chu Shih-chieh	20
Cramer, Gabriel	121

D

d'Alembert, Jean le Rond	348
Dedekind, J. W. Richard	58

de Moivre, Abraham	318
Descartes, René	4
Diophantos	37(Ex. 19d)

E

Eisenstein, Ferdinand Gotthold	125, 155
Enriques, Frederigo	354
Euclid of Alexandria	27, 32, 37(Ex. 21b)
Euler, Leonhard	52, 53, 179(Ex. 7), 318, 348–349

F

Fermat, Pierre de	37, 52, 53, 179(Ex. 7), 305, 318, 334
Ferrari, Ludovico	304, 310
Fibonacci, Leonardo	31, 51(Ex. 2d)
Frobenius, Ferdinand Georg	71(Ex. 14), 90, 247(Ex. 23a), 272
Furtwängler, Philip	330

G

Galois, Évariste	178, 274, 288–292, 310
Gantmacher, F. R.	157
Gauss, Carl Friedrich	37, 54(Ex. 6), 84, 125, 154, 216(Ex. 11), 304, 318, 341, 348–349
Gilmer, Robert	358
Girard, Albert	348
Grassmann, Hermann	260

H

Hamilton, William Rowan	70(Ex. 14), 121, 124, 157, 160
Heisenberg, Werner	110
Helmholtz, Hermann L. F. von	169
Hensel, Kurt	141
Herbrand, Jacques	263(Ex. 12b)
Herstein, Israel N.	351

Index of Mathematicians

Hilbert, David	94, 314, 319, 328, 330, 331
Hölder, Otto	212, 248, 251, 259
L'Hospital, G. F. A. de	121

I

Iskovskibh, V.	354

J

Jacobson, Nathan	322, 340
Jordan, Camille	248, 251, 259

K

Khayyam, Omar	20
Klein, Christian Felix	189(Ex. 6)
Kowa, Seki	121
Kronecker, Leopold	58, 74, 125, 142, 266, 318
Kummer, Ernst Eduard	58, 305, 334, 338

L

Lagrange, Joseph Louis	150, 169, 178, 304, 310, 348–349
Lang, Serge	300
Laplace, Pierre-Simon de	121, 123
Legendre, Adrien-Marie	41
Leibniz, Gottfried Wilhelm von	21(Ex. 16), 121, 169
Leonardo of Pisa (*see* Fibonacci)	
Lie, Marius Sophus	207
Lüroth, Jacob	354, 356

M

Maclaurin, Colin	121
Manin, Y.	354

N

Newton, Isaac	21, 141, 169
Noether, Emmy	94, 319, 328, 330

P

Pascal, Blaise	21
Perron, Oskar	313
Plato	170, 314
Poincaré, Jules Henri	179(Ex. 11)

R

Roth, Peter	348
Ruffini, Paolo	310

S

Schmid, Herman Ludwig	314
Schreier, Otto	334
Schur, Issai	263(Ex. 9)
Smith, David Eugene	348
Steinitz, Ernst	103–104
Sun-Tsu	49
Sylow, Ludwig	221, 241–246

T

Takakusa, Seki	
(*see* Kowa, Seki)	
Tartaglia, Nicolo	304, 310

V

Vandermonde, Alexandre Theophile	121, 329
van der Waerden, B. L.	310
Venn, John	3
Viète, François	348

W

Wallis, John	21
Waring, Edward	153
Wedderburn, J. H. M.	304, 351
Weierstrass, Karl	121
Wessel, Caspar	349
Weyl, Hermann	169, 170
Wilson, John	153
Witt, Ernst	304, 340, 351

Y

Yih-Ling	51(Ex. 2b)

Z

Zariski, Oscar	354
Zassenhaus, Hans	248
Zorn, Max August	266, 299, 300

Index of Notation

Symbol	Meaning	Introduced on page
Set related and logic symbols:		
\	difference of sets	3
$\cap$, $\bigcap_{i=1}^{n}$	intersection (of sets)	3, 4
$\triangle$	symmetric difference of sets	3
$\cup$, $\bigcup_{i=1}^{n}$	union (of sets)	3, 4
$\Rightarrow$	"implies"	2
$\Leftrightarrow$	"is equivalent to"	2
$\in$	"is an element of"	2
$\supset$, $\subset$	proper (strict) containment	2
$\supseteq$, $\subseteq$	containment, with possible equality	2
$\emptyset$	null (empty) set	2
Set operational symbols:		
$\sim$	denotes a relation	6
$\approx$	denotes an equivalence relation	6
$\nsim$	"is not in relation to"	6
$\cong$	isomorphic	66, 186
$R \times S$	cartesian product of sets R, S	4
$\oplus$	internal direct sum	72
$\dotplus$	external direct sum	71
$\otimes$	internal direct product	213

Index of Notation

$\dot{\times}$	external direct product	214
∘	denotes composition of mappings	194

Labels for specific sets:

C	set of complex numbers	2
N	set of natural numbers	2
Q	set of rational numbers	2
R	set of real numbers	2
R^n	set of n-tuples of real numbers	5
U$_m$	(multiplicative group of) units in **Z**$_m$	52
Z	set of integers	2
Z$_m$	residue class ring of integers modulo m	39, 41

Miscellaneous symbols:

$\{\cdots\}$	denotes a set	2
$\langle g \rangle$	group generated by g	186
(m)	ideal generated by m	24
$\langle\!\langle g \rangle\!\rangle$	set of conjugates of g	238
$b \mid a$	"b divides a"	25
$b \nmid a$	"b does not divide a"	26
$\sigma \mid K$	restriction of mapping σ to K	277
$\equiv$	congruence	38
$\lvert a \rvert$	absolute value of $a \in \mathbf{C}$	15
$[s]$	equivalence (congruence) class of s	6
(a, b)	GCD of $a, b \in \mathbf{Z}$ (equivalently, ideal in **Z** generated by a, b)	24, 26
$[a, b]$	LCM of $a, b \in \mathbf{Z}$	34
$o(g)$, $[\langle g \rangle : e]$	order of an element g in a group	173
$o(G)$, $[G:e]$	order of the group G	177
$[G:H]$	index of the subgroup H in the group G	178
$[K:F]$	degree of the extension field K/F	269
$[K:F]_i$	degree of inseparability of K/F	283
$[K:F]_s$	degree of separability of K/F	283

Alphabetized notation:

A_n	alternating group on n elements	200 (Ex. 6c)
$A_G(L)$	annihilator of a subset L in a group G	232

Index of Notation

^tA	transpose of the matrix A	112
$\mathrm{Adj}(A)$	adjoint of the matrix A	124
$\mathrm{Af}(\mathbf{Z}_m)$	affine group over $\mathbf{Z}_m$	171 (Ex. 10)
$\mathrm{Aut}(G)$	group of automorphisms of the group G	205
C	(commonly) center of a group	177 (Ex. 10), 207
$C_G(H)$	centralizer of the subgroup H in the group G	183 (Ex. 12)
C_n	(multiplicative) cyclic group of order n	186
$\mathrm{Card}(S)$	cardinality of the set S	177 (Ex. 13)
$\mathrm{char}\, F$	characteristic of the field F	88
$\deg f(x)$	degree of the polynomial $f(x)$	128
$\det(A) = |A|$	determinant of the matrix A	121
$\dim V = \dim_F V$	dimension of the vector space V over the field F	104
$\mathrm{End}_F(V)$	ring of endomorphisms of the vector space V over the field F	108
$F_{m,n}$	set of $m \times n$ matrices with components in the field F	111
$F[x]$	polynomial ring in the indeterminate x over the field F	127
$F(x)$	field of rational functions (field of quotients) in the indeterminate x over the field F	128
G^1	derived (or commutator) group of the group G	207, 254 (Ex. 9)
G^*	the dual group of the group G	230
$G(K/F)$	galois group of the field extension K/F	273, 286
G/N	quotient (factor) group	181
gH	(left) coset of H by g in a group	171
Hg	(right) coset of H by g in a group	171
$GL(n, \mathbf{R})$	general linear group	208
GCD	greatest common divisor	26
$H_K(M)$	stabilizer of the subgroup M in a group with respect to the subgroup K	237
$H_K(s)$	stabilizer of the element s in a group with respect to the subgroup K	236
$\mathrm{Hom}(G, \Gamma)$	group of homomorphisms $\varphi : G \to \Gamma$ where G, Γ are abelian groups	217
$\mathrm{Hom}_F(U, V)$	vector space of linear transformations	

Index of Notation

	$\varphi: U \to V$ of vector spaces U, V over the field F	107
I_n	$n \times n$ identity matrix	113
$I(G)$	group of inner automorphisms of the group G	206
$\operatorname{Im} \varphi = \begin{Bmatrix} \varphi(G) \\ \varphi(R) \end{Bmatrix}$	image of a homomorphism φ	63, 185
K/F	extension K of the field F	266
$\ker \varphi$	kernel of the homomorphism φ	67, 185
$\operatorname{mod} m$	modulo the integer m	38
LCM	least common multiple	34
$m_a(x)$	minimal polynomial of $a \in K/F$	267
M_φ	matrix representing the linear transformation φ	114
$N(a)$	norm of the field element a	320
$N(K) = N_G(K)$	normalizer of a subset (subgroup) K in the group G	176 (Ex. 8)
$\operatorname{Orb}_G(s)$	orbit of the element s with respect to the group G	236, 237
$PL(2, F)$	projective linear group	184 (Ex. 19), 356
$Q(D)$	field of quotients of the integral domain D	85
$\operatorname{Rad} A$	radical of the ideal A	61 (Ex. 8)
$SL(2, F)$	special linear group	275 (Ex. 6)
$\operatorname{sgn}(\pi)$	sign or signature of the permutation π	122, 200 (Ex. 6)
$\operatorname{Span}(S)$	vector space generated by the set S	98
$T(a)$	trace of the field element a	320
$T(A)$	trace of the matrix A	158
$T(\varphi)$	trace of the linear transformation φ	160
UFD	unique factorization domain	342

Alphabetized greek notation:

$\chi_A(x)$	characteristic polynomial of the matrix A	158
$\chi_a(x)$	characteristic polynomial of the field element a	320
$\chi_\varphi(x)$	characteristic polynomial of the linear transformation φ	159

Index of Notation

δ_{ij}	Kronecker delta	74
Γ_n	dihedral group of order $2n$	168 (Ex. 6f)
$\Gamma(L)$	group of automorphisms of G which leave fixed the elements of the intermediate field L	288
$\Phi(H)$	fixed field of a subgroup H of a group G	288
$\Phi_n(x)$	cyclotomic polynomial of degree $\varphi(n)$	305
$\varphi(m)$	(usually) Euler φ-function of $m \in \mathbf{Z}$	52
σ_x	(usually) inner automorphism by x	206
Σ_n	symmetric group on n elements	194
$\Sigma(S)$	group of permutations of the set S	194
ζ	(commonly) primitive nth root of unity	230
ω	(commonly) a primitive cube root of unity	271 (Ex. 7c)

Index of Mathematical Terms

A

Abelian extension, 292(Ex. 1c), 335–337, 339
  galois group of, 340
Abelian groups, 163, 241
  annihilator of, 232
  characters of, 230, 231
  dual, 230
  dually paired, 235(Ex. 7)
  finitely generated, 222
  free, 229(Ex. 12, 18)
  Fund. Th. of Fin. Gen., 222–227
  homomorphisms of, 217–219
  order of, 241
  rank of, 229(Ex. 12)
Absolute value, 15
Addition, 11, 127
Additive inverse:
  for integers, 11–13
  for rings, 56, 57
  for vector spaces, 95
n-adic expansion, 29, 30, 141(Ex. 15)
  example of, 29, 42
  of integers, 29, 41–42
  of polynomials, 141(Ex. 13–15)
Adjoint matrix, 124
Adjunction of elements, 269
Admissible homomorphism, 257
  isomorphism theorem of, 257
  kernel of, 257
  Ω-admissible, 257
Admissible subgroup, 256
  normal, 256–257

Affine group, 171(Ex. 10), 254(Ex. 5a, 6), 341(Ex. 7)
Algebraic closure (*see also* Closure):
  of fields, 268, 301
  separable, 303
Algebraic element, 267–269
  characteristic polynomial of, 320, 321
  conjugate, 277, 278, 321, 325–326
  degree of, 267
  different of, 328
  discriminant of, 328
  minimal polynomial of, 267, 321
  norm of, 320–324
  separable, 270
  trace of, 320–324
Algebraic extension (*see also* Field extension(s)):
  equivalent, 277
  example of, 143–144
  finite, 269
  primitive element of, 275
  of **Q**, 286
  simple, 275
Algebraically closed, 300
Alternating group, 200(Ex. 6c), 252(Ex. 3)
  not solvable, 253
  simple, 253, 254(Ex. 13)
Annihilator:
  double, 232
  of an ideal, 70(Ex. 9)
  of a subset of a group, 232
Archimedean Principle, 19, 91(Ex. 23b)

Associates, 342
Associativity, 6
  for coset arithmetic, 41
  general law of, 11, 57, 164–165
  for groups, 163
  for ideals, 62(Ex. 11)
  for integers, 11
  for matrices, 119(Ex. 3)
  of permutations, 194
  for rings, 56, 57
  for vector spaces, 95
Automorphism(s) (*see also*
    Endomorphism; Isomorphism;
    Automorphism group):
  of **C**, 67(Ex. 13)
  of **Q**, 280(Ex. 2)
  of cyclic groups, 206(Ex. 1)
  of a field, 274, 277, 286
  Frobenius, 90, 274
  of a group, 186, 205, 229(Ex. 20)
  group of inner, 207, 238–239, 256
  inner, 206
  of a ring, 67
Automorphism group, 205, 211(Ex. 2)
    (*see also* Galois group)
  of **Z**, 212(Ex. 7)
  of cyclic groups, 206(Ex. 1)
  of Γ_4, 212(Ex. 9)
  of Klein's Four-group, 208(Ex. 2)
  of quaternions, 212(Ex. 8)

B

Basis:
  normal, 326
  of $\mathbf{R}^n$, 120(Ex. 12)
  of a vector space, 102–104, 111, 112
Bijections (*see also* Isomorphism;
    Permutations):
  of sets, 235–236
Bijective mapping (*see* Isomorphism)
Binomial coefficients, 20–21(Ex. 8–10),
    36 (Ex. 12)
Binomial Theorem, 19, 20(Ex. 9), 53, 147
Butterfly Lemma, 248
Cancellation, Law of:
  for addition, 12, 57
  for congruences, 38, 39(Ex. 5), 41

  for groups, 164
  for an integral domain, 83–84
  for multiplication, 11, 14, 56
  for rings, 57, 83–84
  for **Z**, 11
  for $\mathbf{Z}_m$, 38, 41

C

Canonical projection, 68, 139, 185,
    201
Cardinality, 177(Ex. 13), 236
Cartesian plane, 5, 9(Ex. 12), 96(Ex. 1),
    314
Cartesian product, 4, 91 (Ex. 21a),
    166(Ex. 5), 258
  nonassociativity of, 6(Ex. 6)
Casting out nines, 42(Ex. 2)
Cauchy Convergence Criterion,
    36(Ex. 6b)
Cauchy's Theorem, 241, 247(Ex. 19)
Cayley's Theorem, 200(Ex. 10)
Cayley-Hamilton Theorem, 124, 157,
    160–161, 320
Center:
  of dihedral groups, 209(Ex. 3)
  of a division ring, 351
  of a group, 177(Ex. 10), 179(Ex. 2),
    183(Ex. 3), 207, 210–211(Ex. 6),
    239
Centralizer, 183(Ex. 12)
Chains of groups, 250, 254(Ex. 9)
    (*see also* Composition series)
  isomorphic, 250
  normal, 250, 251
  refinement of, 250
Characteristic (of a field), 88
Characteristic polynomial:
  of a field element, 320, 321
  of a linear transformation, 160,
    264(Ex. 18)
  of a matrix, 125, 157, 158, 160–161
Characters (*see* Abelian groups)
Chinese Remainder Theorem, 10,
    48–51, 81, 325
Class equation, 238, 240
  application of, 239, 242, 352–353
Closure:
  additive, 15

algebraic, 268, 301, 359
  integral, 344
  multiplicative, 15
  perfect, 360(Ex. 6)
  separable, 303
Cofactor, 124
Commutative diagram, 201
Commutativity:
  for coset arithmetic, 41
  for groups, 163
  for ideals, 62(Ex. 11)
  for integers, 11
  for rings, 56, 57
  for vector spaces, 95
Commutator:
  of group elements, 179(Ex. 3), 207
  of permutations, 199
Commutator subgroup, 207
  higher, 254(Ex. 9, 10)
Complete inverse image, 66, 67, 202
Complex of products, 176, 180
Complex numbers, 2, 144
  of absolute value 1, 166(Ex. 3a)
  automorphisms of, 67(Ex. 13)
  as a residue ring, 144, 349
Composite number, 32
Composition series, 248, 250–252
    (*see also* Solvable group)
  of factor groups, 254(Ex. 10), 255(Ex. 19)
  of groups, 250
  length of, 250, 255(Ex. 19), 260
  of modules, 259
  of subgroups, 253(Ex. 2a), 255(Ex. 19)
Congruence(s):
  general systems of, 48–51
  modulo m, 37–38
  of polynomials, 136, 137
  solutions of simultaneous systems, 48–51, 52(Ex. 6), 138–139
  solvability of, 40(Ex. 11, 12), 43, 46, 52(Ex. 6), 140(Ex. 3)
Congruence classes, 1, 38
Congruence relations, 38, 176(Ex. 5a)
Conjugacy class, 238 (*see also* Equivalence class; Equivalence relations)
Conjugacy relations, 236

Conjugate field elements, 277, 278, 321, 325–326
Conjugate field extensions, 277, 278, 281, 289
  example of, 281
Conjugate group elements, 179(Ex. 2), 238, 239, 281–283
Conjugate roots of a polynomial, 270, 281
Conjugate subgroups, 176(Ex. 6), 239, 289
Constructions with ruler and compass, 54(Ex. 5, 6), 314–318
Content, 343
  p-content, 343
Coset(s), 1, 7 (*see also* Residue classes)
  double, 247(Ex. 23a)
  examples of, 7
  of integers, 38
  left, 171
  mapping of, 62–63, 205
  of polynomials, 137
  product of, 40–41
  representative of, 39, 172
  right, 171
  of a subgroup, 171
  sum of, 40–41
Coset decomposition, 177–178, 247–248(Ex. 23)
Cycle(s):
  commutator of, 199
  r-cycle, 197
  disjoint, 197
  even, 200(Ex. 6)
  examples of, 198–199
  inverse, 197
  length of, 196
  notation for, 195
  odd, 200(Ex. 6)
  order of, 200(Ex. 5)
  product of, 195
  transposition, 196, 198
Cyclic group, 191–193, 214, 219(Ex. 3), 272–273
  automorphisms of, 206(Ex. 1)
  dual group of, 230
  generators of, 186, 192
  homomorphic image of, 192

number of generators, 192
  subgroup of, 191
  sufficient condition for, 192
Cyclotomic extension, 305
Cyclotomic fields, 305
  examples of, 307-308
  galois groups of, 307
Cyclotomic polynomial, 157(Ex. 8a),
    305, 308(Ex. 2, 3), 352–353

D

Defining relations (of a group), 167
Degree:
  of an algebraic element, 267
  of a field extension, 269
  of inseparability, 149, 283
  of a polynomial, 128
  reduced, 149, 283
  of separability, 283
  of a transcendental element, 354
Delian Problem, 317
de Moivre's formula, 318
Derivative (see Formal derivative)
Derived group (see Commutator
    subgroup)
Determinant, 121–122, 158 (see also
    Norm)
  Laplace expansion by minors, 123
  of a linear transformation, 160
  properties of, 122–123
  Vandermonde, 329
Different (of an element), 328
Dihedral group, 168(Ex. 6f),
    169(Ex. 7), 174–175, 252(Ex. 2)
  automorphisms of, 212(Ex. 9)
  center of, 209(Ex. 3)
  commutator subgroup of, 210(Ex. 4)
  quotient group of, 210(Ex. 4)
  solvability of, 252(Ex. 2)
Dimension:
  of a field extension, 269
  of a vector space, 104, 105(Ex. 11),
    260
Diophantine equation, 37(Ex. 19d)
Direct product:
  external, 213, 214
  of groups, 212–215
  internal, 213

Direct sum:
  external, 72
  of groups, 213
  of ideals, 72
  internal, 72
  of rings, 72
  of vector spaces, 98, 99, 264(Ex. 18)
Discriminant, 327, 328
Distributivity:
  for coset arithmetic, 41
  general law of, 20(Ex. 6, 7), 57,
    70(Ex. 11), 126
  for ideals, 62(Ex. 11)
  for infinite vectors, 126–127
  for integers, 11
  for matrices, 119(Ex. 4)
  for polynomial functions, 126
  for rings, 56, 57, 70(Ex. 11)
  for vector spaces, 95
Divisibility, 25, 32, 33, 130–132
Division Algorithm (see also
    Euclidean Algorithm):
  application of, 24, 134, 174, 191
  for integers, 21–23
  for polynomials, 130, 131
Division ring, 351
  center of, 351
  finite, 351
Divisor, 26, 131, 342 (see also
    Greatest common divisor)
Divisor of zero (see Zero divisor)
Domain:
  integral (see Integral domain(s))
  of operators, 255, 256
  principal ideal, 131, 347(Ex. 2)
  unique factorization, 342
Double coset, 247(Ex. 23a)
Doubling the cube, 317
Dual group, 230–233, 336, 337
  annihilators, 232
  double, 231
Dual space, 110(Ex. 14), 235(Ex. 5)

E

Eisenstein's Criterion, 125, 155–156,
    308(Ex. 2)
Elementary symmetric functions,
    290, 346

Index of Mathematical Terms

algebraically independent, 346–347
  Theorem of, 347
Elements (in a domain):
  associated, 342
  divisor, 342
  integral, 344
  irreducible, 342
  unit, 46, 61, 85, 131
Elements (in a field):
  algebraic, 267, 361(Ex. 13)
  conjugate, 270, 277, 321, 325–326
  degree of, 267, 354
  inseparable, 283
  primitive, 275
  purely inseparable, 359, 361(Ex. 13)
  separable, 270, 284, 361(Ex. 13)
  transcendental, 271(Ex. 6), 354
Elements (in a group):
  commutator, 179(Ex. 3), 199, 207
  conjugate, 179(Ex. 2), 238
  generators, 186, 222
  identity, 163
  orbit of, 184(Ex. 15), 238
  order of, 173, 174, 178, 186
  self-conjugate, 179(Ex. 2), 238
  torsion, 225
Elements (in a ring):
  idempotent, 45, 61
  identity, 56, 57
  nilpotent, 43, 61
  orthogonal idempotent, 73, 138–139
    (see also Orthogonal
    idempotent)
  unipotent, 61(Ex. 5)
  unit, 46, 61 (see also Unit(s))
  zero divisor, 43, 45, 46(Ex. 10), 61, 84
Elements (in a set):
  equivalence classes of, 7
  least, 16
  maximal, 300
  upper bound, 300
Embedding, 67, 85
Endomorphism(s), 67, 160
  ring of, 57, 108, 116, 220(Ex. 11)
Epimorphism (see Surjective mapping)
Equality:
  of mappings, 194
  of polynomials, 127
  of sets, 3

Equivalence classes, 1, 7
  disjoint, 7
Equivalence of field extensions, 277, 280–283
Equivalence relation(s), 1, 6–8
  examples of, 6, 8(Ex. 10)
Euclidean Algorithm, 21, 27 (see also Division Algorithm)
  applications of 28–29
Euler φ-function, 10, 52–54, 81
  applications of, 54(Ex. 5), 192
  properties of, 54(Ex. 3,4), 83(Ex. 11,12)
Euler's Theorem, 179(Ex. 7)
Exact sequence, 260, 262(Ex. 5), 264(Ex. 15,16)
Expansion by minors, 123
Exponent:
  of a field extension, 335
  of a group (see Minimal exponent)
  of inseparability, 149
  minimal (see Minimal exponent)
  of nilpotency, 43
Exponents, Law of, 167(Ex. 6a), 170(Ex. 4)
Extensions (see Field extensions and Isomorphism)

F

Factor group (see Quotient group)
Factorization:
  of integers, 33
  in integral domains, 342–344
  of polynomials, 132
  unique, 33, 132, 342
Fermat's Last Theorem, 334
Fermat's Little Theorem, 10, 52–53, 69(Ex. 6), 179(Ex. 7)
Fermat primes, 318
Fibonacci sequences, 31(Ex. 15)
Field(s), 85–90, 153 (see also Field extension(s); Subfield)
  algebraically closed, 300, 353(Ex. 1), 358
  characteristic of, 88
  composite of, 293
  conjugate, 277, 289

cyclotomic, 305
  examples of, 84–85, 142
  finite, 271–274, 303(Ex. 1), 313
  fixed, 274, 288
  ground, 266
  intermediate, 274, 282, 288
  intersection of, 293
  perfect, 358, 359, 361(Ex. 14)
  prime, 88–89
  product of, 293, 296
  quotient, 85–88, 128
  of rational functions, 128–129
  separable, 270, 284
  separable closure, 303
  splitting, 145, 270, 280, 285
Field extension(s), 266
  abelian, 292(Ex. 1c), 335–337, 339
  algebraic, 143–144, 267, 269, 277(Ex. 2), 301, 311–312, 318
  algebraic closure, 268, 301
  conjugate, 277, 278, 289
  cyclic, 292(Ex. 2), 311, 331, 332, 334, 335
  cyclotomic, 305
  degree of, 269
  equivalent, 277, 278, 289
  exponent of, 335
  galois, 285, 289, 296, 311, 330, 346
  inseparable, 285
  Kummer, 335
  normal, 285, 289, 332
  quadratic, 316-318
  radical, 335
  separable, 270, 277(Ex. 2), 284, 285, 330
  simple, 275
  splitting, 145, 270, 280, 281, 285
  transcendental, 355
Field homomorphism, 92(Ex. 31)
Finite fields, 271–274
  examples of, 144
  Frobenius automorphisms of, 272
  generator of, 272
  multiplicative group of, 271–272
  prime field of, 88
  subfield, of, 274
  units of, 271
Fixed field, 274
Fixed point, 237(Ex. 4c)

Formal derivative, 146–148
Formal power series, ring of, 129(Ex. 6)
Frobenius automorphism, 90, 272–274, 333
Fundamental Theorem of:
  Algebra, 56, 125, 145, 266, 341, 348–351
  Arithmetic, 33, 48
  Elementary Symmetric Functions, 341, 347
  Finitely Generated Abelian Groups, 163, 222–227, 264(Ex. 18d)
  Galois Theory, 266, 271, 284, 289–291, 292

G

Galois extensions, 285, 330
  abelian, 292(Ex. 1c), 335
  cyclic, 292(Ex. 2), 331, 332, 334, 335
Galois groups, 286, 346, 355
  of abelian extensions, 335–337, 339
  example of, 286–287, 307–308
  for finite fields, 272–274
  solvable, 312–313
Galois Theory, 288–292
  for finite fields, 274
Gauss, Lemma of, 125, 343
Gaussian integers, 84(Ex. 4)
General associative law, 11, 57, 164–165
General distributive law, 20(Ex. 6,7), 57, 70(Ex. 11), 126
General linear group, 208–209(Ex. 2)
Generalized quaternion group, 167–168(Ex. 6e)
Generator:
  of a group, 167, 186, 222
  of an ideal, 24, 59
Grassmann's relation, 105(Ex. 11), 205(Ex. 6), 260
Greatest common divisors:
  computational examples, 28–29
  of ideals, 35(Ex. 5c), 59, 70(Ex. 10)
  of integers, 26, 27, 35
  in integral domains, 342

Index of Mathematical Terms

of polynomials, 131
uniqueness of, 27
Group(s), 162–234 (*see also* Abelian groups; Affine group; Groups, examples of; Homomorphism(s); Subgroup(s))
  abelian, 163
  affine, 171(Ex. 10)
  of automorphisms, 205–206
  center of, 177(Ex. 10) (*see also* Center)
  of characters, 230
  commutative, 163
  commutator, 207
  cyclic, 167(Ex. 6a,b), 186, 191–192
  defining properties of, 163–164
  defining relations of, 167–168
  derived, 207 (*see also* Commutator group)
  direct product of, 213–214
  double dual, 231
  dual, 230–233
  dually paired, 235(Ex. 7)
  exponent of (*see* Minimal exponent)
  factor, 181
  finite, 177
  Fund. Th. of Finitely Generated Abelian, 222–227
  galois, 274 (*see also* Galois group)
  generated by elements, 167–169 (Ex. 6a–i)
  generator(s) of, 167(Ex. 6a), 186, 222
  of homomorphisms, 217–218
  identity in, 163–164
  inverse in, 163, 164, 170(Ex. 1)
  isomorphism theorem of, 201–202
  law of exponents for, 167(Ex. 6a), 170(Ex. 4)
  minimal exponent of, 179(Ex. 6), 217(Ex. 12), 225, 335
  order of, 177, 241
  p-group, 225, 239, 248(Ex. 24)
  p-primary group, 225
  quotient, 181, 182, 202, 203, 207, 210, 233
  simple, 250
  solvability property of, 164
  solvable, 251, 313
  Sylow (*see* Sylow subgroups)
  symmetric, 194 (*see also* Symmetric groups)
  of symmetries (*see* Symmetric groups; Symmetries)
  torsion, 225
  torsion free, 225
Groups, examples of, 52, 165–169, 182, 186–188, 210–211 (*see also* Automorphism group)
  affine group, 171(Ex. 10) (*see also* Affine group)
  alternating, 200(Ex. 6), 252–253, 254(Ex. 13)
  cyclic, 167(Ex. 6a,b), 188 (*see also* Cyclic group)
  dihedral, 168(Ex. 6f), 169(Ex. 7) (*see also* Dihedral group)
  general linear, 208–209(Ex. 2)
  generalized quaternion, 167–168(Ex. 6e)
  group of the square, 169(Ex. 7b)
  group of the triangle, 169(Ex. 7a), 172–174, 201(Ex. 11)
  Klein's Four-group, 189(Ex. 6b) (*see also* Klein's Four-group)
  of matrices, 166(Ex. 4)
  of order p^2, 240(Ex. 4)
  of order p^m, 248(Ex. 24,26,27), 251–252, 254(Ex. 11), 255(Ex. 18)
  of order pq, 247(Ex. 12, 16)
  of permutations, 194
  projective linear, 184(Ex. 19), 355, 356
  quaternion, 167–168(Ex. 6e), 189(Ex. 6) (*see also* Quaternion group)
  semidihedral, 168(Ex. 6h)
  special linear, 275(Ex. 6)
  symmetric, 194, 313 (*see also* Symmetric groups)
  of translations, 254(Ex. 5a)
Groups with operators, 255–258 (*see also* Modules)
  admissible subgroups, 256
  examples of, 256, 263(Ex. 12)
Group action, 237(Ex. 4)
Group characters, 230

Group homomorphism (*see* Homomorphism(s))
Group ring, 263(Ex. 10)

H

Herbrand Quotient, 263(Ex. 12b)
Hilbert's Theorem Ninety, 328, 331–333, 335, 339
Homomorphism(s), 62–71, 184–191, 259 (*see also* Linear transformations)
  of abelian groups, 186–187(Ex. 1), 217–219
  admissible, 257
  automorphism, 67, 186
  bijective, 66, 186
  canonical, 68, 185
  embedding, 67
  endomorphism, 67, 221(Ex. 11)
  epimorphism, 65, 186
  examples of, 64–65, 186–188
  of groups, 185
  image of, 63, 185
  injective, 66, 186
  inverse image of, 66, 67
  isomorphism, 66, 186
  kernel of, 67, 185, 257, 259
  of modules, 259
  monomorphism, 66, 186
  one-one, 66, 186
  onto, 65, 186
  of rings, 62
  surjective, 65, 186
  trivial, 68

I

Ideal(s), 23, 58, 67
  annihilator of, 70(Ex. 9)
  dense, 69(Ex. 7)
  direct sum of, 72, 73, 82(Ex. 7,8), 83(Ex. 10)
  extended, 93(Ex. 39)
  generators of, 24, 59
  GCD of, 35(Ex. 5c), 59, 70(Ex. 10)
  of integers, 23–25, 76
  intersection of, 25, 34, 59
  inverse image of, 67
  LCM of, 35(Ex. 5a), 59, 60
  left, 59
  maximal, 91(Ex. 15)
  of nilpotents, 61(Ex. 7)
  of polynomials, 131
  power of, 59
  primary, 91(Ex. 12)
  prime, 71(Ex. 17), 91(Ex. 11)
  principal, 24, 75, 131
  product, 59
  proper, 300
  radical of, 61
  right, 59
  in a ring, 58
  semiprime, 91(Ex. 22a)
  sum of, 35(Ex. 5c), 59
  trivial, 23
  two-sided, 59
Idempotent elements:
  irreducible, 74(Ex. 7)
  orthogonal, 73, 74(Ex. 7), 75
  in a ring, 61
  in $\mathbf{Z}_m$, 45–46, 47(Ex. 9)
Idempotent linear transformations, 110(Ex. 16,17)
Identity element:
  for coset arithmetic, 41
  for groups, 163–164
  for groups of permutations, 194
  for integers, 11–13
  for rings, 56, 57, 63
  for vector spaces, 95, 96
Image (of a mapping), 63, 65, 185, 187, 188, 192
  inverse, 66
  of a subgroup, 188
Indeterminate element (*see* Transcendental element)
Index:
  of nilpotency, 43
  of a subgroup, 178
Index set, 4
Induction, Principle of, 14, 17–19
  alternate form, 17
  application of, 11, 19, 22, 164–165
Infinite vectors, 126, 152
Injective mapping, 66, 186

Inner automorphism, 206, 207, 238–239, 256
Inseparability:
 degree of, 149, 283
 exponent of, 149
Inseparable:
 extension, 285
 polynomial, 149
Integers, 2, 14, 92(Ex. 32) (*see also* Ring of integers)
 absolute value of, 15
 algebraic properties, 10–14, 55
 analytic properties, 14–19
 arithmetic of, 11
 axioms for, 11
 composite, 32
 factorization of, 33
 GCD of, 26, 27
 ideals in, 23–25
 LCM of, 34
 ordering of, 14–15, 92(Ex. 33–35)
 prime, 31–33, 37(Ex. 21), 153
 residue classes of, 38, 41 (*see also* Residue class rings of integers)
 solvability in, 13
 well-ordering of, 16
Integral closure (of a domain), 344, 345
Integral dependence, 341, 342
Integral domain(s), 14, 57, 83–88
 direct sum of, 91(Ex. 21)
 divisor in, 25–26, 342
 examples of, 84, 128, 153
 finite, 85
 GCD in, 26
 irreducible element in, 31, 342
 unique factorization in, 33, 342
 well-ordered, 93(Ex. 35)
Integral elements, 344, 345
Integral Root Theorem, 135(Ex. 9), 348(Ex. 8a)
Intermediate field, 274, 282, 288
Intermediate Value Theorem, 349
Inverse:
 for addition, 11–13, 56, 57
 for coset arithmetic, 41
 for groups, 163, 164
 for integers, 11
 for multiplication, 11, 46
 for rings, 56, 57
 for vector spaces, 95, 96
Inverse permutation, 197
Irreducibility Criterion, 155–156
Irreducible element of a domain, 342
Irreducible polynomial, 132, 148, 149, 155
 examples of, 305, 308(Ex. 2)
Isomorphic chains of groups, 250
Isomorphic fields, 273, 277, 280
Isomorphic groups, 186
Isomorphic rings, 66
Isomorphic vector spaces, 106
Isomorphism:
 admissible, 257
 extension of, 277
 of fields, 277, 278, 280–284
 of groups, 186
 prolongation of, 277, 281, 283
 restriction of, 277
 of rings, 66–68
 Theorem of, 201–202
 of vector spaces, 105–106
Isomorphism Theorem:
 applications of, 203–205
 for groups, 201–202
 for groups with operators, 257–258
 for rings, 68, 204(Ex. 1)
 for vector spaces, 109(Ex. 13)

J

Jordan-Hölder Theorem, 201, 251, 258, 259

K

Kernel (of a homomorphism):
 group, 185, 257
 module, 259
 ring, 67
Klein's Four-group, 189(Ex. 6b), 201(Ex. 13), 222
 automorphisms of, 208(Ex. 2)
Kronecker, Theorem of, 318
Kronecker's construction, 125, 142–143

Kronecker delta, 74
Kummer, Theorem of, 338
Kummer extensions, 334–341
Kummer theory, 334

L

Lagrange, Theorem of, 178
Lagrange Interpolation Formula, 135(Ex. 11), 150–151, 326
Laplace expansion by minors, 123
Lattice diagram, 173
Leading coefficient, 128
Least common multiple:
 of ideals, 35(Ex. 5a), 59, 60
 of integers, 34–35
Linear algebra, 94
Linear combinations, 97–98
Linear dependence, 101, 102
Linear functionals, 110(Ex. 14), 235(Ex. 5)
Linear independence, 101, 102
Linear space, 95
Linear transformations, 105–108, 169(Ex. 10), 264(Ex. 17)
 characteristic polynomial of, 160, 264(Ex. 18)
 composition of, 108, 115
 determinant of, 160
 examples of, 106, 109(Ex. 3), 118
 idempotent, 110(Ex. 16)
 isomorphism, 105–106
 kernel of, 109(Ex. 12)
 matrix representation of, 114, 117, 118, 159
 orthogonal, 110(Ex. 17)
 ring of, 57, 108
 set of, 107
 trace of, 160
Luroth's Theorem, 354, 356

M

Mathematical induction (*see* Induction, Principle of)

Matrices, 110–114
 adjoint, 124
 characteristic polynomial of, 125, 158
 cofactor of, 124
 determinant of, 121–124, 158
 groups of, 166(Ex. 4)
 inverse of, 117
 minor, 123
 multiplication of, by scalars, 111
 nonsingular, 117, 120(Ex. 13), 123, 159, 161(Ex. 1), 166(Ex. 4a)
 product of, 112
 ring of, 57
 trace of, 158–159
 transpose, 112
 vector addition of, 111
 vector space of, 111–114, 116–117
Matrix representation:
 examples of, 114–115, 118
 of field elements, 319–321
 of linear transformations, 114, 117, 159
Maximal element, 300
Maximal ideal, 91(Ex. 15), 300
Maximal subgroup, 247(Ex. 22)
Minimal exponent:
 of a field extension, 335
 of a group element, 179(Ex. 6), 217(Ex. 12), 225, 246(Ex. 2), 335
Minimal polynomial, 141(Ex. 11), 267, 321
Module(s), 94, 221, 258–261, 264(Ex. 17,18), 342
 composition series of, 259–260
 direct sum of, 262(Ex. 5)
 finitely generated, 260
 homomorphism of, 259
 irreducible, 263(Ex. 9)
 quotient, 259
 R-left, 258
 R-module, 259
 R-right, 258
 sequences of, 261
 unitary, 258
Monic polynomial, 128
Monomorphism (*see* Injective mapping)

Multiple roots, 149
  definition of, 148
  of an irreducible polynomial, 148
Multiplication:
  cancellation, law of, 11, 14, 57
  of cosets, 40–41
  of integers, 11, 13, 14
  of natural numbers, 15
  rule of signs for, 13, 57
  by scalars, 95
Multiplicative identity:
  for groups, 163
  for integers, 11, 13
  for rings, 56, 57
  for scalar multiplication, 95
Multiplicative inverse (*see* Inverse)
Mutually disjoint sets, 4

N

Natural Irrationality, Theorem on, 294–295
Natural numbers, 2, 15, 16
Natural projection (*see* Canonical projection)
Newton's formulas, 348(Ex. 10)
Nilpotency:
  exponent of, 43
  index of, 43
Nilpotent elements:
  in a ring, 61
  in Z_m, 43–45, 47(Ex. 12)
  nontrivial, 45, 47(Ex. 12a)
Noether's Equations, 330, 332
Norm (*see also* Determinant):
  of a field element, 320–322
  p-norm, 36(Ex. 16)
Normal basis, 326
  Theorem of, 151, 325–326
Normal chains:
  isomorphic, 250
  refinement of, 250
Normal field extensions, 285
Normal subgroups, 180, 182, 207, 214, 289
  example of, 182
  direct product of, 213–214

Normalizer, 176(Ex. 8,9), 183(Ex. 1,2), 239, 240
Null set, 2, 3

O

One-one mapping (*see* Injective mapping)
Onto mapping (*see* Surjective mapping)
Operator domain, 255–256
Oracle of Apollo, 317
Orbit, 236, 237
  of an element, 184(Ex. 15a), 238
Order:
  of an element, 166(Ex. 4c), 173, 174, 264(Ex. 18)
  of a group, 177, 241
  properties of, 173–174, 178
Order function:
  for polynomials, 343
  for rational numbers, 36(Ex. 14–16)
  for ring elements, 342
Ordered pairs, 4
Ordered n-tuples, 4, 96
Ordering, 91(Ex. 23)
  inductive, 300
  Law of Trichotomy, 15
  partial, 299
  total, 300
  well-ordering, 16, 93(Ex. 35)
Orthogonal idempotents, 73, 74(Ex. 7), 75, 110(Ex. 17)
  examples of, 76, 138–139

P

Partition(s), 7, 8
  examples of, 9(Ex. 11,12)
Pascal's Triangle, 21(Ex. 9)
Perfect closure, 360
Permutations, 194–199
  array notation for, 195
  commutator of, 199
  cycle notation for, 195
  cycles, 196, 197
  even, 200(Ex. 6)

examples of, 195, 197–199
group of, 194, 286
inverse of, 197
odd, 200(Ex. 6)
order of, 200(Ex. 5)
signature (sgn) of, 200(Ex. 6)
transposition, 196
Permutation groups (*see* Symmetric groups)
p-groups, 225, 247(Ex. 22), 248(Ex. 26)
center of, 239
solvability of, 251–252
structure of, 248(Ex. 24)
Platonic solids, 170
Polygons:
constructible, 54(Ex. 5), 317–318
regular, 54(Ex. 5,6), 169(Ex. 7)
Polynomials, 126–155 (*see also* Characteristic polynomial; Polynomial equations)
m-adic expansion of, 141(Ex. 13)
congruence of, 136, 137
constant, 127
content of, 343
p-content of, 343
cyclotomic, 157(Ex. 8a), 305, 308(Ex. 2)
degree of, 128
division algorithm for, 130, 131
factorization of, 130–132, 155, 156
GCD of, 131
inseparable, 149
irreducible, 132, 148, 149, 155, 305, 308(Ex. 2)
irreducible cubic, 286, 314(Ex. 1)
irreducibility criterion for, 155–156
leading coefficient of, 128
of linear transformations, 264(Ex. 17)
minimal, 141(Ex. 11), 267
monic, 128
order function of, 343
prime, 132
primitive, 154–155
properties of, 127–128
reduced degree of, 149
reducible, 155
relatively prime, 132
residue class ring of, 136–137

ring of (*see* Polynomial rings)
root of (*see* Root of a polynomial)
separable, 149, 270
in several indeterminates, 129
symmetric, 347
unique factorization of, 132
zero, 127
zero of (*see* Root of a polynomial)
Polynomial equations:
cubic, 310, 313, 314(Ex. 1)
of degree greater than or equal to 5, 310, 313
quartic, 310, 313
quintic, 310, 313
solvable by radicals, 310–314
Polynomial functions, 126, 152–153, 275(Ex. 7)
Polynomial rings, 127–129, 155, 228, 268
factorization in, 130–132, 343, 344
ideals in, 131
properties of, 129(Ex. 9), 153
in several indeterminates, 129
units in, 140(Ex. 3)
Primary ideal, 91(Ex. 13)
Prime field, 88
Prime ideal, 71(Ex. 17), 91(Ex. 11)
Prime numbers, 31–33, 37(Ex. 21), 153
Prime residue, 38
Prime residue classes, 43, 165(Ex. 2c)
Primitive element, 275, 276, 277(Ex. 3)
Theorem of, 276, 276–277(Ex. 1)
Primitive polynomials, 154–155
Primitive root of unity, 189(Ex. 3), 230
Principal ideal, 24, 75, 131
Principal ideal ring, 131, 347(Ex. 2)
Projection, 110(Ex. 16)
canonical, 68, 139, 185, 201
onto quotient structure, 68, 139, 185, 201
orthogonal, 110(Ex. 17)
Projective linear group, 184(Ex. 19), 355, 356
Prolongation of an isomorphism, 277–279, 281–283
Proper subset, 3
Purely inseparable element, 359

Q

Quadratic Law of Reciprocity, 318
Quaternions, 353(Ex. 2) (*see also* Quaternion group)
  generalized group of, 167–168(Ex. 6e)
  ring of, 70(Ex. 14)
Quaternion group, 167–168(Ex. 6e), 189(Ex. 6), 193(Ex. 8)
  automorphisms of, 212(Ex. 8)
Quotient field, 85–88, 128
Quotient group, 181, 202, 203, 207
  dual group of, 233
  example of, 182, 210
Quotient ring, 68
  of polynomials, 128–129
Quotient space, 109(Ex. 13a)

R

Radical:
  of an ideal, 61(Ex. 8)
  of a ring, 61(Ex. 7)
Radical extension, 335
Rational functions, field of, 162
Rational numbers, 2
  automorphisms of, 280(Ex. 2)
  embedding of integers, 85–87
  geometric constructions of, 315
  p-norm of, 36(Ex. 16)
  as a prime field, 88
Rational Root Theorem, 135(Ex. 8), 348(Ex. 8)
Real numbers, 2
  geometric constructions of, 315–316
Reduced degree (of a polynomial), 283
Refinement, 250
Regular polygon, 54(Ex. 5,6), 169(Ex. 7)
  constructibility of, 317–318
Relation:
  defining, 167–168
  equivalence, 6–8
  inductive ordering, 300
  ordering, 91(Ex. 23)
  partial ordering, 299
  reflexive, 6
  symmetric, 6
  total ordering, 300
  transitive, 6
Relatively prime integers, 27
Relatively prime polynomials, 132
Representative of a coset, 39
Residue classes, 1 (*see also* Coset)
  of integers, 38, 41, 137
  of polynomials, 137
  prime, 38
Residue class rings, 68, 69(Ex. 3), 82(Ex. 7) (*see also* Residue class rings of integers and of polynomials)
Residue class ring of integers, 41–46, 55, 74–77, 139
  examples of, 44–45
  as an external direct sum, 78–81
  ideals in, 76
  idempotents in, 45–46, 47(Ex. 9), 75–76
  as an internal direct sum, 78, 79
  multiplicative inverses in, 46
  nilpotent elements in, 43–45, 47(Ex. 12)
  units in, 46, 52
  zero divisors in, 43–45
  Z_p, 88
Residue class ring of polynomials, 137
  examples of, 138–139, 143–144
  of irreducible polynomials, 142–144
  units in, 140(Ex. 3)
  as a vector space, 140
Restriction of a mapping, 139, 277
Ring(s), 14, 56–73 (*see also* Residue class rings; Ring of integers; Subring)
  axiomatic properties, 56
  cancellation law, 57, 84
  commutative, 57
  of continuous functions, 69(Ex. 8), 91(Ex. 20)
  defining properties, 56–57
  direct sum of, 73, 82(Ex. 7)
  of endomorphisms, 108, 220(Ex. 11)
  examples of, 57, 61(Ex. 1a)

of formal power series, 129(Ex. 6)
group, 263(Ex. 10)
ideals in (see Ideal(s))
identity in, 56, 57, 63
isomorphism theorem for, 68
of matrices, 57
noncommutative, 57
of polynomials, 127, 129
principal ideal, 131, 347(Ex. 2)
of quaternions, 70(Ex. 14)
quotient, 68, 91(Ex. 11, 13, 15)
radical of, 61(Ex. 7)
of sets, 61(Ex. 1a)
subring (see Subring)
units of, 61, 131, 165(Ex. 2)
Ring homomorphism, 62
  examples of, 64–65, 67
  kernel of, 67, 141(Ex. 11)
  multiplicative identity under, 63, 65
Ring of integers, 14, 57
  construction from natural numbers, 92(Ex. 32–35)
  ideals in, 23–25, 63–64
  isomorphisms of, 91(Ex. 24)
Ring of polynomials (see Polynomial rings)
Root of a polynomial, 133, 267, 320
  α-fold, 148
  conjugate, 270, 281
  constructible, 316–318
  of an irreducible polynomial, 148–149
  multiple, 148
  multiplicity of, 148, 149
  number of, 133–134, 144–145
  primitive, 305
Roots of unity, 309(Ex. 4)
  primitive, 305
Rule of signs, 13, 57
Ruler and compass constructions, 54(Ex. 5,6), 314–318

S

Schur's Lemma, 263(Ex. 9)
Self-conjugate elements, 179(Ex. 2), 238
Self-conjugate fields (see Galois extensions)

Self-conjugate subgroups, 289, 291
  (see also Normal subgroups)
Semidihedral group, 168(Ex. 6h)
Semiprime ideal, 91(Ex. 22a)
Separable algebraic closure, 303
Separable element, 270, 284
Separable extension, 270, 277(Ex. 2), 284, 285, 330
Separable polynomial, 149, 270
Sequences:
  exact, 260, 262(Ex. 5), 264(Ex. 15, 16)
  induced, 261
  of modules, 261
  of module homomorphisms, 261
Set(s) (see also Subset):
  cardinality of, 177(Ex. 13), 236
  cartesian product of, 4
  difference of, 3
  disjoint, 4
  empty, 2
  equal, 3
  index, 4
  inductively ordered, 300
  intersection of, 3, 4
  maximal element of, 300
  null, 2, 3
  partially ordered, 299
  partition of, 7
  ring of, 61(Ex. 1a)
  symmetric difference of, 3
  totally ordered, 300
  union, 3, 4
  upper bound of, 300
  well-ordered, 16, 93(Ex. 35)
Set notation, 2
Short Five Lemma, 261–262
Simple extension, 275, 355
Simple group, 250
Simple vector space, 259
Simultaneous systems of congruences:
  examples of, 48–51, 74, 138–139
  solvability conditions, 52(Ex. 6)
Solvability axiom, 13
Solvability of congruences, 43, 46, 52(Ex. 6)
Solvable group, 248, 251–253, 312
Solvable polynomial, 310–314
Span of a set, 98, 100(Ex. 10), 102, 175

Index of Mathematical Terms

Special linear group, 275(Ex. 6)
Splitting field, 145, 270, 280, 285
   construction of, 144–145
   galois group of, 312
Stabilizer, 236, 237, 239, 240
Stable subgroup, 256, 263(Ex. 12)
Steinitz Exchange Lemma, 103–104
Subfield, 87
   conjugate, 281, 289
   linearly disjoint, 360(Ex. 8)
   prime, 88
Subgroup(s), 171, 202 (*see also* Sylow subgroups)
   admissible, 256
   alternating, 200(Ex. 6)
   annihilator of, 232
   center, 177(Ex. 10), 207 (*see also* Center)
   centralizer, 183(Ex. 12)
   commutator, 207
   conjugate, 176(Ex. 6), 239, 289
   of a cyclic group, 191
   dual group of, 233
   first derived group, 207
   generators of, 175
   higher commutator, 254(Ex. 9,10)
   index of, 178
   intersection of, 175, 180, 232
   invariant, 207
   maximal, 247(Ex. 22)
   nontrivial, 171
   normal, 180, 182, 207, 214
   normalizer, 176(Ex. 8)
   order of, 178
   p-Sylow, 242
   product of, 175, 180, 232
   stable, 207, 256, 263(Ex. 12)
   stabilizer, 236, 237
   Sylow, 242
   torsion, 176(Ex. 7), 225
Subring(s), 58, 62
   examples, 58
   identity element in, 58
   integrally closed, 345
   nontrivial, 58
Subset:
   nontrivial, 3
   proper, 3
   trivial, 3

Subspace, vector, 97
   cyclic, 264(Ex. 17)
   direct sum of, 98–99, 264(Ex. 18)
   generators of, 98
   intersection of, 98
   invariant, 157
   $p(x)$-primary, 264(Ex. 18)
   sum of, 98
   union of, 100(Ex. 5)
Surjective mapping, 65, 186
Sylow subgroups, 242
   conjugate, 244, 245
   normal, 244
   normalizer of, 246(Ex. 4,5)
   number of, 244
Sylow Theorems, 242(Th. 2), 244(Th. 3), 247(Ex. 23)
Symmetric groups, 194, 286, 313, 345–346
   alternating subgroup, 200(Ex. 6), 253
   not solvable, 253
   solvable, 252
Symmetric polynomial, 347
Symmetries, 169–171

T

Torsion element, 225
Torsion free group, 225
Torsion group, 225
Torsion subgroup, 176(Ex. 7)
Trace:
   of a field element, 320–322, 330
   of a linear transformation, 160
   of a matrix, 158–159
Trace function, 158–159 (*see also* Trace)
Transcendental element, 271(Ex. 6), 354
   degree of, 354
Transcendental extensions, 355
Transitive relation, 6
Transpose of a matrix, 112
Transposition, 196, 198
Triangle inequality, 16, 351(Ex. 1)
Trichotomy, Law of, 15
Trisection of an angle, 317

U

Unipotent element, 61(Ex. 5)
Unique factorization, 342
   of integers, 33–34
   of polynomials, 132
   Theorem of, 33
   UFD, 342–344
Unit(s):
   group of units in Z_m, 52, 165(Ex. 26), 193(Ex. 12), 206(Ex. 1), 216–217(Ex. 11), 307
   in a ring, 61, 131, 165(Ex. 2a)
   in a ring of polynomials, 131
   in Z_m, 46, 47(Ex. 10)
Unitary module, 258
Upper bound, 300

V

Vectors, 95
   as a group, 169(Ex. 9)
   linearly dependent, 101
   linearly independent, 101
Vector addition, 95
Vector spaces, 95–108, 256, 267, 269 (*see also* Vector spaces, examples of; Subspaces)
   bases of, 102–104, 111–112
   cartesian *n*-space, 96(Ex. 1)
   dimension of, 104, 105(Ex. 11)
   direct sum of, 98, 99, 264(Ex. 18)
   dual, 110(Ex. 14), 235(Ex. 5)
   finite dimensional, 104, 107
   finitely generated, 99, 102
   generators, 99, 101
   intersection of, 98
   isomorphic, 106
   linear functionals on, 110(Ex. 14), 235(Ex. 5)
   linearly dependent subset, 101
   linearly independent subset, 101
   order of an element, 264 (Ex. 18)
   polynomial action on, 264 (Ex. 17, 18)
   quotient space, 109(Ex. 13)
   ring of endomorphisms, 108, 116
   simple, 259
   span, 98, 102
   sum of, 98
Vector spaces, examples of, 96
   of complex numbers, 71
   finite fields, 271
   of linear transformations, 107, 113–115
   of matrices, 111–117
   of polynomials, 126
   of quaternions, 70(Ex. 14)
   R^n, 96(Ex. 1), 100
   of real-valued functions, 96(Ex. 3)
Venn diagrams, 3–4, 206

W

Wedderburn's Theorem, 304, 351–353
Well-ordered integral domain, 93(Ex. 35)
Well-ordered set, 16
Well-Ordering, Principal of, 14, 16, 18
   application of, 16–17, 23, 24, 33, 191
Wilson's Theorem, 153
Witt vectors, 340

Z

Zassenhaus Lemma, 248
Zero (*see also* Root):
   in Z, 11
   of a polynomial, 133, 148
Zero divisor, 43, 45, 47(Ex. 10), 61, 84
Zero vector, 95
Zorn's Lemma, 300